Polymers in Concrete

Polymers in Concrete

Proceedings of the Second East Asia Symposium on
Polymers in Concrete (II-EASPIC)

College of Engineering, Nihon University, Koriyama, Japan
May 11–13, 1997

EDITED BY

Yoshihiko Ohama
*Department of Architecture, College of Engineering,
Nihon University, Koriyama, Japan*

Makoto Kawakami
*Department of Civil Engineering, Akita University,
Akita, Japan*

AND

Kimio Fukuzawa
*Department of Urban and Civil Engineering,
Ibaraki University, Hitachi, Japan*

CRC Press
Taylor & Francis Group
Boca Raton London New York

CRC Press is an imprint of the
Taylor & Francis Group, an **informa** business
A CHAPMAN & HALL BOOK

First edition 1997 published by E & FN Spon

Published 2020 by CRC Press
Taylor & Francis Group
6000 Broken Sound Parkway NW, Suite 300
Boca Raton, FL 33487-2742

First issued in paperback 2020

ISBN 13: 978-0-367-65952-3 (pbk)
ISBN 13: 978-0-419-22330-6 (hbk)

Visit the Taylor & Francis Web site at
http://www.taylorandfrancis.com

and the CRC Press Web site at
http://www.crcpress.com

A catalogue record for this book is available from the British Library

Publisher's Note This book has been prepared from camera-ready copy provided by the individual contributors in order to make the book available for the symposium.

CONTENTS

PART THREE   PROPERTIES

PART SEVEN   DURABILITY 379

PROCEEDINGS OF THE SECOND EAST ASIA SYMPOSIUM ON POLYMERS IN CONCRETE (II-EASPIC)

International Advisory Committee

Chairman   Kiyoshi OKADA (Japan)
Members   David W. FOWLER (USA)
    Katsuro KAMIMURA (Japan)
    Yoshio KASAI (Japan)
    Kazusuke KOBAYASHI (Japan)
    Shigeyoshi NAGATAKI (Japan)
    H. Reiner SASSE (Germany)
    Rongxi SHEN (China)
    Yang-Seob SOH (Korea)
    R. Narayan SWAMY (UK)

Regional Organizing Committee

Members   Yoshihiko OHAMA (Japan)
    Kyu-seok YEON (Korea)
    Zhiyuan CHEN (China)

Organizing Committee

Chairman   Yoshihiko OHAMA (Nihon University)
Co-chairman   Kimio FUKUZAWA (Ibaraki University)
Secretary   Katsunori DEMURA (Nihon University)
Members   Toshio FUKUSHIMA (Building Research Institute)
    Toru HIRANO (Hokkaido Industrial Research Institute)
    Hiroshi IBE (Onoda Corporation)
    Makoto KAWAKAMI (Akita University)
    Akio KAWAMURA (Kumagai Gumi)
    Toshio KAWANO (Maeta Techno-Research)
    Wataru KOYANAGI (Gifu University)
    Hirozo MIHASHI (Tohoku University)
    Toyoaki MIYAGAWA (Kyoto University)
    Naohiro NISHIYAMA (Nishimatsu Construction)
    Minoru SAWAIDE (Shimizu Corporation)
    Atsushi SHIRAI (Tokyo Kasei Gakuin University)
    Mikio WAKASUGI (Sumitomo Osaka Cement)
    Takehiro YAMASAKI (Kyushu Institute of Technology)

Reviewers Committee

Chairman	Yoshihiko OHAMA (Nihon University)
Co-chairman	Makoto KAWAKAMI (Akita University)
Members	Katsunori DEMURA (Nihon University)
	Kimio FUKUZAWA (Ibaraki University)
	Toshio KAWANO (Maeta Techno-Research)
	Wataru KOYANAGI (Gifu University)
	Hirozo MIHASHI (Tohoku University)
	Toyoaki MIYAGAWA (Kyoto University)
	Takehiro YAMASAKI (Kyushu Institute of Technology)

Sponsors
College of Engineering, Nihon University
International Congress on Polymers in Concrete (ICPIC)
Japan Chapter of ICPIC
Japan Technology Transfer Association (JTTAS)

Co-sponsors
Architectural Institute of Japan (AIJ)
Japan Cement Association (JCA)
Japan Concrete Institute (JCI)
Japan Society of Civil Engineers (JSCE)
Japanese Society of Irrigation Drainage and Reclamation Engineering
Society for the Advancement of Material & Process Engineering
    (SAMPE)
The Adhesion Society of Japan
The Ceramic Society of Japan
The Japan Reinforced Plastics Society
The Society of Inorganic Materials, Japan
The Society of Materials Science, Japan (JSMS)
The Society of Polymer Science, Japan
China Civil Engineering Society
The Architectural Society of China
The Chinese Ceramic Society
Korean Society of Agricultural Engineers (KSAE)
Korean Society of Civil Engineers (KSCE)
Korea Concrete Institute (KCI)
Research Center for Advanced Mineral Aggregate Composite Products,
    Korea (CACP)
Prestressed and Precast Concrete Society, Singapore
American Concrete Institute (ACI)
American Society for Testing and Materials (ASTM)
International Ferrocement Society
The International Union of Testing and Research Laboratories for
    Materials and Structures (RILEM)

PREFACE

The history of the research and development of concrete-polymer composites is considerably different in various countries because of a difference in the background. The active research and development of the concrete-polymer composites have been performed all over the world, particularly in the United States, United Kingdom, Japan, Germany and Russia for the past 70 years. At present, the concrete-polymer composites are drawing a great attention as high-performance or multi-functional materials in the construction industry as well as the mechanical, electrical and chemical industries. The applications of the concrete-polymer composites have been developing in East Asian countries in recent years. The First East Asia Symposium on Polymers in Concrete was held in Korea in 1994.

Against such a background, the Second East Asia Symposium on Polymers in Concrete (II-EASPIC) is to be held in Koriyama, Japan on May 11-13, 1997, and is co-sponsored by the College of Engineering of Nihon University, the International Congress on Polymers in Concrete (ICPIC), the Japan Chapter of ICPIC and the Japan Technology Transfer Association (JTTAS), under the auspices of the Architectural Institute of Japan (AIJ), the Japan Cement Association (JCA), the Japan Concrete Institute (JCI), the Japan Society of Civil Engineers (JSCE), the Japanese Society of Irrigation Drainage and Reclamation Engineering, the Society for the Advancement of Material and Process Engineering (SAMPE), the Adhesion Society of Japan, the Ceramic Society of Japan, the Japan Reinforced Plastics Society, the Society of Inorganic Materials, Japan, the Society of Materials Science, Japan (JSMS), the Society of Polymer Science, Japan, the China Civil Engineering Society, the Architectural Society of China, the Chinese Ceramic Society, the Korean Society of Agricultural Engineers (KSAE), the Korean Society of Civil Engineers (KSCE), the Korea Concrete Institute (KCI), the Research Center for Advanced Mineral Aggregate Composite Products, Korea (CACP), the Prestressed and Precast Concrete Society, Singapore, the American Concrete Institute (ACI), the American Society for Testing and Materials (ASTM), the International Ferrocement Society, and the International Union of Testing and Research Laboratories for Materials and Structures (RILEM). Financial support was provided by the Kishitani Foundation and JTTAS.

The main objective of this international Symposium is the continuous development and dissemination of knowledge and technology of concrete-polymer composites in East Asian countries. The Call for Papers for this Symposium was responded by a great number of specialists from industries, universities and research institutes. More than 50 papers were accepted for presentation at the Symposium and publication in the Proceedings after strict review by the Reviewers Committee for them. The authors of the papers came from 13 countries including the United States and European countries. The Symposium plans to make clear a comprehensive overview in the areas of the concrete-polymer composites, and offers a forum for the exchange of information, ideas and research results as well as an opportunity for private discussions. The following topics were selected for the Symposium, and rearranged for eight chapters of the Proceedings:

- New developments in theory and practice of concrete-polymer composites
- Performance of concrete-polymer composites
- Manufacturing techniques and materials selection
- Structural design of concrete-polymer composites
- Precast products of concrete-polymer composites
- Restoration and conservation using polymers
- High-strength concretes using polymers
- Super-workable concretes
- Overlays
- Adhesives and coatings used in concrete works
- Special applications and new innovative developments of concrete-polymer composites
- Recycling of concrete-polymer composites

In particular, editorial work for the preparation of this Proceedings was done in close cooperation with a working group comprising the following members:

Professor Yoshihiko Ohama (Chairman of Organizing and Reviewers Committees)
Professor Makoto Kawakami (Co-chairman of Reviewers Committee)
Professor Kimio Fukuzawa (Co-chairman of Organizing Committee)
Dr. Katsunori Demura (Secretary of Organizing Committee)
Dr. Atsushi Shirai (Member of Organizing Committee)

As a result, the editorial group consists of Professor Yoshihiko Ohama, Professor Makoto Kawakami and Professor Kimio Fukuzawa.

On behalf of the Organizing Committee, I would like to thank all the authors of the papers included here for their cooperation. I wish to acknowledge the national and international organizations or institutions which supported the Symposium. I would also like to express my sincere appreciation to the members of the International Advisory Committee and Regional Organizing Committee for their useful advice and suggestions. Special thanks are due to the members of the Reviewers Committee and the Symposium Secretariat.

Yoshihiko Ohama
Chairman
Organizing Committee
II-EASPIC

PART ONE
RESEARCH TRENDS

PART ONE

RESEARCH TRENDS

RESEARCH AND DEVELOPMENT OF POLYMERS IN CONCRETE IN CHINA

Z.Y. Chen and S.Y. Zhong
The State Key Laboratory of Concrete Materials Research,
Tongji University, Shanghai, China

Abstract
This paper deals with the polymer concrete composites developed in China among which polymer cement concrete (PCC) has gotten faster development. Application of polymer impregnated concrete (PIC) declined in China because of complicated technology and high cost. Also polymer resin concrete (PC) has been used in less amount because its high cost. Therefor, research and application of polymer concrete are focused on PCC. The performance, structure and application of PCC are described. The high strength cement based composites are discussed. This paper reviews the situation of research and development of polymer concrete in China based on domestic literature recently.
Keywords: Polymer cement concrete (PCC), performance, structure, application.

1 Introduction

Polymer concrete has developed rapidly in its research and application in recent years because of its outstanding performance. For example, in Japan the polymer emulsion used in polymer cement mortar has amounted 100,000 tons each years; in US the amount of polymer cement mortar used in old and new building structure has reached $360,000 \text{ m}^3$ annually. Research in this field also developed in China and has resulted in much progress. Some international terminology is adopted in China, that is, polymer concrete is divided into polymer resin concrete (PC), polymer cement concrete (PCC), and polymer impregnated concrete (PIC). Application of PIC declined in China because of complicated technology and high cost[1]; also PC has been used in less amount because its high cost. Therefor, research and application of polymer concrete are focused on PCC[1, 2].

Polymers in Concrete, edited by Y. Ohama, M. Kawakami and K. Fukuzawa. Published in 1997 by E & FN Spon, 2–6 Boundary Row, London SE1 8HN, UK. ISBN: 0 419 22330 4.

This paper reviews the situation of research and development of polymer concrete in China based on domestic literature recently.

2 Polymer cement concrete (PCC)

2.1 Performance of PCC
In addition to mechanical properties, other properties like adhesion, corrosion resistance, Cl⁻ diffusion resistance, permeability resistance, freezing and thawing resistance, creep, shrinkage and weathering resistance are also studied.

2.1.1 Adhesion
Wang Xin-you et al. [3] studied fracture behavior of bonding interface between carboxyl SB (Styrene butadiene) latex modified cement mortar and old concrete using ordinary mechanical test and also wedge splitting test methods, showing that the splitting tensile strength of bonded sample decreases at first then increases slightly with increasing amount of polymer although it is lower than that without latex modification; however, fracture energy increases remarkably (Table 1).
   Guan Daqing et al. [4] found by shearing and permeability resistance test that the bonding strength of the cement modified by SB and SA (Styrene acrylate) emulsion to old concrete is lower than that of cement paste or mortar without modification. However, the shear strength is much higher (reaching 80% of concrete strength) and the permeability resistance improves much (S16) with the SB latex cement concrete as new concrete. The reason for that is as the figured out that the polymer particles in polymer cement mortar used as adhesives are larger than cement particles. They tend to occupy the pores on the surface of old concrete preventing cement paste that has better adhesion to concrete from permeating into the pores. The polymer particles in polymer cement concrete tend rather to distribute in concrete than concentrate in bonding interface as those in polymer cement mortar do. Tan Muhua et al. [2] studied the bonding strength between carboxyl SB latex modified cement mortar and ordinary cement mortar by direct tensile test and found that the bonding strength increases with increasing amount of polymer (P/C=0~0.16) whereas freezing resistance is best at P/C=0.08 and lower than ordinary cement mortar at P/C=0.16. Lu Xiaoyu[5] used SB latex as modified repair materials for fast grouting showing that strength and permeability are improved substantially.

Table 1. Mechanical properties of carboxyl SB latex modified cement mortar

P/C	Compressive strength (MPa)	Splitting tensile strength (MPa)	Fracture behavior K_{1C} (MN/M$^{3/2}$)	G_F(N/M)
0	27.5	2.97	0.338	30.5
0.05	26.5	2.04	-	-
0.10	24.0	2.00	0.367	61.4
0.15	23.6	2.55	0.390	54.6
0.20	22.7	2.61	0.408	56.5
0.25	21.1	2.71	0.408	70.9

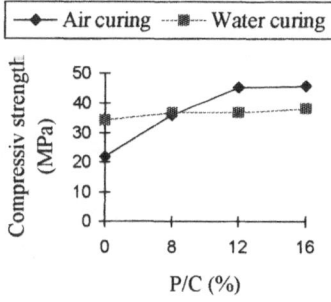 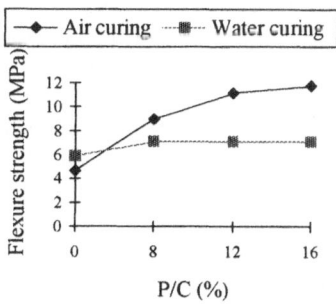

Fig. 1 The mechanical properties of modified cement mortar

2.1.2 Mechanical properties

Tan Muhua et al. [2] determined W/C in carboxyl SB latex modified cement mortar based on same fluidity, found that both tensile strength and flexure strength increase much more when curing in air than in water with increasing amount of polymer added (P/C=0~0.16) (Fig. 1). Whereas Xu Zhiqing et al. [6] prepared carboxyl SB latex modified cement mortar with different P/C but same slump as base concrete, found that the strength had highest value at certain amount of polymer added in the P/C range of 0~0.15 and the change in strength did not depend on W/C. Xong Zuoyun[7] found that compressive strength increases slightly with EVA emulsion added in amount less than 5% and decreases rapidly when amount of EVA emulsion added exceeds 5%.

With same W/C the compressive strength and elastic modules decrease and flexural strength and flexural fracture energy increase remarkably with increasing amount of polymer added ([3,7] for carboxyl SB latex, [8,9] for SB latex, [10] for acrylate copolymer emulsion and EVA emulsion). Curing condition does not influence much strength and flexural fracture energy[9]. K_{1C} and G_F has highest value when SB latex is used with P/C=0.1[8]. Ding et al. [10] found also that addition of Epoxy (EP), EVA and acrylate copolymer emulsion with dosage of below 5% improves flexural strength but lowers it with dosage above 5%, that is, an optimum dosage exists; compressive strength decreases in all cases but only slightly for EP.

The strength of polymer cement materials using unsaturated polyester (UP) and EP also has maximum at P/C=40/60 in weight[11] or 26~28% of UP in volume[18]. Brittless index decreases, that is toughness improves with increasing amount of polymer.

In addition the properties before fracture such as strength of initial crack, deformability and toughness of steel fiber concrete modified by SA emulsion improves remarkably[12].

2.1.3 Cl⁻ diffusion resistance

Improvement of Cl⁻ diffusion resistance is of great importance to improve salt corrosion of steel reinforcement concrete. Cl⁻ permeability resistance of cement mortar increases almost linearly with the increase of amount of SB latex added[2,6].

2.1.4 Corrosion resistance
Carboxyl SB latex can be used to improve greatly acid and oil resistance of cement mortar. (Refer to table 2 and 3[2].)

2.1.5 Effect of polymer on creep and shrinkage
In contrast to PIC the creep of PCC increases with increasing amount of polymer (Table 4)[14], and the effect differs with different type of polymer used (Table 5)[13]. Furthermore, the creep of PCC changes substantially with temperature humidity, it increases in the range of 18~49 ℃ to several percentages.

   Shrinkage of PCC changes only slightly with change of amount of polymer added but remarkably with different type of polymer[13].

Tab. 2 Weight Loss of Modified Cement Mortar by Acid Corrosion, Unit: %

Corrosive Media	P/C (%)			
	0	8	12	16
HCl (10%)	63.5	10.2	5.8	5
Lactic Acid (10%)	10.6	1.5	0.4	0.2
Acetic Acid (10%)	17.5	4.6	0.9	0.4

Tab. 3 Permeation Depth of Oil of Modified Cement Mortar, Unit: mm

Media	P/C (%)			
	0	8	12	16
Machine oil	>20	4.70	2.80	1.00
Kerosene	>20	6.30	4.60	2.60

Tab. 4 Change of Creep with Polymer Content

Polymer content, based of cement	Curing condition	Creep ($\times 16^{-6}$)
0	Dry	769
15	Dry	1054
25	Dry	1224
0	Wet	406
15	Wet	887
25	Wet	875

Tab. 5 Effect of Polymer Type on Creep

Type of polymer	Curing Condition	Creep ($\times 10^{-6}$)
Acrylate	Dry	1224
Modified acrylate	Dry	887
Styrene-Acrylate	Dry	863
SB	Dry	694

Tab. 6 Weathering Resistance

Weathering way	Adhesive Strength (MPa)	Tensile Strength (MPa)	Flexure Strength (MPa)	Compressive Strength (MPa)
Ultraviolet 1000 hr	4.21	4.29	7.53	22.65
Ultraviolet 2000 hr	4.10	4.32	11.05	30.10
Ultraviolet 4000 hr	3.64	4.73	11.67	41.0
Dry and Wet 100 Cycles	3.60	3.73	10.98	40.7
Dry and Wet 300 Cycles	4.19	3.69	8.83	36.6
Outdoor 18 Months	4.06	3.95	6.94	29.1
Indoor 18 Months	3.97	3.58	10.1	32.8
Immersion in water 18 Months	4.82	4.12	8.80	38.5
Indoor Curing 28 days	4.23	4.75	8.40	36.3

2.1.6 Weathering resistance
Ultraviolet, cycling of dry and wet and immersion in water (18 months) have slightly effect on properties of PCC but the strength of PCC decreases slightly by outdoor and indoor weathering (Table 6)[15].

2.2 Structure of PCC

2.2.1 Morphology
Gao Xiufeng[11] studied the morphological structure of EP and UP cement materials by SEM indicating that in this system there exists an interpenetrating network (IPN) formed by two separate phases, which are however itself continuos. At the P/C ratio approximately equal 15/85 the polymer phase transforms its structure from separate to continuo's phase; at P/C=40/60 an IPN with optimal interpenetrating degree is obtained which has smallest domain size.   All samples observed have "cellular" morphological structure in which polymer acts as cell wall with cement stone being cytoplasm and aggregate being nuclear.   The interpenetrating capability can be improved by raising solidified agent/water ratio.

Liang Naixing[16] also investigated the microstructure of cement stone modified by SB latex finding its morphology was related to P/C. When less SB latex is added (P/C=0.1) styrene-butadiene polymer does not really form a film although it has network structure.   The network is built up merely by SB latex particles (approximately 0.5 mm) sticking each other because of less amount of polymer added. The particles are bonded only by intermolecular force, therefor, the network is not completely continuous film but a network with many defects. When more SB latex is added (P/C=0.5) SB latex forms a film quickly in hardened cement paste, which is integral continuous network, no particle is observed. In some area sheet and lump are even found causing that cement stone is divided into separate islands, which is detrimental to the capability of bearing load. When proper amount of SB latex is added (P/C≈0.3) both SB and cement form three dimensional networks, which are interpenetrated to each other to form a distinctive integral so that the loading capability of hardened cement paste, especially tensile loading capability, is greatly improved.

2.2.2 Pores
With increasing amount of polymer (chloroprene and Carboxyl SB latex) added the volume percentage of the pores with diameter of above 500 Å or 1000 Å is substantially reduced[1].   Ye Zhiyong[17] determined the porosity of reactive UP cement material finding that the porosity lower with increasing amount of UP added, but levels off at P/C ≥ 1/1.5.

2.3 Interaction between polymer and hardened cement

2.3.1 Formation of chemical bond between polymer and hardened cement
Feng Xiujie et al. [19] studied the composition of interface between polymer and calcium aluminate with XPS showing that chemical reaction occurs between PVA or PAM and cement and the amount of reaction resultant from PAM is more than that from PVA. PAM reacts with cement to form ~C(O)$-$O$-$M bond with C$-$N bond

broken. In addition it is shown that the reactivity of Ca is higher than that of Al by depth analysis, the higher reactivity is indicated by higher diffusion concentration to organic phase.

Ye[17] found that a calcium carboxylate complex was formed in UP cement using infrared analysis. The resultant increases with age and decreases with increasing P/C ratio (P/C=1/7~1).

Long Jun et al[20] found also the reaction between acrylate copolymer emulsion and $Ca(OH)_2$ forming a macromolecular system built up by a interlinked network bonded by ionic bond. In the range of P/C=0~0.12 the higher the P/C the more reaction occurs, the more the amount of resultant and the higher the degree of hydration.

Taking into consideration of the fact that polyvinyl acetate (PVAc) and cement does not react[19]. It is not sure that the polymer containing ester or amide group reacts certainly with cement.

2.3.2 Effect of polymer on hydration of cement

Acrylate copolymer emulsion slows down hydration of cement, prolongs set time[21]. SB latex also slows down the process of hydration of cement especially at early stage of hydration, but has no effect on final hydration degree.

2.4 Application of PCC

PCC is mainly used for repair, anticorrsion and protection of concrete in China, for example, permeation of protection of floor in power plant, treatment of cracks and joints in concrete surface, protection and anticorrosion of water gate, bonding of old and new concrete, anticorrosion of floor, etc. [15,22,23]. Some practical examples are shown in Table 7.

Table 7 Examples of application of PCC

Project	Date	Area m^2	Situation after renovation
Permeation protection of floor slab in Jiangxi Tonggu Daduan Power Station	1990.12	2885	25~28 m no separation, good result
Joint in fast filtering pond in water plant	Beginning from 1991	~10,000	Good
Floating gate protection in Hebei Xinhe Gaoshou	1993.5	1535	Good
Wool washing workshop in Yimeng Sweat Factory, Inner Mongolia	1989	3000	Good

2.5 High strength cement based composites

Beginning from 80's MDF material was developed which was prepared by forced process using water soluble polymer like PAM and PVA mixed cement at very low W/C ratio. Recently the research on MDF material has gotten much progress. Feng Xiujie et al. [19,24] studied the structure of interface between polymer and cement using XPS technique showing that chemical bond exists between cement and PAM or PVA, which provides direct evidence of chemical reaction mechanism. Academician Wu Zhongwei[25] proposed Central Substance hypothesis to explain and modify the structure and composition of concrete for improving its performance. According to the hypothesis so called "center substance" is dispersing phase and the medium is interfacial zone. Center substance, medium and interface all have 3 levels (Fig. 3). He emphases that interfacial zone shell be not weak part in cementitous materials, in contrast it is advantageous to the forming of center substance network and development of center substance effect. He maintains that if right mix P/C ratio and proper process are used so called interpnetration network of two phases[24] is formed, which should be a sub-center substance, micro-center substance network.

Regarding the preparation of high strength cement Huang Congyun[26] determined a set of technological parameter with practical importance, that is, W/C=0.14~0.20, PVA 7~12%, process pressure 5~15 MPa, temperature 50~150 °C.

Moisture sensitivity is main problem of MDF, which drew much attention. Cracks are formed in the interface between cement particle and polymer caused by initial expansion of polymer when immersed in water. The cracks are easy to develop in hydrated product area causing permanent reduction of mechanical strength of MDF material. Al_2O_3 can improve MDF's moisture sensitivity in an extent. The best way to improve moisture sensitivity is to treat the surface of MDF material so that the surface is isolated from water[27].

Regarding the application of MDF material examples using MDF material with flexural strength of above 50 MPa are antistatic floor and woodlike materials, etc. They show low cost and high performance [28]. At present making the size of cement based high strength material cannot meet the requirement of the construction of civil engineering. Thus for developing the application in the field of civil engineering , Prof. Wu[29] proposed an image of mount element.

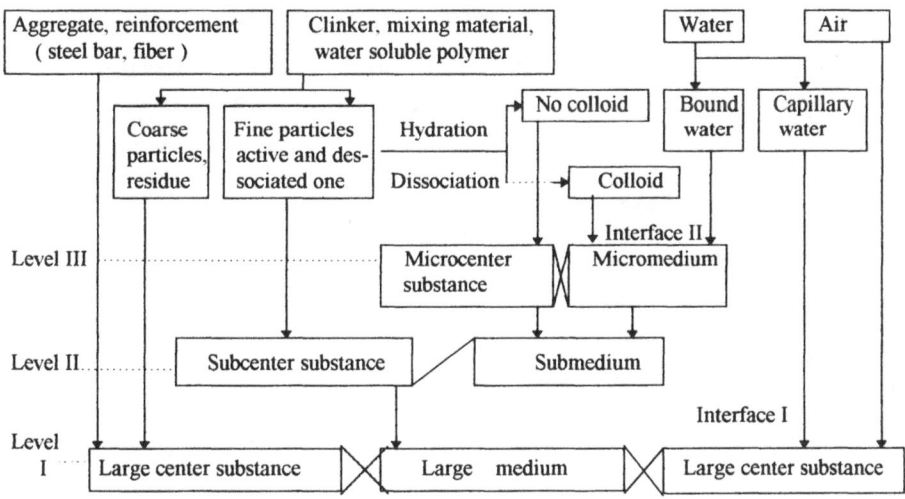

Fig. 3 Schematic description of composition and structure with center substance hypothesis[25].

3 Polymer concrete (PC)

Main polymer for PC is EP and UP that are ordinary commercial products. There are many reports [30-32]reviewing development of new additives and process for meeting special requirement and special construction condition such as repair of concrete structure at low temperature in winter.

Because of good damping property and very high strength in addition to repair EP concrete is used as vibration protection foundation and beam for machine[33] and machine tool[1] case. Table 8[33] shows the comparison between EPC, OPC and cast iron.

Sun Jiaying[34] prepared repair mortar from PUR, which meets the requirement for repair of road surface. It was used in road surface repair of cracks on Tailai highway in Tai'an City, Shandong Province and Maomin road South in Shanghai. The road was open to traffic only 6 hours after the repair was completed. This creates a new way of fast repair and maintenance of cement concrete road in China.

4 Conclusion

Polymer concrete composite has developed greatly in China among which PCC has gotten faster development. Overall and deep study has been carried out of PCC, but still not sure is the mechanism of modification by polymer, especially the mechanism of the effect of polymer on mechanical properties. Many research programs remain to be carried out concerning the application of polymer cement composites.

Table 8 Characteristics of EPC at Room Temperature Compared with OPCC and Iron

Characteristic	Unit	EPC	Plain OPCC 400#	Reinforced OPCC 400#	Cast Iron
Modules of elasticity E	KN/mm^2	26.72	31.58	36.84	88.00 ~ 113.00
Compressive strength R_C	KN/mm^2	107.22	45.31	47.32	720.00
Bending strength Rb	N/mm^2	26.15	6.67	17.53	290.00
Density γ	g/cm^3	2.35	2.39	2.56	7.17
Loss factor h	/	0.044	0.025	0.017	0.003
Bending to compressive strength ratio R_b/R_C	/	1:4.1	1:6.79	1:2.70	1:2.48
Specific compressive strength RC/γ	$N \cdot cm^3/g \cdot mm^2$	45.63	18.96	18.48	100.42
Specific bending strength Rb/γ	$N \cdot cm^3/g \cdot mm^2$	11.13	2.79	6.85	40.45
Specific stiffness E/γ	$KN \cdot cm^3/ g \cdot mm^2$	11.37	13.21	14.39	12.27 ~ 15.76

5 References

1. Chen Zhiyuan and Tan Muhua, (1994) Progress of Polymer Concrete Composite in China, *Proceeding of the lst. East Asia Symposium*, Korea, pp. 25.
2. Tang Muhua, Lu Jinping, Wu Keru, (1995) Performance of Cement mortar modified by Carboxyl SB Latex, *Journal of Tongji University*, Vol. 23 supple, pp. 60-5.
3. Wang Xinyou, Wu Keru, (1995) Study on Fracture Property of bonding Interface between PCC and Concrete, *Concrete and Cement Products*, No. 1. pp. 8-10,13.
4. Guan Daqing, Chen Zhanghong, Sho Yunzhu, (1994) Effect of Interface Treatment on Bonding Property of Old and New Concrete, *Concrete*, No. 5. pp. 16-22,11.
5. Lu Xiaoyu, (1993) Inorganic Fast Grouting Repair Material and Its Bonding to Old Concrete, *Thesis of graduate Student in Tongji University*.
6. Xu Zhiqing, Wang Peiming, (1994) Strength and Chloride Permeation Resistance of SB modified concrete, *Journal of Shanghai Construction Material Institute*, Vol. 7, No. 4. pp. 351-8.
7. Xiong Zuoyun, (1993) Properties and Application of VAE Polymer Cement Mortar with low P/C Ratio, *Chemical Construction Material*, No. 2. pp. 70-3.
8. Chen Meixin, (1991) Test and Analysis of Fracture Mechanics Property of Polymer Cement Concrete, *Thesis of graduate Student in Tongji University*.
9. Liang Naixing, (1994) SB Latex Modified Aluminate Cement, *Concrete and Cement Products*, No. 5. pp. 11-3.
10. Ding Jixin, Ji Lizhu, (1995) An Experimental Study on Repair Materials for Sandstone Statue Rockmass, *Proceeding of New Development in Concrete and Technology*, Published by Polytechnic University of South East China, pp. 1034-9.
11. Gao Xiufeng, (1994) Study on the Morphology of IPCN Polymer Concrete, *Application of Engineering Plastics*, Vol. 22, No. 5. pp. 50-4.
12. Wu Guoqiang, (1992) Study on Axial Extension of Polymer Steel fiber Concrete, *Concrete*, No. 3. pp. 30-3.
13. Yang Zhiqiang, Wang Dong, (1995) Study on the Effect of Polymer on Structure and Property of Concrete, *Concrete and Cement Products*, No. 1. pp. 11-3.
14. Yang Dianwen, Zhu Hequan, Zhao Shuqin, (1990) Some Properties of Polymer Concrete, *Chemical Construction Material*, No. 3. pp. 30-2.
15. L Yalian, (1995) Development and Application of Polymer Cement Mortar PCCM, *Concrete*, No. 2. pp. 31-7.
16. Liang Naixing, (1994) Effect of SB Latex on Hydration of Cement and Microstructure of Hardened Cement, *Journal of Silicate*, Vol. 22, No. 4. pp. 340-6.
17. Ye Zhirong, (1995) Research on the Mechanical Behavior and the Structure of Reactive Polymer Cement Materials, *Proceeding of New Development in Concrete and Technology*, Published by Polytechnic University of South East China, pp. 221-6.
18. Ye Zhirong, (1992) Strength and Brittleness of Active Unsaturated Polyester Cement Material, *Concrete and Cement Products*, No. 6. pp. 16-8.

19. Feng Xiujie, Hu Shuguang, (1991) XPS Study of Interface Composition and Structure of Polymer-Aluminate System, *Journal of Silicate*, Vol. 19, No. 6. pp. 481-7.

20. Long Jun, Yu Ke, Li Guoding, (1995) Interaction between Polymer and Cement Hydrated Product, *Concrete*, No. 3. pp. 35-7, 41.

21. Long Jun, Yu Ke, Li Guoding, (1995) Effect of addition of Polymer on Hydration Process of Cement, *Silicate Building Product*, No. 2. pp. 14-7.

22. Hong Shengshang, Yang Xigong, Jia Shuguang et al. (1993) Development and Application of BJ-894 Protection and Decoration Multilayer Coating Made from Polymer cement, *Concrete*, No. 5. pp. 37-43.

23. Hong Shengshang, Yang Xigong Jia Shuguang et al. (1992) Coating Made from Polymer cement, Concrete, Development and Application of BJ Polymer Cement Product Series, *Concrete*, No. 1. pp. 24-9.

24. Hu Shuguang, (1991) Study On High strength Mechanism and Some Interfacial Property of Polymer Cementitous Material with Low w/c, *Doctoral Thesis in Wuhan Industry University*.

25. Wu Zhongwei, (1992) Development and Application of High and Extrahigh Strength Cement-Based Material, *Concrete and Cement Products*, No. 5. pp. 4-9.

26. Huang Congyuan, (1992) Study On Hydration and Preparation of New High Strength Cement Composite, *Letter of Silicate*, No. 5. pp. 8-14.

27. Wei Guoquan, (1990) Study On Preparation and Moisture Sensitivity of PMC Material, *Thesis of graduate Student in Tongji University*.

28. Anonym, (1994) Research Project on High Strength Special Cement Material Passed Appraisal, *Information of Construction Material Industry*, No. 9. pp. 7

29. Wu Zhongwei, (1994) High Tech Concrete, Letter of Silicate, Vol.13, No. 1. pp. 41-5.

30. Zhang Yi, Zhang Shuming, (1994) Investigation On Repair of Cement Pipe with Epoxy Material, *Concrete*, No. 3. pp. 57-8, 48.

31. Le Minping, (1995) Method of Repairing Defects in Winter Thin Wall Beam With Epoxy Resin Concrete, *Construction Technique*, Vol. 22, No. 10. pp. 618-9.

32. Wang Baoting, (1993) New Field of Treating the Weaken Surface of Concrete by Using H80 Epoxy Coating, *Low Temperature Construction Technique*, No. 1. pp. 19-20.

33. Song Fangzhen et al., (1995) Analysis of Vibration Reduction Ability of Epoxy Concrete (EPC) Beam Mounted on an Elastic Foundation, in *New Development in Concrete Science and Technology*, (ed. Wu Zhongwei et al.), Southeast University Press, Nanjing, China.

34. Sun Jiaying, (1995) Development and Application of PG-308 Polymer Concrete for Road, *Concrete*, No. 2. pp. 38-40.

RESEARCH TRENDS OF CONCRETE-POLYMER COMPOSITES IN KOREA

K.S. Yeon, K.W. Kim, J.D. Choi and K.S. Kim
Dept. of Agricultural Engineering, Kangwon
National University, Chunchon, Korea

Abstract
Concrete-polymer composites have been developed lately and are used in place of conventional cement and concretes (including mortar) by virtue of their excellent characteristics. This study analyzed research and development trends of concrete-polymer composites in Korea through literature search.
Keywords: Polymer, polymer concrete, concrete-polymer composite, research trend.

1 Introduction

Recent years in Korea, construction industry has great interests in concrete-polymer composites, and therefore research and development (R&D) for these materials are under progress in various levels. Early deterioration of cement concrete structures have created many problems, and plant-precast productions of concrete members are becoming popular in stead of field casting. Actually, field casting of concrete structure in Korea becomes more and more difficult due to not only its poor durability, but also high labor and transportation cost. Therefore, there is a need for developing a new materials in place of the conventional concretes.

Some polymer composite products, such as, synthetic marble and polymer mortar bath tub, were produced in 1970s. But those products are not based on systematic research and development result. Major research efforts began

Polymers in Concrete, edited by Y. Ohama, M. Kawakami and K. Fukuzawa. Published in 1997
by E & FN Spon, 2–6 Boundary Row, London SE1 8HN, UK. ISBN: 0 419 22330 4.

in early 1980s in Korea. Therefore, since there have been a little history in research and application of polymer concretes, the developing speed of polymer concrete in Korea is relatively slower than that in other advanced countries. It is true that significant research in concrete-polymer composites has been performed by a limited number of researchers, and several international conferences were held in Korea by their active efforts.

2 Polymer-cement concrete (mortar)

Polymer-cement concrete (mortar) is the material in which a liquid or powder monomer or polymer is partially used as binder in addition to water. Currently in Korea, polymer-cement concretes are becoming widely used. The major reason is that polymer cement mortar is the most appropriate material for repairing old cement concrete structures built in decades ago. Therefore, since it is widely applied on repairing and finishing of concrete structures, research and demand are highly increased.

The research on polymer-cement mortar in Korea are as follows. Soh[27] used acrylic latex for polymer-cement mortar and conducted research on mechanical characteristics of the mortar. Some research dealt with durability of polymer- cement mortar using SBR latex, EVA, PAE and/or MMA/EA emulsion[29, 30]. A research was carried out for fundamental characteristics of polymer-cement mortar using EVA powder and SBR latex.

There are several research which dealt with application of the materials, including application for concrete slab water proof layer and for rust protector of concrete structures. Park conducted a study in which polymer-cement mortar was used as a repair material of distressed pavements[36]. Kwak et al.[25] dealt with PVDC (poly vinylidene chloride) latex polymer mortar, and Shim et al.[38] evaluated performance of concrete structural beams repaired using polymer-cement mortar.

On the other hands, filed application becomes popular and the case includes the application onto factory floors for multi purpose, slab for water proof, as a binder for tile placement, in place of concrete under water, as a binder of joint between old and new concretes, inside sealing of waste chemical tank, and so on. Many companies are established with the specialty to deals with these application on the particular purpose.

3 Polymer Concrete (Mortar)

Polymer concrete or mortar is a composite material in which the binder consists entirely of polymer resin. Up to these days in Korea, the polymer mortar has been mainly used for repairing old concrete structures, and the polymer concrete for making precast products.

Literatures regarding research in polymer concrete in Korea are summarized below. Many experimental studies were focused at fundamental properties, such as, various strengths, bond capacity, impact resistance, hardening shrinkage and creep, of epoxy resin mortar[2,3,4,8,17]. Many other studies were contributed to evaluation of physical properties, mechanical properties and maturity of unsaturated polyester polymer concretes in different mix proportions of materials used[13,14,23,32].

Other types of polymer concretes, such as, light weight polymer concrete[28], reinforced polymer concrete[5,6] and fiber reinforced polymer concrete[11], were also studied. Lately, aggregates from waste resources, such as, coal mine wastes and lime stone wastes, are used in polymer concrete study[34].

Applications of polymer concrete for some other used were studied. These include use of it for reinforcing existing load-bearing wall[9,18,31], for repairing concrete pavement[10], and for reinforcing asphalt pavement[39]. Polymer concrete was studied for use as a material of many precast products in place of conventional concrete, ceramic and iron. The precast items for research and development projects include sandwich panels[7,20], underground structures, such as, manholes[16,21] and sewage pipes[22,23], and electric insulators[19].

Some items are practically used in the field already. These include usages in factory floor finish, bridge deck repair, anchor bolt filler, RC structure repair, and water proofing for water tank and building slab.

4 Polymer Impregnated Concrete (Mortar)

Polymer impregnated concrete (PIC) or mortar is a concrete or mortar into which a monomer or polymer is impregnated to achieve polymerization. In overseas, PIC was used in field application and for precast products. In Korea, however, only limited research was carried out regarding application of fundamental impregnation techniques, and R&D level is far from practical usage. The reason is that the polymerization process is highly complicated, methodology and quality control is difficult.

Some research in Korea were contributed to evaluation of fundamental properties of PIC[1], and of strength, durability and fracture characteristics[12,24]. Byun et al[31] evaluated mechanical properties of carbon fiber reinforced PIC. These research are mostly focused at fundamental studies of polymer impregnated concretes.

5 Summary and Conclusions

During last 10 years, research and development on concrete-polymer composites have been relatively active in Korea. The R&D for polymer-cement mortar was focused at repair and finish of concrete structures. Those polymer concrete or mortar was partially used for repair and rehabilitation, and mainly applied for precast products. R&D for Polymer impregnated concrete are in primitive stage and no field application was achieved yet.

Lately, disasters by collapsing of several concrete structures make people begin to believe that new materials are strongly in need to overcome the properties of conventional materials. Also, due to drastic development of nation's economics, construction industry wants appearance of an advance construction materials that can satisfy customer's high standards. Therefore, demands for concrete-polymer composites will be increased continuously as new and advanced construction materials.

Acknowledgement
This study was supported by the **Research Center for Advanced Mineral Aggregate Composite Products** designated by **KOSEF** at **Kangwon National University,** in Chunchon, Korea.

6 Refrences

1. Jung-woo Na (1982), "A Study on the Fundamental Properties of Polymer Impregnate Concrete, M.S. Thesis, Seoul National University, p. 24
2. Kyu-Seok Yeon, Sin-Up kang (1982), "Strength Characteristics of Epoxy Resin Mortar", Journal of the Korea Society of Agricultural Engineers, Vol. 24, No. 3, pp. 92~99
3. Kyu-Seok Yeon, Sin-Up Kang (1983), "A Study on the Bond Stength and Impact Resistance of Epoxy Resin Mortar", Journal of the Korea Society of Agricultural Engineers, Vol. 25, No. 1, pp. 67~74
4. Kyu-Seok Yeon, Sin-Up kang (1984), "Experimental Studies on the Properties of Epoxy Resin Mortars", Journal of the Korea Society of Agricultural Engineers, Vol. 26, No. 1, pp. 52~72
5. Kyu-Seok Yeon, David W. Fowler and Dan. L. Wheat (1987), "Static Flexual Behaviour of Various Polymer Concrete Beam", 5th International Congress on Polymers in Concrete, Brighton, England, pp. 85~90
6. Kyu-Seok Yeon (1988), "Deformation Characteristics of Reinforced Polymer Concrete Beams", Journal of the Korea Society of Agricultural Engineers, Vol. 30, No. 1, pp. 63~72
7. Kyu-Seok Yeon (1989), "Flexural Characteristics of Polymer Concrete Sandwich Constrictions", Journal of the Korean Society of Agricultural Engineers, Vol. 31, No. 2, pp. 125~134
8. Nam-Seok Heo (1990), "Charateristics of Hardening Shrinkage and Creep of Epoxy Resin Concrete", M. S. Thesis, Kangwon National University, p. 27

9. Kyu-Seok Yeon (1990), "Development of Reinforced Wood Beams Using Polymer Mortar", Journal of the Korean Society of Agricultural Engineers, Vol. 32, No. 3, pp. 79~86

10. The Korea Highway Corporation (1992), An Experimental Study on Repairing Methods of Portland Cement Concrete Pavement with Polymer Concrete Material, Research Report, pp. 148~149

11. Byung-Hwan Oh(1992), "Strength and Mechanical Characteristics of Fiber Reinforced Polymer Concrete, Journal of the Korea Concrete Institute, Vol. 4, No. 3, pp. 147-155

12. Keun-Joo Byun, Young-Jin Kim, Sang-Min Lee, Byeng-Cheol Lho (1992), "Stength, Durabilty and Fracture of Polymer Impregnated Concrete". Procee ding of Joint Seminar between Korea and Japan, JeonBuk National Universi ty, pp. 43-52

13. Kyu-Seok Yeon, Kwang-Woo Kim, Ki-Sung Kim, Kwan Ho Kim (1993), "Effects of Filler on Mixing and Mechanical Properties of Polymer Concrete ", Journal of the Korean Society of Agricultural Engineers, Vol. 35, No. 2, pp. 81~91

14. Kwan-Ho Kim (1994), "Mechanical Properties of High Strength Polymer Concrete", M. S. Thesis, Kangwon National University, p. 64

15. Seung-Bum Park, Burtrand I. Lee, Eui-Sik Yoon (1994) "Mechanical Proper ties of Carbon-Fiber-Reinforced Polymer Impregnated Cement Composite" Proceedings of the First East Asia Symposium on Polymers in Concrete, Kangwon National University, Chunchon, Korea, pp. 141-154

16. Dong-Soo Kim, Jin-Keun Kim, Kyu-Seok Yeon, Yoon-Je Park, Tae-Hyung Kim, Myung-Woo Seo, (1990) "Sectional Design of Undergrond Box Structu re Using High Strength Polymer Concrete" Ibid., pp. 185-194

17. Seong-Hwan Yang, (1994) "An Experimental Study on the Mechanical Properties of Epoxy Mortar and Concrete Ibid., pp. 241-256

18. Byung-Hwan Oh, Hyung-Joon Lee, Dong-Hwan Lim, Shin-Won Baik (1994), Behavior of Two-Layered Reinforced Concrete Slab with Polymer Interface " Ibid., pp. 315-324

19. Hyun-Jin Jung, Jong-Ho Keum, Jin-Young Bark (1994), "Development of 22.9KV VCB Insulator Using Polymer Concrete" Ibid, pp. 361-372

20. Kyu-Seok Yeon, Kwan-Ho Kim, Kwang-Woo Kim, Yoon-Su Lee, Seong-So on Kim, Hyung-Gil Ham (1994), "Flexural Behavior of Sandwich Panels with Polymer Concrete Facing" Ibid., pp. 391-400

21. Dong-Soo Kim, Jai-Myung Kim, Yoon-Te Park, Tae-Hyung Kim, Myung-W oo Seo, In-Seop Han (1994), "An Economic Evaluation of Precast Polymer Concrete Manhole for Telecommunication" Ibid., pp. 415-416

22. Kyu-Seok Yeon, Seong-Soon Kim, Kwang-Woo Kim, Yoon-Su Lee, Kwan-Ho Kim, Hyung-Gil Ham (1994), "Mechanical Properties of Polymer Mortar Pipes" Ibid., pp. 425-434

23. Kwang-Ryul Hwang, Hyoung-Seok Soh, Seung-Young Soh, Hong-Shin Park, Yang-Seob Soh (1994), "Effects of Resin Quantity on the Strength Properties of Polyester Resin Concret", Proceedings of Fall Convention, Korea Concrete Institute pp. 235-239

24. Keun-Joo Byun, Sang-Min Lee, Hong-Shik Choi, Byeong-Cheol Lho(1994), "Material Properties of Polymer Impregnated Concrete and Nonlinear Fracture Analysis of Flexural Members, Journal of the Korea Concrete Institute, Vol. 6, No. 2, pp. 97-107

25. Kae-Hwan Kwak, Jong-Gun Park, Hui-Nam Han (1994), "Experiment Study on Shear Behavior of Polymer Concrete Beams, Journal of the Korean Society of Agricultural Engineers, Vol. 36, No. 4, pp. 39~47

26. Chan-Yong Sung (1994), "Engineering Properties of Permeable Polymer Concrete", Proceedings of the First East Asia Symposium on Polymers in Concrete, pp. 271~281

27. Yang-Seob Soh, Hong-Shin Park, Hyong-Seok Soh, Soo-Hyung Lee, Dai-Soo Lee (1994), "Properties of Polymer Cement Modified Mortar by Mixtures of Acrylic Latex" Idid., pp. 121-130

28. Chan-Young Sung (1995), "An Experimental Study on the Mechanical Properties of High Performance Light Weight Polymer Concrete", Journal of the Korea Society of Agricultural Engineers, Vol. 37, No. 3.4, pp. 72~81

29. Yang-Seob Soh (1995), "Durability of Polymer Modified Mortars" Proceedings of Joint Seminar Between Nihon University and JeonBuk National University, pp. 9-18

30. Eui-Hwan Hwang, Takuya Hasegawa, Yuiko Hama, Eiji Kamada and Yoshihiko Ohama(1994) "First Resistance and Physical Properties of Polymer Modified Motars", Proceedings of the First East Asia Symposium on Polymers in Concrete, Kangwon National University, Chunchon, Korea, pp. 53-80

31. K. J. Byun, H. S. Jeong and H. W. Song (1995), "Mechanical Behavior of Steel Fiber Reinforced Polymer Impregnated Concrete", Proceedings of 8th International Congress on Polymers in Concrete, Oostende, Belgium, pp. 643~667

32. Kyu-Seok Yeon, Kwang-Woo Kim, Yoon-Soo Lee and Kwan-Ho Kim (1995), "Maturity of Polyester Polymer Concretes", RILEM, Technical Committee Tc-113, Proceedings of Symposium on Properties and Test Methods for Concrete Polymer Composites, Oostende, Belgium, pp. 81~88

33. Sung-Soon Kim (1996), "Strength Characteristics of Centrifugal Pipes Using Polymer Mortar", M. S. Thesis, Kangwon National University, p. 62

34. Tae-Yeon Jang (1996), "Physical and Mechanical Properties of Polymer Concrete Using Coal Mine wastes, M. S. Thesis, Kangwon National University, p. 64

35. Kyung-Hyun Jeong (1996), "An Experimental Study on Fundamental Properties of Polymer Cement Mortar", M. S. Thesis, Kangwon National University, p. 51
36. Soon-Moo Park (1996), "Patching Concrete Pavement Using Polymer Composite Materials", M. S. Thesis Kangwon National University, p. 49
37. Kyu-Seok Yeon (1996), "Reinforcement of Load Bearing Wall Using Polymer Composite" Seminar Proceedings, Structural Rehabilitation Research Group, pp. 85-100
38. Jong-Sung Shim, Owan-Chol Choi, Kyu-Seok Yeon (1996), "Development of the Repair and Rehabilitation Methods of RC Structures, Research Report, Hanyang University, p. 122
39. Kwang-Woo Kim, Yong-Chural, Kyu-Seok Yeon (1986), "Tensile Reinforcement of Asphalt Concrete Using Polymer Coating", Construction and Materials, Vol. 10, No. 2, pp. 141-146

RECENT PROGRESS IN POLYMER MORTAR AND CONCRETE IN JAPAN

Y. Ohama
Department of Architecture, Nihon University,
Koriyama, Japan

Abstract

In Japan, polymer mortar and concrete already became the dominant construction materials in 1970s. Binders for the polymer mortar and concrete are chiefly epoxy resin, unsaturated polyester resin, vinyl ester resin and acrylic resin (methacrylate-based resin). The polymer mortar is widely used for finish, repair or chemical-resistant lining works, and the polymer concrete is generally employed for precast products. This review deals with the recent progress in the research and development of the polymer mortar and concrete, and their recent applications in Japan.
Keywords: applications, construction materials, polymer concrete, polymer mortar, research and development.

1 Introduction

In Japan, polymer mortar and concrete were developed in the early 1950s, and are currently used as popular construction materials in various applications because of their high performance and multifunctionality compared to conventional cement mortar and concrete. This has also been considered as a result of marked progresses in technical innovations in the Japanese construction industry in recent years. In particular, this trend is remarkable in the new frontiers of the construction industry, i.e., superhighrise buildings, very deep underground space, ocean and lunar base developments in Japan.

The present review deals with the recent progress in the research and development of polymer mortar and concrete, and their recent applications in Japan.

Polymers in Concrete, edited by Y. Ohama, M. Kawakami and K. Fukuzawa. Published in 1997 by E & FN Spon, 2–6 Boundary Row, London SE1 8HN, UK. ISBN: 0 419 22330 4.

2 Recent research and development activities

For 40 years or more, the active research and development of polymer mortar and concrete have been conducted in Japan. As a result, the polymer mortar and concrete became the dominant materials in the construction industry in the 1970s in Japan. Nowadays, the polymer mortar and concrete are employed as popular construction materials. In general, the polymer mortar is used for finishing work in cast-in-place applications, and the polymer concrete for precast products.

Commercially available liquid resins for polymer mortar and concrete include various thermosetting resins, tar-modified resins and methacrylate monomers in Japan. The liquid resins for the polymer mortar are chiefly epoxy resin(EP), unsaturated polyester (UP) resin (i.e., polyester-styrene system), vinyl ester (VE) resin and acrylic resin, and the most common liquid resin for the polymer concrete is the unsaturated polyester resin.

Tables 1 and 2 summarize the recent research and development activities of polymer mortar and concrete for the past several years. Some commentaries on the recent research and development activities of the polymer mortar and concrete in Tables 1 and 2 are described below.

In the structural applications of polymer mortars and concretes, mild steel bars (for conventional cement concrete), high-strength steel bars (for prestressed concrete) and FRP (fiber reinforced plastics) rods using carbon or glass fibers are used to reinforce the members with the polymer mortars and concretes, and steel fibers, glass fibers, etc. are employed to reinforce the polymer mortars and concretes themselves. It is most important to select the effective reinforcements corresponding to the strength of the polymer mortars and concretes.

The low-temperature curability of polymethacrylate mortars and concretes is an exotic, interesting property compared to ordinary cement concrete, and enables repair work for refrigeration warehouses and construction work in winter season or cold districts.

In general, polymer mortars and concretes are mixed by use of forced mixing-type batch mixers, and cast into forms or molds in a manner similar to that for conventional cement mortar and concrete. Refined continuous mixers are also available. It is most important that the developments of mass production systems for precast polymer mortars and concretes and of automated application systems for cast-in-place polymer mortars and concretes cause a cost reduction and good cost-performance balance.

Topic 13 in Table 2 is chiefly related to environment-conscious developments in the field of polymer mortar and concrete. The effective reuse and recycling of large quantities of PET (polyethylene terephthalate) bottles and waste scallop shells are also examined from the viewpoints of environmental protection and resources exploitation. The development of permanent forms as replacements for plywood forms is considered from the viewpoint of the preservation of forest resources as related to Topic 10. Some problems in the polymer mortars and concretes are the toxicity, explosibility, fire hazard and uncomfortable odors of liquid resins, initiators and promoters for their binders, and the ecologically safe disposal of cleaning solvents for the mixers and tools used in their applications.

Table 1. Recent research and development activities of polymer mortar and concrete
(Part 1)

Topic	Outline
1. New liquid resins for PM and PC binders	(1) Development of high-molecular-weight methacrylate, low-odor acrylic monomer and urethane methacrylate for binders [1] (2) Development of composite or combined liquid resins for binders, consisting of UP resins and vinyl monomers [2]
2. Mix design systems for PM and PC [3]	(1) Mix design systems for UP mortar and concrete (2) Mix design systems for polymethyl methacrylate mortar and concrete (3) Development of ready-mixed UP concrete (4) Mix design systems for UP mortar with lightweight aggregates [4] (5) Characterization of fillers for polymethyl methacrylate concrete by solubility parameters [5] (6) Predictions of working life and strength of polymethyl methacrylate mortars at low temperatures [6]
3. Reinforcement of PC and PM	(1) Fiber reinforcement of PC and PM by steel fibers or glass cloths [7] (2) Development of reinforced or prestressed UP concrete with FRP rods using carbon or glass fibers [8]
4. Low-temperature curability of PM and PC	Low-temperature curability of polymethacrylate mortars and concretes using methyl methacrylate, urethane methyl methacrylate and glycerol methyl methacrylate-styrene at 0 to 25 °C [6], [9], [10]
5. Setting shrinkage of PC	(1) Setting shrinkage and restrained stress of UP concrete [11] (2) Setting shrinkage of polymethyl methacrylate concrete bonded on reinforced concrete in a strengthening technique for reinforced concrete bridge decks [12] (3) Development of UP resin with low shrinkage [13] (4) Development of one-component-type UP resin with low shrinkage for PC [14]
6. Thermal properties and temperature dependence of PC	(1) Thermal properties and temperature dependence of mechanical properties of UP and EP concretes [15] (2) Temperature dependence of strength and elastic modulus of polymethyl methacrylate concrete [16] (3) Thermal properties of glycerol methyl methacrylate styrene-type concrete [9]

Table 2. Recent research and development activities of polymer mortar and concrete (Part 2)

Topic	Outline
7. Durability of PM and PC	(1) Long-term durability of UP or polymethyl methacrylate mortars and concretes [17] (2) Chemical resistance of UP and VE concretes [18] and UP and EP mortars [19] (3) Fatigue behavior of polymethyl methacrylate concrete used in an additional thickness strengthening technique for reinforced concrete bridge decks [20] (4) Freeze-thaw durability of polymethyl methacrylate concrete [21]
8. Automated application systems for PM and PC	(1) Development of small-diameter shield tunneling system using a quick-setting UP mortar [22], [23], [24], [25], [26], [27] (2) Shotcreting systems using PM and PC [28]
9. PM and PC for underwater construction work	Development of polymethyl methacrylate mortars and concretes placed and bonded underwater [29], [30], [31]
10. Repairing or durability-improving systems using PM and PC for reinforced concrete structures [32], [33], [34]	(1) Development of repairing systems using PM and PC [35] (2) Repairing systems using polymethyl methacrylate concrete and mortar for concrete pavement [36], [37] (3) Repairing system using FRP-sandwiched UP mortar panels for conduit lines or headrace tunnels [38] (4) Design of PC or PM permanent forms [39] (5) Development of permanent forms using UP or VE concrete [40] (6) Development of highly-durable permanent forms using UP concrete [41] (7) Development of decorative permanent forms using UP concrete with crushed natural stone as an aggregate [42]
11. Artificial marble tiles and panels	(1) Development of artificial tiles and panels made with polymer pastes with flame-retarding fillers such as aluminum hydroxide and magnesium hydroxide [3], [43], [44] (2) Development of VE for artificial tiles and panels [45]
12. PM and PC for machine tool structures	Development of EP concrete for machine tool structures [46]
13. Reuse and recycling	(1) PM and PC using recycled polyester resins produced from PET (polyethylene terephthalate) bottles [47] (2) VE concretes using waste scallop shells as aggregates [48]

3 Applications

In Japan, polymer mortar and concrete already became the dominant construction materials in the 1970s as mentioned above. The polymer mortar is now used as a popular construction material in various applications as listed in Table 3. At present, polyester concrete is most widely used for various structural and nonstructural precast products, and has some cast-in-place applications. Table 4 gives various applications of the polyester concrete.

In recent years, the demand of polymeric artificial stone products using polymer paste, mortar or concrete has been growing year after year in Japan. Fig. 1 gives the classification of the polymeric artificial stone in Japan. In particular, both polymeric marble and terrazzo are called "Artificial Marble". The share of the polymeric marble in the "Artificial Marble" is found to be 95% or more, therefore, the term, "Artificial Marble" in this paper means "Polymeric Marble". The three largest markets of the artificial marble products are systematized kitchen, washstand and systematized bath fields. In the near future, if the weather resistance of artificial marble is improved to a great extent, the artificial marble products will widely be used as exterior finish materials for buildings. Furthermore, if a success in the great cost reduction of the artificial marble products is brought, the share of natural marble will considerably be replaced by the artificial marble. This may be a natural resource saving.

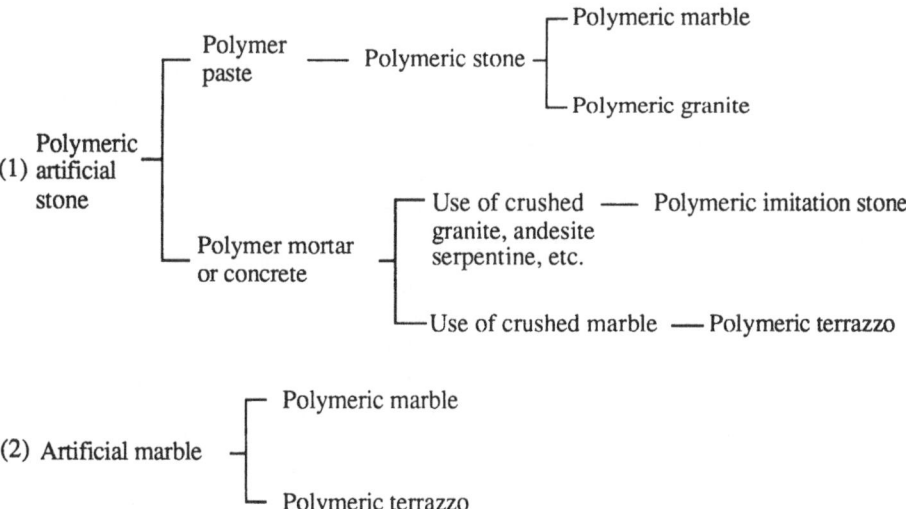

Fig.1. Classification of polymeric artificial stones in Japan.

Table 3. Applications of polymer mortar in Japan

Application	Location of work
Floorings (including decorative finishings)	Floors for houses, warehouses, offices, schools, hospitals, factories and shops, toilets, passages, stairs, garages, railway platforms, train floors,etc.
Pavements	Roads (sidewalks and roadways), bridge decks, foot-bridge decks, parking lots, airport runways, etc.
Anticorrosive linings	Effluent drains, chemical or machinery plant floors, grouts for acid-proof tiles, floors for chemical laboratories and pharmaceutical warehouses, septic tanks, electrolytic baths, hot spring baths, offshore structures, e.g., piers and sea berths,etc.
Adhesives	Adhesives for floorings, walling materials and heat-insulating materials, tile adhesives, adhesives for joining new cement concrete or mortar to old cement concrete or mortar, embedment of anchor bolts, etc.
Repair materials	Grouts for repairing cracks and delamination of concrete structures, patching materials for damaged concrete structures, rustproof coatings for corroded reinforcing bars, resurfacings for damaged concrete pavements, etc.
Integral waterproofings	Concrete roof floors, mortar walls, concrete blocks, water tanks, swimming pools, septic tanks, silos, etc.
Small-diameter automated shield tunneling systems	Telecommunication cable lines, sewageworks, etc.
Precast products	FRP-sandwiched polymer mortar pipes for sewage and irrigation systems, and for penstocks at hydroelectric power stations,highly durable or decorative permanent forms for reinforced concrete structures, FRP-sandwiched polymer mortar panels for conduit lines or headrace tunnels, vibration isolators, artificial marble tiles and panels, etc.

Table 4. Applications of polyester concrete in Japan

Application	Location of work
Structural precast products	Manholes and handholes for telecommunication cable lines, electric power cable lines and gas pipelines, prefabricated cellars, or stack rooms, tunnel liner segments for telecommunication cable lines and sewerage systems, piles for port or hot spring construction, forms for reinforced concrete structures, FRP-reinforced frames or panels for buildings, machine tool structures, e.g., beds and saddles, works of art, e.g., carved statue and objet d'art, tombs for buddhists, etc.
Nonstructural precast products	Gutter covers, U-shaped gutters, footpath panels, terrazzo tiles and panels, and large-sized or curved decorative panels for buildings, partition wall panels, sinks, counters, washstands, bathtubs, etc.
Cast-in-place applications	Spillway coverings in dams, protective linings of stilling basins in hydroelectric power stations, coverings of checkdams, foundations of buildings in hot spring areas, acid-proof linings for erosion control dams with acidic water, etc.

4 Conclusions

To cope with the technical innovations in the construction industry in recent years, useful polymer mortar and concrete as high-performance and multifunctional construction materials have actively been developed in Japan. In particular, a great interest is currently focused on repairing systems for deteriorated reinforced concrete structures, permanent forms (from the viewpoints of the durability improvement of reinforced concrete structures and the preservation of forest resources), automated cast-in-place application systems and artificial marble products. A development competition of the artificial marble products has become cutthroat for the past ten years or more.

5 References

1. Ohama, Y., Demura, K., Kobayashi, K. and Matsueda, H. (1993) Basic Properties of Polyurethane Methacrylate Mortars, in *Proceedings of the Third Japan International SAMPE Symposium Vol.1*, (eds. T. Kishi, N. Takeda and Y. Kagawa), Japan Chapter of SAMPE, Yokohama, pp.190-193.
2. Ohama, Y. and Demura, K. (1994) Basic Properties of Polymer Mortars Using Composite Binders with Unsaturated Polyester Resin and Vinyl Monomers. *Journal of the Society of Materials Science, Japan*, Vol.43, No.491, pp.997-1003.
3. Ohama, Y. (1994) Recent Trend in Research and Development of Polymer Mortar and Concrete in Japan, in *Proceedings of the First East Asia Symposium on Polymers in Concrete*, (eds. K.S. Yeon and J.D. Choi), Kangwon National University, Chuncheon, Korea, pp. 11-23.
4. Lee, Y., Ohama, Y., Demura, K. and Ide, K., Influences of Mix Proportioning Factors on Properties of Lightweight Polyester Mortars, in Press.
5. Kuromoto, M., Kawamura, A., Asai, S. and Sumita, M. (1996) Characterization of Fillers Applied to MMA Polymer Concrete (in Japanese). *Proceedings of the Japan Concrete Institute*, Vol.18, No.1, pp.489-494.
6. Kitagawa, S., Ohama, Y., Demura, K. and Ozaki, T. (1996) Predictions of Working Life and Strengths of Polyurethane Methacrylate Mortars at Low Temperature (in Japanese). ibid., pp.483-488.
7. Ohama, Y. (1994) Recent Research and Development in Concrete-Polymer Composites, in *Advances in Concrete Technology, Second Edition*, (ed. V.M. Malhotra), Canada Centre for Mineral and Energy Technology, Ottawa, pp.753-783.
8. Fukuzawa, K., Numao, T. and Yoshimoto, M. (1994) Flexural Behavior of Polymer Concrete Beams Reinforced with Fiber Reinforced Plastics Rods, in *Proceedings of the First East Asia Symposium on Polymers in Concrete*, (eds. K.S. Yeon and J.D. Choi), Kangwon National University, Chuncheon, Korea, pp. 325-335.
9. Moriyoshi, A., Hirano, T., Itami, K., Tokumitsu, K., Takahashi, M. and Nagata, S. (1995) Thermal Properties of Polymer Concrete Using GM/St Resin at Low Temperatures (in Japanese). *Sekiyu Gakkaishi*, Vol.38, No.1, pp.57-61.

10. Moriyoshi, A., Tokumitsu, K., Hirano, T., Ogasawara, A. and Nagata, S. (1995) Thermal Properties of Polymer Concrete Using Glycerol Methacrylate/Styrene System at Low Temperature, in *Proceedings of the Eighth International Congress on Polymers in Concrete*, (ed. D.Van Gemert), Technological Institute of the Royal Flemish Society of Engineers, Antwerp, pp.509-514.

11. Nguyen, V. L., Uchida, Y., Hayashi, F. and Koyanagi, W. (1994) Stresses Due to Setting Shrinkage in Polyester Resin Concrete (in Japanese). *Proceedings of the Japan Concrete Institute,* Vol.16, No.1, pp.663-668.

12. Tsutsumishita, T., Kurita, A., Tokuoka, F. and Konishi, H. (1994) A Study of the Behavior of Hardening Shrinkage of the Mathacrylic Resin Concrete Bonded to Damaged RC Slabs (in Japanese). ibid., pp.1013-1018.

13. Tsukamoto, T. and Okuno, T. (1996) New Shrinkage-Reduced Unsaturated Polyester Resin for Polymer Concrete (in Japanese). *Reinforced Plastics*, Vol.42, No.5, pp.189-191.

14. Nagai, K. and Ujikawa, N. (1994) Development and Evaluation of a New LPA (in Japanese), in *Proceedings of the 39th FRP CON-EX'94*, Japan Reinforced Plastics Society, Tokyo, pp.358-361.

15. Hayashi, F., Oshima, M. and Koyanagi, W. (1996) Thermal Properties and Temperature Dependence of Mechanical Properties of Resin Concretes for Structural Use (in Japanese). *Journal of the Society of Materials Science, Japan*, Vol.45, No.9, pp.1014-1020.

16. Yamada, Y., Iwai, T., Kawamura, A. and Kuromoto, M.(1994) Temperature Dependence of Mechanical Properties of Polymer Concretes Using Methyl Methacrylate (in Japanese). *Proceedings of the 49th Annual Conference of the Japan Society of Civil Engineers*, No.6, pp.232-233.

17. Yamasaki, T., Idemitsu, T., Watanabe, A. and Miyakawa, K. (1991) A Study on Creep Characteristics of Unsaturated Polyester Resin Concrete (in Japanese). *Journal of the Society of Materials Science, Japan*, Vol.40, No.456, pp.1178-1184.

18. Yamasaki, T., Idemitsu, T., Futajima, K. and Saita, K. (1994) Study on Durability of Polyester and Vinyl Ester-Type Resin Concretes for Chemical Attack (in Japanese). *Proceedings of the Japan Concrete Institute*, Vol.16, No.1, pp.865-870.

19. Tanaka, J., Matsuura, M., Ishida, R. and Suzuki, T.(1993) Study of Durability of Concretes with Polymer Mortar Linings in a Sewage Treatment Plant (in Japanese). *Proceedings of the 48th Annual Conference of the Japan Society of Civil Engineers*, No.5, pp.186-187.

20. Kurita, A., Tsutsumishita, R. and Umeda, M. (1996) Fatigue Behavior of Polymethyl Methacrylate Concrete for Additional Thickness Strengthening Technique for Reinforced Concrete Deck Slabs of Highway Bridges (in Japanese). *Proceedings of the 51st Annual Conference of the Japan Society of Civil Engineers*, No.5, pp.444-445.

21. Kuromoto, M., Iwai, K., Kawamura, A. and Akiyama, Y. (1994) Strength Properties and Durability of Polymer Concretes Using Methyl Methacrylate (in Japanese). *Proceedings of the 49th Annual Conference of the Japan Society of Civil Engineers*, No.6, pp.230-231.

22. Kondou, S., Kuroiwa, M. and Kurahashi, W. (1989) Report on Large Depth, Long Distance Execution of Construction by Small-Diameter Automatic Shield Tunneling System Employing the Cast-in-Place Method, *Presented at the International Congress on Progress and Innovation in Tunneling*, Toronto, pp.1-8.

23. Kimura, Y., Ohtake, S. and Kobayashi, Y. (1995) Influence of Calcium Carbonate Filler on Resin Mortar for Tunnel Construction, in *Proceedings of the Eighth International Congress on Polymers in Concrete*, (ed. D.Van Gemert), Technological Institute of the Royal Flemish Society of Engineers, Antwerp, pp.393-398.

24. Ohtake, M., Tsuchiya, M. and Takigami, T. (1994) Control of Mixing Ratio of Premixed Polymer Mortar and Promoter in Automated Shield Tunneling System for Polymer Mortars (in Japanese). *Proceedings of the 49th Annual Conference of the Japan Society of Civil Engineers*, No.6, pp.76-77.

25. Noro, K., Sakurada, T., Kobayashi, T. and Kawai, K. (1994) Long-Term Storage of Polymer Mortars (in Japanese), ibid., pp.236-237.

26. Ohtake, M. and Kobayashi, Y. (1995) Execution Management Technique for Polymer Mortars for Automated Shield Tunneling System (in Japanese). *Proceedings of the 50th Annual Conference of the Japan Society of Civil Engineers*, No.6, pp.110-111.

27. Maruyama, T., Sakurada, T., Kobayashi, Y. and Takigami, T. (1996) Strength Increase of Polymer Mortars as Linings for Automated Shield Tunneling System (in Japanese). *Proceedings of the 51st Annual Conference of the Japan Society of Civil Engineers*, No.5, pp.442-443.

28. Anon. (1996) "Shot-Rem" Construction Method (in Japanese), in *Nikkei Construction Books*, Nikkei BP, Tokyo, p.199.

29. Bhutta, M.A.R., Ohama, Y. and Demura, K. (1993) Polymethyl Methacrylate Concrete for Underwater Construction, in *Concrete 2000 Economic and Durable Construction through Excellence Vol.2*, (eds. R.K. Dhir and M.R. Jones), E & FN Spon, London, pp.1061-1070.

30. Bhutta, M.A.R., Ohama, Y. and Demura, K. (1995) Cement Paste Primers for Underwater Adhesion of Polymethyl Methacrylate Mortars to Cement Mortar Substrates, in *Proceedings of the Eighth International Congress on Polymers in Concrete*, (ed. D.Van Gemert), Technological Institute of the Royal Flemish Society of Engineers, Antwerp, pp.521-526.

31. Bhutta, M.A.R., Maeda, N., Kawano, T., Ohama, Y. and Demura, K. (1995) Underwater Adhesion of Polymethyl Methacrylate Mortar to Cement Concrete Substrates, in *Proceedings of the Fourth Japan International SAMPE Symposium*, (eds. Z. Maekawa, E. Nakata and Y. Sakatani), Japan Chapter of SAMPE, Yokohama, pp.1364-1369.

32. Kobayashi, S. (1995) Precast Concrete Forms Approved by the Newly Developed Civil Engineering Technology Assessment and Verification System (in Japanese). *Cement & Concrete*, No.582, pp.34-38.

33. Tokunaga, G. (1996) Precast Concrete Panels at the Honshu-Shikoku Bridges (in Japanese). ibid., pp.64-74.

34. Idemitsu, T. (1996) Precast Concrete Formwork Used for Civil Structure (in Japanese). ibid., pp.13-18.

35. Anon. (1994) Filling the Rutting with 200m³ of Polymer Concrete (in Japanese). *Nikkei Construction*, No.105, pp.28-29.

36. Anon. (1996) Filling the Rutting with 200m³ of Polymer Concrete: Pavement Repair Work of Higashi-Kuriko Tunnel (in Japanese), in *Nikkei Construction Books*, Nikkei BP, Tokyo, pp.12-13.

37. Anon. (1996) Raising Method for Expansion Devices of Highway Bridges (in Japanese). ibid., p.113.

38. Yamada, T., Fujita, S., Nakaoka, Y. and Koizumi, S. (1993) Repair Method Using FRPM (Fiberglass Reinforced Plastics Mortar) Panels and Its Characteristics (in Japanese). *Proceedings of the 48th Annual Conference of the Japan Society of Civil Engineers*, No.6, pp.440-441.

39. Ono, S., Kishida, S., Ohnishi, Y. and Yamauchi, K. (1995) Design of Permanent Forms Using Polymer Concrete (in Japanese). *Proceedings of the 50th Annual Conference of the Japan Society of Civil Engineers*, No.6, pp.228-229.

40. Ono, T., Matsuo, K., Kishida, A. and Ono, S. (1994) An Experimental Study on Bending Behavior of RC Beams Using Resin Concrete Embedded Forms. *JCA Proceedings of Cement & Concrete*, No.48, pp.826-831.

41. Ono, T., Matsuo, K., Yamauchi, K., Takenaka, H., Maeda, T. and Yamato, T. (1995) Development of High-Durability Precast Embedded Form Utilizing Resin Concrete and Three-Dimensional Steelmesh (in Japanese). *Journal of Construction Management and Engineering*, Vol.26, No.510, pp.23-30.

42. Kondou, S., Kuriyama, K., Komori, M. and Enomoto, H. (1995) Performance of Decorative Permanent Forms (in Japanese). *Proceedings of the 50th Annual Conference of the Japan Society of Civil Engineers*, No.6, pp.108-109.

43. Anon. (1992) Artificial Marble with Stone Texture (in Japanese). *Nikkei New Materials*, No.108, pp.26-28.

44. Anon. (1992) Artificial Marble (in Japanese). *Engineering Materials*, Vol.40, No.11, pp.114-118.

45. Goh, Y., Izumi, H., Tanaka, K., Tadaoka, E. and Murata, M. (1996) Vinyl Ester Resin for Artificial Marble (in Japanese). *Reinforced Plastics*, Vol.42, No.5, pp.185-188.

46. Tanabe, I. (1993) Development of Ceramic Resin Concrete for Precision Machine Tool Structure (Young's Modulus and Compressive Strength of the Ceramic Resin Concrete). *JSME International Journal*, Series C, Vol.36, No.4, pp.494-498.

47. Kitamura, T. (1994) Market Development for Recycling Thermoset in Japan, in *Proceedings of the 49th Annual Conference of the Composites Institute*, The Society of the Plastics Industry, New York, pp.1-5/Session 15-F.

48. Tsukinaga, T., Sekikawa, S. and Tamaru, S. (1996) Study of Development of Construction Materials by Using Scallop Shells (in Japanese). *Architectural Institute of Japan Tohoku-Shibu Kenkyu-Hokokushu*, No.59, pp.353-356.

PART TWO
MANUFACTURING TECHNICS

CONTINUOUS PRODUCTION OF POLYMER CONCRETE — A NEW GENERATION OF MACHINES

R. Kreis
Respecta–KWM Kunststoffmaschinen GmbH.,
Wülfrath, Germany

Abstract

The properties of polymer concrete are considerably influenced by the preparation method of the mix. To reach optimum properties in the compound best suitable methods are necessary. Manual metering and batch mixing operations have a lot of disadvantages, which can be overcome by using continuously working casting machines, which are available in all sizes and for all kinds of binder systems. Continuous casting machines can also handle foaming binders, which cannot be operated in conventional batch mixers.

The paper describes the latest developments of continuous preparation machinery for polymer concrete and explains some examples of full automatic manufacturing processes for prefabricated products of polymer concrete.

Keywords: casting machinery, compound preparation, metering, mixing, continuous production

Major criteria for preparation machinery

Regardless of the preparation method, whether batchwise or continuous, the working equipment must meet essential requirements, as there are:

- The metering of the single components must be exact and precise.

- The mixing of the ingredients must be homogeneous, degration of materials has to be avoided, even if materials with big differences in gravity are mixed.

- The preparation process should be fast to be able to make use of the short curing times and to avoid the volatilization of reactants necessary for the process.

Polymers in Concrete, edited by Y. Ohama, M. Kawakami and K. Fukuzawa. Published in 1997 by E & FN Spon, 2–6 Boundary Row, London SE1 8HN, UK. ISBN: 0 419 22330 4.

- The cleaning of the equipment must be easy and fast with lowest demands of cleaning agents, particularly as most polymer concrete formulations require chemical solvents for cleaning and not just harmless liquids like water.

- Even big polymer concrete parts should be produced as units, different mixes in one product mean different physical and chemical properties in various areas of the product.

- The preparation process must be economical, it should be arranged as simple as possible but as sophisticated as senseful to minimize cost.

- The preparation process should require as less people as possible, not only because of economical reasons but because of the health risks linked with every chemical process, even if the risk in case of polymer concrete is very small.

- The preparation process should be easy adaptable to the varying production demands. It should be possible to change formulations, capacities, etc. easily and fast in order to reach maximal flexibility in the manufacturing process.

It is quite obvious - the development in the industry during the last 25 years has shown it - the manufacturing processes of polymer concrete parts all over the world are more and more automated by using sophisticated equipment to meet the requirements of modern industries in regard to economical and ecological demands.

Batch mixers with batch meterings disappear more and more out of the factories and are replaced by continuously operating metering-mixing- and casting machines, which satisfy the requirements of industrial mass production as well as custom moulding. The basic disadvantage of such machines compared to batch mixers - the higher investment cost - is compensated by a lot of advantages. Continuous metering and mixing means always uniform conditions, casting machines work precisely and fast, they allow lowest binder contents, and they are easy to clean with lowest amounts of solvents. Continuous casting machines maximize the production rates and they help a lot to avoid strong odorous annoyances, skin irritations, etc., which may effect the health of people.

The development of modern casting machines

Continuous metering- mixing and casting machines have been known for more than 25 years. Although the basic principle of such machines - separate, continuous metering of all single components and then continuous, homogeneous mixing and dispensing - has remained the same, there have been a lot of changes and developments on the machines during the years of which a few are described hereafter.

Premixing of different aggregates or different grain sizes of aggregates before they enter the casting machine is not necessary any more. The RESPECTA-casting machines are able to handle as many separate aggregates as necessary by themselves. Degration is therefore impossible, the formulation of the granulometry can be changed and adapted to the necessities within seconds, special additives like lightweight aggregates, reinforcing material, foaming agents, etc. can be handled, and this new system in total requires less cost as special metering devices and premixers for the aggregates are not needed any more.

Technical development and improvement are particularly evident when the metering systems for the reactants are considered. 20 years ago the materials, even rather dangerous peroxides, had to be stored in pressure tanks and from there to be metered by air pressure. (It is hard to believe, but even today such systems are still in use in single cases.) Later special gear pumps for transportation were developed, which could be used in combination with special designed metering devices, and which allowed to avoid the dangerous pressure tanks. Gear pumps and quantity regulators were then replaced by metering diaphragm pumps and double diaphragm pumps, which meant further improvement and further reduction of danger. Today small, high-speed metering pumps are used, allowing highest metering accuracy in large ranges of adjustment without warming up the flow media even at high speeds. These pumps do not require any dynamic seals and mean absolute safety against leakages.

Modern computer-controlled casting machine, suitable for the production of polymer concrete, mortar, and also marble.

For a long time it was normal to regulate the different metering quantities with the speeds of the pumps or metering devices via mechanical transmissions. By now using frequency transformers the metering accuracy is improved, the ranges of adjustment are remarkably enlarged, and the necessary times for adjustments are reduced to a minimum.

Besides the single metering drives now also the main mixer drive is adjustable, which allows an immediate adaption of the casting capacities to the actual necessities without changing the formulation, when, for example, different moulds are used on the same conveyor system. This development has also led to an essential reduction of the wear of the mixing tools. In many operations the tool lives could be more than doubled.

Today SPS-steerings in combination with computers are normal in continuous casting machines. All single data and complete formulations are preprogrammable and all data are automatically surveyed and registered. This enables the producer to keep his production and even each single product under steady control. The producer is always informed about the situation in the factory and he has all data available even long after the finished product has left the factory, which means a great help in cases of later complaints.

Modern casting machines can be made mobile in all three dimensions and can be equipped with special visual systems, with which barcoded moulds can be recognized and the adjustments of the casting machine, for example the single metering quantities, the output, and/or the machine position, etc. can be changed and adapted to the specific requirements of a particular mould.

Generally it can be said that today each factory can obtain the specific machine for its specific needs according to the rule "as simple and cheap as possible, but as sophisticated as necessary or senseful".

In the following chapters three different operations are described, in which various products of polymer concrete in differnt ways of organization are manufactured.

Continuous production of sewer and agricultural parts

The factory produces various items for the sewer and agricultural industry with average weights of about 15 kgs/-piece. The polymer concrete mix is composed of silica sands and a non-shrinking unsaturated polyester resin with the necessary activators. The production line comprises:

- Storage of raw materials and transport to the casting machine
- Mould filling and compacting station with casting machine and vibrating table
- Mould transport with entrance and exit station
- Release agent application station
- Demoulding station
- Storage for finished products

Depending on the production requirements the necessary moulds circulate on an automatic mould conveying system, electrically powered roller trains and belts transport the moulds to the different working stations in the factory. For changing moulds the conveying system is equipped with special sideways where moulds not needed any more can be taken out of the circulation and new moulds can be added. Each mould is numbered and barcoded. The casting machine is arranged on a two-way-rail system and can also be lifted hydraulically so that it is totally mobile in all three dimensions. Short before a mould reaches the filling station a barcode scanner reads the mould number and transfers it to the computer of the casting machine. The casting machine is automatically provided with all necessary data for filling the concerned mould, as there are:

- Product number
- Mould volume and weight
- Necessary mix formulation
- Filling speed
- Necessary position of the casting machine
- Vibrating frequency and time
- Different other data

The adjustments at the casting machine are made within parts of a second, and - as practical experience shows - it is possible to fill and compact each mould precisely according to its specific requirements, even if two following moulds require rather different mixes and filling conditions.

Two casting machines for different casting capacities, both mobile in three dimensions

Immediately after the moulding process is completed, the curing process in the polymer concrete can start and by the time the mould needs to reach the demoulding station, the polymer concrete mix is cured so far that the product can be demoulded without difficulties. For demoulding the side parts of the outer mould are opened and the finished product is taken from the inner mould by a gripping device, which places the finished product directly into the storage area. An automatic locking device closes the mould again so that it then can be transferred to the release agent station and from there back to the casting machine for the next filling process.

All data of each produced item are kept in the computer data bank and available for controlling and in cases of possible later complaints. With the formulation data the consumption of the raw materials is automatically calculated and these data are taken as basis for reordering the different materials.

In the above described factory two complete production lines are installed. Both work from the same raw material storage but produce independently from each other. The smaller capacity line works with about 10 kgs/min., the larger one with about 30 kgs/min., the total capacity of the factory is about 2,5 tons of polymer concrete products in each hour. As the company works normally four shifts, it reaches a production of more than 50 tons every day. Due to the automatization process the complete factory with the two production lines only requires 4 people per shift to manage and control the operation.

Production of filter pipes

This process was at first developed for vertically installed drainage pipes for lignite beds, later it was found out that the same pipes were also suitable for deep wells in dry areas to gain drinking water. The polymer concrete used in this application is prepared with only coarse silica gravel (grain size about 2-4 mm) without any fines so that the finished products do not have dense structures with closed surfaces but open porous structures with about 1/3 of the pipe surfaces open so that the ground water can penetrate from the outside into the pipes, from where it then is pumped out. The filter pipes are usually produced in pieces of 1 m length, glued in the factory to sections of totally 6 m, and installed in the field section after section in depths down to 200 m. The regulations for drinking water in this case are met by using a polymer concrete with a specially developed hot curing polyurethane resin as binder.

The manufacturing process can be divided into the following sections:
- Storage of raw materials
- Preparation equipment for the polymer concrete
- Mould arrangement
- Mould filling equipment

Overall view into the production hall for filter pipes

- Demoulding and transporting manipulator
- Plant steering
- Gluing division
- Storage of finished products

As the above picture shows the manufacturing process is designed very compact in a rather small area. The whole production hall comprises a surface of about 1.100 sqm, of which about 900 sqm are used for storage of the raw materials and the finished products while only about 200 sqm are needed for the manufacturing process itself.

The continuous casting machine is automatically filled with the necessary raw materials, in this case only one aggregate and two liquids. Due to the chemical requirements of the binder the mixing process is divided into two steps, first the liquids are mixed and then the aggregat is added.

For a continuous production several mould are necessary.Considering the essential parameters - desired total capacity, working time, curing time of the pc, mould filling time, handling time, etc. six equal casting moulds have been set up in an row. Each mould consists of an inner an outer mould and of a base ring, all these parts made of steel. Inner and outer mould, both are heated hydraulically by a heat exchange oil to a temperature of about 100°C in order to cure the polymer concrete properly. The outer moulds are additionally equipped with vibrating motors supporting a fast mould filling and a good compaction of the polymer concrete in the moulds. All six moulds are installed below floor level, inner and outer moulds are stable, the base rings are mobil and can be moved hydraulically over the mould lengths between inner and outer mould. As already said before in this case pipe pieces of 1 m length are produced but in the same moulds shorter pipes, too, can be made just by changing the positions of the base rings.

The moulds standing beside each other with equal distan-

ces are filled one after the other with polymer concrete
directly from the casting machine. This means the moulds do
not travel to the casting machine as usual in most pc-pro-
ductions, but the casting machine must travel to the moulds.
Therefore, the casting machine and all other equipment
necessary for the mould filling are arranged on a platform
which runs on rails over the mould openings on floor level.
The filling time for 1 mould is about 1 minute, then the
casting process is stopped for a short time during which the
platform travels to the next mould, and then the casting
machine starts a new mould filling. During the filling of
the second mould the pipe in the first mould is already
worked by an edge trimming unit also arranged on the platform.

At normal working conditions 1 mould filling cycle, i.e.
the filling of all 6 moulds, takes totally 9 minutes of
which 6 minutes are needed for the real fillings and 3
minutes for the travelling time between the moulds and back
to the starting position before the first mould. After every
5 fillings cycles the mould surfaces must be supplied with
release agent. This is done with a spray unit also arranged
on the platform which is sunken into the moulds and moves
along the mould lengths spraying the necessary release film.

Immediately after a mould filling cycle is finished the
automatic demoulding process of the pipes starts. In the
same sequence as the moulds have been filled they are demoulded by pressing the base rings and so the pipes standing
on them hydraulically from the low position to the top. In
this way the pipes are pushed out of the moulds and they
finally stand above the moulds free to be taken by a special
manipulator.

The complete production process and the single sections
are controlled by a process computer hierarchy supervised by
one person only. Eventually necessary manual interferences,
e.g. changing the cycle times, the pc-formulation, etc., can
be made without any problem at any time.

The total capacity of the described manufacturing process
with 6 moulds the mentioned cycle times, and 2-shifts working time is about 70.000 pipe pieces per year meaning a
consumption of polymer concrete of about 5.000 tons.

Production of sewer pipes

Fully continuous manufacturing processes of pipes - i.e.
processes, in which not only the preparation of the mix is
continuous but the moulding process, too - are known from
the winding technique, with which sewer pipes composed of a
combination of polymer concrete and glass fibre reinforced
plastic are produced. Recently, however, a new technology
has been developed, which allows the fully continuous production of pure polymer concrete pipes without any fibres or
other additives in a simple casting process.

The process is laid out for a polymer concrete using
graded silica sands as aggregates and a standard unsaturated
polyester resin as binding agent with reactants for a warm

curing. Aggregates (three different grain sizes) and resin, both are preheated to about 50°C, metered and mixed with the reactants in a sophisticated casting machine, and the polymer concrete mix runs out of the machine nozzle directly into the moulding tool, which is also heated. Opposite to the process for the filter pipes described above this technology does not require several moulds, one moulding tool is sufficient. In this special technique - details cannot be published yet, worldwide patents have been applied - the pasty polymer concrete passes the moulding tool and is cured during this passing time so that then a moulded pipe sufficienty cured runs out of the moulding tool continuously with a speed of 0,75 - 1,5 m/min. Inside and outside surface of the pipe are absolutely smooth, the pipe structure is dense, the physical and chemical properties meet the requirements. At the exit of the moulding tool the pipe is cut to the required lengths. Today pipes till 2,5 m length are produced, theoretically, however, all kinds of pipe lengths can be made, this is only a question of the geometry of the plant. The cut pipe lengths are picked up by gripping device and taken into a tempering oven, where they stay at about 100°C for about 30 minutes for final curing. Then the hot pipes leave the oven and are cooled down slowly to a temperature of about 40°C. The heat taken from the pipes is transferred back into the manufacturing process and used to preheat the aggregates.

During the complete manufacturing process the polymer concrete is kept in fully closed areas without any emissions of toxic vapors, odors, etc. After final curing the remaining content of styrene is less than 0,1 %, i.e. lower than in the rules for food materials required.

In this fully continuous operation only calibrated pipes can be produced, necessary seal systems must be added to the pipes later. If, for example, sleves or sockets are required

Model of the new continuous production plant for sewer pipes

Major part of the plant - the continuous mix preparation and the continuous moulding

they are cast on the pipes immediately after curing, when the pipes are still hot.

During the manufacturing process all data of the process are registered by the computer of the casting machine, and additionally each pipe is tested mechanically so that absolute quality control is secured.

The complete plant runs automatically with an extreme low percentage of cost for manpower. The producer of the pipes expects a much larger share of the pipe market than before due to the excellent technical properties and the rather low cost of his pipes.

Conclusions

After a long time during which the development of new materials, new formulations, and improvement of the properties of polymer concrete were placed in the foreground, now many attempts are made to improve the manufacturing processes in the sense of more automatization and easier handling and control. More and more people accept that money for sophisticated equipment is a good investment, which is paid back in short time. Producers using the new possibilities expect much bigger market shares so that the acceptance of polymer concrete should be further increased.

APPLICATION OF EPOXY EMULSION CEMENT MORTAR IN CHINA

G.H. Liu, Y.P. Xu and H.Y. Chen
China Building Materials Academy, Beijing, China

Abstract
The paper describes that Epoxy emulsion cement mortar (EECM) has been used in many fields in China recently, such as, waterproof, corrosion resistance, repairing & reinforcing treatment and bond & primer coat engineerings, etc.
keywords: Epoxy emulsion, impermeable, corrosion, patch, reinforcement, bond.

1 Introduction

EECM is one of polymer cement concrete (PCC). It is composed of emulsified diphenol A epoxy resin, cement and fine aggregate. EECM has advantages of good workability, high bond strength to cement mortar, stone, ceramic tiles and cast stone, etc., high inherent strength, water-resistance, alkali-resistance and durability. It can be used for patching & strengthening concrete and concrete products in order to improve impermeability and corrosion resistance. Moreover, it is used as the floor, the liner with corrosion resistance of pipelines and container in Chemical plant as well as the bonding materials for natural and artificial stone, ceramic tiles etc.

Since the late 1980's, we have been making use of EECM in the impermeable engineering (to air and water), corrosion resistance engineering, patching and reinforcement engineering and bonding engineering, etc.. As a result, we have gained the benefits of social and economic very well. The main uses of EECM are summarized as follow.

Polymers in Concrete, edited by Y. Ohama, M. Kawakami and K. Fukuzawa. Published in 1997 by E & FN Spon, 2–6 Boundary Row, London SE1 8HN, UK. ISBN: 0 419 22330 4.

2 Impermeable Engineering

The barn of BingDing wharf on QinHuangDao Port is the largest cylindric silo of concrete construction for bulk grain in Asia. It consists of 48 main silos and 30 star ones which are tangent to main silos. The diameter of the main silos is 8m and the height is 39m. Its operating control equitment is imported from Japan. The barn can store grain of 63500t. Capacity of loading and unloading is 600 t/h. It is the most fully equipped and the top automation of the bulk grain barn in China, shown as Fig.1.

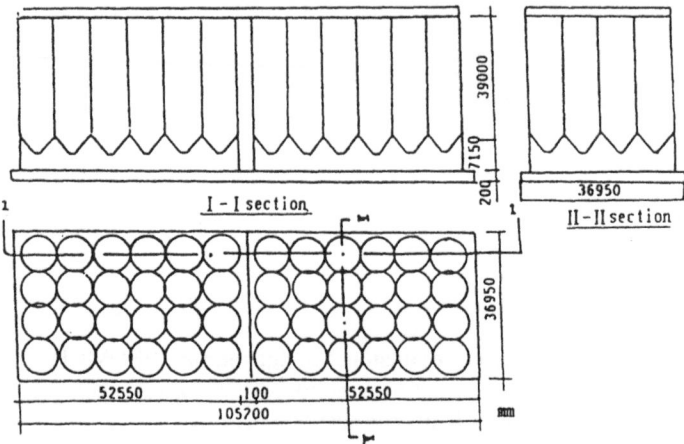

Fig.1 The barn structure drawing

According to the demand of epidemic prevention, imported grain must be fumigated to kill insects. The air-tightness of the barn has to be up to the grade A of JIS, that is, the barn is pressed to 5kPa, the remain of pressure in the barn must be higher than 2kPa after 20 min. Concrete itself is a porous material, so its air-tightness is unsatisfactory, the silos were constructed with slipforms and the beams, columns and panels were precast and assembled on the inner tops of silos which lead to their poor airtightness and fumigation of grain can't be carried out in the silos. The impermeable treatment is necessary to assure requirement of impermeability of silos. All surfaces of the barn were treated with different methods in accordance to their various conditions, then EECM was used for embedding and coating the barn of the silos. After 3d curing, tightness test was conducted.

The fumigation equipment was used for pressurizing the barn, the blast volume was 55 m^3/min.

Before testing, all of the valve and the passages were closed, by starting the fan and opening the vent valve, making the pressure increased gradually. When the pressure of the barn reached 5kPa, the valve was closed. The pressure of the barn and the time were measured. The result of the testing was shown in Fig.2.

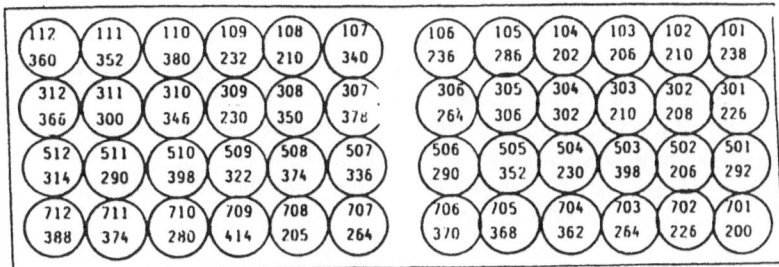

Fig.2 Result of testing for airtightness of the silos

The tested result shows that the airtightness of all the silos satisfied the grade A of JIS, and 66.7% main silos keep residue pressure over 2.5kPa. Thus, conclusion can be drawn as follows.

(1) Airtightness of the barn in Qinhuangdao port treated with EECM is up to the grade A of JIS.

(2) It is effective to use the EECM as airtight sealant.

3 Patching of Concrete Pressure Pipe for Delivery Water

While concrete pressure pipes are in the process of manufacturing, some factors, for instance, improper operation and the flaw of molds may cause some defects such as surface dusting and honeycomb in the spigot joint, and mortar flowing off at the joint of pipewall which cause pipes leakage, make the pipes into waste products. Epoxy resin mortar is usually used for patching the waste pipes, but there are some short-comings in the treatment, such as the surface of the pipes must be dried out,when epoxy resin mortar is coated on it, The large difference in thermal expansion coefficient between epoxy resin and concrete exists, the patching layer may break off from the concrete products in service life. EECM can overcome the shortcomings. Because of the small difference in expansion coefficient between EECM and concrete, patching layer bonds sound & service life lasts longer. In addition, operation of patching is simple without heating epoxy resin and repaired substrate, cleaning the container and tools used for patching is easy without toxic organic solvent resulting in no pollution to enviroment so that cost of repair can be decreased by about 30%, As an example, Guangzhou Panyu cement product plant, qualified ratio of product was low during the early production period and leakages of the pipes were serious because of the inaccurate equipment and improper operating,. When unqualified pipes are repaired with EECM, the patching surface of pipes must be treated first. To a leakage joint, A V groove was made on the place and thin slurry of EECM was brushed into the V groove to increase EECM bonding force to the pipe. Then, the suitable proportion EECM was filled into the V groove and evened up. When patching a pipe wall or a spigot joint, it is cleared with a metal brush and coated with EECM for 2-3 times, then the pipe was put into a water pool and cured for 3d. In consequence, all of 1000 repaired pipes satisfied the quality standard of the plant, namely the pipes can stand the pressure of 1.2 MPa for 20 min.

As other example, Nanping cement pipe plant have used EECM for patching 5000 concrete pipes of $\Phi 500$ mm and $\Phi 300$ mm within ten years, the qualified ratio of the patched concrete pipe reached to 100%. The cement pipe plant (with Beijing water conservancy first department) used to make a $\Phi 600$mm × 5m concrete pipes with a suspension technique (introduced from Australia), the leakages of the pipes was serious, qualified ratio of products was as low as only 40%. After patching with EECM, it got to above 95%.

4 Corrosion Resistance Engineering

Nanping paper-mill in Fujian province is one of the four largest news-paper mill in China, which drains off sewage (with acid sewage of paper pulp) about 40000t/d. The delivery pipelines were corroded seriously, the pipeline needed replacing, so that the production could not be in continuously process. By coating the inner wall of the concrete pipes with EECM, the problem of the corrosion was solved. The treatment method was described as follows: $\Phi 800$ mm self-stressing concrete pipes were formed by means of centrifugal method with ferroaluminate self-stressing cement. After the pipes curing, the inner walls were coated with EECM several times. The first layer of EECM was composed of emulsion : cement : fine aggregate : curing agent = 1:1.8:2.5:0.3, the EECM coating has 3-5 mm in thickness. The second layer of EECM was composed of emlusion:cement:fine:aggregate:hardener = 1:1:0.8:0.3 which was coated on the first coating to form a smooth surface layer. This pipeline is 2.5 km long and the inner surface of the pipeline coated with EECM keeps very well after used for 5 years.

In the diamond workshop of exploring team of GMR in Yanjiao, Hebei, the acid washing floor of the workshop was made of ordinary concrete and corrosion-resistance coating which was attacked seriously and service life of the floor was less than 2 years. After the floor was coated by 10 mm thick with EECM, the wall was coated with EECM. The floor and the wall are yet in good condition after used for several years.

There are many sluice gates along with the coast in Fujian, some of which are RC gates, the others are steel gates. These gates have been corroded seriously by sea water splashing and some was in bad condition. After the gates was treated with EECM by the institute of water conservancy with water electricity department of Fujian, the gates keep in good condition for 4 years.

5 Reinforcement & Bonding Engineering

At the Huadi station of subway in Guanzhou, honeycombs and pocks occurred in the concrete structure elements, which caused local leakage and the reduction of concrete strength as well. Luban waterproof & strengthening Co. of Guangzhou treated above sation with EECM. They gained a good result. In addition, the 1500m^2 of basement of Date Building in Guangzhou seeped seriously, concrete beams and posts of Ganzhou Lushui Village had large area of honeycombs which affected quality of the concrete structure seriously, they used EECM for waterproof and reinforcement treatment, the results satisfied users.

In the bottom slab the 6 # gate of Fujian Beixi diversion engineering, was seriously broken and leaked out because the sand base of the gate was vacated by the surging water, so Fujian water electric department first used rapidly hardening mortar for sealing, then used EECM for reinforcing, The reinforcement effects were satisfactory.

In the water conservancy project of Fujian Minnan area, the joint of the cement pipeline used to be made hole in the pipes with manual labor. the work efficiency was low and the quality could not be assured. Afterward, they used EECM as bonding materials, not only the quality of the joint became satisfactory, but also the productivity increased by a factor of at least 2 and the cost decreased by 2/3 as well, it solved the problem of irrigating 100000mu. The economic effectivity is good. In addition, at Beijing Shisanling pumped storage station and Guizhou Tianshengqiao hydropower station, the engineering of concrete tunnel used EECM for reinforcing and impermeating, the results of use were all very satisfactory.

6 References

1. Xu Yaping, Chen Heyun and Liu Guanghua, (1990). *Study of the epoxy emulsion cement mortar*, Polymere in Concrete, 1990, pp.322~329.
2. Xu Yaping, etc.. *Durability of epoxy emulsion cement mortar*, The Proceeding of the Third National Symposium on Durability of Concrete. pp.398~404.

THE CHALLENGE OF POLYMER CONCRETE FOR FORMATIVE ART: APPLICATION OF POLYMER CONCRETE TO ARCHISCULPTURE AND ADVANCED MATERIAL SCULPTURE

S. Mitsuyoshi
Mitsuyoshi Digital Create Inc., Omiya, Saitama-ken, Japan

Abstract
Polymer concrete have been used as construction materials because of its superior properties such as quick hardening, high strength and excellent durability. Specially, the decorative precast products such as artificial marble tiles and panels using the polymer concrete are given much attention in recent years. The texture of the polymer concrete can be controlled by its mix design and the selection of materials used. The polymer concrete having such properties may be applicable for formative arts. The introduction of the polymer concrete as a material for formative arts may create new category of hyper-expressionism called Hyper Art.

In this paper, polymer concretes were used to produce the archisculpture and advanced material sculpture categorized as Hyper Art. The sculptures of the polymer concretes and their process technology were presented. It is concluded through this work that the polymer concrete may be accepted as a ultimate or the most appropriate material for Hyper Art.
Keywords : Polymer concrete, archisculpture, advanced material sculpture, hyper-expressionism, polyester concrete, polymethyl methacrylate concrete, sculptor, product

1 Introduction

Polymer concrete have been used as construction materials for structural members, flooring, surface finishing and repairing of reinforced concrete structures, and precast products because of its superior properties such as quick hardening, high strength and excellent durability. Specially, the decorative precast products such as artificial marble tiles and panels using the polymer concrete are given much attention in recent years. The polymer concrete having such properties may be applicable for formative arts. Sculptors are always looking for a material having free-formability without size limit, great looking sense, high strength and good weatherbility. Such material can be accepted as a ultimate material or the most appropriate material for formative arts. The creativity of the sculptors will spread out widely without any limitation if such material was obtained. The freedom of expression for products as works of art is promised. The products are categorized as hyper-expressionism called Hyper Art. The author as the sculpture pick out the possibility of archisculpture and advanced material sculpture by using the polymer concrete.

In this paper, sculptures are produced by using polymer concretes. The possibility of the application of the polymer concrete for archisculpture and advanced material sculpture is presented.

Polymers in Concrete, edited by Y. Ohama, M. Kawakami and K. Fukuzawa. Published in 1997 by E & FN Spon, 2–6 Boundary Row, London SE1 8HN, UK. ISBN: 0 419 22330 4.

2 Polymer concrete as material for Hyper Art

Hyper Art includes following subjects:
(1)Research and developments of effective material for formative arts.
(2)Product classified into Hyper Art expresses straightly the imagination of sculptor.
(3)Recognition with strong impact of the products is maintained long period.
(4)Unordinary fantastic impression is expressed by structural design techniques. The feeling of fantastic arts and surrealistic arts will be represented in the existing space.
(5)The product gives unusual expression in physical and structural sense and gives strong impact to human mind.
Therefore, the following nature is required for the material to create the product as a work of arts categorized as Hyper Art:
(1)The material gives no limitation in shape and size of products which express unordinary fantastic impression.
(2)It is possible to produce the product having the scale of 10m or more with lightweight and high strength. The products should be produced without any joints.
(3)Material should have superior weatherbility more than 20 years.
(4)Material should have unchanged or ever-changing texture.
(5)Casting, polishing and chipping should be allowed to produce and treat the product.
It is considered from the above mentioned view points that polymer concrete will be a material for Hyper Art, because the polymer concrete has many advantages such as good castability, high strength and superior durability. The large scale formation and repairing work with no trace of sculpture are possible by using the polymer concrete. Sculptor may also built up original expressionism with the modification of the texture of the polymer concrete in consideration of mix proportion and materials used.

3 Application of polymer concrete for archisculpture and advanced material sculpture

3.1 Product one
Photo 1 shows the sculpture named "BONDAGE HIP".
3.1.1 Materials
Commercially available ultraviolet-curing polyester resin was used as a binder of glass fiber-reinforced plastic phase. The binder system for polyester concrete was commercially available unsaturated polyester resin, together with methyl ethyl ketone peroxide as a catalyst and cobalt octoate as an accelerator. Ground calcium carbonate was used as a filler. Black- and white-granites having three sizes of 3mm (3mm or finer), 5mm (5mm or finer) and 10mm (5-10mm) were employed as aggregates for coloring the polyester concrete. Expanded polystyrene form was also employed to built up sculpture. Glass fiber mat was used for the glass fiber-reinforced plastic phase to strengthen the sculpture.
3.1.2 Procedures for Making Sculpture
The production of the sculpture was done outdoor under the temperature of 30 °C. The working life of binder was adjusted to be 15 minutes with the formulations of unsaturated polyester resin : catalyst : accelerator = 100:0.03:0.1(by mass). Polyester concrete with the mix proportions of binder : filler : 3mm-granite : 5mm-granite : 10mm-granite = 15.0:10.0:10.0:19.5:45.5(by mass) was mixed. The black- and white-colored polyester concretes were prepared by using the black- and white-granites, and the sculpture was produced by following procedures:
(1)Expanded polystyrene form was shaped as the core of the sculpture.

Photo 1 Product one : BONDAGE HIP
(Polyester concrete with height/width/depth:600/300/300cm and weight:1500kg)

(2)The glass fiber mats were faced on the expanded polystyrene core and ultraviolet-curing polyester resin was applied to the mats by hand lay-up method to make the fiber-reinforced plastic phase as the reinforcing core of the sculpture. The surface of the fiber-reinforced plastic was treated by sand paper.

(3)The polyester concrete was trowelled and the shape of the sculpture was roughly formed. Then the exact shape of the sculpture was formed by using electric cutter. The surface of the sculpture was finally polished.

3.1.3 Discussion

The compressive strength of the polyester concrete is 90 MPa and cracks after hardening of the polyester concrete are not observed. The color of the sculpture is successfully created by using the black- and white-granites as aggregates and is not changed at 6 years after outdoor exhibition because the granites are densely compacted. The use of expanded polystyrene as the core of the product considerably decreases its weight. The broken-down at the portion of the leg of the product during processing may be occurred if stone was used because the slender leg supports upper part of the sculpture with heavy weight. However, the glass fiber-reinforcing plastic phase successfully strengthens the slender leg. The product expresses texture like stone and gives impact against the gravity. It is considered that the polyester concrete creates a new category of formative arts.

3.2 Product two
Photo 2 shows the sculpture named "BONDAGE TORSO".

3.2.1 Materials
The employed materials for polymer concrete and glass fiber mat were the same as

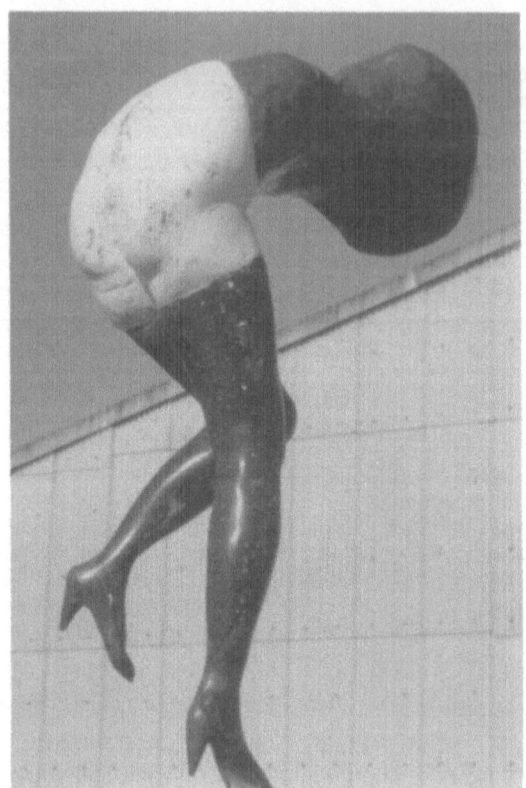

Photo 2 Product two : BONDAGE TORSO
(Polyester concrete with height/width/depth:400/180/180cm and weight:800kg)

the product one.   Steel bars with diameter of 5 mm and wire cloth with an opening dimension of 10 mm were used as reinforcements.   Clay and commercial molding gypsum were used to prepare a mold for the lower part of the sculpture.
3.2.2 Procedures for Making Sculpture
The binder having the working life of 15 minutes at 20 ℃ was formulated with the formulations of unsaturated polyester resin : catalyst : accelerator = 100:0.8:1.0(by mass). Polyester concrete with the mix proportions of binder : filler : 3mm-granite : 5mm-granite : 10mm-granite = 16.0:10.0:9.0:19.5:45.5(by mass) was mixed.   The black- and white-colored polyester concrete was prepared by using black- and white-granites and the sculpture was produced by following procedures:
(1)The model of the lower part (leg) of the sculpture was craved by using the clay, and a gypsum mold was formed from the clay model.
(2)The mixed polyester concrete was casted into the gypsum mold embedded steel reinforcements to make the leg of the sculpture as a supporting part.
(3)The wire cloth was shaped for the upper or the main part of the sculpture having a hollow structure, and then glass fiber mats were faced on the mesh. Then the binder of the polyester concrete used was applied to the mats by hand lay-up method to make the fiber-reinforced plastic phase as the reinforcing core of the sculpture. The surface of the fiber-reinforced plastic was treated by sand paper.
(4)The polyester concrete was trowelled and the shape of the sculpture was roughly formed. Then the exact shape of the sculpture was created by using electric

cutter. The surface of the sculpture was finally polished.
3.2.3 Discussion
The compressive strength of the polyester concrete is 95 MPa and cracks after hardening of the polyester concrete are not observed. The color of the sculpture is successfully created by using the granites and is not changed at 5 years after outdoor exhibition. The steel reinforcements for the leg of the sculpture, and steel mesh and glass fiber-reinforced plastic phase strengthen the products. The hollow structure of the upper part considerably decreases the weight of the sculpture. The profound impact of the sculpture against the gravity is expressed as compared to the product one of "BONDAGE HIP". Process technology for casting of the polymer concrete and techniques for built up the hollow structure for the sculpture using thepolymer concrete are successfully established through this work.

3.3 Product three
Photo 3 shows the sculpture named "A TOWER OF ARCHISCULPTURE".

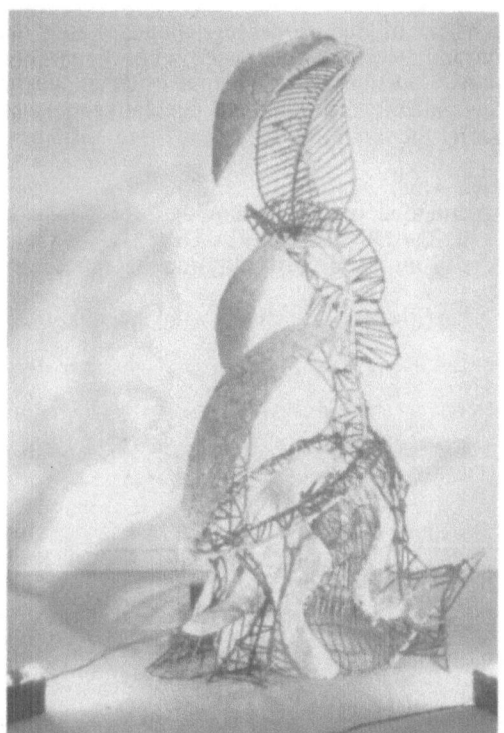

Photo 3 Product three : A TOWER OF ARCHISCLUPTURE
(Polyester concrete with height/width/depth:175/70/70cm and weight:20kg)

3.3.1 Materials
The employed materials for polyester concrete and glass fiber mat were the same as the product one. White-granite was used for coloring polyester concrete. Steel bars with a diameter of 5 mm were also employed. Artificial ruby, aluminum chips, ceramic and glass beads were used as decorative materials.
3.3.2 Procedures for Making Sculpture
The binder having the working life of 15 minutes at 20 ℃ was formulated with unsaturated polyester resin : catalyst : accelerator = 100:0.8:1.0(by mass).

White-colored polyester concrete with the mix proportions of binder : filler : 3mm-granite : 5mm-granite : 10mm-granite = 16.0:10.0:9:19.5:45.5(by mass) was mixed.   The mixture consisted of binder : artificial ruby : aluminum chips : ceramic beads : glass beads = 20.0:20.0:20.0:20.0:20.0(by mass) was also prepared as a decorative polyester composite. The sculpture was produced by following procedures:

(1)The steel bars were welded together to form basic shape of sculpture as a steel structure.

(2)The glass fiber mats were applied on the parts of steel fabric which required polymer concrete casting, and then the binder of the polyester concrete used was applied on the mats by hand lay-up method to built up the fiber-reinforced plastic phase.   The surface of the fiber-reinforced plastic was treated by sand paper.

(3)The polyester concrete and the decorative polyester composite were trowelled on the fiber-reinforced plastic phases and the shape of the sculpture was formed. The steel structure parts were painted.

3.3.3 Discussion

The compressive strengths of the polyester concrete and decorative polyester composites are 95MPa and 90MPa respectively.   The polyester concrete normally has the texture like stone , but the decorative polyester composite creates unknown texture.   The polyester concrete including the decorative polyester composite may become a original material for formative arts.

3.4 Products four to six

Based on the above mentioned process technology for products one to three, the sculptures named "A FORM OF WHIRL FUTURE", "A CRISIS" and "JAGUAR" were successfully produced by using polymethyl methacrylate concretes as shown in Photos 4 to 6.

Polymethyl methacrylate(PMMA) concretes with the mix proportions of PMMA

Photo 4 Product four : A FORM OF WHIR FUTURE
(Polymethyl methacrylate concrete with height/width/depth:70/70/50cm
and weight:30kg)

binder : 10mm-brack-granite = 16.0:84.0 (by mass) and PMMA binder : aluminum hydroxide : 3mm-white-granite : 5mm-white-granite : 10mm-white-granite = 16.0:10.0:9:19.5:45.5(by mass) were used for the product four. The product four was produced by cast molding method by using commercial molding gypsum mold.

The product five was produced by almost the same techniques applied to the product four by using polymethyl methacrylate concrete with the mix proportions of PMMA binder : ground calcium carbonate : 3mm-white-granite : 5mm-white-granite : 10mm-white-granite = 16.0:10.0:9:19.5:45.5(by mass).

Photo 5 Product five : A CRISIS
(Polymethyl methacrylate concrete with height/width/depth:350/110/110cm
and weight:300kg)

PMMA concrete with the mix proportions of PMMA binder : ground calcium carbonate : 3mm-granite : 5mm-granite : 10mm-granite = 16:10:9:19.5:45.5(by mass) was used to make the product six. The same size of black- and white-granites were mixed to produce gray-colored concrete. The body of a car was made from the concrete by trowelling. The black colored surface of the body was painted. The PMMA concrete body of the car gives unknown impression.

Photo 6 Product six : JAGUAR
(Polymethyl methacrylate concrete with height/width/depth:150/2300/600cm and weight:3500kg)

4. Conclusions

Conclusions obtained through the above mentioned works are summarized as follows:
(1) The polymer concretes are easily handled and molded to make any shapes without size limit. The polymer concrete products are also chipped, polished and repaired with no traces.
(2) The polymer concrete not only gives the texture like stone, but also creates its own texture which express unknown impression.
(3) Process technology of polymer concretes for archisculpture and advanced material sculpture was established. Sculptors freely express their imagination if polymer concrete was chosen, and the polymer concrete may be accepted as a ultimate or the most appropriate material for hyper-expressionism called Hyper Art.

Acknowledgment
The author wish to thank Mr.Sachio Yamada, president of REX Daito Co., Ltd. who made a chance to meet polymer concretes and gave technical suggestions for producing archisculpture and advanced material sculpture presented in this paper.

CHARACTERIZATION OF POLYMER–FILLER INTERACTION IN MMA POLYMER CONCRETE

M. Kuromoto and A. Kawamura
Technical Research and Development Institute,
Kumagai Gumi Co., Ltd., Tsukuba, Japan
T. Iwai
Civil Engineering Technology Dept., Kumagi Gumi Co., Ltd.,
Tokyo, Japan
M. Sumita and S. Asai
Dept. of Organic and Polymer Materials,
Tokyo Institute of Technology, Tokyo, Japan

Abstract
For a system of polymer concrete containing alumina as filler and methyl methacrylate (MMA) as binder, several coupling agents were investigated for their effect on improving the affinity and dispersion capability, as well as the influence exerted by their different use conditions.   The investigation was performed by both, dynamic-mechanical approach and statistical analysis of SEM images.   The results demonstrated that surface modification of alumina with silane having methacrylic functional groups provided uniform distribution and strong interaction for the MMA polymer binder. Thus the surface modification of filler results in significantly improved properties such as workability, strength and volume stability of polymer concrete.
Keywords: Affinity, Alumina, Coupling agent, Dispersion, Filler, Interaction, Methyl methacrylate, Polymer concrete.

1   Introduction

Polymer concrete (PC) using methyl methacrylate (MMA) as a binder is less viscous, offering greater workability than unsaturated polyester or epoxy PC's.   Besides, MMA-PC is more superior in weathering resistance and setting property at low temperatures ranging to -20℃ [1].   However, since it is more expensive and shows more marked setting shrinkage, it has not been widely used in large-scale civil works.

PC is composed of synthetic resin serving as binder as well as aggregate and inorganic filler. The essential functions of filler are filling voids between aggregates, providing thereby a dense structure, improving the workability and preventing segregation [1].   Moreover, recent studies have revealed that fillers not only reduce the liquid resin proportion, leading to cost saving,   but exert significant influence upon the

Polymers in Concrete, edited by Y. Ohama, M. Kawakami and K. Fukuzawa. Published in 1997 by E & FN Spon, 2–6 Boundary Row, London SE1 8HN, UK. ISBN: 0 419 22330 4.

mechanical properties and durability of PC [2-4]. Such behavior of fillers is one of important subjects of study for the authors aiming at improvement of properties and wider application of MMA-PC. The properties of PC depend upon the physical characteristics of the filler such as grain size or specific surface area, as well as affinity of the filler for polymer and it's surface chemistry determining the affinity. Coupling agents are well known for enhancing the affinity of filler for polymer binder.

This paper, focuses upon the interaction between MMA resin and filler, discusses its quantitative evaluation and reports the results of its improvement by coupling agents. More specifically, the investigation discusses here involves evaluation by the dynamic-mechanical approach of the interfacial affinity between filler and MMA polymer binder, and research on the dispersion of filler in the polymer binder through statistical analysis of electron microscopic images. Furthermore, this paper specifies, a system with alumina as filler and MMA as binder, the best coupling agent for improving the affinity and dispersion, as well as their applications, and finally demonstrates a filler excellent in affinity and dispersion improves the workability, strength and volume stability of PC.

2 Materials

Two kinds of alumina shown in Table 1 were used as filler. These aluminas are different in physical properties such as particle size, but almost the same in chemical properties. A smaller particle size aluminaA, was employed to conduct precise evaluation of the interaction and the dispersion of filler in polymer pastes. By contrast, aluminaB, which had a particle size adequate to the preparation of MMA-PC, was employed to study the fluidity or mechanical properties of PCs.

For improving the affinity between filler and binder, four typical silane coupling agents were used, which had different organic functional groups or hydrolysis groups. Table 2 summarizes the formulas and physical properties of the coupling agents.

As binder, a liquid resin available on the market, the main component of which is MMA monomer, was used. Besides MMA monomer, the liquid resin includes MMA polymer (PMMA), crosslinking monomer, polymerization accelerator and a small amount of polymerization prohibitor. The liquid resin has a density of 0.965 g/cm^3 and a viscosity of 2 mPa·s at 20℃. The polymerization initiator was a compound of one

Table 1. Physical properties of fillers

Filler	Specific gravity	Particle size (μ m)	BET surface area (m^2/g)
aluminaA	3.95	2.2	3.0
aluminaB	3.95	3.7	1.4

Table 2. Formula and physical constants of silane coupling agents

Name	Abbreviated Name	Molecular formula	Molecular weight	Density (g/cm^3)
γ -Aminopropyltriethoxysilane	Silane-APE	$NH_2C_3H_6Si(OC_2H_5)_3$	221.3	0.951
Vinyltriethoxysilane	Silane-VE	$CH_2=CHSi(OC_2H_5)_3$	190.3	0.903
Vinyltrimethoxysilane	Silane-VM	$CH_2=CHSi(CCH_3)_3$	148.6	0.970
γ -Methacryloxypropyl-trimethoxysilane	Silane-MPM	$CH_2=CCH_3COOC_3H_6$ $\cdot Si(OCH_3)_3$	248.3	1.045

part of benzoyl peroxide and one part of PMMA. In principle, an amount of the compound equal to 6.5% by weight of the liquid resin was added

3. Experimental

3.1 Dynamic-mechanical analysis

Many studies on particle-loaded polymer compounds such as polymer pastes have been implemented for not solely determining their physical properties but also evaluating the interaction between polymer and particles by the dynamic-mechanical technique [4,5].

The investigation reported here used polymer pastes prepared by adding 10 to 30 vol. % of aluminaA and a variety of coupling agents to liquid resin, to measure the real part (storage modulus G') and imaginary part (loss modulus G") of the dynamic complex modulus. On the basis of the measurement results, the affinity and reactivity of coupling agents for MMA resin as well as interaction between filler particles and polymer binder were investigated. The measurements were performed with a forced vibration dynamic tester available on the market. The vibration frequency was 10 Hz and testing temperature was 40°C. Test specimens were thin pieces 1 cm wide, 3 cm long, 0.05 to 0.07 cm thick prepared by using Teflon molds.

Tensile tests were also conducted using specimens of the same geometry to determine the strength of different polymer pastes with aluminaA.

3.2 Statistical analysis of SEM images by Morishita's quadrate method

As a general rule, the properties of a particle-loaded polymer compound is significantly influenced by not only the volume fraction of particles, but also their dispersion state. Observations using a scanning electron microscope (SEM), etc. have been frequently conducted to know the dispersion of particles in a matrix.

For the quantitative evaluation, it is necessary to statistically process the images by a SEM in order to obtain appropriate parameters which represent the dispersion state. In the present study, back-scattering electron images (BEI) of the section of the polymer pastes, obtained by a SEM, were analyzed to study the dispersion state of the filler by Morishita's quadrate method [6].

Let us describe the summary of Morishita's quadrate method. Each of (a), (b), (c) in Fig. 1 illustrates the typical dispersion modes respectively. The number of particles and the area are the same in every diagram. The diagram (a) represents a Poisson's distribution where all points (particles) are distributed randomly i.e., independent of each other. The diagram (b) shows a regular mode of distribution which is likely to occur when some reaction is present in a certain radius from each point (particle). The diagram (c) shows an agglomerated (cluster) distribution which is found when affinity or cohesion exists between particles.

In Morishita's quadrate method, Morishita's index $I\delta$ is an essential parameter for characterizing the dispersion state. This index $I\delta$ is expressed by [6]

$$I\delta = q \cdot \delta \tag{1}$$

$$\delta = \left\{ \sum_{i=1}^{q} ni(ni-1) \right\} / N(N-1) \tag{2}$$

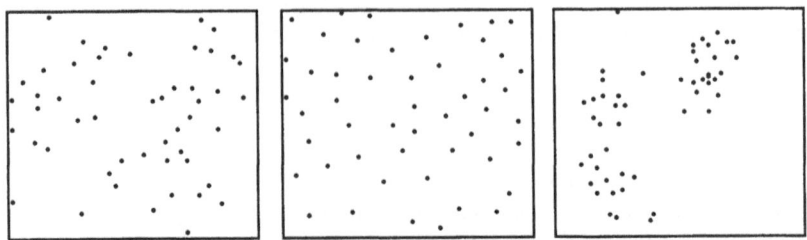

(a) Poisson's distribution   (b)regular distribution   (c)agglomerated distribution

Fig. 1.   Representative distribution modes of particles [6].

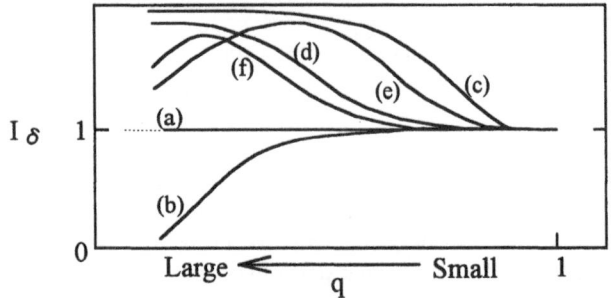

Fig.2.   Schematic representation of the relationship between Morishita's index, I δ, and quadrate number, q, for various distribution modes of particles [6].

where q is the number of quadrates, n_i the number of particles in the i-th quadrate, N the total number of particles, δ represents the probability of presence of two random particles observed in a same quadrate. I δ is given by multiplying δ by the total number of quadrates q.

Fig. 2 shows, for the typical distribution states, the variation of I δ with q. If, with increasing q, I δ remains unity, the distribution is in Poisson's mode (curve a in Fig. 2); if I δ < 1, the distribution is in the regular mode (curve b); if I δ > 1, the distribution is in the agglomerated mode (curves c to f). By the size of cluster and dispersion of particles in each cluster, the agglomerated distribution is classified into four categories: (c) with large size clusters, Poisson's distribution in each cluster, (d) with small size clusters, Poisson's distribution in each cluster, (e) large size clusters, regular distribution in each cluster, (f) small size clusters, regular distribution in each cluster.

3.3   Viscosity of polymer pastes

The absolute viscosity of the polymer pastes with aluminaB was measured by a viscometer.   Besides, the effect of reducing the viscosity was investigated using a silane coupling agent under different conditions.   In all cases, the measurements were done at room temperature (16 to 18℃), without using polymerization initiator.

3.4   Strength and shrinkage of PCs

An immersed type strain gauge and a thermocouple were installed at the center of the mold (10 x 10 x 40 cm), into which PC was placed.   The strain and   internal

Table 3. Mix proportion of MMA polymer concrete

Mix proportion	(wt.%)			Binder./Filler	s/a
Binder	Filler (AluminaB)	Fine agg.	Coarse agg.	(wt.%)	(%)
7.5	14.1	37.2	41.2	53.5	48.7

temperature at various time intervals were recorded from just after placement of PC at a constant ambient temperature of 20℃.   In addition, cylindrical specimens of 10 cm in diameter, 20 cm long were made to determine their compressive strength and static elastic modulus at the age of 28 days (cured at 20℃, 60% R.H.).   The mix proportion of PC with aluminaB is shown in Table 3.   The fine and coarse aggregates were sand (oven dry specific weight = 2.56) and crushed stones (oven dry specific weight = 2.69).

4   Results and discussion

4.1   Characterization by the dynamic-mechanical technique
The storage modulus (G') and loss modulus (G") determined by measurements for different filler proportions and coupling agents are shown in Figs. 3 and 4.   The mechanical loss (tan δ = G"/G') and tensile strength of polymer pastes are plotted in Figs. 5 and 6 respectively.   The mix proportion of the coupling agents was kept constant to 2 wt.% of the filler.   In every test mix, an amount of liquid resin equal to

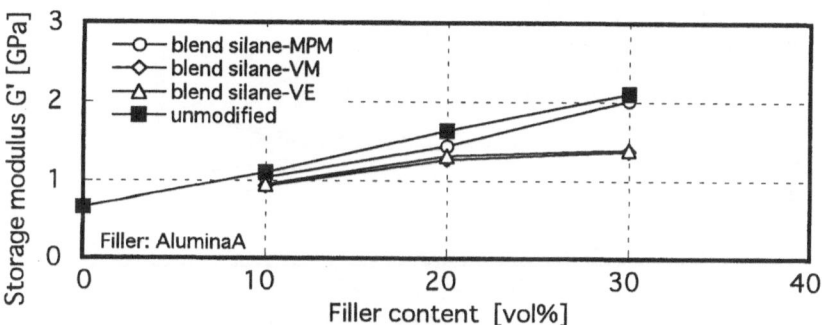

Fig. 3.   Storage mudulus of polymer pastes blended with different coupling agents.

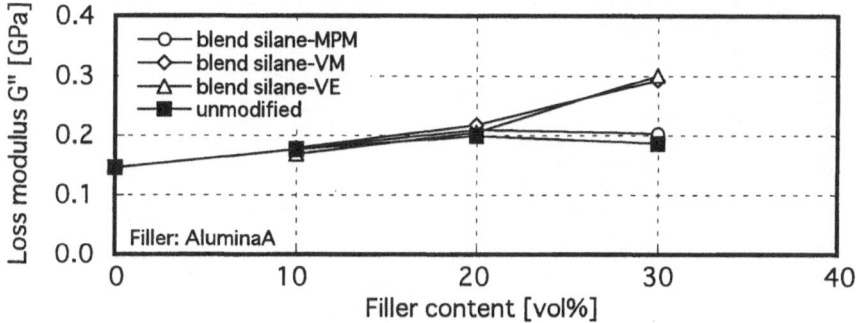

Fig. 4.   Loss mudulus of polymer pastes blended with different coupling agents.

the volume of the silane to be blended was removed so that the volume of liquid components (silane and liquid resin) remained constant for the same proportion of filler.

Polymer pastes blended with silane-APE did not set satisfactorily, so it was impossible to measure their elasticity and strength.

As revealed from Figs. 3 and 4, the variation in storage modulus (G') and loss modulus (G") of the polymer pastes blended with silane-MPM is almost the same as that of the specimens without using silane. The specimens blended with silane (VE or VM) containing vinyl as organic functional groups, regardless of the kind of hydrolytic groups they contain, show almost the same G' and G". When the filler content is 30 vol.%, the influence of the coupling agent, in other words, the influence of the organic functional groups of the coupling agent appears most significantly. With this level of mix proportion, the value of G' of polymer paste with silane having vinyl functional groups is smaller than that of the specimen with silane having methacrylic groups (MPM) or the specimen without silane. G" of polymer paste mix with silane-VE or -VM is greater than that with silane-MPM or that without silane.

Thus, as revealed by Fig. 5, the specimens blended with silane having vinyl groups gives greater loss factors than that blended with acrylic groups or that without silane; when the filler loading is 30 vol.%, the difference between them reaches about twice. The loss factor of the paste with silane-MPM is slightly greater than that without silane.

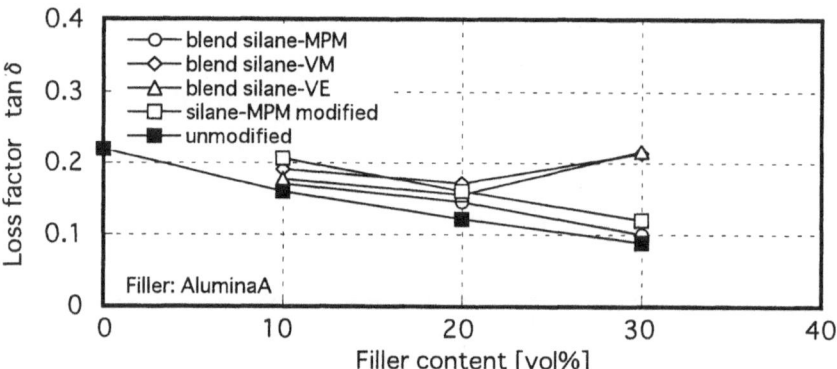

Fig. 5. Loss factor of various polymer pastes.

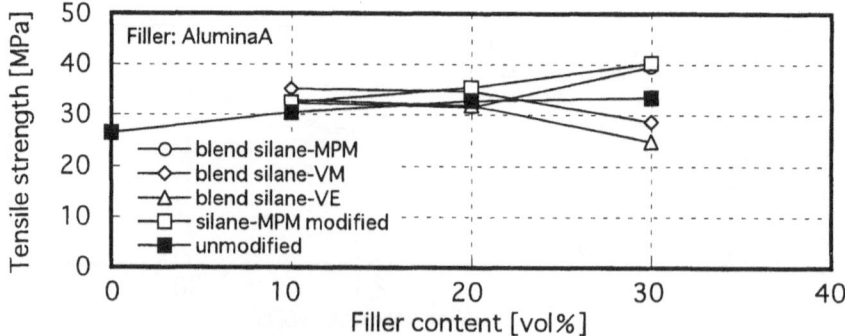

Fig. 6. Tensile strength of various polymer pastes.

Generally, the increment of loss factors can be explained by these facts: (a) filler particles agglomerate and friction is generated by contact between filler particles, (b) in the system where adhesion does not occur between filler particles and polymer binder, friction occurs in the filler-binder interface, and (c) due to morphological change of polymer binder, its loss factor near the interface becomes very large [5]. Besides, the loss factor becomes larger (d) when there are voids in the polymer paste, or (e) strong adhesion occurs between polymer binder and filler due to specific interaction (e.g. coupling agent), to create immobile interphase in the surrounding of filler particles [4,5].

In the present investigation, the value of tan δ of the paste with silane having vinyl groups was larger than the paste without such silane. It can be explained not by the reason (e) above, but by (b) or (c), because as shown in Fig. 6 the strength of polymer pastes decreases with increasing tan δ. Silane-VE or -VM may show reactivity to filler particles, but, far from exhibiting affinity or reactivity to MMA resin, it may prevent adhesion between filler and polymer binder. Furthermore, it is quite probable that the polymer binder near the interface with filler particles was changed in property due to blending of such coupling agent.

By contrast, as for the system blended with silane-MPM having methacrylic functional groups, the tensile strength is greater as shown in Fig. 6. It can be explained, as mentioned above, by the fact that the coupling agent offers strong interaction between polymer binder and filler to form immobile interphase around the filler particles, which may be suggested by the slight increase of loss factor by blending of silane-MPM.

Figs. 5 and 6 are the plots of loss factor and tensile strength of the polymer paste loaded with aluminaA the surface of which had been modified with silane-MPM respectively. The modified alumina was a commercial product obtained by a dry process. In the process, alumina was treated with alcohol-water solution of the silane-MPM (1 wt.% for filler) and was then heated to complete the reaction of hydrolysis and vaporize the excess solution.

The paste loaded with modified alumina ($\square$) shows greater loss factors and tensile strengths, compared with silane-MPM blended one ($\bigcirc$). This result shows that the surface modification of filler may strengthen the interaction with the polymer binder.

From the above discussion, it is inferred that, among the four types of silane coupling agents, silane-MPM is the most remarkable for producing interaction and adhesion in the interface between filler and MMA polymer binder.

4.2 Characterization of the dispersion state

Fig. 7 illustrates the dispersion state of aluminaA in the MMA polymer binder and improvement of dispersion provided by the coupling agents. Three types of specimens were used: (1) specimen with unmodified aluminaA ($\blacksquare$), (2) silane-MPM equal to 2 wt.% of the filler was blended with liquid resin; then the blend was loaded with unmodified aluminaA ($\bigcirc$), and (3) specimen using aluminaA whose surface was modified with silane-MPM ($\square$). For every specimen, the filler content was 20 vol.%.

As shown in Fig. 7, the specimen with unmodified alumina and that blended with silane-MPM, the filler particles are dispersed in Poisson's mode, with a slight sign of agglomerated distribution. By contrast, the specimen with surface-modified alumina with silane-MPM shows a regular distribution.

These results can be explained as follows: in the cases with unmodified alumina and silane-MPM blended one, repulsion does not act between particles, resulting in random

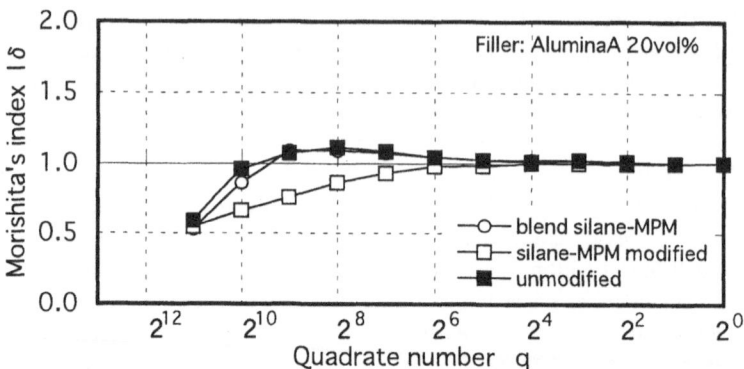

Fig. 7.   Morishita's I δ index for aluminaA as a function of quadrate number, q.

distribution where the particles are distributed independent to each other. Besides, cohesion operates between particles, forming small agglomerations.

When some repulsion is produced between filler particles, prohibiting the formation of clusters, and, at the same time, good affinity is present in the interface between filler and binder, the distribution state must be changed from Poisson's mode or agglomerated mode to regular mode.   Surface modification with silane-MPM satisfies these conditions for changing the distribution mode, by preventing cohesion of filler and improving the distribution in the MMA polymer binder.

The results of this experiment reveal the fact that the dispersion state varies with method of use of the coupling agent (blending or surface modification).   It may be correctly explained not by the affinity between resin and filler but by the affinity and cohesion between filler particles, because, as mentioned in the previous section, blending silane-MPM, like the case conducting surface modification by silane-MPM, produces interaction between filler and polymer, increasing the strength of polymer paste.

Thus the presence of a coupling agent and its method of application are the essential factors which determine the dispersion state of filler particles in the polymer binder.   In the following sections, we will study the influence exerted by fillers excellent in dispersion capability and affinity in the physical properties of polymer pastes and PCs.

4.3   Effect of silane coupling agent on the viscosity of polymer pastes

The proportion of liquid resin in PC can be reduced most easily by substituting filler for a part of liquid resin.   But, careless increase of filler proportion makes greater the viscosity of polymer paste, worsening the workability of PC.   Thus we tried to reduce the viscosity of polymer paste by loading fillers excellent in both affinity for MMA polymer binder and dispersion in it.

As known from Fig. 8, the absolute viscosity of polymer paste becomes greater with increasing proportion of aluminaB where surface is modified with silane-MPM (referred to as aluminaB-MPM) or unmodified aluminaB.   AluminaB-MPM is also available on the market.   When the filler proportion is 10 vol. % or more, the viscosity of polymer paste with aluminaB-MPM is lower than that with unmodified aluminaB.   The difference of viscosity between these two polymer pastes becomes greater with increasing proportion of filler.   As inferred clearly from these facts, surface modification by silane-MPM reduces the viscosity of polymer paste.

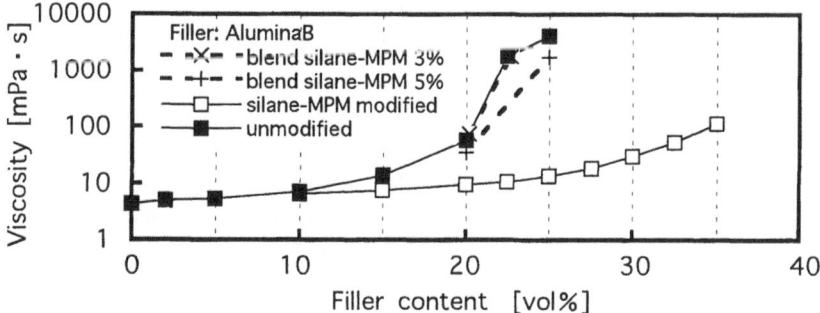

Fig. 8. Viscosity of polymer pastes with different use conditions of silane-MPM.

Fig. 8 shows the viscosities of the polymer pastes made of liquid resin blended with silane-MPM equal to 3 or 5 % by weight of filler, to which unmodified aluminaB was added. Addition of silane-MPM, whatever its mix proportion, does not reduce remarkably the viscosity unlike mixture of aluminaB-MPM.

For improving the workability of PC or reducing the mix proportion of liquid resin, it is therefore necessary to increase the affinity of filler for resin and, especially, the dispersion capability of filler. From this standpoint, simple blending with silane does not satisfy the requirements, so modification of the filler surface by silane is necessary.

4.4 Influence of filler upon setting shrinkage and strength of PCs

Fig. 9 shows the shrinkage curve of PC with aluminaB-MPM and that of PC with unmodified aluminaB. These PCs have a same mix proportion shown in Table 3. As illustrated by the graphs, PC with aluminaB-MPM shows greater stability of volume, that is, less variation both in positive and negative strains (expansion and shrinkage) than PC with unmodified aluminaB. It is noteworthy that the shrinkage of PC with aluminaB-MPM is approximately 70% of that of PC with unmodified aluminaB at 100 hours. These phenomena may be due to the fact that, because of uniform distribution of aluminaB-MPM in the MMA polymer binder and strong interaction between filler and binder, change in volume at PC caused by polymerization is significantly restricted.

Table 4 shows the properties of PC with aluminaB-MPM and PC with unmodified aluminaB. Surface modification enhances the fluidity and workability, increases about

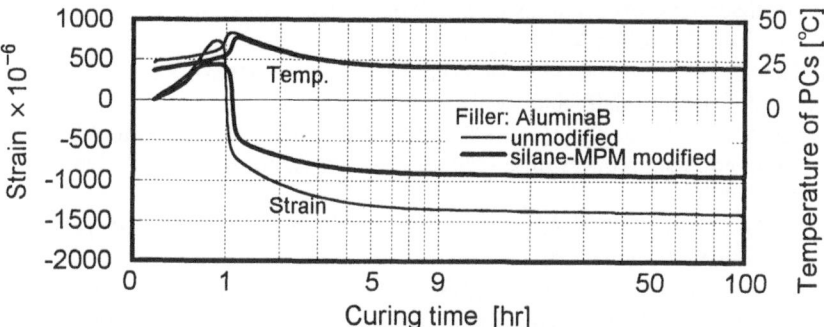

Fig. 9. Curing time vs. strain and internal temperature of PCs.

Table.4   Properties of MMA-PCs using unmodified or modified aluminaB as a filler

AluminaB	Slump (cm)	Compressive strength (MPa)	Static elastic modulus (GPa)
unmodified	18.0	66.5	26.0
silane-MPM modified	21.0	72.7	24.3

10 % the compressive strength and slightly diminishes the static elastic modulus of PC.

As described above, it is clear that fillers with dispersion capability and interaction with polymer, improved by surface modification with silane-MPM, can not only ameliorate the physical properties of polymer pastes but enhances the workability, strength and volume stability of PC.

5 Conclusions

Polymer concrete is a versatile material prepared by replacing water and cement of conventional cement concrete, with liquid resin and filler respectively. Therefore, it is essential for research on polymer concrete to fully utilize knowledge obtained not only on the field of cement concrete but also on that of polymer composites.

From this point of view, the authors investigated the fundamental properties of filler-loaded polymer compounds (polymer paste) as reported in this paper. The dynamic-mechanical approach, though it is an indirect technique, offered information useful for evaluating the interaction between filler and polymer binder, and for assessing the affinity or reactivity between coupling agent and resin. In addition, a direct approach, that is, statistical analysis of SEM images was performed. The analysis revealed that the dispersion state is determined, rather than by the affinity between filler and resin, by the chemical properties of the filler surface which governs the repulsion between filler particles. It was also demonstrated that information on the interaction of fillers obtained in polymer pastes can be directly translated into the properties of polymer concrete. This is among others an important results of this investigation.

6 References

1.   Koyanagi, W. (1993) Resin concrete. *Concrete Journal*, Vol. 31, No. 4. pp. 5-13.
2.   Kuromoto, M. et al. (1993) Study on fillers in resin concrete using methyl methacrylate, *Proceedings of the 48th Annual Conference of the Japan Society of Civil Engineers*, Vol. 5. pp. 62-63.
3.   Kawamura, A. et al. (1993) Study on mix proportion of resin concrete using methyl methacrylate, ibid., Vol. 5. pp. 60-61.
4.   Kuromoto, M. et al. (1996) Characterization of fillers applied to MMA polymer concrete, *Proceedings of the Japan Concrete Institute*, Vol. 18., No. 1. pp. 489-494.
5.   Nielsen, E. L. (1975) *Mechanical properties of polymer and composite*, Marcel Dekker Inc., New York.
6.   Karasek, L. and Sumita, M. (1996) Characterization of dispersion state of filler and polymer-filler interaction in rubber-carbon black composites, *Journal of Materials Science*, No. 31. pp. 281-289.

EFFECT OF FILLER ON THE MECHANICAL PROPERTIES OF UNSATURATED POLYESTER RESIN MORTAR

Y.S. Soh
Dept. of Architectural Engineering, Chonbuk National University, Chonju, Korea
Y.K. Jo and H.S. Park
Building Inspection Division, Korea Infrastructure Safety & Technology Corporation, Anyang, Korea

Abstract
The purpose of this study is to find out the effect of fillers on the mechanical properties of unsaturated polyester resin(UP) mortars. Fillers such as ground calcium carbonate, silica powder and fly ash are used to extend polymeric binders and to supplement the very fine particles for aggregates. The UP matrixes and mortars are prepared with various contents of the calcium carbonate and fly ash as fillers, and tested for compressive, flexural, direct tensile and splitting tensile strengths. From the test results, the strengths of UP matrixes and mortars decrease with increasing filler content. This tendency is markedly shown with the filler contents of 60 to 70%. The strengths of UP matrixes and mortars using fly ash as a filler are a little higher than those using ground calcium carbonate.
Keywords: Mechanical properties, unsaturated polyester resin, fly ash, filler, compressive strength, flexural strength, tensile strength.

1 Introduction

Polymer concrete(mortar) is a composite material produced by fully replacing the cement hydrate binders of conventional cement concrete with a polymeric binder, and is dealt with the category of concrete-polymer composites(1). The polymer concrete(mortar) shows excellent mechanical properties and chemical resistance compared with the conventional cement concrete. Therefore, the polymer concrete(mortar) draws a strong interest as high performance or multifunctional materials in the construction industry(1,2,3). UP concrete(mortar) is made from unsaturated polyester resin and inorganic aggregates such as gravel, sand and filler. At present, UP concrete(mortar) is most widely used very effectually for various structural and nonstructural precast products, the repair of structural member or the overlay of pavements, bridges, industrials floors and dams. The properties of UP concrete are effected from various factors such as types of resin and filler, resin and filler

Polymers in Concrete, edited by Y. Ohama, M. Kawakami and K. Fukuzawa. Published in 1997 by E & FN Spon, 2–6 Boundary Row, London SE1 8HN, UK. ISBN: 0 419 22330 4.

contents, binder-filler ratio, curing condition and temperature, etc. In this study, for optimum mix design of UP matrixes and mortars, UP matrixes and mortars are prepared with various binder-filler ratios of ground calcium carbonate and fly ash as filler, and the effect of fillers on their mechanical properties is discussed.

2 Materials

2.1 Materials for binder

Binder for UP matrixes and mortars were based on unsaturated polyester resin, together with styrene monomer(SM) as a diluent, methyl ethyl keton peroxide(MEKP) as an initiator and cobalt octoate(CoOc) as an accelerator. The properties of UP are indicated in Table 1.

Table 1 Properties of unsaturated polyester resin.

Type	Specific Gravity	Acid Value	Viscocity (Pa· s)	Gel Time (min.)
Orthophtalate	1.102	19.9	1.80	1.34

2.2 Filler and aggregate

Commercial ground calcium carbonate and fly ash were used as fillers, and crushed granite, river sand and silica powder as fine aggregates. The water contents of the filler and fine aggregates were controlled to be less than 0.1%. The physical properties of fillers and aggregates are shown in Table 2, 3 and 4.

Table 2 Physical properties of calcium carbonate.

Specific Gravity	Unit Weight (t / m³)	Blain's Specific Surface (cm² /g)	Water Content (%)
2.7	0.984	2500	< 0.1

Table 3 Physical properties of fly ash.

Specific Gravity	Unit Weight (t / m³)	Blain's Specific Surface (cm² /g)	Water Content (%)
2.16	0.994	3200	< 0.1

Table 4 Physical properties of aggregate.

Type of Aggregate	Size(mm)	Specific Gravity	Content (wt.%)
Crused granite	2.5 - 5.0	2.43	21.7
River Sand	0.6 - 1.2	2.64	62.0
Silica Power	0.3	2.69	16.3
Unit Weight (kg / l)			1.67
Specific Gravity			2.60
Solid Volume Ratio (%)			64.3
Fineness Modulus			2.7

3 Test procedures

3.1 Preparation of specimen

According to KS F 2419(Method of making polyester resin concrete specimens), UP matrixes consisted of binder and filler, and mortars were prepared with the mix proportions given in Table 5. Specimens 40x40x160mm were molded, then cured at 20℃ and 50% R.H. for 7 days.

Table 5 Mix proportions of UP matrixes and mortars by mass.

Type of Specimen	Binder %				Binder-Filler Ratio, B/F	Aggregate Content (%)
	UP	SM	CoOc	MEKP		
Matrix					3 / 7	
					4 / 6	
					5 / 5	—
					6 / 4	
	55.44	42.56	1.0	1.0	7 / 3	
Mortar					3 / 7	
					4 / 6	
					5 / 5	70
					6 / 4	
					7 / 3	

3.2 Viscosity

Viscosity of UP matrixes was measured by brookfield viscometer at 20℃ and 60% R.H.

3.3 Strength test

According to KS F 2482(Method of test for flexural strength of polyester resin concrete), KS F 2483(Method of test for compressive strength of polyester resin concrete using portions of beams broken in flexure), KS F 2480(Method of test for splitting tensile strength of polyester resin concrete) and ASTM C 190(Tensile strength of hydraulic cement mortars), UP matrixes and mortars were tested for strengths.

4 Test results and discussion

This experimental study is to find out the effect of filler content on the strengths of UP matrixes and mortars using a ground calcium carbonate and fly ash as filler to decide optimum mix design for UP matrixes and mortars.

Fig.1 represents the effect of filler content on the viscosity of UP matrixes. Regardless of types of filler, the viscosity of UP matrixes tends to increase with an increase in filler content. This tendency is markedly shown with filler contents of 40% or more. The viscosity of UP matrixes having fly ash is higher than that having ground calcium carbonate. This reason seems to be due to the physical properties of a low specific gravity, unit weight and high blaine's specific surface of fly ash compared with those of ground calcium carbonate.

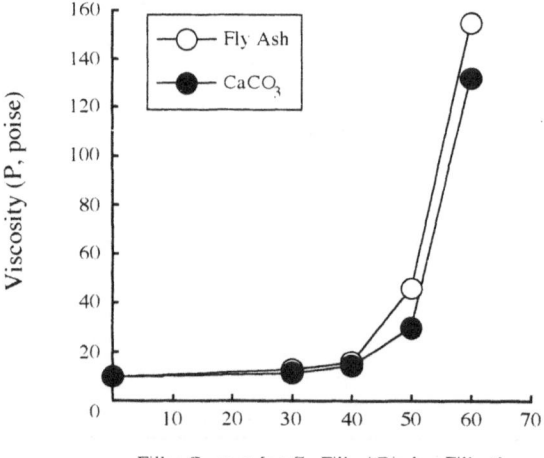

Fig.1 Effect of filler content on viscocity of UP matrixes.

Fig.2 exhibits the effect of filler content on the compressive and flexural strengths of UP matrixes having ground calcium carbonate and fly ash as filler. The compressive strength of UP matrixes tends to decrease with increasing filler content irrespective of types of filler. The compressive strength of UP matrixes having fly ash is higher than that having ground calcium carbonate. However, the flexural strength of UP matrixes decreases with increasing filler content as the above-mentioned compressive strength test results. And the change in flexural strength of UP matrixes according to types of filler is not nearly recognized.

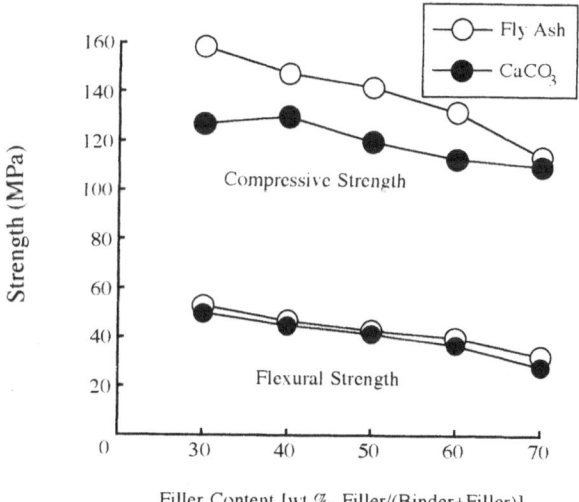

Fig.2 Effect of filler content on compressive and flexural
strengths of UP matrixes.

Fig.3 shows the effect of filler content on the direct tensile and splitting tensile strengths of UP matrixes having ground calcium carbonate and fly ash as filler. The tensile strength of UP matrixes tends to markedly decrease with an increase in filler content. This tendency is remarkable with the splitting tensile strength of UP matrixes having fly ash.

Fig.4 illustrates the effect of filler content on the compressive and flexural strengths of UP mortars having ground calcium carbonate and fly ash as filler. As above-mentioned, the compressive and flexural strengths of UP matrixes tend to decrease linearly with increase in filler content. On the other hand, the compressive strength of UP mortars reaches maximums with the filler content of 50% and the flexural strength is nearly constant with filler content 30 to 60%, and then markedly decreases with the filler content of 70%. The maximum compressive strengths of UP mortars having fly ash and ground calcium carbonate are 109.0MPa and 97.8MPa respectively. However, the difference in compressive strength between UP mortars having ground calcium carbonate and having fly ash is bigger than that of flexural strength. That is, types of filler affect the compressive strength rather than flexural strength.

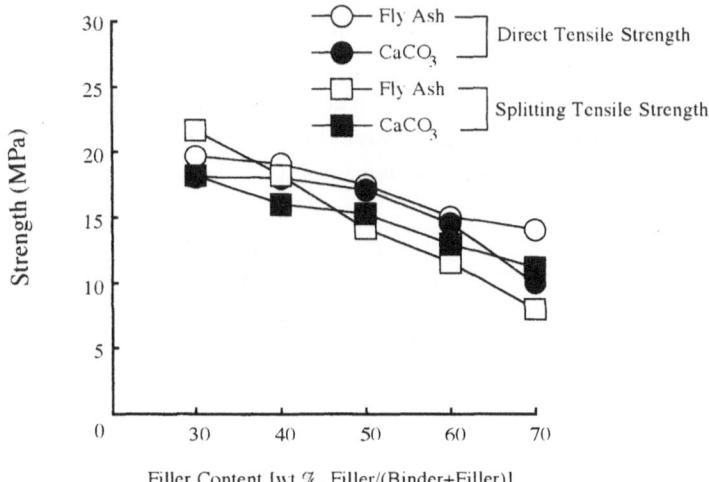

Fig.3 Effect of fliier content on direct tensile and splitting tensile
strengths of UP matrixes.

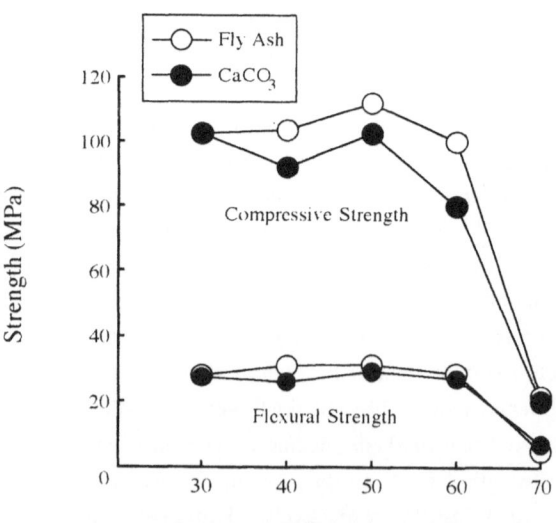

Fig.4 Effect of filler content on compressive and flexural
strengths of UP mortars.

Fig.5 indicates the effect of filler content on direct tensile and splitting tensile strengths of UP mortars having ground calcium carbonate and fly ash as the filler. The direct tensile and splitting tensile strengths of UP mortars having fly ash as filler increase with increasing filler content, and the maximum strengths attain with the filler content of 50%. This maximum strength is higher than those having ground calcium carbonate. The direct tensile and splitting strengths of UP mortars having ground calcium carbonate and fly ash with filler content of 70% are about one-sixth and one-third of those with filler content of 60%. In UP mortars, it is evident that the maximum filler content is limited to 60% to make the most of the excellent strengths of UP mortars.

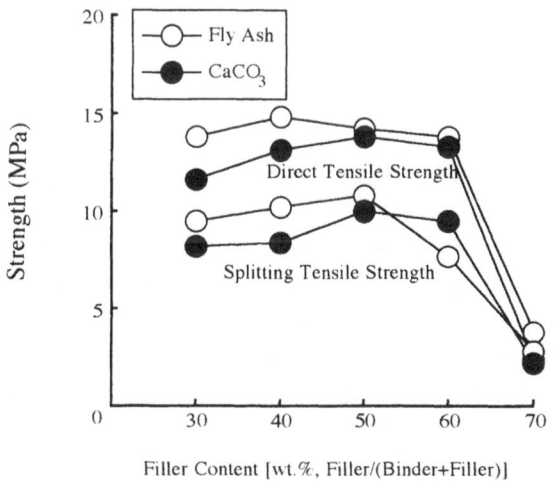

Fig.5 Effect of filler content on direct tensile and splitting tensile strengths of UP mortars.

From the above-mentioned test results, it is obvious that the strengths of UP matrixes and mortars are improved when fly ash rather than ground calcium carbonate as filler is mixed with filler content of 50%, and proper viscosity of binder for UP mortars is 30 poise to 50 poise. The viscosity of binder matrixes for UP matrixes and mortars could be controlled and the properties of resin are improved by using filler. It is found out that the filler content is one of the important factors to decide an optimum mix design for UP mortars.

5 Conclusions

The conclusions obtained from the test results can be summarized as follow:
(1) Regardless of types of filler, the viscosity of UP matrixes is effected from filler

content. The viscosity of UP matrixes having fly ash is higher than that having ground calcium carbonate as filler.

(2) The strengths of UP matrixes tend to decrease with raising filler content irrespective of types of filler. Except for compressive strength, the difference in strength of UP matrixes according to types of filler is not nearly recognized.

(3) The compressive strength of UP mortars reaches maximums with filler content of 50%, and the flexural strength is nearly constant with filler content 30 to 60%, and then markedly decrease with the filler content of 70%.

(4) The direct tensile and splitting tensile strengths of UP mortars having fly ash as filler increase with raising filler content, the maximum strengths attain with filler content of 40% and 50%.

(5) The maximum filler content is limited to 60% to make the most of the excellent strengths of UP mortars.

6   References

1. Ohama, Y. (1994) Recent Trend in Research and Development of Polymer Mortar and Concrete in Japan, Polymer in Concrete, Proceeding of the First East Asia Symposium, Kangwon National University, Chuncheon, Korea, pp.11-23.

2. Fowler, D. W. (1990) Status of Concrete-Polymer Materials, Polymers in Concrete, Proceedings of the Sixth International Congress on Polymers in Concrete, International Academic Publishers, Beijing, pp.10-27.

3. Ohama, Y. (1994) Recent Research and Development in Concrete-Polymer Composites, Advances in Concrete Technology (2nd Edition, Edited by V.M Malhotra) Canada Centre for Mineral and Energy Technology, Ottawa, pp.753-783.

AN EMPIRICAL APPROACH FOR PREDICTING COMPRESSIVE STRENGTH OF LIGHTWEIGHT POLYESTER MORTARS BY THE MATURITY METHOD

Y.S. Lee, Y. Ohama and K. Demura
Dept. of Architecture, Nihon University, Koriyama, Japan
K.S. Yeon
Dept. of Agricultural Engineering, Kangwon National University, Chuncheon, Korea

Abstract
The maturity method in which the strength increase of cement concrete is expressed as a function of an integral of the curing period and temperature of the concrete has often been applied to its strength prediction. For the purpose of the application of the maturity method to the compressive strength prediction for lightweight polyester mortars using an unsaturated polyester resin as a binder, the lightweight polyester mortars with various catalyst and accelerator contents, are prepared, tested for compressive strength, and the datum temperatures for the maturity equations are estimated. The maturity is calculated by using the maturity equations with the estimated datum temperatures. The compressive strengths of the lightweight polyester mortars are predicted from the maturity-compressive strength relationships.
Keywords : Accelerator content, catalyst content, compressive strength, datum temperature, maturity equation, maturity method, lightweight polyester mortar

1 Introduction

In recent years, technical innovations in the construction industry have progressed to a great extent, and the development of high-performance or multifunctional construction materials has actively been conducted to cope with such innovations. In particular, this trend is remarkable in the new frontiers of the building industry, i.e., superhighrise buildings and long-span structures in advanced countries. There are high-strength lightweight concretes in the building materials toward which a great interest is oriented in such background. The high-strength lightweight concretes are divided to two types : high-strength lightweight cement concrete and polymer concrete. This paper is related to high-strength lightweight polyester mortar as a model of high-strength lightweight polymer concrete. The objective of this investigation is to try the compressive strength

Polymers in Concrete, edited by Y. Ohama, M. Kawakami and K. Fukuzawa. Published in 1997 by E & FN Spon, 2–6 Boundary Row, London SE1 8HN, UK. ISBN: 0 419 22330 4.

prediction for lightweight polyester mortars with high specific strength by the maturity method, which has often been used for strength prediction for cement concrete.

In the present paper, lightweight polyester mortars using an unsaturated polyester resin are prepared with various catalyst and accelerator contents at different temperatures, and tested for compressive strength. The estimation of the datum temperatures for the maturity equations from the ambient temperature-compressive strength relationships, and the prediction of the compressive strengths from the maturity-compressive strength relationships are discussed for the lightweight polyester mortars.

2 Materials

2.1 Materials for binder systems
Binder systems were based on an unsaturated polyester resin(UP), together with styrene monomer(St) as a diluent, methyl ethyl ketone peroxide(MEKPO) as a catalyst and 8% mineral turpentine solution of cobalt octoate(CoOc) as an accelerator. The properties of the unsaturated polyester resin are listed in Table 1.

Table 1. Properties of unsaturated polyester resin

Specific gravity (20° C)	Viscosity (20 ° C, mPa·s)	Acid value	Styrene content (%)
1.13	325	16.9	38.0

2.2 Materials for filler and lightweight fine aggregate
Commercially available ground calcium carbonate($CaCo_3$) was used as a filler. Expanded shales(ES-1 and ES-2) were employed as materials for a lightweight fine aggregate. The properties of the filler and lightweight aggregates are given in Table 2.

Table 2. Properties of filler and lightweight aggregates

Type of material		Size (mm)	Specific gravity (20 ° C)	Water content (%)	Organic impurities
For filler	Ground calcium carbonate ($CaCo_3$)	2.5×10^{-3}	2.70	<0.1	Nil
For lightweight fine aggregate	Expanded shale (ES-2)	2.5-5.0	1.35	<0.1	Nil
	Expanded shale (ES-1)	<2.5	1.62	<0.1	Nil

3 Testing procedures

3.1 Preparation of mortars
Prior to the mixing of lightweight polyester mortars, the compositions of a lightweight

fine aggregate were prepared as follows:
    ES-1 : ES-2 = 2 : 1 (by volume)
In accordance with JIS(Japanese Industrial Standard) A 1181 (Method of Making
Polyester Resin Concrete Specimens), the lightweight polyester mortars were mixed
with the binder formulations and mix proportions as shown in Tables 3 and 4
respectively. Beam specimens 40 x 40 x 160 mm were molded, and then cured at
different temperatures of 0, 10, 20, 30 and 40°C for 6, 12, 24, 72, 120 and 168 hours.

Table 3. Formulations of binders for lightweight polyester mortars

Formulations (phr)*			
UP	St	MEKPO	CoOc
		0.25	0.25
100	12	0.50	0.50
		0.75	0.75

Note, *phr : parts per hundred parts of resin by mass.

Table 4. Mix proportions of lightweight polyester mortars

Mix proportions (vol%)				Filler-binder ratio, Vf/Vb	Specific gravity of lightweight polyester mortar
Binder, Vb	Filler, Vf	ES-1	ES-2		
25.2	15.6	19.7	39.5	0.62	1.72

3.2 Compressive strength test
According to JIS A 1183 (Method of Test for Compressive Strength of Polyester Resin
Concrete Using Portions of Beams Broken in Flexure), beam specimens were tested for
compressive strength.

4 Test results and discussion

Table 5 lists the ambient temperature vs. compressive strength of lightweight polyester
mortars with various catalyst and accelerator contents.

4.1 Estimation of datum temperatures for maturity equations
Maturity for ordinary cement concrete is defined as an integral of curing period and
temperature of cement concrete above a datum temperature[1][2], and is expressed by
the following equation:

$$M_c = \int_0^t (T - T_0)dt \qquad (1)$$
$$= \sum (T - T_0) \cdot \Delta t$$

where
    Mc = maturity [(°C · h) or (°C · d)] of cement concrete at curing period or age, t or
        Δt (h or d)

T   = temperature (°C) of cement concrete during curing, or ambient temperature

To = datum temperature (°C) below which no compressive strength gain of cement concrete takes place regardless of curing period or age (with additional curing period or age), -10 °C for cement concrete [3]

Table 5. Ambient temperature vs. compressive strength of lightweight polyester mortars with various catalyst and accelerator contents.

Catalyst and accelerator contents (phr)*		Ambient temperature	Compressive strength (MPa)					
			Curing period (h)					
MEKPO	CoOc	(° C)	6	12	24	72	120	168
		0	Not to test					
		10	0.0**	0.0	0.0	2.1	2.3	2.6
0.25	0.25	20	0.9	5.0	24.7	49.5	59.9	67.8
		30	5.0	9.0	57.7	68.5	70.3	73.2
		40	5.9	10.2	60.9	70.5	72.0	74.5
		0	2.1	7.2	9.8	26.0	32.0	32.8
		10	8.8	34.7	38.1	68.0	69.2	71.1
0.50	0.50	20	9.8	24.5	53.8	82.6	83.2	85.4
		30	15.0	60.3	70.8	82.9	84.2	87.6
		40	36.5	75.8	86.6	91.2	94.1	97.0
		0	6.0	12.5	15.8	36.0	39.0	41.5
		10	13.3	35.1	40.7	69.1	71.7	73.6
0.75	0.75	20	17.0	53.6	67.0	90.9	91.2	93.5
		30	50.5	72.7	80.9	91.8	92.0	96.1
		40	75.9	84.9	89.5	94.2	99.8	102.8

Notes, *phr : parts per hundred parts of resin by mass.
   **   : not to harden.

According to the law of compressive strength with maturity, in which cement concrete with the same mix proportions at the same maturity provides approximately the same compressive strength [4], the compressive strength of cement concrete can be expressed as a function of maturity. This is the concept of the maturity method for compressive strength prediction for cement concrete.

In the application of the maturity method to the compressive strength prediction for lightweight polyester mortars, the processing of $\triangle$t data in Eq. (1) and the estimation of a true datum temperature (To) should be sufficiently considered because the setting or hardening process of the lightweight polyester mortars differs considerably from that of ordinary cement concrete. From the results of some preliminary studies, it was found out that the use of $\sqrt{\triangle t}$ in replacement of $\triangle$t in Eq.(1) is desirable because of more rapid progress in the setting or hardening process of the lightweight polyester mortars than the cement concrete. In general, the setting or hardening process of the lightweight polyester mortars is controlled by the catalyst and accelerator contents for the unsaturated polyester resin as a binder. Therefore, maturity (Mp) for the lightweight polyester mortars is expressed as follows :

$$M_P = \sum (T - T_0) \cdot \sqrt{\Delta t} \qquad (2)$$

To estimate the datum temperatures(T_0) for the maturity equations for the lightweight polyester mortars, the ambient temperature-compressive strength relationships of the lightweight polyester mortars with different combined catalyst and accelerator contents at various curing periods are represented in Figs.1 to 3. As seen in these figures, the ambient temperature-compressive strength curves at different combined catalyst and accelerator contents become almost the same or duplicate at curing periods of 72,120 and 168h, and can be expressed as empirical equations by applying regression analysis regardless of curing periods of 72h or more. By extrapolations using the empirical equations, the datum temperatures for the lightweight polyester mortars are estimated as shown in Table 6.

Table 6. Datum temperatures and equations for their estimation

Identification symbol	Catalyst content (phr)*	Accelerator content (phr)	Equation for datum temperature estimation	Coefficient of correlation	Datum temperature (°C)
0.25 / 0.25	0.25	0.25	Y=22.12√X-59.83	0.95	7.3
0.50 / 0.50	0.50	0.50	Y=10.31√X+30.06	0.94	-8.5
0.75 / 0.75	0.75	0.75	Y=9.76√X+38.45	0.91	-15.5

Note, *phr : parts per hundred parts of resin by mass.

Table 7. Maturity of lightweight polyester mortars

Catalyst and accelerator contents (phr)*		Ambient temperature (°C)	Maturity (°C·h) Curing period (h)					
MEKPO	CoOc		6	12	24	72	120	168
		0	Not to test					
		10	6.6	9.4	13.2	22.9	29.6	35.0
0.25	0.25	20	31.1	44.0	62.2	107.8	139.1	164.6
		30	55.6	78.6	111.2	192.6	248.7	294.2
		40	80.1	113.3	160.2	277.5	358.1	423.8
		0	20.8	29.4	41.6	72.1	93.1	110.2
		10	45.3	64.1	90.6	157.0	202.7	239.8
0.50	0.50	20	69.8	98.7	139.6	241.8	312.2	369.4
		30	94.3	133.4	188.6	326.7	421.7	499.0
		40	118.8	168.0	237.6	411.5	531.3	628.6
		0	38.0	53.7	76.0	131.5	169.8	200.9
		10	62.5	88.3	124.9	216.4	279.3	330.5
0.75	0.75	20	87.0	123.0	173.9	301.2	388.9	460.1
		30	111.5	157.6	222.9	386.1	498.4	589.8
		40	136.0	192.3	271.9	470.9	608.0	719.4

Note, *phr : parts per hundred parts of resin by mass.

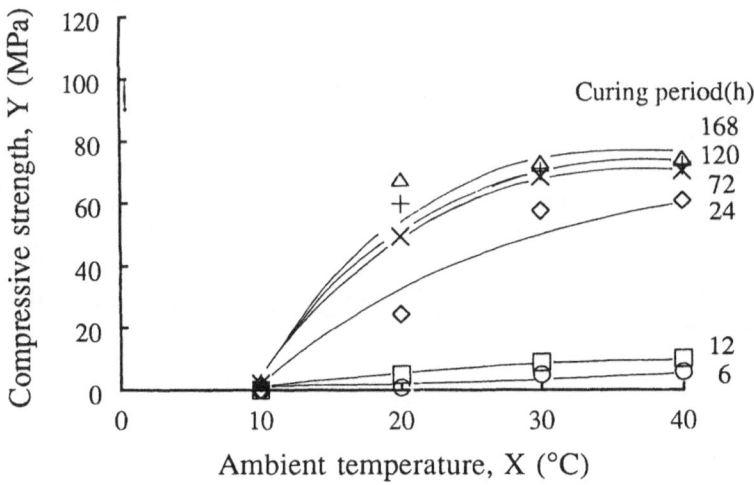

Fig. 1. Ambient temperature vs. compressive strength of lightweight polyester mortars with MEKPO content of 0.25phr and CoOc content of 0.25phr.

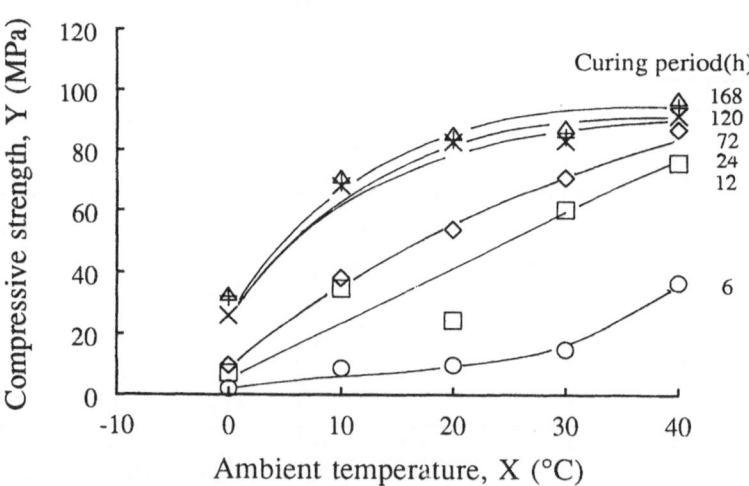

Fig. 2. Ambient temperature vs. compressive strength of lightweight polyester mortars with MEKPO content of 0.50phr and CoOc content of 0.50phr.

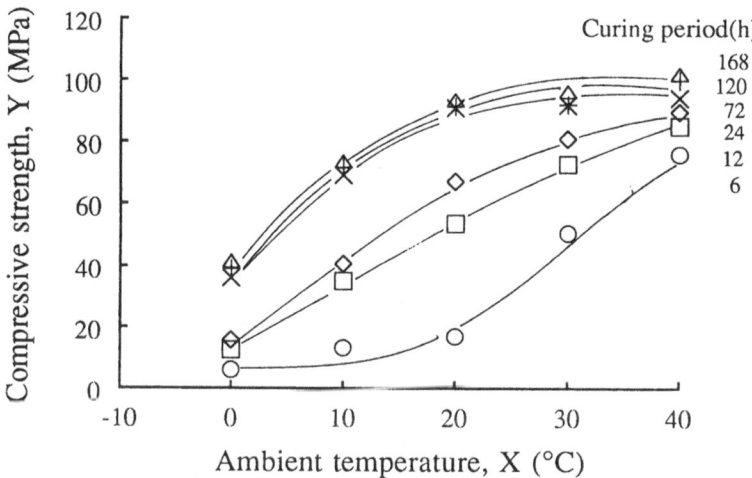

Fig. 3. Ambient temperature vs. compressive strength of lightweight
polyester mortars with MEKPO content of 0.75phr and CoOc
content of 0.75phr.

4.2 Prediction of compressive strengths by the maturity method

Table 7 gives the maturity of lightweight polyester mortars, calculated by using the
maturity equations with the estimated datum temperatures. Figs. 4 to 6 illustrate the
maturity in Table 7 vs. compressive strength of the lightweight polyester mortars with
different combined catalyst and accelerator contents as shown in Table 5. As seen in
these figures, the regression analysis for the maturity-compressive strength relationships
was conducted for each combination of the catalyst and accelerator contents, and
provides three empirical equations expressing such relationships with high coefficients
of correlation. It is suggested from the above results that the compressive strengths of
the lightweight polyester mortars can be predicted from the maturity-compressive
strength relationships.

5   Conclusions

The following conclusions are drawn from the test results in this research work:

(1) The datum temperatures for maturity equations for lightweight polyester mortars
    with different combined catalyst and accelerator contents can be estimated by using
    the ambient temperature-compressive strength relationships.

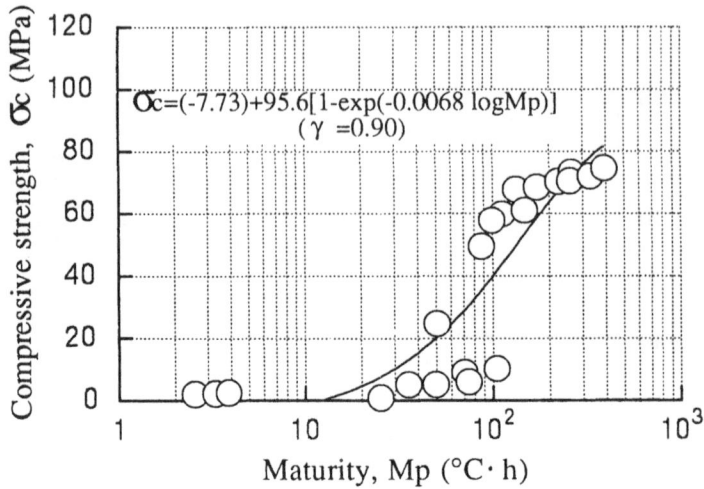

Fig. 4. Maturity vs. compressive strength of lightweight polyester mortars with MEKPO content of 0.25phr and CoOc content of 0.25phr.

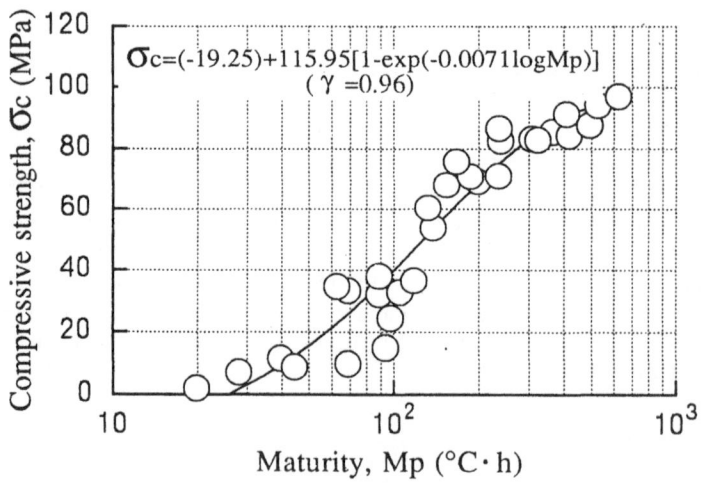

Fig. 5. Maturity vs. compressive strength of lightweight polyester mortars with MEKPO content of 0.50phr and CoOc content of 0.50phr.

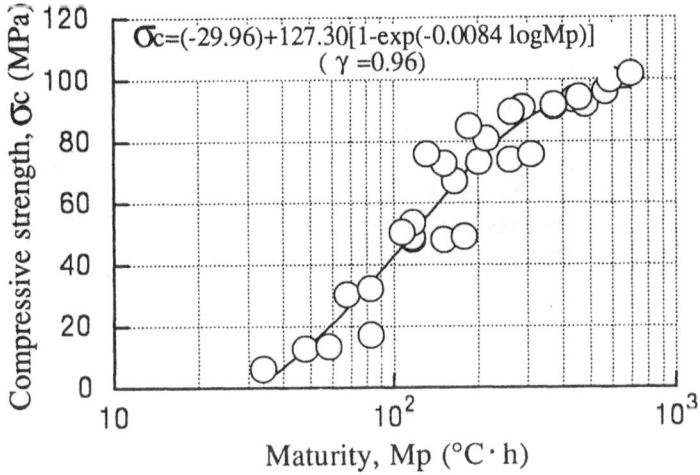

Fig. 6. Maturity vs. compressive strength of lightweight
polyester mortars with MEKPO content of 0.75phr
and CoOc content of 0.75phr.

(2) The maturity for lightweight polyester mortars is expressed as follows:

$$M_p = \sum (T - T_0) \cdot \sqrt{\Delta t}$$

where

Mp = maturity [(°C·h) or (°C·d)] of lightweight polyester mortars at curing
period or age, Δt (h or d)

T = temperature (°C) of lightweight polyester mortars during curing

To = datum temperature (°C) below which no compressive strength gain of
lightweight polyester mortars takes place regardless of curing period or age
(with additional curing period or age)

(3) The compressive strength of lightweight polyester mortars can be predicted by using
the following empirical equation expressing the maturity-compressive strength
relationships.

$$\sigma_c = A + B[1 - \exp(-C \log Mp)]$$

where

σ_c = compressive strength(MPa) of lightweight polyester mortars

Mp = maturity [(°C·h) or (°C·d)] of lightweight polyester mortars

A,B and C = constants

6 References

1. Oluokun,F.A., Burdette,E.G. and Deatherage,J.H. (1990) Early-Age Concrete Strength Prediction by Maturity-Another Look. *ACI Materials Journal*, Vol.87, No.6, pp. 565-572.
2. Carino, N. J. (1984) The Maturity Method : Theory and Application. *Cement, Concrete, and Aggregates*, Vol. 6 , No. 2, pp. 61-73.
3. Bergstörm, S.G.(1953) Curing Temperature, Age and Strength of Concrete. *Magazine of Concrete Research*, Vol.5, No.14, pp.61-66.
4. Plowman, J.M. (1956) Maturity and the Strength of Concrete. *Magazine of Concrete Research*, Vol.8, No.22, pp.13-22.

EFFECTS OF MIX PROPORTIONING FACTORS ON CONSISTENCY OF LIGHTWEIGHT POLYESTER MORTARS

Y. Ohama, K. Demura, K. Ide and Y.S. Lee
Dept. of Architecture, Nihon University, Koriyama, Japan

Abstract
This paper deals with the consistency of fresh lightweight polyester mortars as high-performance or multifunctional building materials. Fresh lightweight polyester mortars using an unsaturated polyester resin, lightweight fillers and aggregate are mixed with various mix proportions, and tested for slump as a measure of consistency. The influences of mix proportioning factors on the slump of the lightweight polyester mortars are examined, and the empirical equations for the slump predictions are successfully proposed.
Keywords :   Artificial lightweight aggregate, binder content, consistency, filler-binder ratio, lightweight polyester mortars, slump, slump predictions, unsaturated polyester resin.

1 Introduction

Recently, polyester concrete has been widely used in the construction industry because of its quick setting, high strength, excellent adhesion, watertightness and chemical resistance compared to ordinary cement concrete. Its applications are also developed increasingly [1]. In particular, this trend is marked in the new frontiers of the building industry, i.e., superhighrise buildings and long-span structures in advanced countries. There are high-strength lightweight concretes in the building materials toward which a great interest is oriented in such background. The objective of this study is to get basic information about the consistency of lightweight polyester mortars using unsaturated polyester resin and artificial lightweight aggregates.

In the present paper, lightweight polyester mortars using an unsaturated polyester resin, lightweight fillers and aggregate are prepared with various mix proportions,

Polymers in Concrete, edited by Y. Ohama, M. Kawakami and K. Fukuzawa. Published in 1997 by E & FN Spon, 2–6 Boundary Row, London SE1 8HN, UK. ISBN: 0 419 22330 4.

and tested for slump. The effects of mix proportioning factors on the slump of the lightweight polyester mortars are examined, and the empirical equations for predicting the slump are successfully proposed.

2 Materials

2.1 Materials for binder systems
Binder systems were based on an unsaturated polyester resin (UP), together with styrene monomer (St) as a diluent, methyl ethyl ketone peroxide (MEKPO) as a catalyst and 8% mineral turpentine solution of cobalt octoate (CoOc) as an accelerator. The properties of the unsaturated polyester resin are listed in Table 1.

Table 1. Properties of unsaturated polyester resin

Specific gravity (20°C)	Viscosity (20°C,mPa·s)	Acid value	Styrene content (%)
1.13	325	16.9	38.0

2.2 Materials for lightweight fillers and fine aggregate
Commercially available hollow glass (HG), hollow fly ash (HF) and ground calcium carbonate (CaCO₃) were used as materials for lightweight fillers. Hollow fused alumina (HFA) and hollow mullite (HM) were employed as materials for a lightweight fine aggregate. The properties of the materials for the lightweight fillers and fine aggregate are given in Table 2.

Table 2. Properties of materials for lightweight fillers and fine aggregate

Type of material		Size (mm)	Specific gravity (20°C)	Water content (%)	Organic impurities
For lightweight filler	Hollow glass (HG)	$10\text{-}120 \times 10^{-3}$	0.45	<0.1	Nil
	Hollow fly ash (HF)	$10\text{-}350 \times 10^{-3}$	0.40	<0.1	Nil
	Ground calcium carbonate (CaCO₃)	$<2.5 \times 10^{-3}$	2.70	<0.1	Nil
For lightweight fine aggregate	Hollow fused alumina (HFA)	0.1-5.0	0.70	<0.1	Nil
	Hollow mullite (HM)	$150\text{-}300 \times 10^{-3}$	0.40	<0.1	Nil

3 Testing procedures

3.1 Preparation of mortars
Prior to the mixing of lightweight polyester mortars, the compositions of lightweight fillers and a lightweight fine aggregate were prepared as follows :

Compositions of the fillers
    (1) $HG:CaCO_3=2:1$ (by volume)
    (2) $HF:CaCO_3=2:1$ (by volume)
    (3) $CaCO_3$ (as a control)
Compositions of the fine aggregate
    $HFA:HM=4:1$ (by volume)

In accordance with JIS (Japanese Industrial Standard) A 1181 (Method of Making Polyester Resin Concrete Specimens), the lightweight polyester mortars were mixed with the binder formulations and mix proportions as shown in Tables 3 and 4 respectively.

Table 3. Binder formulations for lightweight polyester mortars

Formulations (phr)*			
UP	St	MEKPO	CoOc
100	12	0.5	0.5

*phr : parts per hundred parts of resin.

Table 4. Mix proportions of lightweight polyester mortars

Mix proportions by volume			Filler-binder ratio, V_f/V_b (by volume)
Binder, V_b	Filler, V_f	Fine aggregate	
23.6	8.7	67.7	0.37
	10.9	65.5	0.46
	14.6	61.8	0.62
24.4	9.0	66.6	0.37
	11.2	64.4	0.46
	15.1	60.5	0.62
25.2	9.3	65.5	0.37
	11.6	63.2	0.46
	15.6	59.2	0.62

3.2 Slump test

According to JIS A 1173 (Method of Test for Slump of Polymer-Modified Mortar), fresh lightweight polyester mortars were tested for slump. Plate glass as prescribed in JIS R 3202 (Float and Polished Plate Glass) was used as a flat plate. The slump of the mortars on the plate glass was measured at 60 seconds after the removal of the slump cone.

4 Test results and discussion

Figs.1, 2 and 3 show the relation between the binder content and slump of lightweight polyester mortars. The slump of the lightweight polyester mortars is increased with increasing binder content and filler-binder ratio regardless of type of filler. This may chiefly be explained by the improved flowability of the lightweight polyester mortars due to the increased binder content. On the other hand, the effect

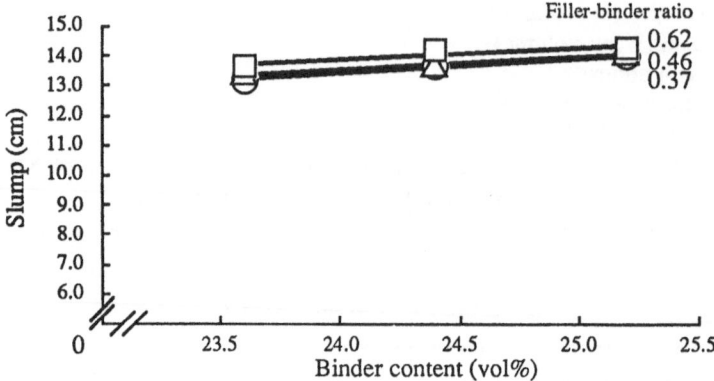

Fig.1. Binder content vs. slump of lightweight polyester mortars using Filler HG/CaCO₃.

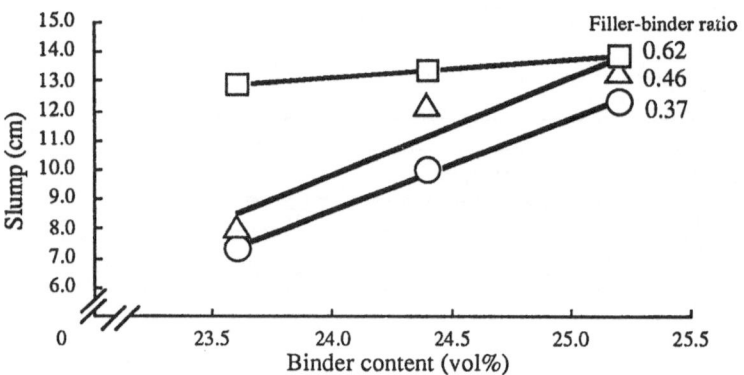

Fig.2. Binder content vs. slump of lightweight polyester mortars using Filler HF/CaCO₃.

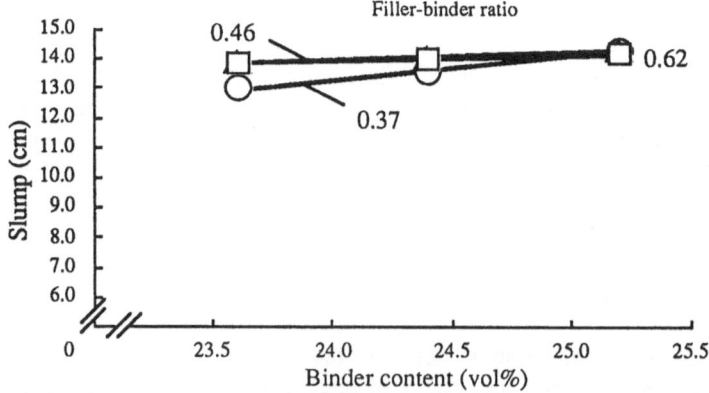

Fig.3. Binder content vs. slump of lightweight polyester mortars using Filler CaCO₃.

of the filler-binder ratio on the slump is marked for the lightweight polyester mortars using Filler $HF/CaCO_3$, but is negligible for the mortars using Fillers $HG/CaCO_3$ and $CaCO_3$. It seems that the lightweight polyester mortars using Fillers $HG/CaCO_3$ and $CaCO_3$ give a good workability even at low binder content. The consistency of a concrete-like mixture of a liquid and solid particles can generally be expressed as the functions of the viscosity of the liquid, the shape, size and content of the solid particles, etc.[2]. In this case, as Filler $HF/CaCO_3$ contains much larger particles from HF compared to Fillers $HG/CaCO_3$ and $CaCO_3$ as seen in Table 2, and the smaller particle volume-total filler volume ratio of the lightweight polyester mortars is changed, the slump of the lightweight polyester mortars using Filler $HF/CaCO_3$ is varied by similar effect to sand-aggregate ratio effect on the slump of ordinary cement concrete. It is considered from Figs. 1 to 3 that the slump of the lightweight polyester mortars is proportional to the binder content. The empirical equations for predicting the slump are expressed for each lightweight polyester mortar as follows:

$HG/CaCO_3$

$V_f/V_b=0.37,$	$Sl=0.50V_b+1.43$	$(\gamma=0.99)$	(1)
$V_f/V_b=0.46,$	$Sl=0.44V_b+3.05$	$(\gamma=0.99)$	(2)
$V_f/V_b=0.62,$	$Sl=0.38V_b+4.92$	$(\gamma=0.93)$	(3)

$HF/CaCO_3$

$V_f/V_b=0.37,$	$Sl=3.12V_b-66.3$	$(\gamma=0.99)$	(4)
$V_f/V_b=0.46,$	$Sl=3.31V_b-69.7$	$(\gamma=0.95)$	(5)
$V_f/V_b=0.62,$	$Sl=0.62V_b-1.85$	$(\gamma=0.99)$	(6)

$CaCO_3$

$V_f/V_b=0.37,$	$Sl=0.81V_b-6.19$	$(\gamma=0.99)$	(7)
$V_f/V_b=0.46,$	$Sl=0.25V_b+7.97$	$(\gamma=0.96)$	(8)
$V_f/V_b=0.62,$	$Sl=0.19V_b+9.46$	$(\gamma=0.98)$	(9)

where Sl is the slump (cm) of the lightweight polyester mortars, V_b and V_f are the binder and filler contents (vol%) respectively, and γ is coefficient of correlation. The above equations can be rewritten in the following general form:

$$Sl= \alpha \cdot V_b + \beta$$

α, $\beta =f(V_f/V_b)$ are anticipated from equations (1) to (9), and the following equations are obtained for constants α and β:

$HG/CaCO_3$

$\alpha =-0.49 \cdot V_f/V_b+0.67$ $(\gamma=0.97)$, $\beta =13.7 \cdot V_f/V_b-3.47$ $(\gamma=0.98)$ (10)

$HF/CaCO_3$

$\alpha =-10.8 \cdot V_f/V_b+7.57$ $(\gamma=0.95)$, $\beta =277 \cdot V_f/V_b-180$ $(\gamma=0.97)$ (11)

$CaCO_3$

$\alpha =-2.23 \cdot V_f/V_b+1.50$ $(\gamma=0.94)$, $\beta =56.4 \cdot V_f/V_b-23.5$ $(\gamma=0.94)$ (12)

Fig. 4 represents the relation between the filler-binder ratio (V_f/V_b) and empirical constants α and β in the general equation of the slump of the lightweight polyester

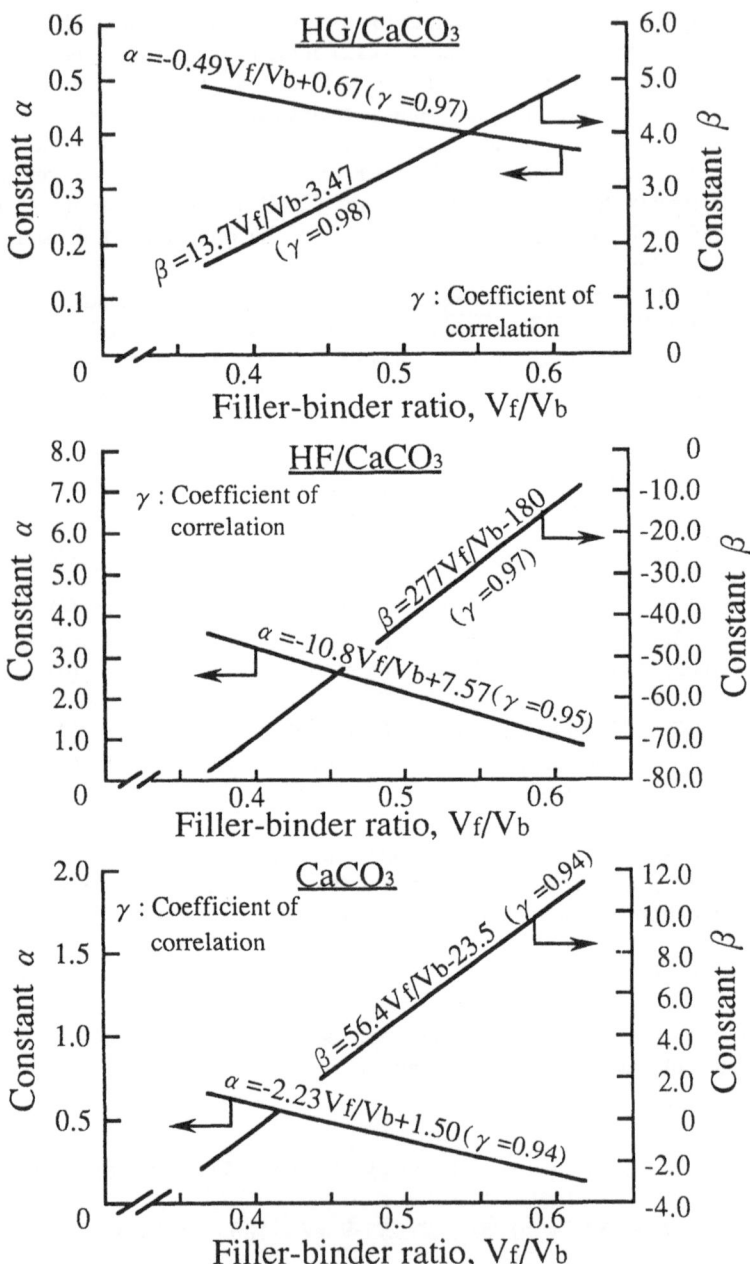

Fig. 4. V_f/V_b-Constants α and β relationships for prediction of slump of lightweight polyester mortars.

mortars. By using the relationships in Fig. 4 or equations (10), (11) and (12), the slump of the lightweight polyester mortars can easily be predicted.

5 Conclusions

The following conclusions are drawn from the test results in this research work:
(1) The slump of lightweight polyester mortars is increased with increasing binder content and filler-binder ratio.
(2) The slump of lightweight polyester mortars can be predicted as follows:

$$Sl = \alpha \cdot V_b + \beta$$

HG/CaCO$_3$
$\quad \alpha = -0.49 \cdot V_f/V_b + 0.67$ ($\gamma = 0.97$), $\beta = 13.7 \cdot V_f/V_b - 3.47$ ($\gamma = 0.98$)
HF/CaCO$_3$
$\quad \alpha = -10.8 \cdot V_f/V_b + 7.57$ ($\gamma = 0.95$), $\beta = 277 \cdot V_f/V_b - 180$ ($\gamma = 0.97$)
CaCO$_3$
$\quad \alpha = -2.23 \cdot V_f/V_b + 1.50$ ($\gamma = 0.94$), $\beta = 56.4 \cdot V_f/V_b - 23.5$ ($\gamma = 0.94$)

where Sl is the slump (cm) of the lightweight polyester mortars, V_b and V_f are the binder and filler contents (vol%) respectively, and γ is coefficient of correlation.

6 References

1. Ohama, Y., Demura, K. and Shimizu, A. (1986) Effects of mix proportioning factors on consistency of fresh polyester resin concrete. *Proceedings of the Transactions of the Japan Concrete Institute*, vol.8, pp.51-56.
2. Iwasaki, N. (1975) *Properties of Concrete* (in Japanese). Kyoritsu Shuppan, Tokyo, pp.25-30.

PART THREE
PROPERTIES

ENGINEERING PROPERTIES OF PERMEABLE POLYMER CONCRETE USING STONE DUST AND HEAVY CALCIUM CARBONATE

C.Y. Sung, S.W. Kim, J.K. Min, Y.J. Song, H.J. Jung and K.T. Kim
Dept. of Agricultural Engineering, Chungnam National University, Taejon, Korea

Abstract

Permeable polymer concrete can be applied to roads, sidewalks, river embankment, drain pipes, conduits, retaining walls, yards, parking lots, plazas, interlocking blocks, etc. This study was to explore a possibility of using stone dust and heavy calcium carbonate as fillers for the permeable polymer concretes. Different mixing proportions were tried to find an optimum mixing proportion of the permeable polymer concretes. The tests were carried out at $20 \pm 1°C$ and $60 \pm 2\%$ relative humidity. At 7 days of curing, compressive, flexural and splitting tensile strengths and water permeability ranged between 209-246kgf/cm², 101-121kgf/cm², 36-52kgf/cm² and 3.076-4.390 ℓ /cm²/hr, respectively. It was concluded that the stone dust and heavy calcium carbonate could be used in the permeable polymer concretes.
Keywords : elastic modulus, heavy calcium carbonate, permeable polymer concrete, stone dust, strengths, stress-strain curve, water permeability.

1 Introduction

Demand for concrete material supply has been widening with the rapid growth of construction works such as structures. Supply of natural materials from river beds and mountains are not sufficent, environmental problems associated with the material collection from river beds by dredging and from mountains by excavation have caused strong protests from environmentalists. And with the growth of construction industry, the supply of materials in construction industry have been pressing question to solve in near future. Application of stone dust and heavy calcium carbonate as on

Polymers in Concrete, edited by Y. Ohama, M. Kawakami and K. Fukuzawa. Published in 1997 by E & FN Spon, 2-6 Boundary Row, London SE1 8HN, UK. ISBN: 0 419 22330 4.

additive or filler of concrete has widely increased and recent studies have found to the excellent compatibility between those industrial byproducts and polymers[1][2]. The use of polymer concrete as on alternative to cement concrete products has been increasing because of its superior mechanical properties, chemical resistance, durability, strong adhesion and rapid curing[3].

This study initiated to find a way to reuse an industrial byproducts as a precious resource. A way to reuse the industrial byproducts is to use it as a construction material. The objectives of this study were (1) to find a way to reuse industrial byproducts as filler in permeable polymer concrete that has properties of high flexural strength, long durability and corrosion resistivity, (2) to test the permeable polymer concrete that uses an industrial byproducts as filler with respect to engineering properties, and (3) to provide the test data for factory production of the permeable polymer concrete products.

2 Materials and methods

2.1 Materials

2.1.1 Unsaturated polyester resin
An ortho-type unsaturated polyester resin was used in this study and its general properties are shown in Table 1.

Table 1. General properties of unsaturated polyester resin.

Specific gravity at 20℃	Viscosity at 20℃ (poise)	Styrene content (%)	Acid value
1.12	3.5	37.2	26.5

2.1.2 Hardner
Hardner used in this study is for normal hardening and its general properties are shown in Table 2.

Table 2. General properties of harder.

Component		Specific gravity at 20℃	Active oxygen (%)
MEKPO	55%	1.13	10.0
DMP	45%		

2.1.3 Aggregates

Fine and coarse aggregates were natural aggregate collected at a river bed. The river was near Tacjon, Korea. Also, aggregates washed and dried at 100 ± 5°C for one day before use[4]. Physical properties of natural aggregates are shown in Table 3.

Table 3. Physical properties of natural aggregates.

Classification	Size (mm)	Specific gravity (20°C)	Absorption (%)	F.M	Unit weight (kg/m')
Coarse aggregate	4.75~10	2.63	2.15	6.00	1,560
Fine aggregate	0.597~4.75	2.63	1.32	2.96	1,530

2.1.4 Filler

Common fillers are powders of fly ash, calcium carbonate, alumina, silica, cement, stone dust and so on. Among the fillers, stone dust and heavy calcium carbonate were used in this study because it is relatively cheap and easy to buy. Fillers were dried at 100 ± 5°C for one day before use[4]. Chemical composition and physical properties of fillers are shown in Table 4.

Table 4. Chemical composition and physical properties of fillers

	Item	Stone dust	Heavy calcium carbonate
	$CaCO_3$	–	100
	SiO_2	73.40	–
	Al_2O_3	11.30	–
Chemical	K_2O	4.77	–
composition	Fe_2O_3	3.00	–
(%)	Na_2O	3.63	–
	MgO	0.25	–
	CaO	0.98	–
	Ig.loss	0.61	–
Specific gravity (20°C)		3.18	2.20
Specific surface (Blain)(cm/g)		3,054	3,126
Bulk density (kg/m')		1,682	620
Grain size (mm)		< 0.595	< 0.297

2.2 Mixing proportion of permeable polymer concrete

After many preliminary tests, 5 mixing proportions were tried to determine an optimum mixing proportion of permeable polymer concretes using stone dust and heavy calcium carbonate contents of unsaturated polyester resin, hardner, fillers and natural aggregates were fixed as in Table 5.

Table 5. Mixing proportions of permeable polymer concretes

| Mix type | Mixing proportion (wt. %) | | | | | | Total (wt. %) |
| | Resin | Hardner | Aggregate | | Filler | | |
			Fine	Coarse	Stone dust	CaCO$_3$	
G$_1$	7.20	0.14	16.00	68.26	–	8.40	100
G$_2$	7.28	0.15	16.05	68.27	2.06	6.19	100
G$_3$	7.33	0.15	16.07	68.25	4.10	4.10	100
G$_4$	7.40	0.15	16.10	68.25	6.08	2.02	100
G$_5$	7.64	0.15	16.00	68.21	8.00	–	100

2.3 Specimen

Specimens were prepared according to the Korean Standard Testing Methods, KS F 2419 (Specimen preparation methods for strength measure of polyester resin concrete). Permeable polymer concretes were mixed by using a high performance concrete mixer. Two types of specimen, i.e., cylindrical and block specimens, were made depending on test. Specimens were formed by putting permeable polymer concrete into a cylindrical and block mold, the mold was put on a table vibrator and compacted sufficiently by vibration. All the specimens were demolded after cured at 20 ± 1℃ that was the room temperature for three hours, and cured again at 20 ± 1℃ for up to 7 days.

2.4 Test methods

2.4.1 Water permeability

Water permeability was measured by volumn of permeated water(ℓ /cm²/hr). The size of block specimen was 20cm×20cm×6cm. The permeability test apparatus of permeable polymer concrete tested in this study is shown in Fig.1.

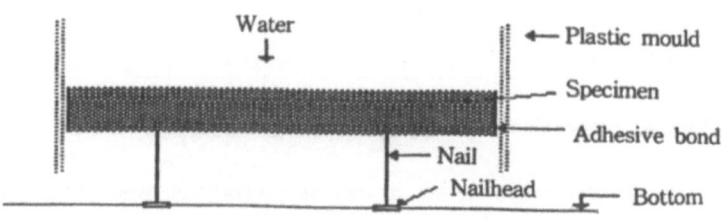

Fig.1. Schematic drawing of permeability test apparatus

2.4.2 Strengths

Compressive, flexural and splitting tensile strength tests were carried out according to KS F 2481 (Compressive strength test method for polyester resin concrete), KS F 2482 (Flexural strength test method for polyester resin concrete), and KS F 2480 (Splitting tensile strength test method for polyester resin concrete), respectively.

2.4.3 Static modulus of elasticity, poisson's ratio and strain

Tests for static modulus of elasticity and poisson's ratio were performed by using the strain gauge method which is in KS F 2438 (Testing method for static modulus of elasticity and poisson's ratio in compression of cylindrical concrete specimens). Static modulus of elasticity and poisson's ratio were obtained from the secant modulus that was computed from the tangent of stress-strain curve. Loads were applied up to 40% of the failure load. The size of cylindrical specimen was ϕ15cm×30cm.

2.4.4 Dynamic modulus of elasticity

Dynamic modulus of elasticity test was carried out according to BS 1881 (Methods of testing hardened concrete for other than strength). The dynamic modulus test was conducted by excitation in the longitudinal mode of vibration. The size of cylindrical specimen was ϕ15cm×30cm.

3 Results and discussion

3.1 Water permeability

Table 6 shows the results of water permeability tests of permeable polymer concretes. The water permeability increased with the decrease of the content of stone dust. When the content of stone dust was more 8% of the total permeable polymer concrete material weight, the water permeability seemed to decrease rapidly. It was thought that the large amount of stone dust absorbed more resin than calcium carbonate did and thus, the compactability of the permeable polymer concrete got raised, resulting in less pores in the permeable polymer concrete. Measured water permeability of the permeable polymer concrete ranged from 3.067 to 4.390 ℓ/cm²/hr and were more 100 times as large as in the world maximum rainfall, and it was largely dependent upon the mixing proportion. It proved that the permeability of the permeable polymer concrete is superior. Accordingly, these permeable polymer concretes can be used to the members and structures which needs appropriate strength and water permeability.

Table 6. Results of water permeability tests of permeable polymer concretes

Type	Water permeability (ℓ /cm²/hr)
G_1	4.390
G_2	4.130
G_3	3.923
G_4	3.669
G_5	3.076

3.2 Strengths

Strength developments of permeable polymer concrete were assumed to be related to the content of fillers as stone dust and calcium carbonate. And thus, strength tests were performed with respect to the different content of fillers. Table 7 shows the results of compressive, flexural and splitting tensile strengths. Measured strength decreased with the increase of the content of heavy calcium carbonate. The highest strength was achieved by stone dust filled permeable polymer concrete, and the choice of filler is very important[5]. It was increased 17% by compressive, 188% by flexural and 148% by splitting tensile strength than that of normal cement concrete, respectively.

Table 7. Strengths of permeable polymer concretes

Type	Strength ($kgf/cm²$)		
	Compressive	Flexural	Splitting tensile
G_1	209	101	36
G_2	217	109	42
G_3	230	115	44
G_4	237	117	50
G_5	246	121	52

Table 8 shows the strength ratio of permeable polymer concretes with different mixing proportion. The strength ratio is one of the important properties of the permeable polymer concretes. The ratio of flexural strength to compressive strength of the permeable polymer concretes was between 0.483 and 0.502, and it was 2.5 times higher than that of normal cement concrete. Also, the compressive strength of the permeable polymer concretes was higher, and the flexural strength was much higher than that of normal cement concrete. It could be explained that the polymer has a particular property and thus, the toughness of polymer concrete was higher than that of normal cement concrete.

Accordingly, the application of permeable polymer concretes in the structures for bending would be very useful.

Table 8. Strength ratio of permeable polymer concrete

Type	σ_t/σ_c	σ_b/σ_c	σ_t/σ_b
G_1	0.172	0.483	0.356
G_2	0.194	0.502	0.385
G_3	0.191	0.500	0.383
G_4	0.211	0.493	0.427
G_5	0.211	0.492	0.429

3.3 Static modulus of elasticity, poisson's ratio and strain

The effects of stone dust and heavy calcium carbonate on the static modulus of elasticity, poisson's ratio and strain of permeable polymer concretes are shown in Table 9. The static modulus of elasticity was in the range of $1.17 \times 10^5 - 1.32 \times 10^5$ kgf/ cm', which was approximately 53-56% of that of normal cement concrete. Stone dust filled the permeable polymer concrete was showed relatively higher elastic modulus, poisson's ratio and strains. The poisson's number and strain of the permeable polymer concretes were smaller and larger than that of normal cement concrete, respectively. Also, static modulus of elasticity increased with the increase of strength[6].

Table 9. Static modulus of elasticity, poisson's ratio and strain of permeable polymer concretes

Type	Static modulus of elasticity ($\times 10^3$ kgf/cm')	Poisson's		Strain ($\times 10^{-3}$)	
		Ratio	Number	Longitudinal	Horizontal
G_1	117	0.167	5.981	0.640	0.107
G_2	122	0.174	5.738	0.637	0.111
G_3	126	0.176	5.686	0.671	0.118
G_4	130	0.178	5.612	0.724	0.129
G_5	132	0.196	5.106	0.725	0.142

Compressive stress-strain curves of permeable polymer concretes tested in this study is shown in Fig.2. Tangents of stress-strain curve increased with the increse of the content of stone dust. However, it was thought that the pore of the permeable polymer concretes was closely related to the static modulus of elasticity.

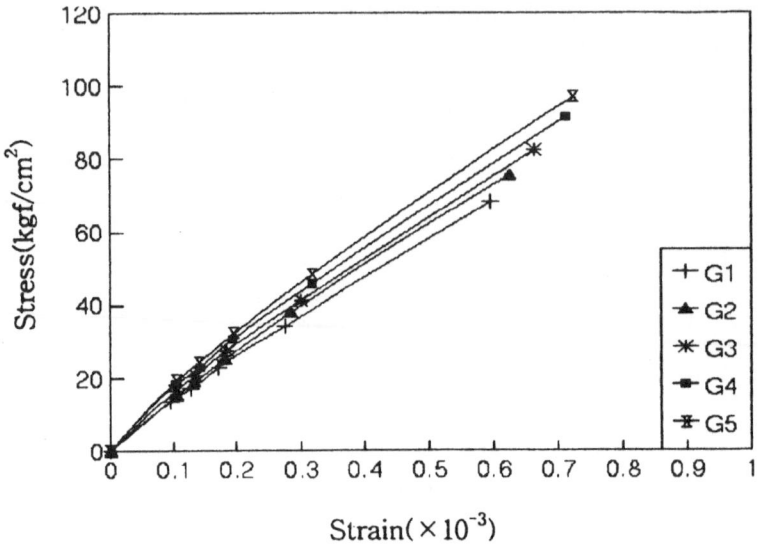

Fig. 2. Compressive stress-strain curves of permeable polymer concrete
by type of filler

3.4 Dynamic modulus of elasticity

Dynamic modulus of elasticity of permeable polymer concretes is shown in Table 10. Dynamic modulus of elasticity increased with the increase of the content of stone dust. Also, dynamic modulus of elasticity was interrelated with the compressive, flexural and splitting tensile strengths and water permeability[7]. This technique offers potentially useful nondestructive test methods to identify the physical condition of concrete.

Table 10. Dynamic modulus of elasticity of permeable polymer concretes

Type	Dynamic modulus ($\times 10^3$ kgf/cm^2)
G1	130
G2	135
G3	140
G4	143
G5	150

4 Conclusions

This study was performed to evaluate the engineering properties of permeable polymer concretes. An unsaturated polyester resin was used as binder material, and stone dust and heavy calcium carbonate were used as fillers. The following conclusions were drawn.

1) Water permeability was in the range of 3.076 ~ 4.390 ℓ/cm/hr and was largely dependent upon the mixing proportion and it was inversely dependent upon the magnitude of the concrete strengths. These permeable polymer concretes can be used to the members and structures which need appropriate strength and water permeability.

2) Strength increased with the increase of the content of stone dust. At 7 days of curing, compressive, flexural and splitting tensile strengths ranged between 209-246 kg f/cm, 101-121 kgf/cm and 36-52 kgf/cm, respectively. The highest strength was achieved by stone dust filled permeable polymer concrete, it was 17% by compressive, 188% by flexural and 148% by splitting tensile strengths than that of normal cement concrete, respectively. The increase of the flexural strength of the permeable polymer concrete was higher than the increase of the compressive strength compared to that of normal cement concrete, and those concrete to the structures that is subject to flexure can be used.

3) σ_t/σ_c and σ_b/σ_c of permeable polymer concrete were 0.172-0.211 and 0.483-0.502, respectively, and were larger than those strength ratios of normal cement concrete. σ_t/σ_b of the permeable polymer concrete was 0.356-0.429, and was smaller than those strength ratio of normal cement concrete. The ratio of flexural strength to compressive strength of the permeable polymer concrete was 2.5 times higher than that of normal cement concrete. It suggests that the permeable polymer concrete have a big advantage for the design and production of concrete structures.

4) Static modulus of elasticity, poisson's ratio and strain tended to increase with the increase of the content of stone dust. Static modulus of elasticity and poisson's ratio ranged between 1.17×10^5-1.32×10^5kgf/cm and 0.167-0.196, respectively. Static modulus of elasticity and poisson's ratio of permeable polymer concrete were smaller and larger, respectively, than those of normal cement concrete. Also, permeable polymer concrete was showed higher strain than that of normal cement concrete.

5) Dynamic modulus of elasticity was in the range of 1.3×10^5-1.5×10^5kgf/cm, which was approximately less compared to that of normal cement concrete. Stone dust filled permeable polymer concrete was showed higher dynamic modulus. The dynamic

modulus of elasticity were increased approximately 10-13% than that of static modulus of elasticity.

5 References

1. Sung,C.Y. Jung,H.J. (1996) Engineering properties of permeable polymer concrete with stone dust and fly ash, Journal of the Korean society of agricultural engineers, vol.38, No.4, pp.147-154. (in Korean)

2. Yeon,K.S., et al., (1993) Effects of fillers on mixing and mechanical properties of polymer concrete, Journal of the Korean society of agricultural engineers, Vol.35, No.2, pp.81-91. (in Korean)

3. Fowler,D.W. (1987) Current status of polymers in concrete, Proceedings of the 5th ICPIC, Brighton, U.K, pp.3-7.

4. Paturoyer,V.V. (1986) Recommendations on polymer concrete mix design, NIZHB, Moscow, p.18.

5. Aguado,H. Martinez,A. (1984) Effects of different factors in mixing and placing of polymer concrete, Proceedings of the First ICPIC, London,U.K, pp.299-303.

6. Neville,A.M. (1981) Properties of concrete, Pitman publishing limited, London, U.K, pp.605-635.

7. Sung,C.Y. (1995) Mechanical characteristics of permeable polymer concrete, Proceedings of the '95 Japan and Korea Joint Seminar, Tottori University, Japan, pp.32-35. (in Korean)

ON THE SIZE EFFECT IN FLEXURAL STRENGTH OF RESIN CONCRETE

M. Oshima and F. Hayashi
Sunrec Co., Ltd., Aichi, Japan
W. Koyanagi and Y. Uchida
Dept. of Civil Engineering, Gifu University, Gifu, Japan

Abstract
When resin concrete is applied to structures, the design strength should be established, taking into account the scale of the structural member. It is very important to evaluate the size effect quantitatively .

In this study, flexure tests of resin concrete beams were conducted with various cross-sectional sizes and with various maximum aggregate sizes. The effect of the maximum aggregate size on the size effect in flexural strength was investigated by examining the relationship between the depth of cross section and the apparent flexural strength.
Keywords: Flexural strength, maximum aggregate size, resin concrete, size effect.

1 Introduction

When the size of conventional cement concrete specimens increases, their apparent strength decreases. This phenomenon is the so-called size effect in strength, and is one of the inherent mechanical properties of concrete and other quasi-brittle materials.

The effect is also found in resin concrete, which consists of coarse aggregate, fine aggregate, and a filler bound together with polymers. The size effect in flexural strength of resin concrete is known to be more evident than in cement concrete, because it has a higher strength and more brittle mode of failure than ordinary cement concrete[1]. In addition, the flexural strength of resin concrete tends to decrease as the

Polymers in Concrete, edited by Y. Ohama, M. Kawakami and K. Fukuzawa. Published in 1997 by E & FN Spon, 2–6 Boundary Row, London SE1 8HN, UK. ISBN: 0 419 22330 4.

maximum size of aggregate increases, due to the higher strength of the binder than the bond strength between the binder and aggregate.

When resin concrete is applied to structural members in Japan, Type I application is more frequent, where resin concrete is used for flexural members to utilize its high flexural strength, than Type II, where resin concrete only bears the compression while the tension is borne by the reinforcement as in the case of ordinary reinforced concrete.   The scale of Type I applications also tends to increase[2].

The design strength of resin concrete should, therefore, be established taking into account the scale of the structure for its proper design as a structural material.   In this respect, it is very important to evaluate quantitatively the characteristics of the size effect.

In this study, flexure tests of resin concrete beams with five kinds of cross-sectional sizes and having three different maximum aggregate sizes were conducted.   The effect of the maximum aggregate size on the size effect in flexural strength was investigated by examining the relationship between the depth of cross section and the apparent flexural strength.

2   Outline of experiment

2.1   Mix proportions of resin concrete

The mix proportions of resin concrete used in this experiment are given in Table 1.   Unsaturated polyester resin (ortho-phthalic acid type) was used as the binder for the resin concretes.   Coarse aggregate was crushed stone from the same quarry having three different gradings, varying with the maximum sizes as 10 mm, 15 mm, and 25 mm and with the fineness moduli as 5.57, 5.88, and 6.23, respectively.   Their grading curves are shown in Fig. 1.   Natural sand with a fineness modulus of 1.3 was used as the fine aggregate.   Calcium carbonate with an average grain diameter of 40 μ m was used as the filler.

As for the hardener for the polyester resin, methyl ethyl keton peroxide (MEKPO) and cobalt naphthenate were used as the catalyst and accelerator, respectively.   Their dosage was adjusted so that the working life of the resin concrete was 45 ±10 min.

Table 1.   Mix proportions of resin concrete

Name of mixture	Resin (wt%)	Filler (wt%)	Fine aggregate (wt%)	Coarse aggregate (wt%)	Remark
Mix 1	10	20	20	50	M.S.=10 mm
Mix 2	10	20	20	50	M.S.=15 mm
Mix 3	10	20	20	50	M.S.=25 mm

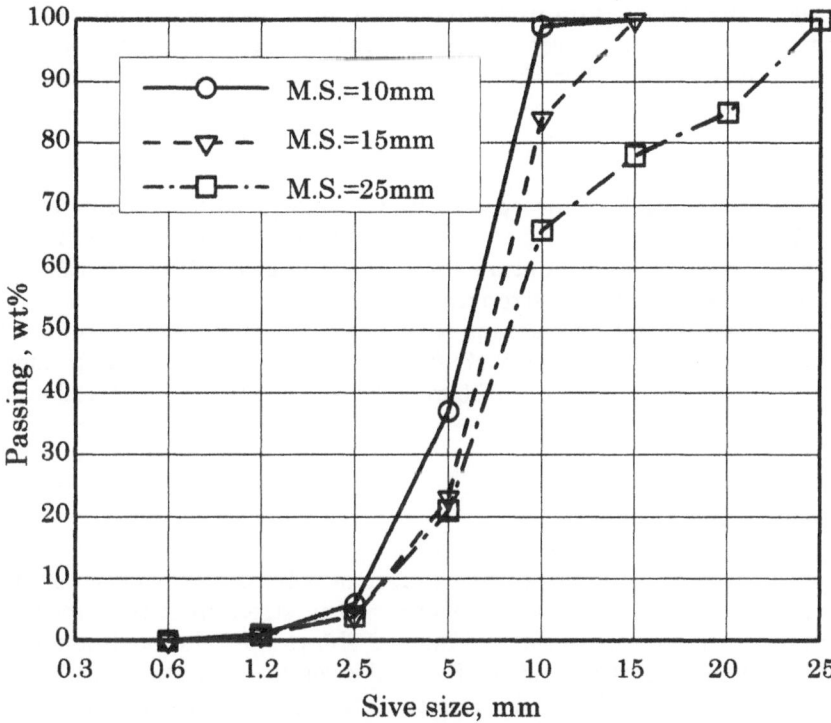

Fig. 1. Grading of coarse aggregates

2.2 Specimens

In this experiment, by employing resin concrete of three mix proportions as given in Table 1, three beam specimens with each of five cross-sectional sizes were cast. The sizes and designations of the specimens are given in Table 2.

The procedure of casting the specimens was as follows:
A forced mixer with a capacity of 1200 kg was used for mixing resin concrete. The batches, 800 kg in size, were mixed for 5 min. Freshly mixed resin concrete was placed in molds and was compacted with a vibrating table.

Table 2.   Sizes and designations of the specimens

Name of specimen	Depth (mm)	Width (mm)	Length (mm)
A	60	60	240
B	100	100	400
C	150	150	530
D	200	200	800
E	250	250	1000

The resin concretes set up in 40 to 50 min, and specimens were removed from the molds 1 hr after set up. The demolded specimens were cured in a heating oven of 80 ℃ for 15 hr and then in a room with a temperature of 20 ℃ for 10 days.

2.3 Strength testing

The flexural strength of the specimens was tested by third-point loading with the span length being 3 times the depth of cross section. The specimens were loaded to rupture with the loading speed being adjusted so as to reach the maximum strength in 2 to 3 min.

A compression testing machine with a capacity of 490 kN and universal testing machine with a capacity of 980 kN were used for loading specimens with a depth of cross section of 150 mm or less (A, B and C) and 200 mm or more (D and E), respectively.

3 Results and discussion

Table 3 gives the average flexural strength of specimens from each mixture and their percent changes with respect to the strength of Specimen B of 100 mm depth of cross section. Figure 2 shows the relationship between the depth of cross section and the flexural strength. As the depth of cross section increases, the flexural strength of the mixes tends to decrease and then comes to be the same. This may result from the fact that the ratios of the maximum sizes of aggregate to the sectional sizes of the specimens decrease and thus the strength of the specimens approach to that of resin mortar.

In Fig. 2, the flexural strength of resin concrete of Mix 1 and Mix 2 with the maximum aggregate size of 10 and 15 mm linearly decreases as the depth of cross section increases. The flexural strength of Mix 3 with the maximum aggregate size of 25 mm, also linearly decreases as the depth of cross section increases, excepting Specimen A, whose depth of cross section is 60 mm. This could result from nonuniform distribution of aggregate particles in Specimen A of Mix 3, because its sectional size is as small as about 2 times the maximum aggregate size.

Table 3. Flexural strength and their percent changes to B

Name of specimen	depth of cross section (mm)	Flexural strength Mix 1 (MPa)	Mix 2 (MPa)	Mix 3 (MPa)	Percent changes to B Mix 1 (%)	Mix 2 (%)	Mix 3 (%)
A	60	24.4	25.3	20.5	+16	+29	+22
B	100	21.0	19.6	16.9	---	---	---
C	150	18.8	18.2	15.5	−10	− 7	− 8
D	200	18.2	17.5	15.2	−13	−11	−10
E	250	16.2	16.4	14.6	−23	−17	−13

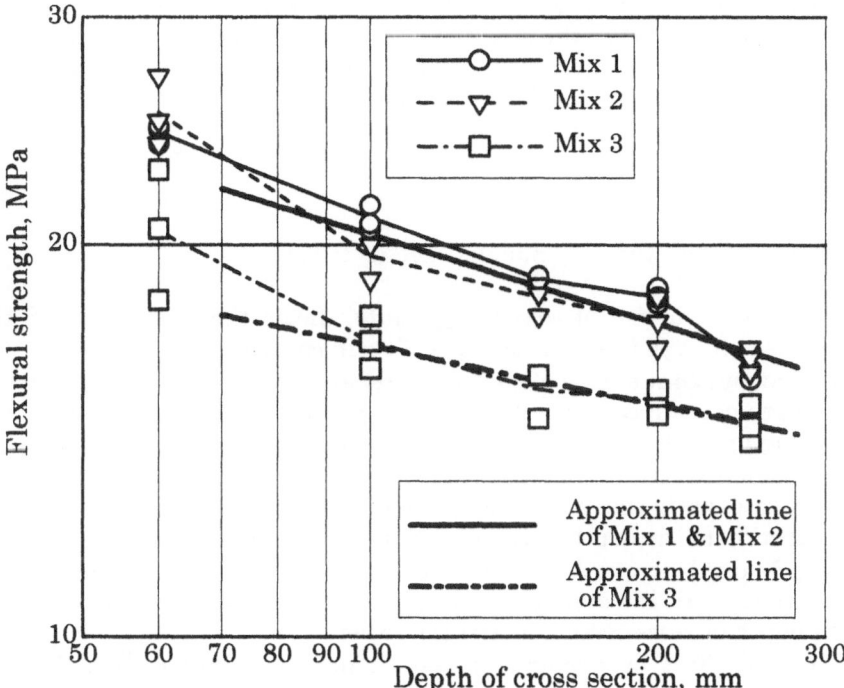

Fig. 2. Depth of cross section vs. flexural strength

The relationships between the depth of cross section and the flexural strength of specimens concerning the case of depth of cross section of 100 mm or more (Specimens B to E) in Fig. 2 were approximated by the least square method.   Results obtained are as follows :

The flexural strength of resin concrete of Mix 1 decreases in proportion to the depth of cross section to the 1/4 power ; that of Mix 2 decreases in proportion to the depth of cross section to the 1/5 power ; and that of Mix 3 decreases in proportion to the depth of cross section to the 1/7 power.

The specimens of resin concrete of Mix 1 and Mix 2 are considered to behave similarly throughout the depth of cross section range of 60 to 250 mm.   In the case that they were combined, it was found that their flexural strength decreases in proportion to the depth of cross section to the 1/4.5 power.

It has been reported that the flexural strength of conventional cement concrete decreases in proportion to the depth of cross section to the 1/7 power[3].   This corresponds to the reduction rate of resin concrete of Mix 3 with the maximum aggregate size of 25 mm.   The reduction rates for resin concrete of Mix 1 and Mix 2 with the maximum aggregate size of 10 and 15 mm are larger than this reported value.

Table 3 and Fig. 2 suggest that the larger maximum aggregate size leads to the lower flexural strength of resin concrete, although it leads to a smaller rate of strength reduction with the increase in the depth of cross section.   It is found that the size effect in flexural strength is affected by the maximum size of aggregate, because the strength of the aggregate crushed from rock of an identical origin is the same.

4   Conclusions

The results of this study are summarized as follows:

① The maximum size of coarse aggregate affects the flexural strength and size effect in flexural strength of resin concrete.
② A larger maximum size of coarse aggregate leads to the lower flexural strength of resin concrete, being independent of the specimen size, but leading to a lower rate of strength reduction due to size effect.
③ The flexural strength of resin concrete beams containing coarse aggregate with the maximum size of 10 and 15 mm decreases in proportion to the depth of cross section as the 1/4.5 power.
④ The flexural strength of resin concrete beams containing coarse aggregate with the maximum size of 25 mm decreases in proportion to the depth of cross section as the 1/7 power.
⑤ The size effect in strength of resin concrete is larger than that of ordinary cement concrete.

5   References

1.   W.Koyanagi,   K.Rokugo   and   Y.Uchida.(1983)   Stability   and Measurement in Concrete failure. Trans. of JCI, Vol.5, pp.149-160.
2.   Polymers-in-concrete   committee(1985)   Recommendation   for   the design of polyester resin concrete structures. Jour. Society of Materials science, Vol.34, pp.1110-1114. (in Japanese)
3.   Uchida,Y.,Rokugo,K.,Koyanagi,W.(1992)   Application   of   fracture mechanics to size effect on flexural strength of concrete. Proc. of JSCE,   V ol.16,No.442,pp.101-107. (in Japanese)

MECHANICAL PROPERTIES OF TRADITIONAL CERAMICS-FILLED POLYMER CONCRETE

D. Suraatmadja, D.R. Munaf and B. Lationo
Dept. of Civil Engineering, Institut Teknologi Bandung,
Bandung, Indonesia

Abstract
Compressive and tensile properties of polymer concrete filled with traditional ceramics, made from brick's powder, were studied under various material sources, grain size and volume fraction of ceramic filler. Galunggung sand (Volcanic Sand) and the Banjaran split stone (Andesite Stone) were used as aggregates. The "Dark resin" matrix or unsaturated polyester made from recycled PET has been used together with Cobalt Naphtenath as a promoter. Styrene monomer was used to reduced the high viscosity of the dark resin. MEKPO curing agent was used as the latest additive. The polymer concrete filled with traditional ceramics was cured at room temperature of 24° C, and tested for compressive and tensile strengths. At seven days age, with 20% filler (passing 200 MESH) by weight of concrete, the brick's powder from "Garut" produced 62 MPa for compressive strength and 3.13 MPa for tensile strength, "Nagrek" produced 51 MPa for compressive strength and 2.29 MPa for tensile strength and "Sapan" produced 61 MPa for compressive strength and 2.73 MPa for tensile strength. With 13% filler (passing 0.3 mm sieve) by weight of concrete, "Garut" brick's powder gave 40 MPa for compressive strength and 4.15 MPa for tensile strength, "Nagrek" gave 28 MPa for compressive strength and 3.12 MPa for tensile strength, "Sapan" gave 35 MPa for compressive strength and 4.12 MPa for tensile strength.
Keywords: Building bricks, clay brick, filler, mechanical properties, polymer concrete, powdered bricks, traditional ceramics.

Polymers in Concrete, edited by Y. Ohama, M. Kawakami and K. Fukuzawa. Published in 1997 by E & FN Spon, 2–6 Boundary Row, London SE1 8HN, UK. ISBN: 0 419 22330 4.

1 Introduction

Commonly, polymer concrete (PC) produced from inorganic aggregates (such as sand and gravel) bonded together by thermosetting (such as epoxies, polyesters, and recycled PET resin) or thermoplastic (such as methyl methacrylates) polymer binder and filled with fly ash.[1] - [5]   Other filler, such as calcium carbonate, have also been used very successfully.[3]

The polymer concrete has a high-strength, cures in a few minutes or hours and its current applications warrant mainly fine-aggregate fillers.[6] Also, this concrete can be used successfully as a very thin overlay (6-24 mm thick), repairing spalled or damaged structural components (such as building, bridges and hydraulic structures), precast components and industrial floor applications.[1] - [6]   To develop for its materials, the objective of this study is to evaluate the application of traditional ceramic materials as inorganic filler in polymer concrete using recycled PET or "dark resin".

2 Materials and mix proportioning

The following coarse and fine aggregates were used in the experiment: Galunggung volcanic sand with a bulk specific gravity of 2.225, 9.04% absorption, 1,516 kg/m^3 dry rodded unit weight, 5.82% clay/silt matter, 2.57 fineness modulus, and its gradation recommended by ASTM C33-78 specification. The 20 mm Banjaran split stone with a bulk specific gravity of 2.44, and 4.475% absorption, 1,44 kg/m^3 dry rodded unit weight, 6.56 fineness modulus, and its gradation recommended by ASTM C33-78 specification. The fine and coarse aggregates were oven-dried for 24 hours at 130° C to reduce their moisture content to less than 4.35% by weight and ensuring good bond to the polymer matrix. The aggregates were then cooled at room temperature and stored in closed packs. The traditional ceramic materials made from building bricks were used as inorganic filler. The brick's sources are from Garut (G specimens), Nagrek (N and A specimens), and Sapan (S specimens).

The bricks was oven-dried for 24 hours at 130° C and then crushed using Los Angeles
abrasion machine to produce passing 200 MESH, 0.150 and 0.3 mm grains. The brick's powder has a reddish brown color. Table 1 shows the physical and mechanical properties of traditional ceramics.

The "dark resin" which made by recycled Poly Ethylene Therephtalate (PET) were supplied from a commercial source. This resin was less than three months age unsaturated polymer with high viscosity. It was therefore diluted with 40% styrene monomer ($CH_2CHC_6H_5$) by weight of resin to reduce its viscosity. Cobalt Naphtenath (CoNp) promoter of 0.5% by weight of resin added into unsaturated polymer. Table 2 gives the mix design proportion of traditional ceramics filled polymer concrete.

Styrene monomer, CoNp were added to the dark resin immediately prior to mixing. Mixing was done using horizontal mixer for a period of about three minutes, continually MEKPO added to the mixing about six minutes.   Specimens were then cast in PVC (for compression specimens) or steel molds (for direct tensile, Young modulus of elasticity and Poisson's ratio specimens) and allowed to cure at room temperature of 24° C.

Tabel 1. Physical and mechanical properties of traditional ceramics.

Sources	Properties	Test results
Garut	Color	Reddish brown
	Specific gravity	2.304
	Unit weight (kg/m^3)	1,046
	Compressive strength (MPa)	2.44
Nagrek	Color	Reddish brown
	Specific gravity	2.53
	Unit weight (kg/m^3)	857
	Compressive strength (MPa)	4.91
Sapan	Color	Redish brown
	Specific gravity	2.551
	Unit weight (kg/m^3)	1,516
	Compressive strength (MPa)	6.57

Table 2 Mix design proportion of traditional ceramics filled polymer concrete.

Mix Code	Split Stone (%)	Volc. Sand (%)	Filler G-200 (%)	Filler G-0.15 (%)	Filler G-03 (%)	Filler N-200 (%)	Filler S-200 (%)	Dark Resin + Styrene Mon. (%)	Styrene Mon. (%) Dark Resin	CoNp (%) Dark Resin	MEKPO (%) Dark Resin
G-1	45	27	13	-	-	-	-	15	40	0.5	5
N-1	45	27	-	-	-	13	-	15	40	0.5	5
S-1	45	27	-	-	-	-	13	15	40	0.5	5
G-2	45	20	20	-	-	-	-	15	40	0.5	5
N-2	45	20	-	-	-	20	-	15	40	0.5	5
S-2	45	20	-	-	-	-	20	15	40	0.5	5
A-11	45	27	13	-	-	-	-	15	40	0.5	5
A-12	45	27	-	13	-	-	-	15	40	0.5	5
A-13	45	27	-	-	13	-	-	15	40	0.5	5

3 Testing procedure

After demolding, the 70 mm x 140 mm cylinders compression's specimens were cutted with diamond saw to make its top and bottom parallel. The ages at testing of compressive specimens were 3, 7, 14 and 28 days after moulding at a constant loading rate of 45 kN per minute. The ages at testing of 100 mm x 200 mm cylinders compression modulus, Poisson's ratio specimens was 28 days after molding at a constant loading rate of 45 kN per minute. Both longitudinal and transversal deformations were measured at every 1 N load increased, using extensometer which connected to the data logger. The age at testing of cylinders direct tensile specimens was 7 days after molding at a constant loading rate of 4.5 kN per minute. The cylinders are 150 mm x 300 mm cylinders completed with 400 mm x 100 mm steel reinforced caps both in top and bottom of the cylinders. The direct tensile equipment constructed from steel and recommended by the Indonesian National Standard (SNI).

The testing completed with scanning electronics microphotograph to evaluate the structure and fracture of composite material. Quantitative microscopy techniques to determine the distribution of the ceramic grains from a fracture surface can be used to derive important parameters which relate to mechanical properties. In the particle filled composite, like this, adhesion of the filler is quite important to improve the mechanical properties, and therefore fillers are usually surface treated with coupling agents to enhance the interfacial bonding and wetting.[5] Using γ methacryloxypropyltrimethoxy silane as coupling agent in polymer concrete has reported by Vipulanandan.[6] In this research, coupling agent was not used.

4 Results and discussion

4.1 Workability

Practically, using the horizontal mixer, the G-2, N-2 and S-2 (contain 20% of total weight ceramics filler) mixes have a better workability than the other. The G-1, N-1, S-1, A-11, A-12, and A-13 (contain 13% of total weight ceramics filler) showed a liquid mixes. The higher proportion of the ceramic filler will decrease the liquid condition, because the ceramic grains surfaces must be wetted by the resin. However, ceramic grains must be lubricated by the resin sufficiently. Over proportion of ceramic filler addition gives a low workability.

The mixes being hardens by polymerization two hours after MEKPO , 5% by weight of resin, were added into the mix. Polymerization of resin is initiated by a free radical action. Free radicals are formed at room temperature by breaking the O-O bond of the relatively unstable MEKPO as an initiator in the presence of CoNp as a promoter. The free radicals react with the monomer molecules and break the carbon double bonds to produce monomers with unpaired electrons. $CH_2CHC_6H_5$ is the cross linking agent reacts with the free radical and attaches itself to the unsaturated region of the polyester chain to form three-dimensional network of cross linked polymer. The rate of polymerization can be controlled by varying the concentration of the MEKPO and/or the temperature.[6]

However, it is difficult to form polymerization before MEKPO addition. Using MEKPO as the latest additive method, it is possible to construct many cast in place structure .

4.2 Compressive strength and unit weight

Fig. 1 represents the curing period vs. compressive strength of polymer concretes. Compression test results show that there is no correlation between the mechanical properties of the bricks and its concrete , and also, no much strength increasing after seven days. At seven days, the compressive strength is about 90% from its 28 days strength.

The compressive strength and unit weight varied with the source of the bricks, and a higher volume fraction of the ceramics filler (20% of total weight) gives a higher compressive strength and unit weight (G-2 = 2330.0 kg/m^3 > G-1 = 2232.00 kg/m^3, N-2 = 2211.00 kg/m^3 > N-1 2215.34 kg/m^3, S-2 = 2243.00 kg/m^3 > S-1 = 2245.20 kg/m^3). However, it is less than of the matrix (70.55 MPa) and its ceramics-resin composite (74.70 MPa) compressive strength values (see Fig. 4). The finer filler (passing 200 MESH) produces a higher strength and unit weight (A-11 = 2055.73 kg/m^3 > A-12 = 2050.95 kg/m^3 >A-13 = 2015.92 kg/m^3), but between the A-11 (passing 200 MESH) and the A-12 (passing 0.15 mm sieve) relatively have the same value for its compressive strength.

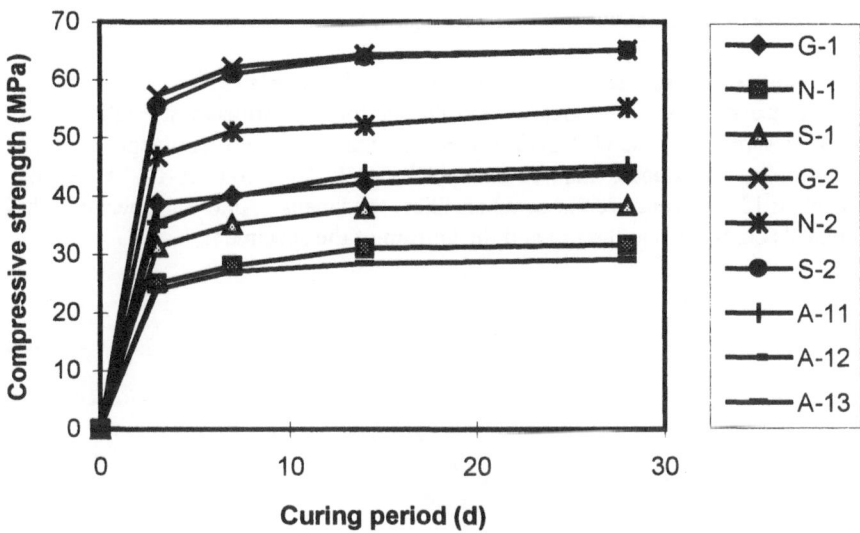

Fig. 1 Curing period vs. compressive strength

4.3 Tensile strength

Fig. 2 illustrates the tensile strength at seven days age. The tensile strength test results below (only for passing 200 MESH) describes that the tensile strength varied with the bricks sources and the volume fraction of ceramics fillers.

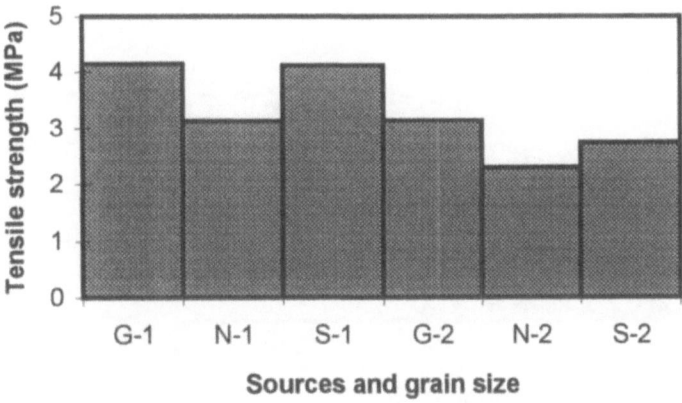

Fig. 2 Tensile strength at seven days age

Converse with the compressive strength, the higher ceramics volume fraction (20% of total weight) gives a lower tensile strength. This phenomena describe the mechanical properties of the ceramic materials which have brittle behavior.

4.4 Failure modes, modulus of elasticity and Poisson's ratio

There are two failure-modes shown during the compressive strength test. Brittle failures showed by the G-2, N-2, and S-2 (contain 20% of total weight ceramics filler) specimens, and ductile failures showed by G-1, N-1, S-1, A-11, A-12, and A-13 (contain 13% of total weight ceramics filler) specimens. All of the specimens have two conicals crack both in the top and the bottom of the specimens.

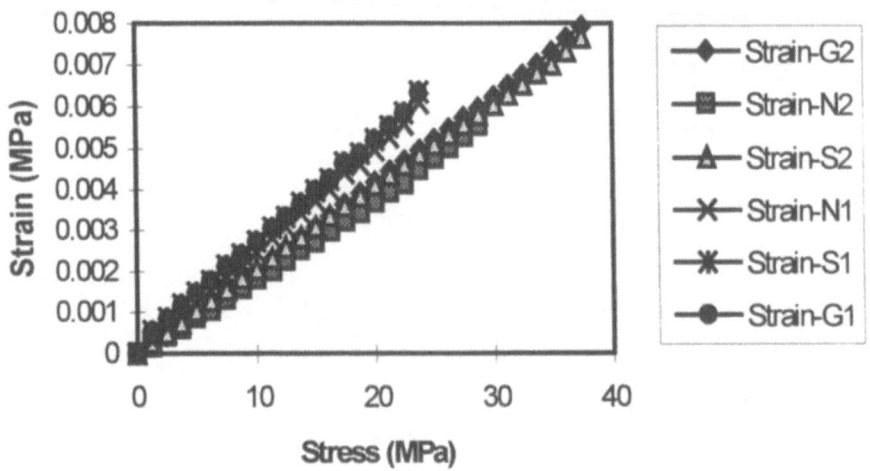

Fig. 3 Stress vs. strain

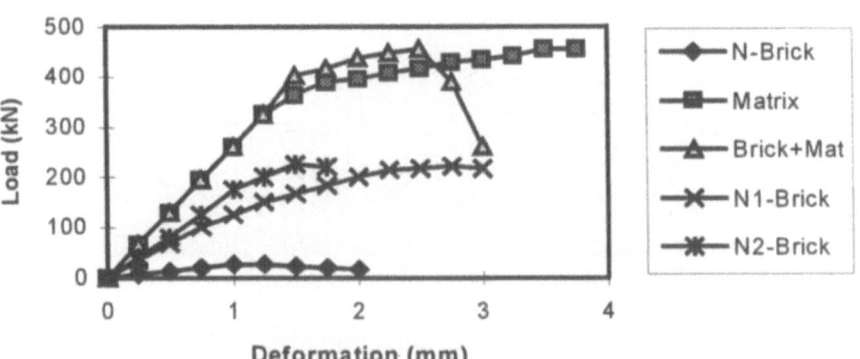

Fig. 4 Deformation vs. load for N specimens

Fig. 3 and Fig. 4 show that increasing the filler is increases the modulus of elasticity (E), but decreases the materials ductility. Poisson's ratio test results exhibit that higher filler has a lower Poisson's ratio (G-2 = 0.12 < G-1 = 0.13, N-2 = 0.18 < N-1 = 0.19, S-2 = 0.22 < S-1 = 0.23).

4.5 Microphotograph

Fig. 5 and Fig. 6 show the SEM microphotograph of interface zone for a lower volume fraction(13% of total weight)and higher volume fraction(20% of total weight) ceramics filler, respectively. G-1, N-1, and S-1 cracks surfaces are smoother than G-2, N-2, S-2 and exhibited a dense matrix. In the G-2, N-2, and S-2 specimens, the ceramics filler is dispersed fairly especially in the interface area. This condition shows that surface wetting between the aggregates and the matrix is not reached excellently. So, the initial cracks possibly begin to grow from this interzone area. Inspite of this, the voids also give contribution to produce the initial cracks. However, the G-2, N-2, and S-2 produce higher strength than the others, because the grains of ceramics filler will give its energy to stop the cracks.

Fig. 5 SEM microphotograph (x1000) of interface zone for a lower
volume fraction (13% of total weight) ceramics filler.
Showing a denser matrix but it produces
a lower compressive strength.

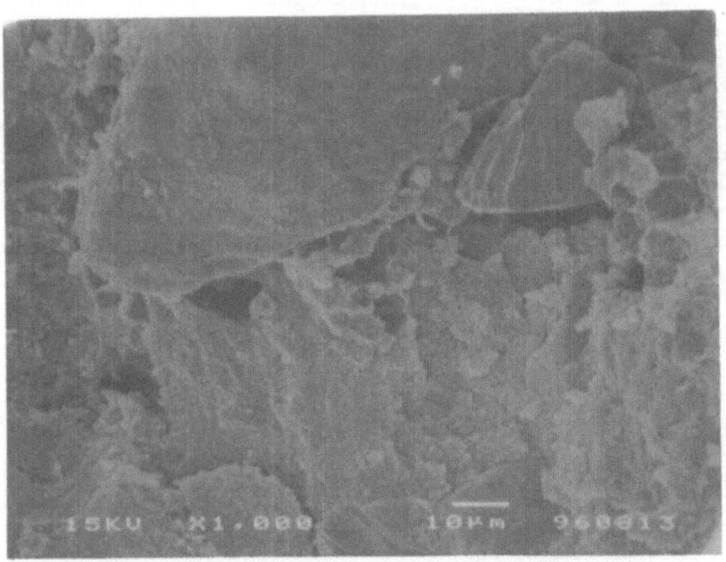

Fig. 6 SEM microphotograph (x1000) of interface zone for a higher
volume fraction (20% of total weight) ceramics filler.
Showing a porous matrix but it produces
a higher compressive strength.

5 Conclusions

From results of the present set of investigations the following conclusions were drawn :

- Traditional ceramics filler made from powdered bricks can be use to produce high strength (more than 60 MPa), low modulus of elasticity and Poisson's ratio polymer concrete.
- There is no much different strength produced by   passing 200 MESH or 0.15 mm grain size of ceramics filler, but 0.30 mm grain size produces lower compressive strength.
- Inspite of its source, compressive and tensile strength of the ceramics filled polymer concrete depend more on the volume fraction of ceramics filler.   The higher ceramics filler volume fraction produces a higher compressive strength, unit weight, and modules of elasticity, but a lower tensile strength, Poisson's ratio and ductility.

6 Acknowledgements

The authors would like to express their appreciation for the research grants made by the National Research Council Government of Indonesia through contract no SPK/005/RUK/BPPT/X/1996 to carry out the reported work, which forms part of the major research program on High Performance Concrete and Cement Based Materials at the Institut Teknologi Bandung.

7 References

1. Rebeiz, S. K., Fowler, D.W., Paul, D.R. (1993) *Recycling Plastic in Polymer Concrete for Construction Application*, Journal of Materials in Civil Engineering, Vol. 5, No. 2, pp. 273 - 248.
2. Rebeiz, S.K., Fowler, D.W., Paul, D.R. (1994) *Mechanical Properties of Polymer Concrete Systems Made with Recycled Plastic*, ACI Materials Journal, Vol. 9, No. 1, pp. 40 - 45.
3. Rebeiz, S.K. ,Serhal, P.S., Fowler, D.W., Structural Behavior of Polymer Concrete Beams Using Recycled Plastic, Journal of Materials in Civil Engineering, Vol. 6, No. 1, pp. 150 - 164.
4. Rebeiz, S.K. , Yang, S., Fowler, D.W. (1994) *Polymer Mortar Composites Made with Recycled Plastics*, ACI Materials Journal, Vol. 91, No. 3, pp. 313 - 319.
5. Sawyer, L.C., Grubb, D.T., (1987) *Polymer Microscopy*, Chapman & Hall, p. 215.
6. Vipulanandan, C., Paul, E. (1993) *Characterization of Polyester Polymer and Polymer Concrete*, Journal of Materials in Civil Engineering, Vol. 5, No.1, pp. 62 - 82.

DEVELOPMENT OF POROUS POLYMER CEMENT CONCRETE FOR HIGHWAY PAVEMENTS IN BELGIUM

A. Beeldens and D. Van Gemert
Civil Engineering Dept., Katholieke Universiteit Leuven, Heverlee, Belgium

Abstract

Porous concrete is a polymer modified cement concrete, designed to obtain high porosity. The addition of polymer emulsion to the mixture provides the required workability and strength. The paper presents the results of the experimental study and discusses the structure of the porous concrete and its performances.

Keywords: Durability, mechanical characteristics, mixture proportioning, porous concrete, structure

1 Introduction

Porous concrete, also referred to as very open concrete (VOC) is a polymer modified concrete, designed to obtain high porosity (25% to 30% accessible porosity). This open structure makes the VOC well suited to be used as top layer on roads since it provides noise reducing capacities and good drainage properties. The durability and low maintenance, proper to concrete, allow to presume a long service life and little or no rutting under heavy traffic, which makes the open concrete preferable over open black topics.

The porosity is achieved by means of a gap-graded aggregate distribution in which there is a very low proportion of fine aggregates. The strength of porous concrete is provided by the interaction polymer-cement-aggregate. Preliminary research accentuated the importance of aggregate distribution, amount and kind of polymer, compaction method and pore structure. Out of the results of the laboratory tests, the mixture proportions are derived to realize a test highway pavement with 40 mm porous concrete as top layer.

Polymers in Concrete, edited by Y. Ohama, M. Kawakami and K. Fukuzawa. Published in 1997 by E & FN Spon, 2–6 Boundary Row, London SE1 8HN, UK. ISBN: 0 419 22330 4.

The paper discusses the mixture proportioning and the properties of the fresh and hardened concrete. It will also discuss shortly the specific structure of the porous concrete and reveal the application of the porous concrete as top layer in a highway pavement.

2 Mixture Proportioning

The basic mixture proportions are given in Table 1. The coarse aggregate size is chosen to be 4/7 mm and occupies 50 % of the volume of the fresh mixture. Larger aggregate sizes produce larger porosity but will diminish the workability of the VOC and the evenness of the surface, which is required for the noise reducing properties [1]. A small amount of fine sand 0/1 mm is used. The cement used is CEMIII/A/42,5/LA. This cement is used because it is less subjected to shrinkage and it provides good resistance against environmental actions, primarily frost-thaw and thawing salts.

Four different types of polymer emulsions are used: two acrylic emulsions, one styrene-acrylic emulsion and one styrene-butadiene emulsion. The polymer-cement ratio is equal to 0,10 (weight solid mass of the polymer emulsion/weight cement). The water-cement ratio is equal to 0,3 (water of the emulsion included).

Table 1   Mixture proportioning for the VOC

Coarse aggregate 4/7	1350 kg/m^3
Fine aggregate 0/1	90 kg/m^3
Cement CEM III/A/42,5	280 kg/m^3
Polymer emulsion (50% solids)	56 kg/m^3
Water (emulsion water not included)	56 kg/m^3

The mixing of the VOC is done as described by Ohama [2]: first the coarse aggregates, the sand and the cement are placed in the mixer. During mixing, first the water and consequently the polymer emulsion are added. Afterwards, mixing continues during 2 minutes.

The curing conditions were taken similar to the outdoor conditions: the samples were covered by a polyethylene sheet during two days. After demoulding the samples were stored under laboratory conditions, sheltered from air current.

3 Properties of the Fresh Mixture

The workability of the fresh mixture is tested by means of the flow-test, according to NBN B15-233. The flow-index varies between 1,42 and 1,55 for the different mixtures, as indicated in Table 2.

The results show that the influence of polymer type is negligible. Only the first acrylic emulsion used gave a lower workability-index. This lower performance could be observed through all the tests. This can be explained by the type of polymer emulsion used: the first acrylic emulsion contains 50% solids, the second acrylic emulsion on the contrary only 30%. It seems that especially the larger amount of water, present in the second case improves the workability of the VOC. Also the type of polymer is important:

the styrene-acrylic and the stryrene-butadiene emulsion both contain 50% of water. Even though they show a good workability.

Table 2: Results of the flow-test for the different mixtures

Polymer emulsion	Flow-index	Total solids of the emulsion
Acrylic emulsion 1	1,42	50%
Acrylic emulsion 2	1,53	30%
Styrene-Acrylic emulsion	1,55	51%
Styrene-Butadiene emulsion	1,53	52%

An important property of the VOC for the practical application as top layer is the behavior at compaction. Due to the irregular form of the aggregates and the gap-graded distribution, there will always be a large amount of pores in the concrete. The porosity depends on the compaction method used (shocking, vibrating, with or without an extra weight placed on top of the sample), on the amount, the composition and the viscosity of the cement-polymer paste and on the type of polymer modification. Table 3 reveals the density and the degree of compaction for different polymer emulsions using different compaction methods. The degree of compaction is equal to the density of the fresh mixture divided by the density, calculated from the mixture proportioning, equal to 1830 kg/m^3. The mixture proportioning is made to obtain approximately 27 % of air volume into the sample at a degree of compaction equal to 100%, calculated by means of the specific volume of each of the components.

Table 3: Degree of compaction

	Vibrated (during 30 s)		Shocked		Fresh mixture[*]	
	density (kg/m³)	degree of compaction (%)	density (kg/m³)	degree of compaction (%)	density (kg/m³)	degree of compaction (%)
Acrylic emulsion 1	1703	93%	1355	74%	-	-
Acrylic emulsion 2	-	-	1827	100%	1847	101%
Acrylic-Butadiene emulsion	1977	108%	1718	94%	1842	101%
Styrene-Butadiene emulsion	1981	108%	1699	93%	1821	99%

(*) The density of the fresh mixture is calculated by means of a cylinder with a content equal to 8 liter, after compacting the fresh mixture by shocking the sample 8 times on the shock table, used for the ABRAMS flow-test according to the standard NBN B15-233

Similar results as in the flow test appear: the first acrylic emulsion reaches a much lower degree of compaction than the mixtures made with the other polymer emulsions, even though an identical compaction method is used. Differences between the other polymer emulsions are small, especially for the fresh mixture, where the degree of compaction is a reflection of the workability. Remarkable however is the good performance of the second acrylic emulsion: in spite of the low content of solids in the polymer emulsion, the mixture obtains a high degree of compaction and also produces good mechanical properties and a high durability.

4 Properties of the Hardened Porous Concrete

The test program for the hardened porous concrete concentrated on the relationship compressive/tensile strength versus total/accessible porosity, related to the dry density of the samples. The test program included also performance and durability tests, such as freezing and thawing resistance, resistance against abrasion and against alkaline environment and the determination of the adhesion of the VOC on a concrete sublayer.

The relation total porosity-compressive strength is investigated by measuring the total porosity and the strength after 28 days on cubes of 150x150x150mm^3. The total porosity is measured after vacuum saturation of the sample. The accessible porosity takes into account all the pores accessible for water. The accessible porosity is measured in the following way: the test specimen is dried at 105°C and put in a vessel with a known volume, the vessel is slowly filled with water. The volume of the water in the specimen divided by the volume of the specimen is equal to the accessible porosity. The difference between the accessible porosity and the total porosity is approximately 6%. Also the dry density of the cubes is measured. The relation total porosity-compressive strength and the relation total porosity-dry density are both linear (Fig. 1 and Fig. 2). A similar relation can be found between the tensile strength and the total porosity.

The influence of the type of polymer is not directly visible in the relation strength-porosity. As said before, the type of polymer plays a role in the compaction degree, obtained with the addition of a certain amount of energy, and consequently in the obtained porosity. Therefrom one observes the high porosity and the low strength for the samples made with the first acrylic polymer emulsion, since the workability and the possibility to compact was much lower for this mixture.

The durability of the VOC is tested on samples, made with the styrene-acrylic emulsion, which simulate the pavement conditions: VOC is laid on a concrete sublayer according to the wet-in-wet method. First the concrete sublayer (150 mm thick) is placed. After approximately three hours, the VOC layer (40 to 80 mm thick) is placed on top of the not completely hardened sublayer. Depending on the time between the placing of the two layers and depending on the compaction method of the VOC, the VOC and the concrete sublayer will penetrate into each other. The adhesion strength is measured by means of a direct tensile test. Two kind of samples are tested: one, where the VOC is compacted through vibration and another where the VOC is compacted by tamping. In the first case, the two layers were penetrated into each other, in the second case a clear line was visible between the two layers. The results of the direct tensile test are given in table 4.

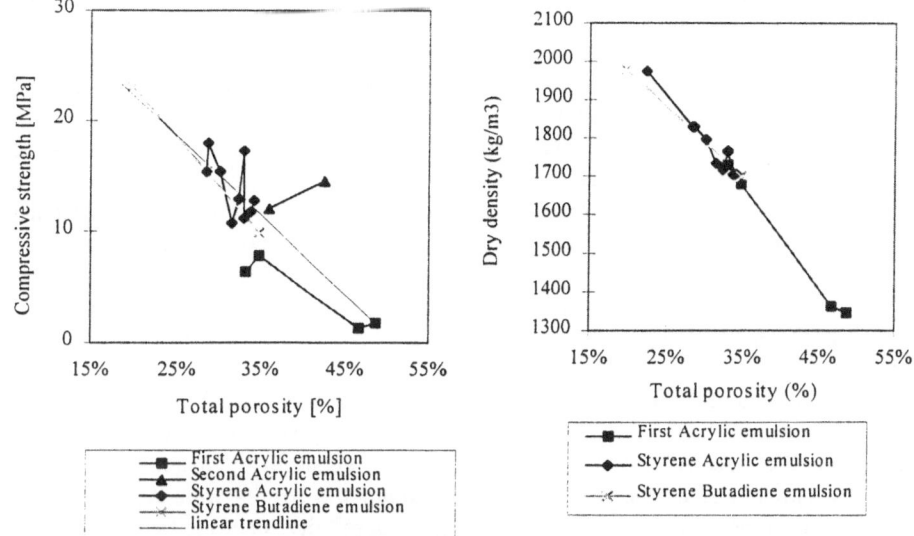

Fig. 1: Total porosity-compressive strength   **Fig. 2**: Total porosity-dry density

The results indicate the importance of the penetration of the two layers into each other and of the compaction method used: the penetration improves the adhesion of the VOC to the sublayer: in the vibrated samples the failure always occurred in the VOC, but the penetration will lower the porosity over the height of the layer: the average porosity is equal to 21%, the porosity of the part of the VOC which is not penetrated into the sublayer is equal to 32%. This lower porosity is advantageous for the strength of the concrete, but will diminish the noise reducing and drainage properties of the VOC.

Table 4: Adhesion strength of the VOC on a concrete sub layer

Height of VOC (mm)	Adhesion strength (MPa)	Porosity (%)	Place of failure
vibrated - 40 mm	3,02	21% - 32%[*]	VOC
vibrated - 70 mm	2,91		VOC
tamped - 40 mm	1,73	33%	2/3 VOC 1/3 transition layer
tamped - 80 mm	1,84	31%	2/3 VOC 1/3 transition layer

(*) average porosity is equal to 21%, porosity of the upper part is equal to 32%.

The Amsler abrasion test (NBN B15-223) revealed promising results: no damage was visible and the abrasion was limited to 5% loss in thickness. The road simulation test showed similar results: after 100 000 cycles with a tire, which was inclined over 5° and submitted to a force equal to 5,5 kN, only a few aggregates came out of the VOC. These tests indicate that the VOC can withstand loads imposed by heavy traffic.

The freezing and thawing resistance is determined according to the European standard prEN 1340 (1993) on samples made with different polymer emulsions. Cores were taken from the two layer samples and were at the sides completely surrounded with an impermeable rubber cloth, in such a way that the VOC is completely submerged in water or ice during the whole test. 28 cycles are carried out, varying the temperature between -20°C and +20°C. To test the influence of a freezing medium, some of the samples were submerged with a 3% NaCl-solution. At the end of the test, the samples were unpacked, cut in two and visually inspected.

The samples made with the second acrylic emulsion and with the styrene-butadiene emulsion, tested without freezing medium were cracked at the transition of the VOC to the sublayer. For the samples, were the two layers flowed into each other, the crack was capricious; for the samples where the transition line was clearly visible, the crack formed a straight line as well. The samples made with the styrene-acrylic emulsion showed no damage.

The samples tested with the 3% NaCl-solution were not damaged in the transition zone between the porous concrete and the sublayer, but the upper surface of these samples was attacked by the salts. All the samples in this case showed similar damage: the aggregates from the upper surface were ripped of. Although the results were not completely satisfying it can be said that the VOC showed good durability, since the freezing-thawing test is very severe: in real circumstances, the water can flow out of the VOC before freezing due to the large porosity.

To ameliorate the freezing-thawing resistance, a new mixture proportioning is introduced with 50 kg/m³ fine aggregate and 330 kg/m³ cement. Samples, made with the new mixture proportioning showed no significant damage after the freeze-thawing test.

A test road is realized with the last mixture composition in cooperation with the Ministry of the Flemish Community, Environment and Infrastructure Department, Flemish Brabant Road Division. The test road consisted out of two adjacent strokes, each 900 m long and 3 m wide. The porous concrete is poured on the fresh continuously reinforced concrete sublayer by means of a slip-form paver. The compaction took place through a high frequency vibration beam. A test program is running to reveal the properties of the in situ laid porous concrete.

5 Structure

VOC is characterized by a specific structure with a large porosity. At 100% compaction it is composed of 50 volume-% coarse aggregate, 23 volume-% mortar (cement, polymer emulsion, water and sand), and 27 volume-% air.

Macroscopically, three phases can be distinguished, which are homogeneously spread over the material (Fig. 3). One phase consists of coarse aggregates and is discontinuously divided over the material. The second phase is continuously spread over the material and consists of the cement-polymer paste. This paste forms a matrix in which the coarse aggregates are embedded: the paste forms a fine layer (0,5 to 1 mm thick) around the aggregates and binds them together by means of 'bridges'. The third phase consists of the large pores, connected with each other through the material. These pores form a second matrix, twinned with the cement-polymer comatrix.

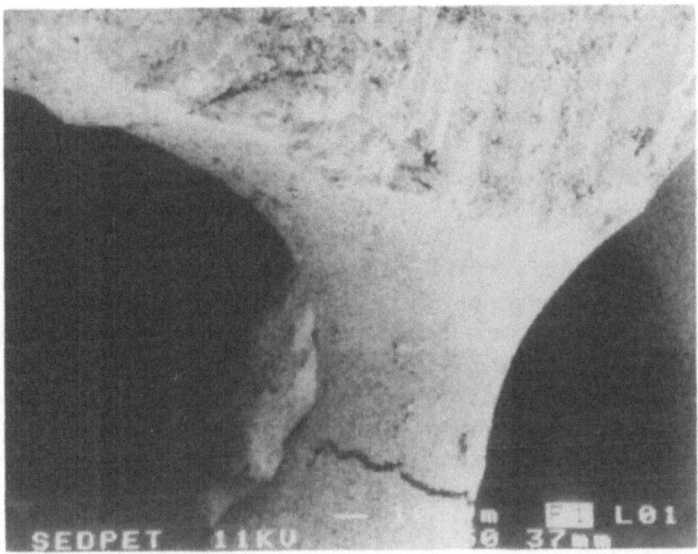

Fig. 3: Cement-polymer comatrix around and in between aggregates (50 x)

Figure 3 shows an enlargement (50 X) of the mortar-film around an aggregate and of a 'bridge' connecting the aggregates, taken by means of a scanning electron microscope (SEM). Ideally the VOC is composed to contain just enough mortar to form a film around the aggregate and to connect the aggregates by means of "bridges", but not enough mortar to fill the open pores between the aggregates.

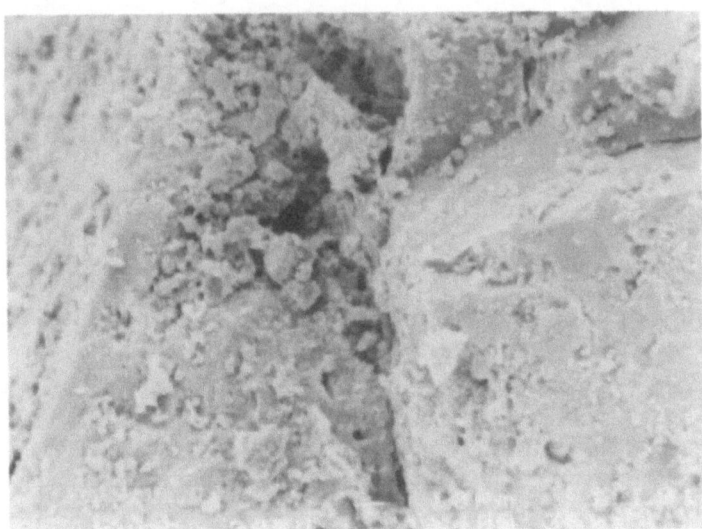

Fig. 4: Transition zone between the polymer-cement and the aggregate (400 x)

Comparison between the samples made with and without polymer emulsion, indicates that the polymer facilitates the adhesion of the cement-polymer paste to the aggregate. A visual control of reference samples without polymer addition showed an incomplete covering of the aggregates by cement paste. With polymer addition, one observes a complete covering of the aggregate by the cement polymer paste. Although the adhesion of the cement-polymer paste to the aggregate is present, it is not uniform. As shown in the central part of figure 4 small pores and small cracks are present along the aggregate. This larger porosity decreases the strength of the transition zone. Ideally the VOC should contain a high amount of large pores, but a very low amount of small pores, which means that the difference between the total porosity and the accessible porosity should be as small as possible. Investigation of a sample treated with HCl (Fig. 5) reveals the formation of the polymer phase at the transition zone. Again larger pores are visible close to the aggregate or to the sand particle than in the bulk polymer phase. The task of the cement-polymer paste is dual: it provides the workability of the fresh mixture and enables a good compaction of the mixture; secondly after combined hydration and polymerization it produces a comatrix in which the aggregates are embedded. This comatrix provides the strength of the VOC.

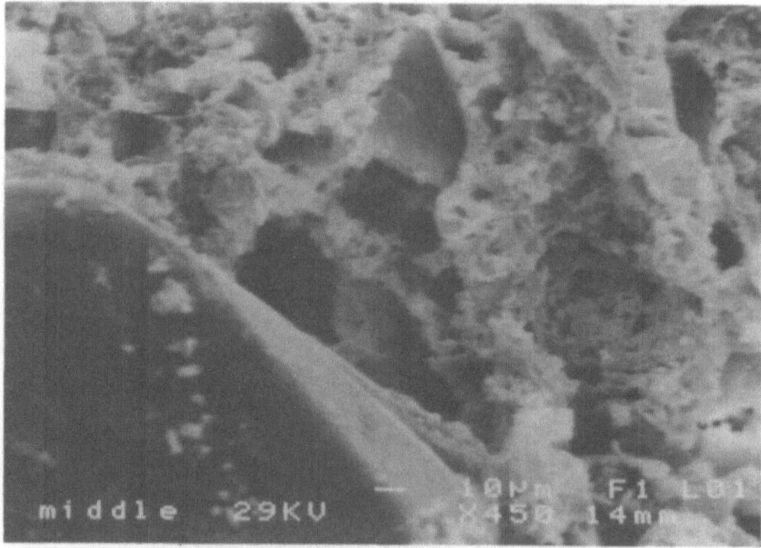

Fig. 5: Transition zone between the polymer phase and the aggregate, after treating the sample with HCl

In general, failure occurs at three different places: in the bulk cement-polymer paste, in the transition zone between the aggregate and the paste and through the aggregate itself. In the second case the cement-polymer paste is ripped off of the aggregate. This is due to the larger porosity of the transition zone and to the non-uniform adhesion of the cement-polymer paste to the aggregate. The addition of the polymer emulsion will contribute to the stiffness and the tensile strength of the very open concrete: the formation of a continuous polymer film through the comatrix will increase the elasticity of the comatrix and will enlarge the resistance to tensile forces. However, care has to be taken to set bounds to the amount of polymer emulsion added. If a large amount of polymer emulsion is added, the cement particles can be completely surrounded by polymer emulsion and further hydration of the cement particles will be prohibited, and the formation of a continuous structure will not be possible anymore. This kind of structure can also arise from a faulty mixing procedure: if the addition of the polymer emulsion to the mixture takes place before the water is added, the polymer emulsion will immediately enclose the cement particles and will prevent all contact with water to allow further hydration. The 'ideal' situation, where the cement hydration and the polymerization act together and where the cement hydrates are covered with polymer films will not be possible in such case. Besides, an excessive amount of polymer will make porous concrete uneconomical.

6 Conclusions

Porous concrete is a polymer cement concrete, built up of three phases: the aggregates, the cement-polymer phase and the open pores. The open pores are connected with each other through the material and form a second matrix besides the matrix, consisting of cement hydrates and polymer films. The addition of the polymer emulsion improves the workability of the fresh mixture, enables a high degree of compaction and ameliorates the mechanical properties of the porous concrete. Care has to be taken at mixing: the addition of the polymer emulsion to the mixture, before the addition of the water can hinder the continuous hydration of the cement particles.

7 References

1.   Onstenck, H.J.C.M. et al. (1994) Concrete Pavements: safe and sound-proofing (in Dutch). *CUR-rapport 94-11*, Centrum voor regelgeving en onderzoek in de grond-, water- en wegenbouw en de verkeerstechniek, Gouda
2.   Ohama, Y. (1995) *Handbook of polymer-modified concrete and mortars, properties and process technology* Noyes Publications, New Jersey.
3.   Brite/Euram State of the art report (1994) Surface properties of concrete roads in accordance with traffic safety and reduction of noise. *Brite/Euram project BE 3415*, October 1994.

PROPERTIES OF HIGH POLYMER CEMENT MORTAR

M. Tamai
Dept. of Civil Engineering, Kinki University, Higashi-Osaka, Japan
K. Yamaguchi
Toa Corporation, Tokyo, Japan

Abstract
In this paper, PC (Polymer Cement) and HPCM (High Polymer Cement Mortar) are made of cement paste and cement mortar, respectively, which contain a new type of acrylic emulsion. Contrary to conventional cement mortar, this type allows for flexible deformation.

The rheological filling conditions for these fresh and hardened materials correspond to slurry range or mixed range composed of slurry range and capillary range when considering the cement or sand particles, polymer emulsion and air contained in this mixture, as a solid phase, a liquid phase and a gas phase, respectively.
The purpose of this study is to show the chemical characterization of HPCM and to clarify the mechanical properties of fresh or hardened HPCM.

As a result, the hardening reaction is complete at 20°C within a short period of time and with little change in volume. The tensile strength of the material is in the range of 0.5 to 3.0 MPa. In particular, the elongation greatly varies depending on the amount of polymer emulsion and becomes 0.1 to 2, which is a superior characteristic for lining material. The purpose of molding the specimen is to indicate the possibility of using it in the making of structural materials, such as for jointing, repairing and other road pavement materials.
Keywords: Acrylic type emulsion, elongation, flexible deformation, high polymer cement mortar, polymer cement, slurry range, strength, volume change.

Polymers in Concrete, edited by Y. Ohama, M. Kawakami and K. Fukuzawa. Published in 1997 by E & FN Spon, 2–6 Boundary Row, London SE1 8HN, UK. ISBN: 0 419 22330 4.

1 Introduction

Conventional polymer cement mortar is comprised of normal cement mortar (NM) mixed and 5~10% polymer emulsion (PE), and is used to improve the properties of bending and tensile strengths.

In the present study, for the purpose of developing a high polymer cement mortar (HPCM) with flexibility by providing the elasto-plastic substance of NM with additional elasticity, a relatively large amount of special polymer emulsion (PE) was mixed with cement to be used as a binder, to investigate the various properties thereof.

In the case of conventional water-based PE, with an ordinary content ratio of polymer : water = 1 : 1, it was difficult to provide the mortar with elasticity by mixing the polymer/cement (P/C), from the viewpoint of strength and drying shrinkage. When using a special acrylic base water-based PE (polymer : water = 3 : 2), it was possible to obtain a slurry range rendering cement particles dispersed in the polymer without increasing W/C. Accordingly, it became possible to provide the HPCM with free deformability as well as elasticity. The present study has aimed to develop a flexible mortar by creating an appropriate mixture of said polymer cement paste (PC) and sand. The content of this paper consists of studies on filling form, mechanical properties, drying shrinkage and microstructure.

2 On the reforming of HPCM taking into account the filling form

2.1 Reforming model of binder phase

Reforming the binder phase with PE involves the fact that PC is fixed as a matrix where cement particles are floated in the PE in a state without mutual contact in the mixing system sand which is dispersed within a slurry range. Among hydration processes for HPCM, schemes at the early stages of hardening and the progress itself are given in Fig. 1(a) and Fig. 1(b), respectively.

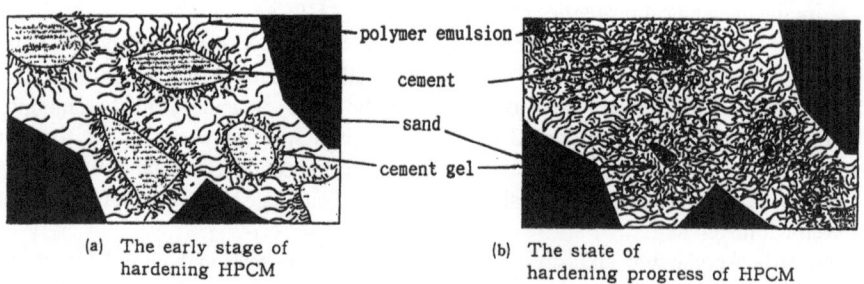

(a) The early stage of
     hardening HPCM

(b) The state of
     hardening progress of HPCM

Fig. 1. Hydrating situation of HPCM

2.2 Hydration reaction of clinker mineral constituting cement

When portland cement is kneaded with a weight ratio of 30%~60%, usually of water, it results in a paste in which $C_3S \cdot C_2S \cdot C_4AF$ and C_3A (clinker minerals) constitute cement reacting with water to produce hydrates.

Table 1. Chemical reaction formula of clinker mineral

Clinker mineral	Chemical reaction formula
C_3S	$2(3CaO \cdot SiO_2) + 6H_2O \rightarrow 3CaO \cdot 2SiO_2 \cdot 3H_2O + 3Ca(OH)_2$
C_2S	$2(2CaO \cdot SiO_2) + 4H_2O \rightarrow 3CaO \cdot 2SiO_2 \cdot 3H_2O + Ca(OH)_2$
C_4AF	$4CaO \cdot Al_2O_3 \cdot Fe_2O3 + 2Ca(OH)_2 + 10H_2O \rightarrow C_3AH_6 + C_3FH_6$
C_3A	$3CaO \cdot Al_2O_3 + 3CaSO_4 + 32H_2O \rightarrow C_3A \cdot 3CaSO_4 \cdot 32H_2O$
	$3CaO \cdot Al_2O_3 + 6H_2O \rightarrow C_3AH_6$

The reaction formula for hydration is given in Table 1[1].

C_3S and C_2S, bring about hydrolysis and via afwillite ($C_3S_2H_3$) finally turn into tobermorite and $Ca(OH)_2$. The tobermorite ($C_5S_6H_5$) covers the surface of the clinker powders with a thin film and $Ca(OH)_2$ dissolves into a liquid phase attaining a saturated state within several minutes. C_4AF reacts with the $Ca(OH)_2$ produced by the hydration of C_3S and C_2S, as well as with water, to produce C_3AH_6 and C_3FH_6. C_3A is hydrated most rapidly, turning into C_3AH_6, and an eventual existence of gypsum ($CaSO_4$) permits ettringite ($3C_3A \cdot 3CaSO_4 \cdot 32H_2O$) to be produced in an amount corresponding to the consumption thereof. Accordingly, in relation to the state of the progress of the hydration of the cement, the volume can be calculated by giving molecular weights to the chemical reaction formula in Table 1 and, on the basis of the generally known specific gravities of respective clinker minerals, those of the compounds and the hydration ratio at various curing days. The contents and specific gravities of the clinker minerals making up the cement, the specific gravities of the chemical compounds[2] and the hydration ratios of the clinker minerals[3] are given in Tables 2, 3 and 4, respectively.

Table 2. Content ratio and specific gravity of clinker mineral

Item	Content ratio (%)	S.g
C_3S	55	3.13
C_2S	25	3.28
C_4AF	9	3.77
C_3A	25	3.04
$CaSO_4$	3	2.98

$C_3S = 3CaO \cdot SiO_2$, $C_2S = 2CaO \cdot SiO_2$
$C_4AF = 4CaO \cdot Al_2O_3 \cdot Fe_2O_3$, $C_3A = 3CaO \cdot Al_2O_3$
S.g = specific gravity

Table 3. Specific gravity of chemical compound

Chemical compound	S.g
$Ca(OH)_2$	2.24
$C_3S_2H_3$	2.44
C_4AH_6	2.52
$C_3A \cdot 3CaO_4 \cdot 32H_2O$	1.73
C_3FH_6	2.6

$C_3S_2H_3 = 3CaO \cdot 2SiO_2 \cdot 3H_2O$
$C_3AH6 = 3CaO \cdot Al_2O_3 \cdot 6H_2O$
$C_3FH_6 = 3CaO \cdot Fe_2O_3 \cdot 6H_2O$

Table 4. Hydration ratio of clinker mineral

Curing age	C_3S (%)	C_2S (%)	C_4AF (%)	C_3A (%)
1 (hr)	9	2	11	37
24 (hr)	60	4	18	38
5 (days)	68	5	29	42
14 (days)	76	15	37	46
28 (days)	77	21	41	47
91 (days)	85	38	43	61

2.3   Relationship between volume ratio of admixture and curing days in HPCM
Each clinker mineral being hydrated, the volume occupied by the former increases according to the lapse of time. The relationship between the volume ratio of the admixture and the number of curing days in the HPCM in connection with mixture type C in Table 5 is shown in Fig. 2. After 91 curing days, the volume of the cement particles has increased by 9.6% due to the hydration reaction of the various clinker minerals. On the other hand, 6.8% of the vaporized water (water within PE and adjusting water) escapes from the inner part of the HPCM to its outer part and crystal water (various clinker minerals) is reduced to 8.1%. Moreover, the reason for the rapid decrease in the volume of the water up to 28 curing days is considered to lie in the fact that C_3A, which has a remarkable influence on the coagulation of cement to the vaporized water, reacts with gypsum within several minutes after kneading to produce ettringite ($C_3A \cdot 3CaSO_4 \cdot 32H_2O$) of needle crystal and that the hydration ratio of C_3S which presents the most rapid hydration among clinker minerals is relatively large up to 28 curing days.

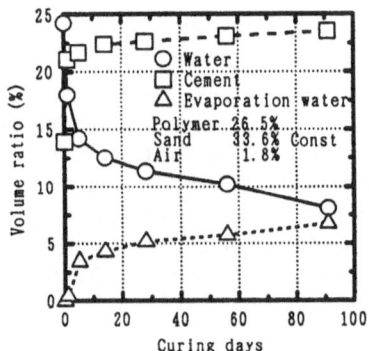

Fig. 2.   Relationship between volume ratio of admixture and curing days

2.4   Relationship between (PE + W) (vol. %) and curing days
Fig. 3 gives the relationship between PE and the adjusting water {(PE + W) vol. %} existing in cement voids within PC and curing days.

Since {(PE + W) vol. %} expresses the volume ratio of voids between cement particles within the PC, it shall be fixed as the void ratio of cement.

In general, a ball filling model with a void ratio of 66% and that with a void ratio of 47.6% are the compression filling with 4 particle contact points to maintain a stable structure model and a cubic lattice filling with 6 particle contact points, respectively, to be a slurry range requiring a void ratio around a compression filling[4]. The cement filling model within the PC showed a void ratio of 77.8%, which is coarser than a compression filling at an early stage of kneading and at a condition of perfect hydration with a void ratio of 66.2% after 7 curing days. It finally obtains a void ratio of 51.6% , approaching a cubic lattice filling via a compression filling. At grain boundaries of cement mingled with this PE, however, the contact force between the particles of cement is thought to be small since hydrates of cement produced from inside to outside of cement particles are in a coarse state. Accordingly, the mixture system of PC shows remarkable flexibility. Moreover, HPCM where PE and sand are regarded as liquid and solid phases, respectively, presents no such great changes from the early stages, the volume ratio {(PE + W + C) vol. %} of voids between sand particles within HPCM being 61.2%~66% which well satisfies the condition for obtaining flexibility.

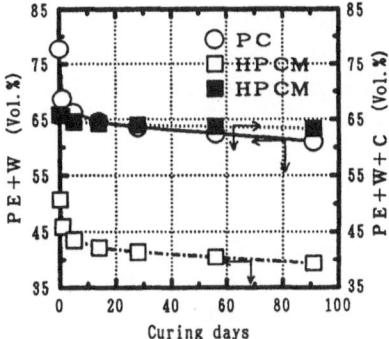

Fig. 3.   Relationship between (PE + W) vol. %, (PE ⊢ W + C) vol. % and curing days

3   Properties of polymer-mixed cement paste and mortar

Tests on the drying shrinkage among volume changes and tests on compression, bending, tensile forces and bending restoration strain among mechanical properties were investigated.

3.1   Experimental method

3.1.1   Applied materials :
Cement : Portland cement as usual (produced by O Cement Co.), sand : standard sand (from Toyoura), polymer : cation base acrylic-emulsion, antifoamer : Adekanate B-940 (made by Chuorika Industry Co.), adjusting water : tap water.

3.1.2   Mixing :
Mixture types of HPCM applied to the present tests are given in Table 5.

The mixing ratio of sand for HPCM was obtained around the compression filling {Vp/(Vp + Vs) = 66%}. It should be mentioned that the kneading water was obtained from the water included in the PE and the adjusting water, 2% of the antifoamer against the PE being mixed.

Table 5. Mixture proportion of HPCM

Mixture Type	C/PE (%)	P/C (%)	W/C (%)	Vp/(Vp+Vs) (%)
A	50	120	110	63.3
B	80	75	69	64.9
C	100	60	55	65.9
D	120	50	46	66.9
E	150	40	37	68.2

3.1.3   Concrete placing and curing method :

The testing samples were prepared in such a manner that the PE was dispersed in an antifoamer, adding water first, and then cement and sand. This was followed by agitations in a JIS type mortar mixer for 1 minute at a low speed and then for 2 minutes at a high speed. The curing method consisted of 24 hours at a temperature of 20℃ and a humidity level of 90~95% RH after the concrete was placed, and was followed by a curing in the air at 20℃ and a humidity level of 60% RH until the predetermined curing days had passed.

3.1.4   Testing method :

For the drying shrinkage test, measurements were taken, according to the contact gauge method, curing in a thermostatic chamber at 20℃, 60% RH. The compression test was made with a ϕ 5 cm x 10 cm cylinder according to JIS A 1108, and the bending test was made with a 4 cm x 4cm x 16 cm rectangular parallelopiped according to JIS R5201. In addition, a tensile test was made with a briquet type test piece according to the tensile test method for polymer cement mortar (proposed) (ASTM C190)[5].

The strain was measured during the bending and tensile tests with a plastic gauge (YL-20) attached to the bottom surface of the pressure plane and the surface of the test piece, respectively.

The bending restoration strain test was made in such a manner that increases and decreases in pressure were repeated in   order of load ratios of 50, 70, 80, 90 and 95% against the breaking load to measure the strain. In addition, in order to compare the stability with NM, the recovery ratio was calculated according to the following formula:

Recovery ratio (%) = (Emax- Δ Ei)/Emax x 100

Emax : maximal strain against load ratio,

Δ Ei : residual strain against load ratio

3.2   Experimental results and considerations

3.2.1   Drying shrinkage test :

Fig. 4 gives the results of the drying shrinkage test on NM, PE and the other   emulsions (EVA : ethylene vinyl acetate and PAE : polyester acrylate) for a base length set at the mold released 1 day after curing at a constant W/C of 55%. Out of all the polymers, the amount of shrinkage is the greatest for NM, and 150 days after the drying begins shrinkage of about 120 x 10^{-5} is seen. On the other hand, the amount of shrinkage for various types of polymers is reduced to 1/6~2/3 of that of NM, the amount of shrinkage for PE being the least. This is thought to be caused by the inhibition of drying shrinkage due to PE replacing the free water in NM. In general, when NM is dried, its capillary water evaporates and, at the same time, its surface tension   increases to cause shrinkage. In the case of HPCM in the present paper, the cement particles are hydrated to absorb the free water within the HPCM and increase its volume. At the same time, the polymer seals the surface of the test piece to inhibit the evaporation of water, which exhibits expansion at an early stage.   The subsequent transition to shrinkage is believed to be caused by the fact that the increase in volume brought about by the hydration of cement particles is overcome, though only by a small amount, by the amount of evaporation of free water.

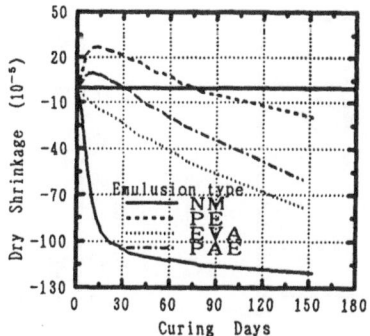

Fig. 4.   Relationship between shrinkage and curing days

3.2.2   Compression test :

For mixture types A~E of HPCM for 28 and 56 curing days, relationship of the compressive strength and mixture types is given in Fig. 5. The compressive strength increases in proportion to C/PE (%) and the number of curing days. It is thought to be caused by the fact that the hydration reaction of the cement unites the PE with the cement and absorbs the free water within the PE, which increases the self-adhesive strength of the PE with cement. Moreover, in comparison to 28 curing days, those of 56 clarify that the compressive strength increases by 1.4~1.8 times inversely as C/PE (%), the strength of mixing E being the greatest.

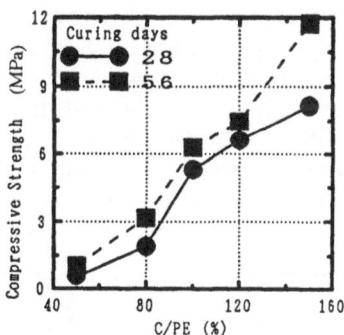

Fig. 5.   Relationship between compressive strength and C/PE (%)

3.2.3   Bending test :

For mixture types A~E of HPCM and PC, the relationship between bending strength and the mixture types is given in Fig. 6 for 28 and 91 curing days. Likewise, the relationship between the lower surface maximally strained edge strain and mixture types is given in Fig. 7. The bending strengths of HPCM and PC increase in proportion to C/PE (%) as well as to the number of curing days, the bending strength of HPCM being 1.3~1.8 times greater than that of PC. Due to the absence of sand, PC is believed to have a larger unit amount of cement, and accordingly, a greater amount of strength than HPCM. However,

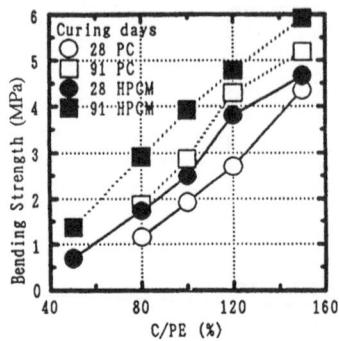

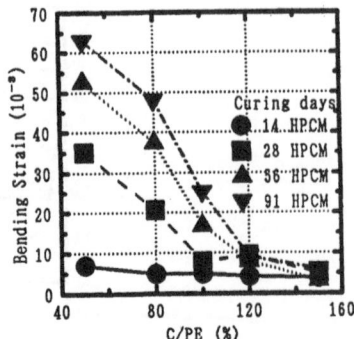

Fig. 6.   Relationship between bending strength and C/PE (%)

Fig. 7.   Relationship between bending strain and C/PE (%)

PC shows such a tremendous amount of drying shrinkage in comparison to HPCM as to be subjected to tensile stress due to the drying shrinkage already at the point of execution in bending test, which is thought to be a cause of the reduction in strength. On the other hand, the lower surface maximally strained edge strain increases in inverse proportion to C/PE (%) and in proportion to the number of curing days.

3.2.4   Tensile test :

For mixture types A~E of HPCM and PC for 28 and 91 curing days, the relationship between the tensile strength and mixture types is given in Fig. 8. And in Fig. 9, the relationship between the tensile strain (elongation capacity) and mixture types A~E of HPCM for 28 and 56 curing days is given. Tensile strengths for PC and HPCM increase with a decreasing ratio of water to cement, similarly to NM, but such a remarkable increase in the compressive strength was not observed. In addition, a comparison of tensile strengths for PC and HPCM of the same mixture type reveals no such changes, HPCM, however, exhibits a tendency to be slightly higher. As for the amount of strain, it increases remarkably in proportion to the number of curing days with an increase in the amount of polymer in the HPCM in comparison to that in the cement.

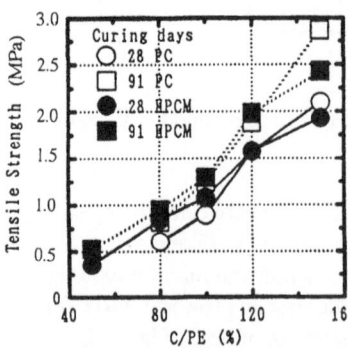

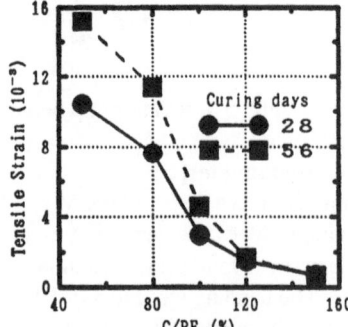

Fig. 8.   Relationship between tensile strength and C/PE (%)

Fig. 9.   Relationship between tensile strain and C/PE (%)

3.2.5 Bending restoration deformation test :

The relationship between the repeated stress and the lower surface strained edge strain in the bending restoration strain test of mixture type A for 28 curing days is given in Fig. 10. In addition, the relationship between the load ratio and the calculated recovery ratio obtained from Fig. 10 is shown in Table 6. Fig. 11 shows that HPCM presents a large deformation capacity against loading and good flexibility. Moreover, although the deformation capacity of NM (as shown in Table 6) is small, such as less than 300 μ , the

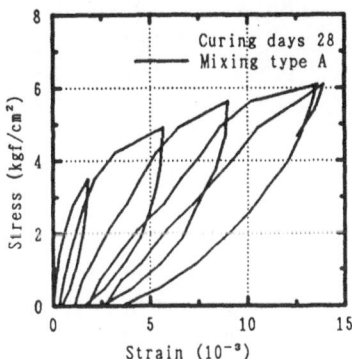

Fig. 10.   Relationship between repetition of stress and strain

recovery ratios are about 100% against any load ratio. Maximal strains against various load ratios of HPCM are dozens of times greater than those of NM. Its recovery ratios are as good as 80.1~83.6% against any load ratios. From the above data, it is clear that HPCM offers good stability against repeated loads. The same results were also obtained with respect to mixture types B~E.

Table 6. ε max and recovery ratio of strain

Type of mortar	Load ratio (%)	50	70	80	90
NM	ε max(x 10^{-6})	113	163	256	------
NM	Recovery ratio (%)	100	100	100	------
HPCM	ε max(x 10^{-6})	1820	5641	9003	13629
HPCM	Recovery ratio (%)	83.6	80.9	80.1	80.4

4   Microstructures of polymer-mixed mortar, obtained through a scanning electron microscope (SEM)

In order to grasp the interfacial condition of PC and aggregates in the mixture system of HPCM, the adhesion of polymers was microscopically investigated by means of a scanning electron microscope (SEM). The samples used for the observation was a mixture of HPCM with PC + fly ash. The photographs of SEM are given in Photos 1 and 2. Photo 1 confirms an aspect of adhesion on the surface of sand particles with a fairly adhesive binder (PC).

In Photo 2, an aspect where PC adheres to spherical particles of fly ash, with a diameter of 1 μ ~5m μ , is also observed.

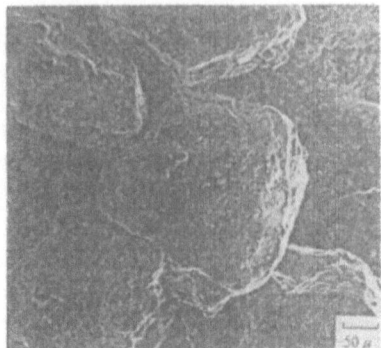

Photo. 1.   SEM Photo of HPCM

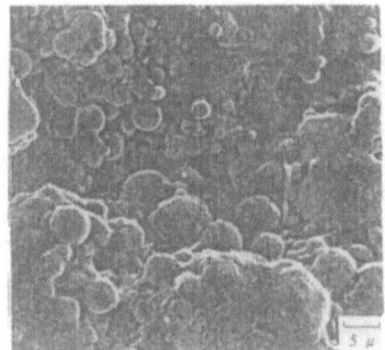

Photo. 2.   SEM Photo of PC + Fly ash

5   Conclusions

(1) For the purpose of providing flexibility, the PC needs the volume ratio of the hydrated products of the cement particles to be around a compressive filling and for HPCM that the proportion of sand to PC be in the slurry range.

(2) It was elucidated that a volume change due to the drying shrinkage of HPCM was, at an early stage, of a slight expansion accompanied by a small amount of shrinkage and presenting a great resistance to drying shrinkage.

(3) It was also clarified that HPCM shows a low strength, but a tendency different from that of NM such that the former is provided with the capacity for dozens of times the elongation that can't be observed with a conventional hardened cement.

(4) HPCM shows very large deformability such that its recovery ratio is as large as 80%.

(5) As a result of the SEM observation of interfaces between various materials in the HPCM, it is shown that the adhesion effect of PC is so strong that it influences the reform of various properties of NM.

(6) It is believed that a high polymer cement mortar such as the material used in the present study shows a relatively weak strength but large free deformability which allows the same to be applicable to the expansion joint requiring elasto-plasticity, shock absorbing materials, etc..

6   References

1. Tokune, Y. (1969) Chemistry of cement concrete, Gyihoudou, Tokyo.
2. Kasai, J. (1984) Cementchemistry (Review), No.6, Concret engineering, Vol.22, No.5,
3. Arai, Y. (1984) Material chemistry of cement, Dainihontosho, Tokyo.
4. Tamai, M., et al. (1990) A study on flexible cement paste and cement mortar containing new acrylic emulsion, Proc. of the 6th Int. Cong. on polymer in concrete, pp. 242-248.
5. Committee of Composit Materials. (1987) Regulation of testing method on polymer cement mortar, Concrete engineering, Vol. 25, No.6.

CHARACTERISTICS OF POLYMER-MODIFIED MORTARS USING REDISPERSIBLE POLYMER POWDERS

T. Kubokawa, H. Matsusato and Y. Yamazaki
Central Research Laboratory, Nihon Cement Co., Ltd.,
Tokyo, Japan

Abstract

Compressive strength, flexural strength, adhesive strength, water absorption, amount of water permeation and drying shrinkage of hardened polymer-modified mortars using redispersible polymer powders were measured. From these results, influence of the charactor of redispersible polymer powders such as kind of polymer, glass transition temperature and kind of protective layer on the properties of the polymer-modified mortars using redispersible polymer powders. The characteristics of polymer-modified mortars using redispersible polymer powders were compared with those using polymer dispersions. The following results were obtained.

(1) Adhesive, flexural strengths and drying shrinkage of polymer- modified mortars using redispersible polymer powders were influenced by the glass transition temperature of redispersible polymer powders.

(2) The kind of polymer influenced on amount of water permeation and compressive strength of polymer-modified mortars using redispersible polymer powders.

(3) The characteristics of hardened polymer-modified mortars using redispersible polymer powders were little influenced by the kind of protective layer of redispersible polymer powders.

(4) The characteristics of polymer-modified mortars using redispersible polymer powders were somewhat inferior compared to that using polymer dispersions.

Keywords :adhesive strength, flexural strength, glass transition temperature, polymer-modified mortar, polymer dispersion, protective layer, redispersible polymer powder

Polymers in Concrete, edited by Y. Ohama, M. Kawakami and K. Fukuzawa. Published in 1997 by E & FN Spon, 2–6 Boundary Row, London SE1 8HN, UK. ISBN: 0 419 22330 4.

1 Introduction

Polymer-modified mortars(PCM) are widely used as surface preparation materials, adhesive of a tile and a brick, floor coating materials and repair materials. In general, PCM are prepared by mixing pre-mixed powder consisting of cement, fine aggregates and some additives with polymer dispersions and /or water. In recent years, polymer pre-mixed products used redispersible polymer powders have been also used. However, few studies have so far been made on the effect of the charactor of redispersible polymer powder on the characteristics of PCM [1] ~ [3] .

The purpose of this study is to make clear the relationship between the characteristics of polymer-modified mortars using redispersible polymer powders and a kind of a polymer composition, glass transition temperature and protective layer of the redispersible polymer powders. For composition, the polymer-modified mortars with the same polymer composition as redispersible polymer powders are also prepared and tested in the same manner.

2 Materials and experimentals

2.1 Materials
2.1.1 Cement and fine aggregate
Ordinary Portland cement (Produced by Nihon Cement Company) and the Toyoura standard sand as specified in JIS R 5201"Physical Testing Methods for Cement" were used for mix proportions in all the mortar mixes.

2.1.2 Polymer
Fifteen types commercial redispersible polymer powders were used. Kinds of polymer, glass transition temperature and kinds of protective layer were shown in table-1. These redispersible polymer powders can be classified into five types by a monomer composition : vinyl acetate(VAC), ethylene-vinyl acetate copolymer (EVA), vinyl-acetate/vinyl-carboxylate copolymer (VAC/VV), acryrate/vinyl-acetate/vinyl-carboxylate copolymer(AC/VAC/VV) and methyl methacrylate-butyl acryrate(AC). Glass transition temperatures of the redispersible polymer powders are in the range of $-10 \sim 35\,^{\circ}\mathrm{C}$. Protective layer were three kinds : non-ionic surfactant, poly-vinyl alcohol(PVA) and carboxylation. Two kinds of polymer dispersions(ethylene-vinyl acetate copolymer(EVA),methyl-methacrylate/butyl-acryrate copolymer(AC)) having the same polymer compositions as redispersible polymer powder were also used.

2.2 Experimental procedures
2.2.1 Specimens preparation procedures
PCM were mixed with a mass ratio of cement to standard sand 1:2, polymer-cement ratios of 0,5,15 mass% and water-cement ratio of 60%. These redispersible polymer powders were pre-mixed in dry state, then mixed with water(W/C=0.6)according to JIS A 6203"Polymer dispersions

and redispersible polymer powders for cement modifiers ".
Beam specimens 40x40x160mm and disk specimens ϕ 150x40mm were moulded and then cured under the conditions as following; a 2-day-20℃-80%-R.H.-moist plus 5-day-20℃-water plus 21-day-60%-R.H.-dry cure. Adhesive substrates 70x70x20mm were fabricated by JIS R 5201. After curing for 28 days, PCM of thickness of 2mm or 10mm were molded on the adhesive substrates, then cured for 14 days at 20℃,60%-R.H.

Table 1. Kinds of polymer, glass transition temperature and protective layer of the redispersible polymer powders and polymer dispersions

Sample[1] No.	Polymer Composition	Tg (℃)	Protective layer	Sample[1] No.	Polymer Composition	Tg (℃)	Protective layer
A－1	VAC	35	Carboxylation	C－2	VAC/VV	5	PVA
B－1	EVA	－10	PVA	C－3	//	10	//
B－2	//	－5	//	C－4	//	30	//
B－3	//	0	//	D－1	AC/VAC/VV	－5	//
B－4	//	0	//	D－2		0	non-ionic
B－5	//	0	//	E－1	MMA/BA	10	//
B－5E	//	0	//	E－1E	//	10	//
C－1	VAC/VV	0	//	F－1	VAC/ET/VC	10	//
				Plain	－	－	－

＊1)B－5E and E－1E are polymer dispersions

2.2.2 Tests
Compressive strength, flexural strength, adhesive strength, water absorption, and drying shrinkage of hardened PCM were tested according to JIS A 6203 "Polymer dispersion and redispersible polymer powders for cement modifiers". Amount of water permeation was tested by JIS A1404 "Method of test for water proof agent of cement for concrete construction".

3. Results and discussion
3.1 Flexural and compressive strengths
Experimental results of the flexural and compressive strengths are shown in Fig.1 and Fig.2 respectively.

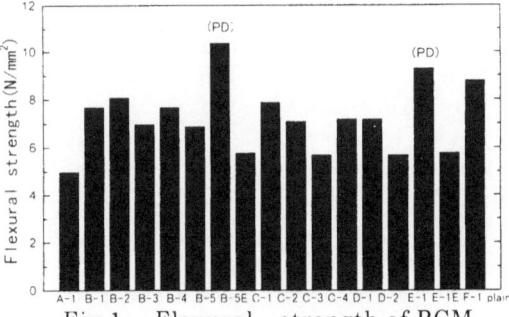

Fig.1. Flexural strength of PCM

Flexural and compressive strengths of PCM using polymer dispersions are higher than those using redispersible polymer powders which have identical compositions. The flexural strength of PCM using redispersible polymer powders were lower than that of plain mortar. On the contrary, the flexural strength of PCM using polymer dispersion were higher than that of plain mortar.

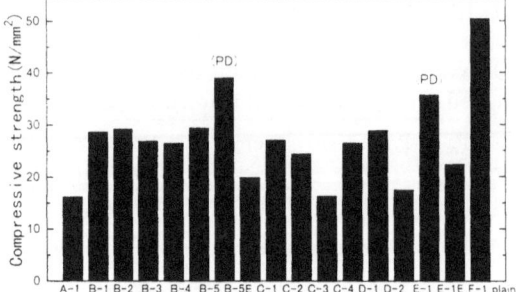

Fig.2.   Compressive strength of PCM

Fig.3 shows polymer-cement ratio vs. the flexural and the compressive strengths of PCM. Both strength were lowered with increasing polymer-cement ratio. The reductions in the compressive strength with increasing polymer-cement ratio are considered as the elastic   modulus of polymers is lower than that of cement mortar.

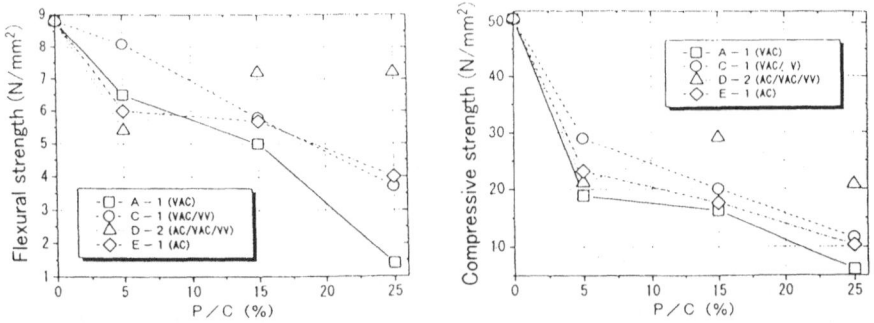

Fig.3.   Polymer-cement ratio vs. flexural and compressive strengths

Fig.4 shows relation between the flexural and the compressive strengths vs. polymer composition. There is no relationship between polymer composition and the flexural strength. Relationship between glass transition temperature and the flexural, the compressive strengths of PCM is shown in Fig.5 . There is the correlation between glass transition temperature and the flexural strength, that is, the flexural strength are lowered with increasing of glass   transition temperature. In the case of the compressive strength, there is no clear relationship between glass transition temperature and strength. It has been reported   that the flexural strength of PCM using acrylic-polymer-dispersions of which polymer-cement ratio were up to 10mass% increased with glass transition temperature [4] . However, an opposite tendency was obtained in this study.

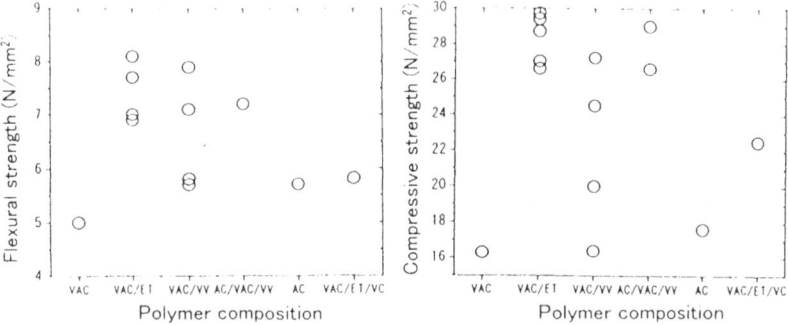

Fig.4. Relation between polymer composition and flexural, compressive strengths of PCM using redispersible polymer powders

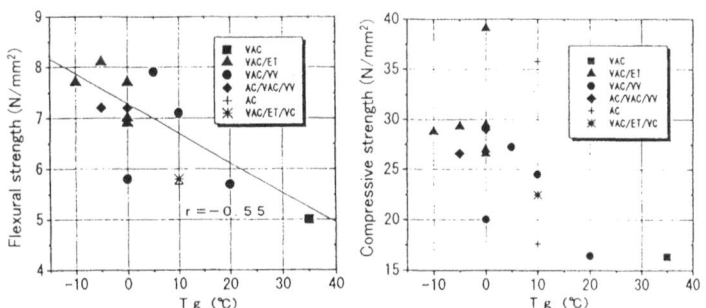

Fig.5. Relation between glass transition temperature and flexural, compressive strengths of PCM using redispersible polymer powders

3.2 Adhesive strength

The adhesive strength of PCM is shown in Fig.6.

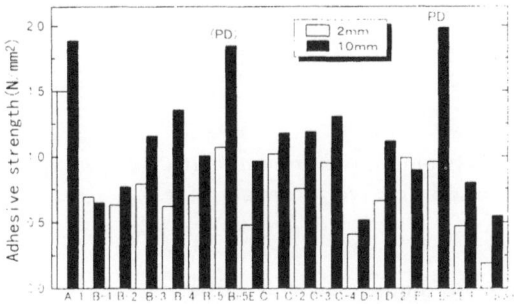

Fig. 6. Adhesive strength of PCM

The adhesive strength of PCM having thickness of 10mm is higher than that having thickness of 2mm, but, this difference is not so much. All of the adhesive strength of PCM are higher than that of plain mortar. A failure modes of PCM are mainly cohesive failure in PCM inside, while, the failure modes of plain mortar are almost adhesive. It is understood that the adhesive strength of PCM is improved due to polymer addition, from the view point of difference in failure modes. The difference of the adhesive strength of thickness of 2mm using between redispersible polymer powders and polymer dispersions is not recognized, but, PCM using polymer dispersions have a high adhesive strength compared with using redispersible polymer powders.

Relation between polymer-cement ratio and adhesive strength is shown in Fig.7. Relation between the adhesive strength of these PCM using redispersible polymer powders and polymer-cement ratio is classified into three types;(1)the adhesive strength increased with increasing polymer-cement ratio, (2)the adhesive strength increased up to polymer-cement ratio of 15%. and (3)the adhesive strength changes slightly at polymer cement ratios of 5% or more.

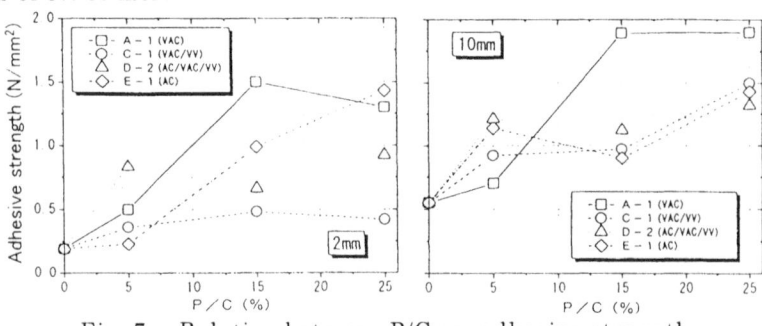

Fig. 7. Relation between P/C vs. adhesive strength

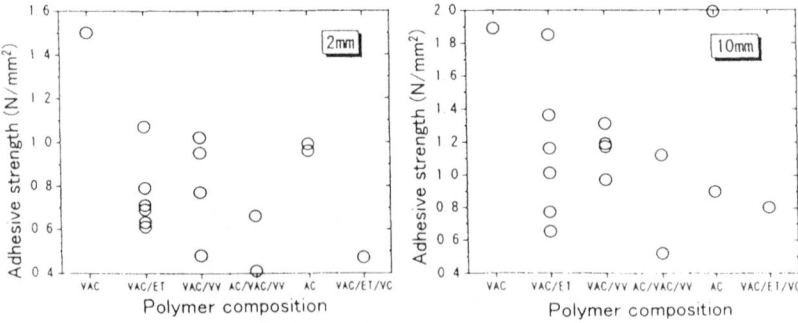

Fig. 8. Relation between polymer composition of redispersible polymer powders and adhesive strength

Fig.8 shows relation between the adhesive strength and polymer composition. There is no relationship between polymer composition and the adhesive strength. Relationship between the adhesive strength and glass transition temperature is shown in Fig.9. It can be seen that there is a correlation between the adhesive strength of PCM using redispersible polymer powders and glass transition temperature, that is, the adhesive strength increases with increasing of glass transition temperature of redispersible polymer powders. It is considered that cohesive force of polymer film in PCM greatly influences the adhesive strength, in addition to cohesive force of cement hydrate.

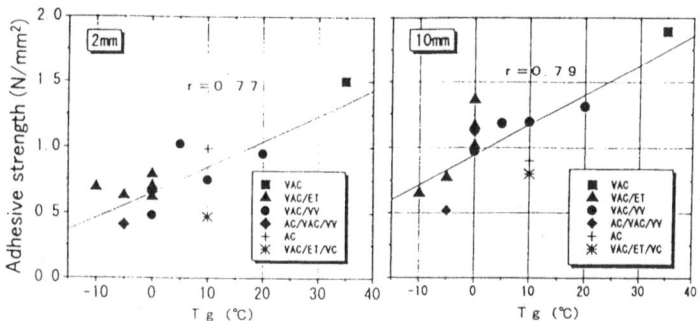

Fig. 9.   Relation between glass transition temperature
and adhesive strength

3.3 Water absorption

Fig.10 illustrates the results of water absorption. Water absorption of PCM using polymer dispersions is lower than that of PCM using redispersible polymer powders of an identical compositions. Water absorption of PCM using redispersible polymer powders is nearly equal to that of plain mortar, except for sample A-1 and C-4. These two samples have a high absorption compared with others.

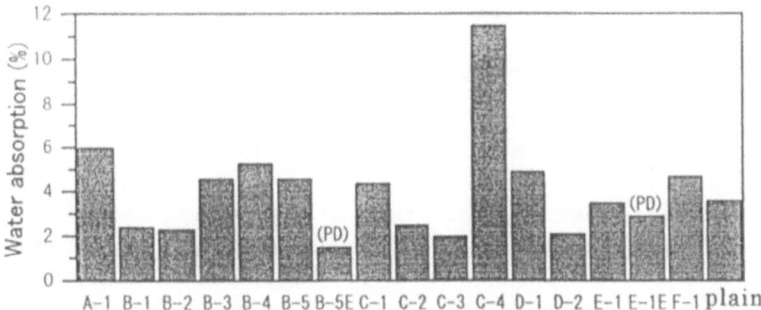

Fig. 10.   Water absorption of PCM

Since glass transition temperature of A-1 and C-4 polymer powders is beyond experimental temperature, it is considered that a continuous polymer film are not formed.

Relation between polymer-cement ratio and the water absorption of PCM using the redispersible polymer powders selected from each polymer composition is shown in Fig.11. The water absorption of PCM decreases with increasing of polymer-cement ratio, except for sample A-1. In the case of sample A-1, the water absorption increases with increasing polymer-cement ratio.

There are no correlation between water absorption of PCM and a polymer composition and/or glass transition temperature.

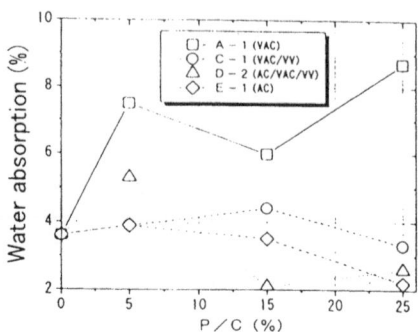

Fig. 11.   Relation between P/C and water absorption

3.4   Water   permeation
Fig.12 shows the water permeation of PCM.

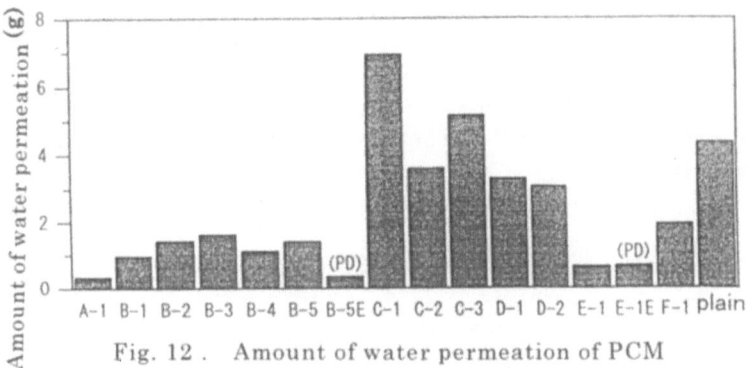

Fig. 12 .   Amount of water permeation of PCM

Amount of water permeation of PCM is much lower than that of plain mortar except for VAC/VV system(C-1,C-3). The tendency that amount of water permeation becomes higher for VAC/VV(C-1 ~ C-3) system and AC/VAC/VV(D-1,D-2) system were shown, but, no correlation was seen with a polymer composition and/or glass transition temperature.

3.5   Drying   shrinkage
Fig.13 illustrate drying shrinkage of PCM according to type of polymer ,and Fig.14 shows relation between drying shrinkage of OCM and age.

Although drying shrinkage after 28 days is bigger than that of plain mortar in case of the sample A-1, B-3 and E-1, the other specimens are almost equal drying shrinkage to the plain mortar. Moreover, the drying shrinkage of PCM using redispersible polymer powders is big in comparison with PCM using polymer dispersions of an identical composition.

Fig.15 represents the polymer-cement ratio vs. 28d-drying shrinkage of PCM.

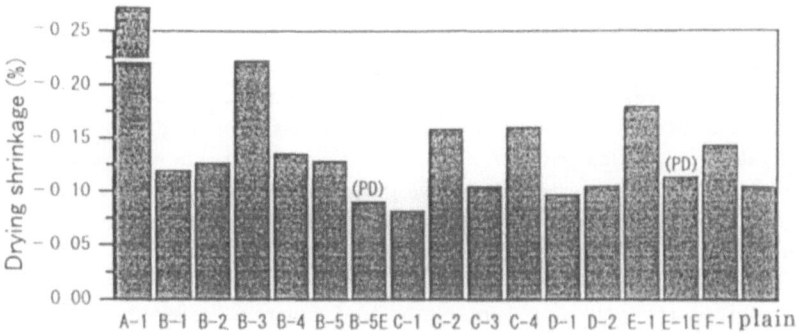

Fig. 13.   Redispersible polymer powders vs. drying shrinkage

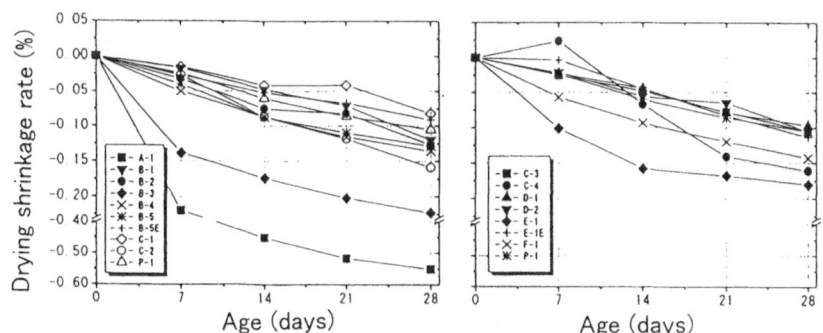

Fig. 14.   Drying shrinkage of PCM cured in room(20℃ and 60%RH.)

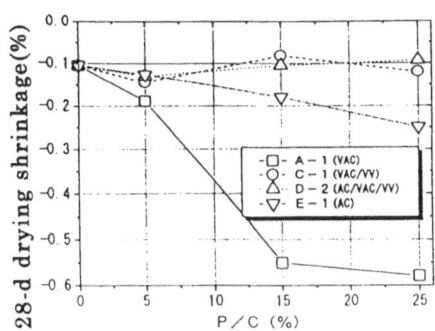

Fig. 15   P/C vs. Drying shrinkage(28days)

The drying shrinkage of PCM using VAC polymer is biggest in other polymer composition. The drying shrinkage of PCM using redispersible polymer powders is almost same manner as that of plain mortar.

4. Conclusions

The relation between the characteristics of polymer-modified mortars using redispersible polymer powders and the charactor of redispersible polymer powders were examined. For comparison, the polymer-modified mortars with the same polymer composition as redispersible polymer powders are also prepared and tested in the same manner.

The following conclusion can be drawn from this experiment.
(1) The adhesive ,the flexural strengths and drying shrinkage of polymer-modified mortars using redispersible polymer powders were strongly influenced by the glass transition temperature of redispersible polymer powders.
(2) The kind of polymer influenced on the amount of water permeation, the compressive strength of polymer-modified mortars using redispersible polymer powders.
(3) The characteristics of hardened polymer-modified mortars using redispersible polymer powders were little influenced by the kind of protective layer of redispersible polymer powder.
(4) The characteristics of polymer-modified mortars using redispersible polymer powders were somewhat inferior to those of polymer-modified mortars using polymer dispersions.

5. References

1. Ohama,Y., Demura,K. and Kim,W.(1994), Proceedings of the First East Symposium on Polymers in Concrete, Chuncheon ,Korea, May 2-3, pp.81-90
2. Sakai,E. and Sugita,J.(1995), Journal of Cement and Concrete Research,Vol.25, No.1 ,pp.127-135
3. Ohama,Y., Demura,K. and Kim,W.(1994) ,JCA Proceedings of Cement & Concrete No.48,pp.796-801
4. Tokumoto,M.(1992), JCA Proceedings of Cement & Concrete No.46, pp.498-503

INFLUENCES OF PROCESS CONDITIONS ON STRENGTH PROPERTIES OF POLYMER-MODIFIED MORTARS USING AN UNSATURATED POLYESTER RESIN

Y. Ohama, K. Demura and K. Kawabata
Department of Architecture, Nihon University, Koriyama, Japan

Abstract
Polymer-modified mortars using an unsaturated polyester resin (UP) are prepared with various mix proportions by two mixing methods, given three types of curings, and tested for flexural and compressive strengths. The influences of process conditions on the flexural and compressive strengths of UP-modified mortars are examined. As a result, increasing polymer-cement ratio chiefly causes an improvement in the flexural strength of UP-modified mortars, and increasing water-cement ratio brings about marked reductions in their flexural and compressive strengths. The highest flexural and compressive strengths are obtained at water-cement ratios of 5% or less regardless of the polymer-cement ratio. Furthermore, the strengths of UP-modified mortars are decreased with an increase in the water-polymer ratio. The production of UP-modified mortars is possible by applying either of Mixing Methods A and B.
Keywords: Curings, mixing methods. polymer-cement ratio, polymer-modified mortars, strengths, unsaturated polyester resin, water-cement ratio, water-polymer ratio.

1 Introduction

In general, polymer-modified mortars are prepared by using polymer dispersions and redispersible polymer powders, and are widely used as high-performance construction materials because of their superior properties such as high flexural strength, good adhesion to ordinary cement mortar or concrete, waterproofness, carbonation and chloride penetration resistance. However, polymer-modified mortars using liquid resins except for epoxy resin have been developed to a lesser extent till

Polymers in Concrete, edited by Y. Ohama, M. Kawakami and K. Fukuzawa. Published in 1997 by E & FN Spon, 2–6 Boundary Row, London SE1 8HN, UK. ISBN: 0 419 22330 4.

now. The purpose of this research is to develop polymer-modified mortars using unsaturated polyester resin which is one of popular low-cost liquid resins.

In this paper, polymer-modified mortars using an unsaturated polyester resin are prepared with various polymer-cement ratios and water-cement ratios by different mixing methods, and subjected to heat, dry and water curings. The cured polymer-modified mortars are tested for flexural and compressive strengths. The effects of process conditions on the strength properties of the polymer-modified mortars using the unsaturated polyester resin are discussed.

2 Materials

2.1 Cement and fine aggregate

Ordinary portland cement and Toyoura standard sand specified in JIS (Japanese Industrial Standards) R 5210 (Portland Cement) and JIS R 5201 (Physical Testing Methods for Cement) respectively were used in all the mortar mixes. The physical properties and chemical compositions of the cement are listed in Table 1.

Table 1.   Physical properties and chemical compositions of cement

Specific gravity (20°C)	Blaine's specific surface (cm^2/g)	Setting time (h-min)		Compressive strength of mortar (MPa)		
		Initial set	Final set	3d	7d	28d
3.16	3250	2-08	3-03	15.8	25.5	42.2

Chemical compositions (%)							
CaO	SiO$_2$	Al$_2$O$_3$	Fe$_2$O$_3$	SO$_3$	MgO	ig.loss	Total
64.0	22.0	5.2	2.6	2.2	1.7	1.8	99.5

2.2 Cement modifier

A commercially available orthophthalate-type unsaturated polyester resin (UP) was used as a cement modifier. The chemical structure and properties of the unsaturated polyester resin are given in Fig. 1 and Table 2 respectively. Ammonium peroxodisulfate [(NH$_4$)$_2$S$_2$O$_8$, molecular weight =228.20] was employed as a catalyst for the unsaturated polyester resin.

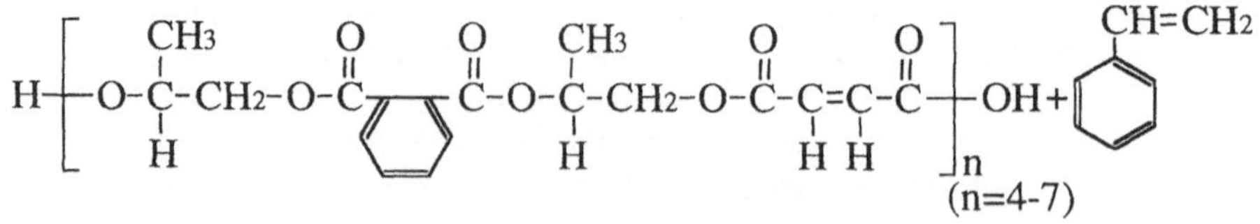

Fig. 1.   Chemical structure of unsaturated polyester resin.

Table 2.   Properties of unsaturated polyester resin

Specific gravity (20℃)	Viscosity (20℃,mPa•s)	Acid value	Styrene content(%)
1.13	325	16.9	38.0

3 Testing procedures

3.1 Preparation of specimens
Polymer-modified mortars using an unsaturated polyester resin were prepared with the mix proportions given in Table 3 by the following two mixing methods:
(1) Mixing Method A: An unsaturated polyester resin was added to the dry mixture of cement and sand, and then the mixing water containing a catalyst was mixed with the mixture of the unsaturated polyester resin, cement and sand.
(2) Mixing Method B: The mixing water containing a catalyst was added to the dry mixture of cement and sand, and then an unsaturated polyester resin was mixed with the mixture of water, catalyst, cement and sand.
The mixing methods for the polymer-modified mortars are illustrated in Fig. 2.

Table 3.   Mix proportions of polymer-modified mortars

Cement : sand (by mass)	Polymer-cement ratio (%)	Water-cement ratio (%)	Catalyst content (phr*)
1 : 3	100 150 200	3 5 10 20	2

Note,*: Parts per hundred parts of resin.

Beam specimens 40x40x160mm were molded with the mortars in accordance with JIS A 1172 (Method of Test for Strength of Polymer-Modified Mortar), and subjected to the following three curings:
(1)2-day-20 ℃ -80%(RH)-moist plus 15-hour-70 ℃ -heat cure (Heat curing)
(2)2-day-20 ℃ -80%(RH)-moist plus 5-day-50 ℃ -dry cure (Dry curing)
(3)2-day-20 ℃ -80%(RH)-moist plus 5-day-20 ℃ -water cure (Water curing)

3.2 Strength tests
The cured beam specimens were tested for flexural and compressive strengths according to JIS A 1172.

4 Test results and discussion

The hardening mechanism of unsaturated polyester (UP)-modified mortars may be summarized as illustrated in Fig. 3 [1]. In general, the hardening of UP-modified

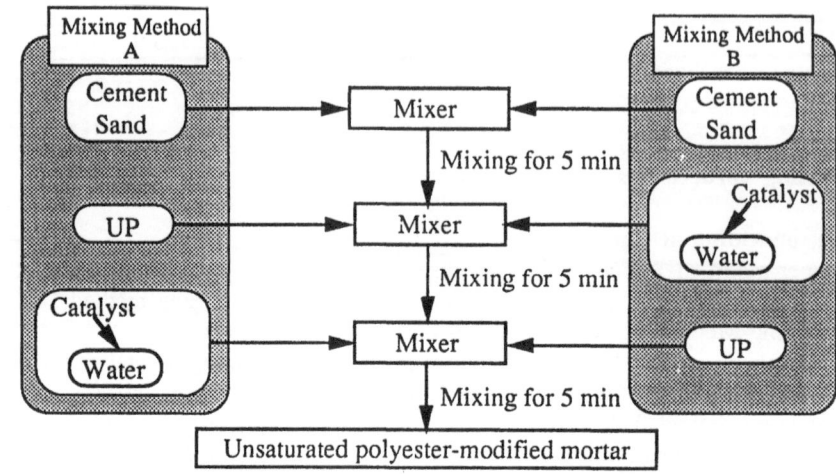

Fig. 2.   Mixing Methods for polymer-modified mortars.

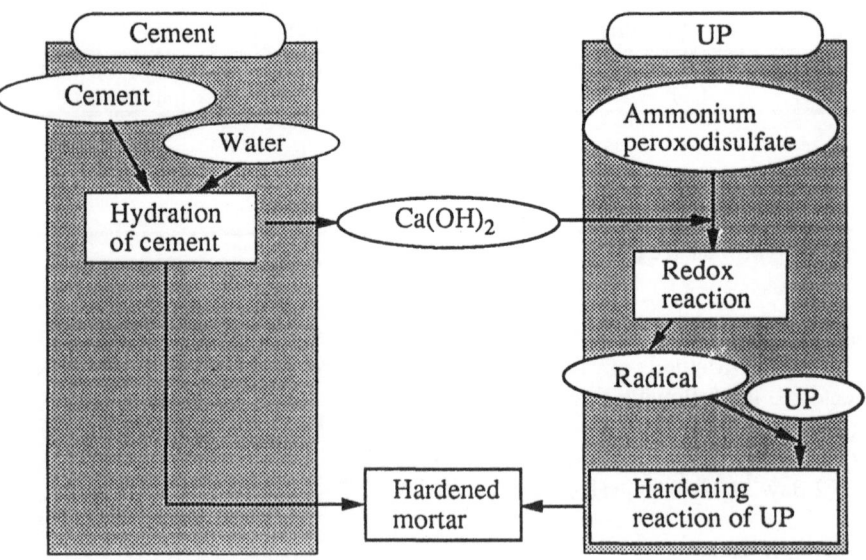

Fig. 3.   Hardening mechanism of UP-modified mortars.

mortars occurs as the result of the hydration of cement and the radical polymerization of unsaturated polyester resin[2]. Ammonium peroxodisulfate as a catalyst is decomposed by a redox reaction with calcium hydroxide [$Ca(OH)_2$] from the hydration of the cement, and generates the radicals necessary for the radical polymerization of the unsaturated polyester resin. The hardening of UP-modified mortars is based on an interaction between the cement hydration and radical polymerization. Therefore, it is very important that both cement hydration and radical polymerization progress simultaneously.

Figs. 4 to 7 represent the effects of polymer-cement ratio and water-cement ratio on the flexural and compressive strengths of UP-modified mortars, prepared by Mixing Methods A and B, and subjected to heat, dry and water curings. The flexural strength of UP-modified mortars tends to increase with an increase in the polymer-cement ratio like polymer-modified mortars using polymer dispersions or redispersible polymer powders. Their compressive strength is reduced or hardly varies with increasing polymer-cement ratio, though there are some exceptions. The flexural and compressive strengths of UP-modified mortars are sharply reduced with increasing water-cement ratio. The highest flexural and compressive strengths of UP-modified mortars are obtained at water-cement ratios of 5% or less regardless of the polymer-cement ratio. It is considered that the strength development of UP-modified mortars with high polymer-cement ratio depends mainly on the polymerization of the unsaturated polyester resin. The strength reduction with an increase in the water-cement ratio may be due to the hydrolysis of the unsaturated polyester resin as a component of a binder for UP-modified mortars [3]. It is difficult to clearly explain the effects of the mixing method and curing condition on the flexural strength development of UP-modified mortars because of the combined effects of various process conditioning factors. By contrast, the effects of the mixing method and curing condition on the compressive strength development of UP-modified mortars are hardly recognized at a water-cement ratio of 20% and a polymer-cement ratio of 100 or 150%. In the application of Mixing Method B, the compressive strength of heat-cured UP-modified mortars is higher than that of dry- and water-cured mortars irrespective of the polymer-cement ratio and water-cement ratio. In other words, Mixing Method B and heat curing are the most suitable for the compressive strength development of UP-modified mortars. However, generally speaking, it is concluded that the production of UP-modified mortars is possible by applying either of Mixing Methods A and B. As in Mixing Method A, the addition of the mixing water and catalyst is the final process of the mixing of UP-modified mortars, Mixing Method A is convenient to produce a prepackaged UP-modified mortar system consisting of the unsaturated polyester resin, cement and sand, which is applied with the addition of the water containing the catalyst in situ.

From the above-mentioned test results, the flexural and compressive strengths of UP-modified mortars are markedly affected by the amounts of the water and unsaturated polyester resin contained in the mortars. Fig. 8 shows the effect of water-polymer ratio on the flexural and compressive strengths of UP-modified mortars. The water-polymer ratio was calculated from the amounts of the mixing water and unsaturated polyester resin used for preparing UP-modified mortars. Regardless of the mixing method, the flexural and compressive strengths of UP-modified mortars are reduced with increasing water-polymer ratio. Close correlations between the flexural and compressive strengths and water-polymer ratio of UP-modified mortars are found out. In particular, high correlation coefficient is obtained for the flexural strength of UP-modified mortars. Consequently, it is considered that the water-polymer ratio can become a mix proportioning factor for UP-modified mortars.

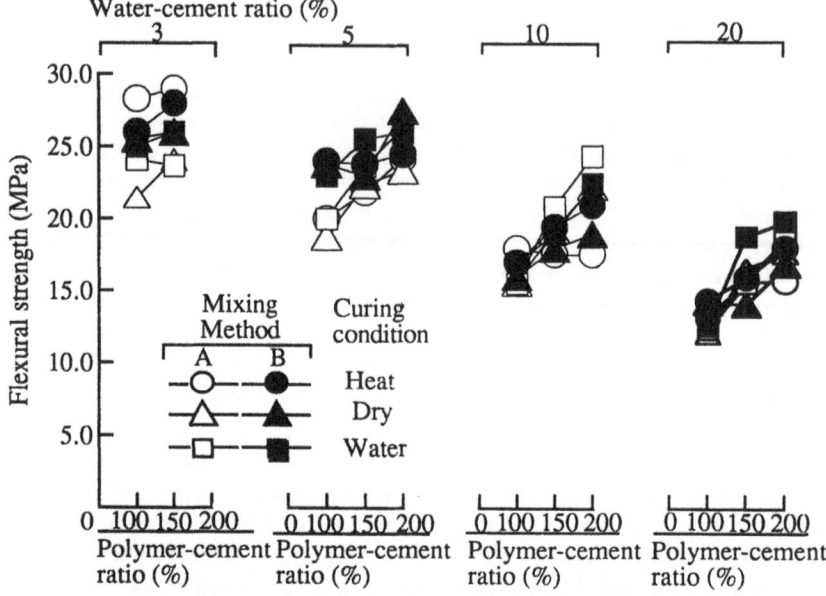

Fig. 4.   Polymer-cement ratio and water-cement ratio vs. flexural strength of heat-, dry- and water-cured UP-modified mortars.

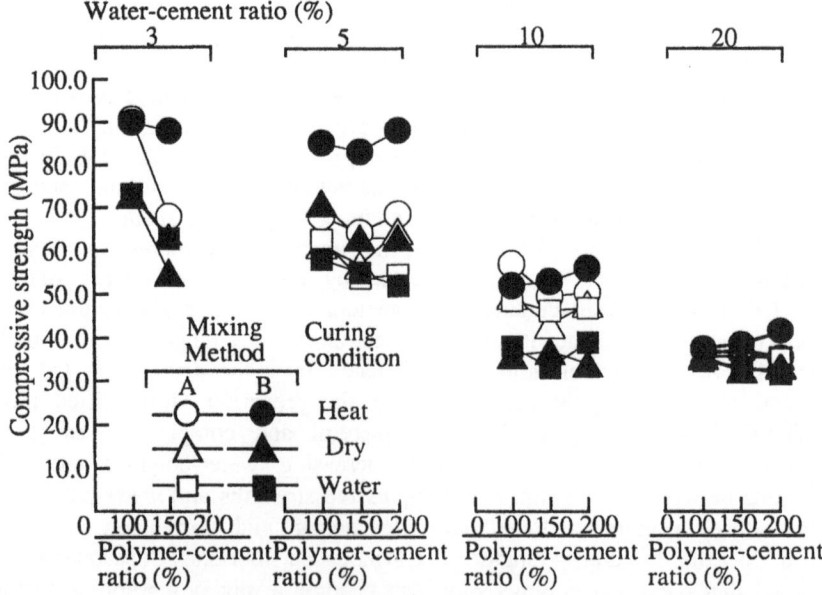

Fig. 5.   Polymer-cement ratio and water-cement ratio vs. compressive strength of heat-, dry- and water-cured UP-modified mortars.

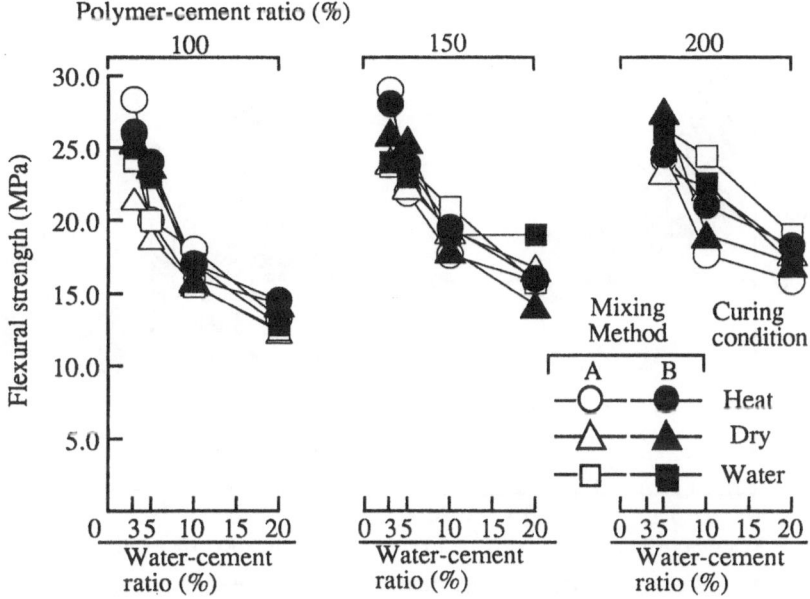

Fig. 6. Water-cement ratio and polymer-cement ratio vs. flexural strength of heat-, dry- and water-cured UP-modified mortars.

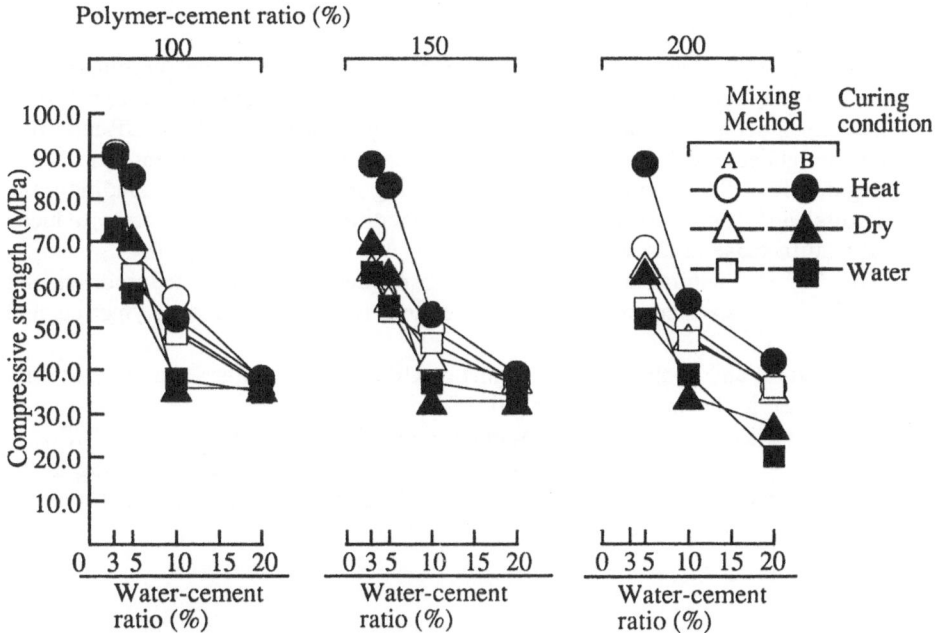

Fig. 7. Water-cement ratio and polymer-cement ratio vs. compressive strength of heat-, dry- and water-cured UP-modified mortars.

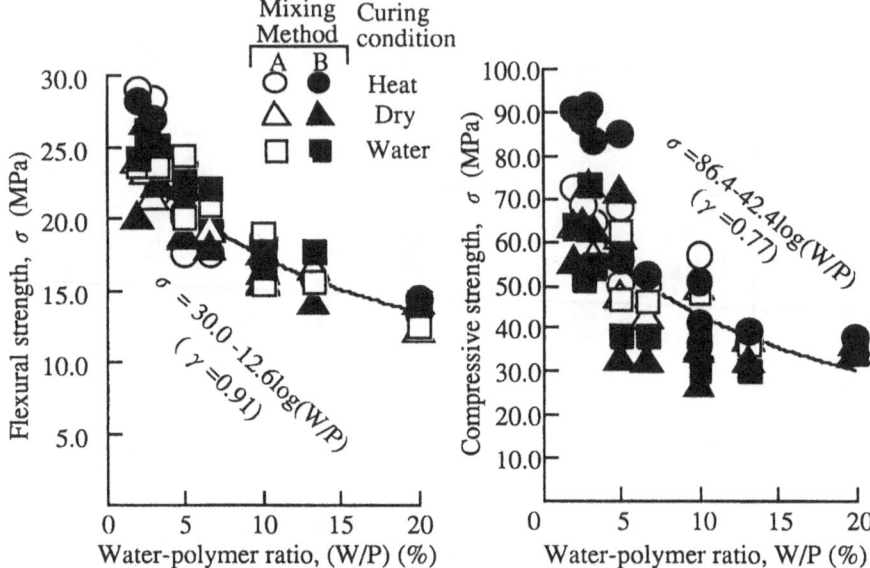

Fig. 8   Water-polymer ratio vs. flexural and compressive strengths of UP-modified mortars.

5 Conclusions

The conclusions obtained in the above research work are summarized as follows:

(1) In general, the flexural strength of UP-modified mortars is increased with increasing polymer-cement ratio, while the compressive strength of UP-modified mortars is decreased or varies with an increase in the polymer-cement ratio.

(2) Regardless of the polymer-cement ratio, mixing method and curing condition, the flexural and compressive strengths of UP-modified mortars are sharply reduced with an increase in the water-cement ratio.

(3) The highest flexural and compressive strengths of UP-modified mortars are obtained at water-cement ratios of 5% or less regardless of the polymer-cement ratio.

(4) The flexural and compressive strengths of UP-modified mortars are reduced with an increase in the water-polymer ratio.

(5) The production of UP-modified mortars is possible by applying either of Mixing Methods A and B. The application of Mixing Method B and heat curing is the most effective for the compressive strength development of UP-modified mortars.

6 References

1.   Ohama, Y. and Demura, K. (1995) Strength properties of unsaturated polyester

resin-modified mortars (in Japanese). *Architectural Institute of Japan Tohoku-shibu Kenkyu-Hokokushu,* No. 58, pp. 229-302.

2.   Nutt, W. O., Stable compositions of hydraulic cement and polymerizable unsaturated polyester resin components copolymerizable by the mere addition of water, and methods of producing same, in *U.S. Patent No. 3437619,* Apr. 8, 1969.

3.   Takiyama, E. (1988) *Handbook of Polyester Resin* (in Japanese), Nikkamkogyo, Tokyo, pp.152-153.

STUDIES OF A NEW NON-DISPERSIVE CONCRETE FOR USE UNDERWATER AND ITS APPLICATIONS

Y.B. Cai, B.Y. Lin and G.L. Shang
Nanjing Hydraulic Research Institute, Nanjing, China

Abstract

This paper presents the research results of mixture on a new non–dispersive concrete admixture for use in underwater concrete. Physical and mechanical properties and durability of the concrete in fresh and hardened were investigated, and applications of the concrete in two underwater repair projects were introduced.Test results showed that the dispersed cement paste is only 6% ~ 16% of control when the concrete poured in underwater, compressive strength ratio of the concrete in underwater to in air reached 83% and over. There are marked improvement in other mechanical properties and durability in contrast to control, furthermore, excellent workability such as no bleeding and never segregation, self–flowability and self–compaction.

Keywords: Applications, Non–dispersive concrete admixture, Physical and mechanical proterties, Underwater concrete.

1 Introduction

Conventionally, the quality of underwater concrete depends on construction technics and quality control of the concrete, the key is separating concrete from environmental water in construction as can as possible. Main placing method for underwater concrete is tremie method. Generally, the covered depth of the vent of a pipeline in concrete pouring must be more than 1 meter, for this reason, pouring underwater concrete with pipepline transmiting is only suitable when water–level is 1.5 meter deep above foundation. Cement paste is easily dispersed for plain

Polymers in Concrete, edited by Y. Ohama, M. Kawakami and K. Fukuzawa. Published in 1997 by E & FN Spon, 2–6 Boundary Row, London SE1 8HN, UK. ISBN: 0 419 22330 4.

concrete when it poured in underwater due to invaded by water, compressive strength of surface concrete contacted with water may be decreased by 50% and over, and poor binding strength with foundation. Moreover, removing 15~ 45 centimeter surface concrete are generally needed when concrete constructed at intervals, so, it is necessary to have leeway of 15 centimeter for each side of some underwater structures, huge waste it is.

Since 1970's , a series of studies for improving quality of underwater concrete have being developed through improving anti−dispersiveness of cement paste in underwater. A non−dispersive concrete (NDC), which cement paste is not disperse even the concrete contacts with water in pouring, was firstly published in former West Germany in 1974. In 1980's, NDC has also being used in underwater concrete projects in Japan, and the quality specification of NDC admixture (NDCA) had been formulated. NDC had also been used successfully in harbor and offshore structures in Europe and America. In China, there is great quantity underwater concrete operation not only in construction but in strengthening of hydraulic and offshore structures, and considerable progress on studies and development of NDC has being achieved. This paper presents partly the test results of a new NDC and its applications.

2   Test methods and materials used

2.1   Test methods

2.1.1   Cast method for underwater concrete specimens

A set of contrast test concerning the influence of water depth of watertank for cast on compressive strength of NDC was carried out. Generally, the longer the distance from water surface to specimen model, the lower compressive strength of concrete. A water depth of 50 centimeter was selected for casting condition and ready fresh concrete was poured continually in underwater model.

2.1.2   Test method for flowability

NDC has the feature of viscous fluidity. Germany standard DIN 1048 specified flow table test method was selected for testing flowability of NDC in the study due to the method is more sensitive for testing flowability than slump.

2.1.3   Test method for resisting dispersity

1. Transparency: the relative transparency of water samples which from watertank for cast was tested by a infrared light−sensitive meter and set the transparency of clear water sample is 100%.
2. pH value: pH value of water samples was tested by pH meter.
3. Washed−sieve analysis: washed−sieve analysis of fresh concrete mixtures

After pouring in underwater was made according to Japan standard JIS A 1112-1975.

2.2 Materials used

1. Cement: ASTM type I of Portland cement
2. Fine aggregate: river sand, γ_f = 2.61, F.M. = 2.60
3. Coarse aggregate: crushed limestone, γ_c = 2.70, D_{max} = 30mm
4. NDCA: to be compound

3 Development of NDCA

Raw material which could be used for developing NDCA and suitable dosage range may be described as folows:

1. Synthetic or natural polymers: such as synthetic ethoxyl cellulose, polyacrylamide and starch gum etc., and dosage used may generally be 0.2% ~0.5% (by weight of cement).
2. Organic emulsion: such as acrylic emulsion and paraffin emulsion etc., and dosage used may generally be 0.1%~1.5%.
3. Mineral materials with high surface area: such as bentonite and silica fume etc., and dosage used may generally be 3%–10%.
4. Inorganic filler: such as fly ash and slaked lime etc., and dosage used may generally be 10%~25%.

A large number of tests were launched for optimizing NDCA use of above various types materials. Test results indicated that ethoxyl cellulose(EC) and polyacrylamide are most suitable as significant ingredient of NDCA. Further, the optimum dosage of EC and polyacrylamide and other supplementary components (including superplasticizer, set–modulating admixture, strength gained admixture and fillers etc.) for developing NDCA were experimental investigated. The parallel test results of optimum mixture of NDCA are listed in Table 1. It showed that from Table 1, when the tested cubes casted in air, there is higher compressive strength for conventional concrete containing superplasticizer and with a lower water cement ratio, but its mechanical properties were remarkably decreased when poured in underwater contrast to NDC containing EC or polyacrylamide. In addition, mechanical properties of the concrete containing EC is superior than the concrete with polyacrylamide. Say, EC is optimum major ingredient for NDCA. Hereafter, the influences of different types EC and various supplementary components on comprehensive properties of fresh and hardened concrete were investigated. A new NDCA which is composed of optimum EC and five supplementary components was conclusively developed, and named NNDC-2 hereafter. The molecular structure of EC of NNDC-2 is showed in Fig. 1.

Table 1.   Comparisoh of propeties of NDC admixture

Main ingredient	Cement (kg / m³)	W / C ratio	Flowability (cm)	Transparency (%)	pH value	Comp. str.(MPa) 7d		28d		Bind. str. 28d (MPa)
						A	U	A	U	U
Superplasticizer	465	0.40	44	3	11.4	24.5	4.7	35.6	6.9	0.44
Polyacrylamide	413	0.64	45	98	9.5	14.3	7.5	29.2	17.9	0.75
Ethoxyl cellulose	421	0.57	44	95	9.7	13.6	9.8	25.1	22.0	1.22

Note: A— Tested cubes casted in air;   U— Tested cubes casted in underwater.

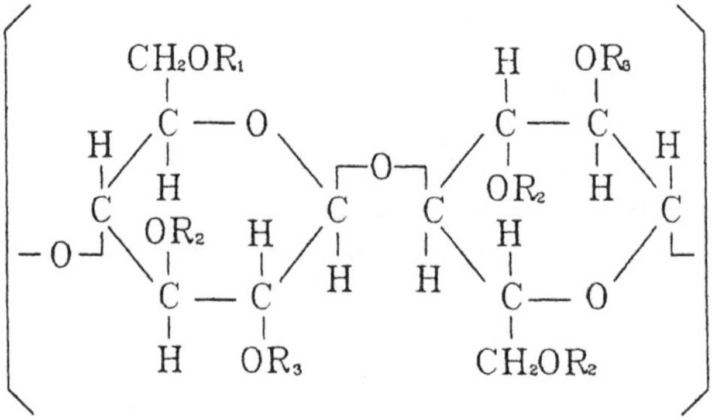

$R_1 \sim R_1$:   alkyl or hydroxyl

Fig. 1. Molecular structure of ethoxyl cellulose.

4   Properties of NNDC−2 concrete

4.1   Properties of fresh NNDC−2 concrete

The test results of fresh NNDC−2 concrete properties are given in Table 2, and the change of mixture of fresh concrete around poured in underwater model are illustrated in Fig. 2.

As showed in Table 2 and Fig.2, fresh NNDC−2 concrete has excellent anti-dispersity in underwater. The cement to aggregate ratio of NNDC−2 concrete was only varied from 1:4.1 in air to 1:5.1 in underwater, in contrast to plain concrete, from 1:4.8 to 1:18.3. Moreover, the fresh NNDC−2 concrete has not bleeding, segregation and rapid flowability loss. It is very favorable the characteristics of fresh NNDC−2 concrete for construction of underwater concrete by pipeline transmiting and pumping.

Table 2. Properties of fresh NNDC-2 concrete

Type	Cement (Kg/m³)	W/C ratio	Flowability (cm)	Transparency (%)
Control	420	0.51	46	24
NNDC-2	430	0.52	45	91

Type	Air content (%)	Bleeding after 3 hr (%)	Flowability change with time (min.) (%) 0 30 60 90		Setting time (hr:min) initial	final
Control	1.2	7.0			12:05	17:40
NNDC-2	2.2	0	100 98 84 70		14:55	19:30

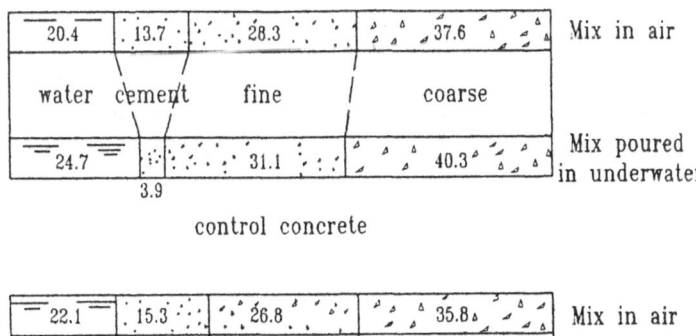

control concrete

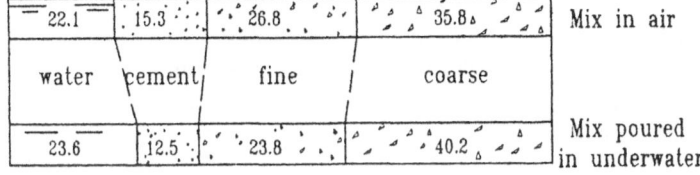

NNDC-2 concrete

Fig. 2. Change of concrete mixtures around poured in underwater.

4.2 Properties of hardened NNDC-2 concrete

The tested properties of hardened NNDC-2 concrete are tabulated in Table 3. It is clear that there are approximate mechanical properties for NNDC-2 and control mix in air condition. However, when poured in underwater, strength and durability of plain concrete were noticeably decreased due to cement paste was excessively dispersed in water. A striking contrast was observed for NNDC-2 concrete in same test conditions.

5 NNDC-2 concrete for In-situ construction

5.1 Strengthening of pier of Xin-An River bridge

Xin-An River bridge which is situated in the western part of ZheJiang province

Table 3. Properties of hardened NNDC–2 concrete

Type	Compr. str. (MPa)		Bending str. (MPa)		Binding str. (MPa)		Elastic modu. ($\times 10^4$MPa)		Weight loss by abrasion
	A	U	A	U	A	U	A	U	U
Control	32.3	7.7	2.8	0.8	6.6	1.6	3.64	2.40	100%
NNDC–2	33.9	28.0	3.4	3.0	6.8	5.0	3.38	3.31	3%

Note: A—Tested cubes casted in air; U—Tested cubes casted in underwater; The age of all cubes is 28d.

was built in 1990. It was inspected that there was a weak binding strength between foundation rock and pier concrete based ȯn the drilling cores. A strengthening concrete circle around the pier basis was needed for safety and durability of the pier, and the concrete circle size was designed as 1 meter high and 0.8 meter wide. According to the field conditions, it seems impossible to guarantee underwater concrete quality in case of pouring conventional concrete. So, NNDC–2 was selected as a suitable replacement. Used typical mixture of NNDC–2 concrete as follows:

ASTM type I cement: 480 kg / m^3
Fine aggregate(sand): 585 kg / m^3, F. M. = 2.80
Coarse aggregate(gravel): 880 kg / m^3, D_{max} = 40 mm
NNDC–2 admixture: 40 kg / m^3
Water: 250 kg / m^3
Slump: 20–22 cm

The tests and construction for In–situ showed that NNDC–2 concrete has good workability and low slump loss, and favorable non–dispersity and self–flowability in underwater. The tested mechanical properties of NNDC–2 concrete in field are given in Table 4. The desired results had been achieved in the strengthening project.

Table 4. Test results of NNDC–2 concrete for In–situ (Xin An River bridge project)

Cast method	Compr. str. (MPa)		Binding str.(MPa)	Clench str. with
	7d	28d	28d	rebar(MPa) 28d
In air	25	36.6		
In underwater	22.7	33.9	4.6	3.3

5.2 Strengthening of spillway apron of Majitang hydropower station

The spillway apron of Majitang hydropower station which is located in HuNan province was critically eroded for abrasion after 2~3 years motion. It is necessary to strengthen the apron as soon as possible for the safety of spillway dam. There is usually 4~5 meter water-level above the apron, it is impossible to pour concrete as conventional method according to ready field conditions. So, underwater concrete by pipeline construction was the most suitable choice. NNDC-2 concrete was selected as major strengthen material in order to increasing strength of underwater concrete and durability for abrasion erosion resistance.

The trial construction was firstly done in small scale in 1991, plain concrete, superplasticized concrete and NNDC-2 concrete were separately poured in an area of apron for comparision. The construction and test results for In-situ showed that NNDC-2 concrete displayed excellent workability in constructing, self-flowability and good non-dispersity in underwater, and that 28 day mean compressive strength of NNDC-2 concrete cores was 30MPa and over.

Based on acquired experiences from the trial construction, strengthening works for the apron was launched in large scall in 1994. Total of 1500 cubic meter NNDC-2 concrete was poured by pipeline. The typical mixture of NNDC-2 concrete is given below:

ASTM type I cement: 500kg / m^3
Fine aggregate(sand): 550kg / m^3 , F.M. = 3.0
Coarse aggreagate(gravel): 900kg / m^3 , D_{max} = 40mm
NNDC-2 admixture: 38.5kg / m^3
Water: 260kg / m^3
Slump: 20-22cm

In the construction, high slump of NNDC-2 concrete was conveyed through tipping tray (15 meter long and 10 meter high difference), and 0.1 cubic meter of hand tipping wagons, but fresh NNDC-2 concrete displayed good workability, never bleeding and segregation from mixer to skip of pipeline. After pouring in underwater, in the range of 3 meter flowing radius which is controlled by a pipeline, the largest high difference of surface of poured the concrete was only 20 centimeter, does excellent self-flowability it is. The test results from drilling cores are given in Table 5. It can be clearly seen that compressive strength of all cores is comparable to, even higher than that cubes strength from mixer (these cubes cured in nature condition). Say, NNDC-2 concrete has outstanding non-dispersity in underwater.

6 Conclusions

1. Ethoxyl cellulose, a synthetic soluble polymer, is the most suitable as significant ingredient of non-dispersive concrete admixture.

Table 5.   Test results of NNDC−2 concrete for In−situ (Majitang apron project)

Cast method	Size of specimens(cm)	Poured area No.	Compr. str.(MPa)28d
In air	$15 \times 15 \times 15$	1	31.5
		2	30.5
		3	31.2
In underwater	$\Phi 9 \times 9$	1	36.9
		2	33.3
	$\Phi 12.5 \times 12.5$	3	26.8

2. Developed   NNDC−2 admixture has excellent all properties for use in underwater concrete, the compressive strength ratio   (in underwater to in air) reached 83% and over, other physical and mechanical properties and durability of the concrete are also markedly increased or improved.

3. Through construction practices of two underwater strengthening projects, it showed that NNDC−2 concrete has satisfied workability, flowability and low slump loss in air, self−flowability and self−compaction in underwater.

7   References

1. Staynes, B. and Corbett, B.(1988). Underwater concreting with polymers, Journal of Civil Engineering, Consulting Engineer International, London.

2. Neeley, B.D.(1988). Evaluation of concrete mixtures for use in underwater repairs, Technical Report of US Army Corps of Engineering, Civil Works Research, Work Unit 32305.

3. Ramachandran, V. S.(1984). Concrete Admixtures Handbook: Properties, Science and Technology, NOYES Publications, New Jersey.

CRACK REPAIR BY RESIN INJECTION: MATERIALS, EQUIPMENT AND METHODS

W.M. Kay
MBT (Singapore) Pte. Ltd., Singapore

Abstract

This paper traces injection resin techniques over the past twenty years. changes in material technology will be discussed showing advances in both physical performances and environmental issues.

Methods of injection will be highlighted including low pressure, high pressure twin line and single line systems.

Injection material types are investigated and listed showing major properties, advantages, disadvantages and uses.

The final part of the paper will outline some case histories of epoxy injection in both repair and rehabilitation of concrete structures and will discuss the practicalities of resin injection today.

Key words : Epoxy, injection, repair.

Introduction

This paper EAS 73 will document some of the changes that have occurred in the resin injection industry over the last twenty years. The definition I have taken in this paper for resin injection is epoxy or polyester crack injection for structural repair or gap filling. It will not cover water stopping ground stabilization or other forms of injection.

Resin Injection Theory

Over the years many ideas have been put forward as to the correct method of grouting or injecting cracks. Today we see two basic methods have evolved and both will be discussed.

Both theories accept that for successful injection the crack should be sealed and have some form of entry port for the injection of the resin. Entry porting will be discussed later in the section. These theories also accept that resin can then be transferred into the crack in a variety of means. The differences come in procedures.

Polymers in Concrete, edited by Y. Ohama, M. Kawakami and K. Fukuzawa. Published in 1997 by E & FN Spon, 2–6 Boundary Row, London SE1 8HN, UK. ISBN: 0 419 22330 4.

Theory A suggests port spacing is equal to the thickness of the member being grouted and the injection starts at the bottom or lowest part of the crack and proceeds upwards. When resin appears at the next highest injection point, the entry port or tube is moved to that point. When all the injection ports show resin the crack must be fully grouted.

Theory B suggests that you start at the widest part of the crack and when the epoxy resin shows at the next port that port is tied off or capped and injection continues. Injection continues until all the ports are filled and the injection pressure is then held until some constant back pressure occurs. Depending on crack width, ambient temperature resin gel time and depth of crack the injection starting point may be moved.

I believe both methods have merit but theory B is in my opinion likely to have greater success in all cases whether the crack is dry or damp. The reason is as follows. If we take two sheets of glass 600mm x 300mm and position them exactly one mm apart and then seal the edge all the way round we now have a gap simulating a crack. if we place 3 injection ports at 0mm, 300mm and 600mm high and then start to inject we will see the following phenomena.

The resin moves both upwards filling the simulated crack. This would than suggest theory A is the preferred method of crack injection. Figures 1 & 2 & 3 illustrate this point.

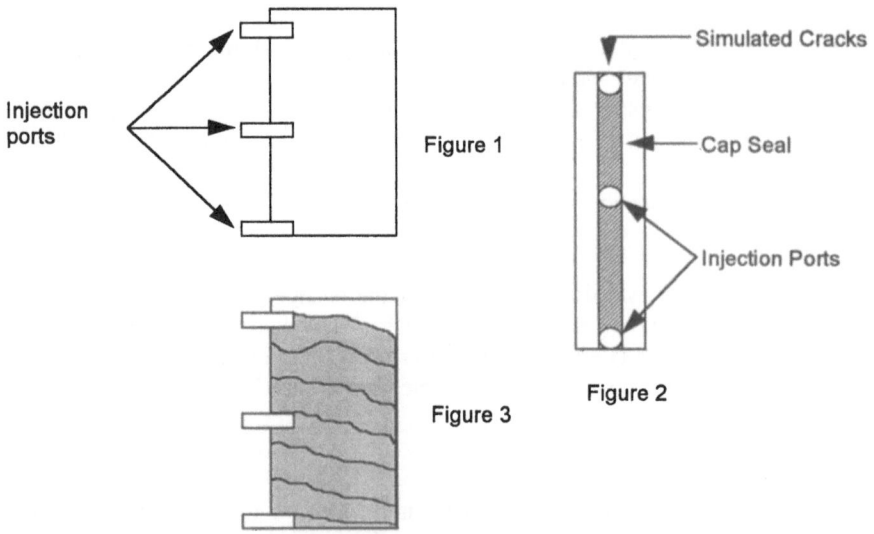

Injection ports

Figure 1

Simulated Cracks

Cap Seal

Injection Ports

Figure 2

Figure 3

Demonstrating the theory of crack injection

Let us now look at a real crack - a beam 4 metres long 300mm cross section with a central crack through the beam. Exaggerated the crack is illustrated in Figure 4 below.

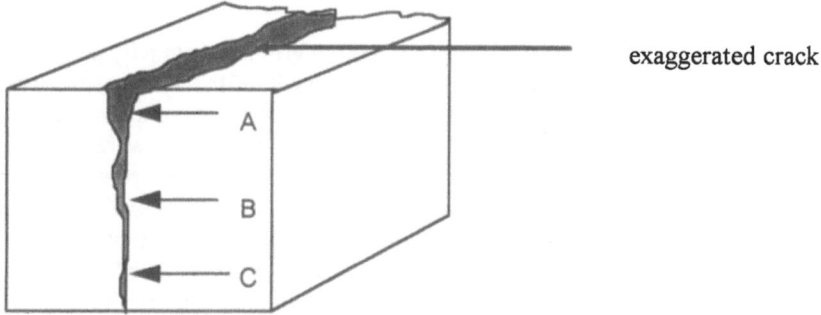

exaggerated crack

Figure 4 Simulated crack in concrete

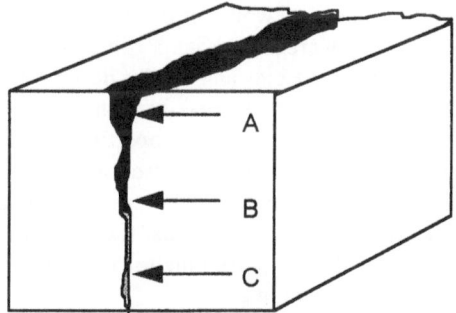

Figure 5 Simulated crack in concrete showing resin flow

When resin is injected resin flows into the crack as well as along the crack. If we then assume the crack has a width of 0.5 mm at point A, 0.3mm at point B and 0.1mm at point C, the following flow pattern will occur. The initial resin will penetrate the crack and as the crack narrows the resin will choose the easier path of a wider crack. Thus as figure 5 shows the resin can and will show at the next port even though it has not filled the crack.

If we simply move onto the next port then we may never fill the crack. this is why I believe we have to adopt the theory B of tying off/or sealing the port and applying pressure until resistance occurs.

Resin Injection Equipment

Early in the history of crack injection people adapted existing equipment to place the resin in the crack. The simplest means were grease guns through standard grease nipples or pressure pots used in the paint industry. The grease nipples were glued on to the surface with a suitable adhesive which then subsequently was used to seal the crack. After hardening the injection resin was mixed, poured into the greased gun and then pumped into the crack through the grease nipple. Pressure pots were used in a similar fashion using steel or copper tubes as injection ports. A system the writer used in 1969 was using steel porting adapters but using a peristaltic pump to push the resin into the crack. The advantages of this pump was that if the resin hardened the tube was discarded and a new piece fitted.

Floor cracks and other horizontal surfaces were often grouted by gravity particularly if the cracks were more than 0.3 mm wide. The underside of the surface (if it was a through crack) was sealed and either "wells" of clay or plastercine were formed or the side of the crack was carefully chipped to create a well. These "wells" were then filled with the mixed injection resin and this flowed into the crack. The well were constantly topped up until the crack was filled, similar methods were adapted on walls but ports were set in to the concrete and each port had a metre head of clear plastic to be. Equipment technology took quantum leaps forward in the mid 70's. The first concrete platforms were being constructed - often by slip form method. Due to the huge size and complexity it was inevitable cracks would occur and also these large gravity structures required complex leak sealing techniques in the construction phase. This demand fueled the development of specific equipment to inject resins under pressure. In U.K. the equipment tended to be high pressure one part whilst in Europe (through American technology) the first twin line equipment appeared. Special porting adapters were manufactured to allow use of these high pressure systems and similar designs are in use today. The reason twin line injection machines were a breakthrough in crack injection was that for crack injection we require two opposite properties in the injection resin material. These are the ability to penetrate very fine crack and yet be reactivity in the resin means the resin has a short pot life and/or working time. The introduction of twin line resin metering and mixing equipment meant that the resin was mixed just prior to placing in the cracks. This then allowed more time for the resin to penetrate into the crack which is essential for the complete filing.

The first twin line equipment used gear displacement to meter the resin. This equipment has the advantages of simplicity, relative low cost and the ability to deliver high volumes of mixed material in a short time frame. Following these developments of twin line which tended to be limited to specialist repair applicators a number of developments occurred using low pressure systems designed for smaller applications. These included spring loaded hypodermic type capsules, balloon type containers of various materials, twin line twin cylinder caulking guns and pneumatically energized reservoirs. These have big advantages in terms of simplicity and low capital cost however the high cost and limitations on volume with some systems can limit the results obtainable - Figure 6 shows Typical Hand Operated System

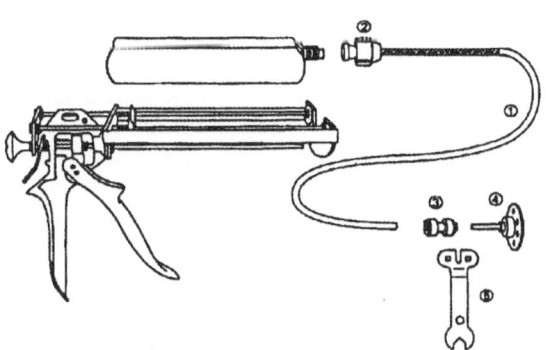

1. Connecting tube of required length.
2. Static disposable mixer
3. Easy connector to join tube to inject on port
4. Injection port.
5. Tool to remove connector

Figure 6 Twin line hand operated

The need to inject small deep cracks - say 0.2 mm - 0.3 mm wide and 600 mm - 1.5 metre deep led to the development of piston displacement equipment. This equipment has excellent ratio accuracy at high pressure but low flow rates. The disadvantages of these systems are cost, relatively complex equipment and of being quite bulky particularly where air driven. To allow this equipment to be cost effective in injecting these large area small volume cracks a system of multi porting or simultaneous injection has been evolved. By taking more than one outlet from the mix head and then leading this to a manifold we can inject many ports at once - The number of injection points can be from four to fifty depending on the crack width, depth etc. Figure 7 shows the Typical System.

Figure 7. Typical example of high pressure twin line equipment

Changes in material technology for crack injection has not been as radical as the change in equipment. Major changes have been in areas of water and moisture tolerances pot life and viscosity.

Many cracks will by definition be damp when injected and it is important that the resin and hardener systems are correctly formulated to displace water and bond to damp concrete. Some formulations claim to adhere even underwater and whilst chemically this is possible practical difficulties may give a reduced bond strength in the crack. To ensure full bonding of the sides of the epoxy injection material has to displace all water in the crack - a difficult task.

I briefly discussed pot life earlier, pot life is the practical working time from when the resin and hardener are fully mixed until this material then starts to gel or thicken. At ambient temperatures of 30 degrees C this can be as short as ten minutes. Advances in hardener technology have produced crack injection resins water thin but having a pot life of one hour at 30 degrees C.

Viscosity is a very important property of a crack injection resin to penetrate very thin cracks. There are limitations on how low a viscosity can be obtained with epoxy resins and therefore we have to use diluents and other means to produce the lowest viscosity with acceptable properties. The range of hardeners and diluents both reactive and non reactive has widened over the years. The selection of correct blend of reactive diluent with specific hardener has to allow the resin technologist to produce lower viscosity materials with specific properties to suit the engineer. Key mechanical properties include :-

Typical Values for an epoxy injection resin
Table 1

Compressive Strength	
Tensile Strength	$100N/mm^2$
Arizona slant Sheer	$55N/mm^2$
H.D.T	$25N/mm^2$
Bond to dry/damp concrete	60°C
Viscosity @25°C	Greater than strength of concrete 350 cps

In the opening part of this paper I outlined two types of resin epoxy an polyester and Table 2 outlines some of their basic properties when used for crack filling
Basic Properties of Epoxy and Polyester Resins
Table 2

Property	Epoxy	Polyester
Strength	Good	Good
Adhesion Dry	Good	Good
Adhesion Wet	Reasonable	Poor
Solvent	No	Yes
Pot Life	Can be varied	Can be easily varied
Toxicity	Low - medium	Medium
Shrinkage	Negligible	Yes

Generally Polyesters are rarely used for crack injection today as we have to allow for shrinkage and a lower adhesive bond. Certain companies however still favour a polyester cap seal. As polyester have a greater rate of gain of strength in the first six hours compared with epoxy they allow injection to proceed in a more timely manner. This increase in speed is at the expense of adhesion so this must be allowed for in the injection process by not having too great an injection pressure that will blow off the cap seal.

One other area for discussion is pressure or more correctly injection pressure. Given that we could have an injection resin with an infinite pot live then we could use any pressure to fully inject a crack say 0.3 mm width regardless of size. The pressure only influences time so if it would take three minutes at 300 psi (20.67bar) to inject a specific crack it might take one hour at 20 psi (1.38bar). so the variable factors in resin crack injection are time versus pressure, with limitation on time due to the pot life of the resin.

Case Histories

Some of the first injection repairs the writer was involved with were from 1968 - 72 in the London and the Home counties. Using a polyester resin technology cracks were grouted using either gravity or peristaltic pumps through stainless steel tubing.

By the mid 70's work was being carried out on some of concrete gravity platforms in both Scotland and Norway.

Case History I

SCB Injection Process Plays Important Role in Andoc Platform

The first production platform for Shell Expro's Dunlin - "A" field is a massive steel-reinforced concrete structure known as Andoc (see figure 8). Anglo-Dutch Offshore Concrete (Andoc) is a consortium comprised of four Dutch and three British construction companies with broad design and construction experience in offshore structures. This North Sea platform supports a payload of modules, drilling equipment and a process plant weighing in excess of 10,500 tons. The platform has 48 working wells and an oil storage capacity of 1.4 million barrels. The structure has a total height of 240 meters and is located where the water depth is 151 meters. Eighty-nine thousand cubic meters of concrete were used in the construction.

During the construction stage of the base cell walls, where the world's largest slip form with an overall length of 2000 m' was jacked up continuously, the 1975 heatwave hit the site with temperatures not experienced of 30 years. The unpredictable combination of these two factors caused shrinkage cracks which appeared in the freshly placed concrete following the jacking of the slip form. Although it was clearly understood that the cracks were not of a structural natures Andoc's owners and engineers wanted to ensure that the cracks would not allow ingress of sea water and as a result might cause corrosion of the reinforcing steel. The thickness of the walls in the 81 cells varied from 31cm to 140cm. The cracks penetrated only to the steel reinforcement. Cracks varied in width from hairline to 1.5cm.

The design engineers determined that the most economical and permanent method of sealing the cracks would be through epoxy injection. Balm B.V., the Adhesive Engineering Company SCB Process in Holland was selected for this work because of their ability to immediately mobilize with injection equipment an trained technicians. Another major reason for the selection of this company was the fact that SCB Process injection system does not require the applicator to drill into the cracks and grout in injection fittings. Balm B.V. selected the appropriate surface seal depending upon the width and condition of the crack and left injection ports spaced at least as far apart as the crack was deep to assure full depth adhesive penetration. Gaskets of PVC hoses were applied against the opening into the crack and CONCRESIVE 1380, an injection resin system which contains resins made by Shell, was pressure injected using two component meter-mix equipment. The system as described is the Structural Concrete

Bonding Process, which is applied in all kinds of structures through licensed applicators in more than 50 countries all over the world. CONCRESIVE 1380 was selected as the injection adhesive for this project because of its low viscosity allowing for full depth penetration of all primary and secondary cracks and its ability to bond wet and dry concrete. Over one hundred 5cm diameter cores were taken from injected cracks in the structure as part of the quality control and inspection procedures on the job. All the cores showed full depth penetration and a bond to the concrete greater than the strength of the concrete itself.

The first injection activities took place in Rotterdam during 1975/76. The structures was then towed from Rotterdam to a fjord near Stavanger, Norway and anchored for further construction and equipment installation. Norwegian SCB licensee Ivar Asting contracted for additional epoxy injection on the anchored Andoc platform which consisted of the filling of voids between steel pipes and concrete as well as voids between steel and steel in the structures, where welding activities were causing these predictable voids. Balm B.V. acted as a subcontractor to Ivar Asting in Stavanger. Over 150 litters of CONCRESIVE 1001 LPL (Long Pot-Life) was pressure injected in voids which were difficult to reach other wise in the concrete walls, towers and the roof structures of the cells. Alternative methods of repair would not have been as thorough or as permanent as the SCB injection process which has been proven in major civil engineering structures throughout the world over 18 years. Other methods would have been disruptive to the flow on construction activity on this project and would have been both more time consuming and expensive.

Figure 8 Concrete Platform under Construction

Case History II

Inverkip, Scotland

This repair was to a chimney at power station. Long vertical cracks were found in the concrete and a repair proposal was made. It was decided to strengthen the chimney by fixing steel plates then grouting the cracks and finally gluing steel plates over the cracks. The adhesive used for the cap seal and gluing the plates was anon pigmented system and thus it was difficult to tell when it was completely mixed. This led to adhesive failures so the base was pigmented white the hardener black so it was very easy to see streaks, a sign of incomplete mixing. The plates were then bonded and the crack was grouted using single line high pressure pumps. After injection the ports were removed and the final steel bonded in place.

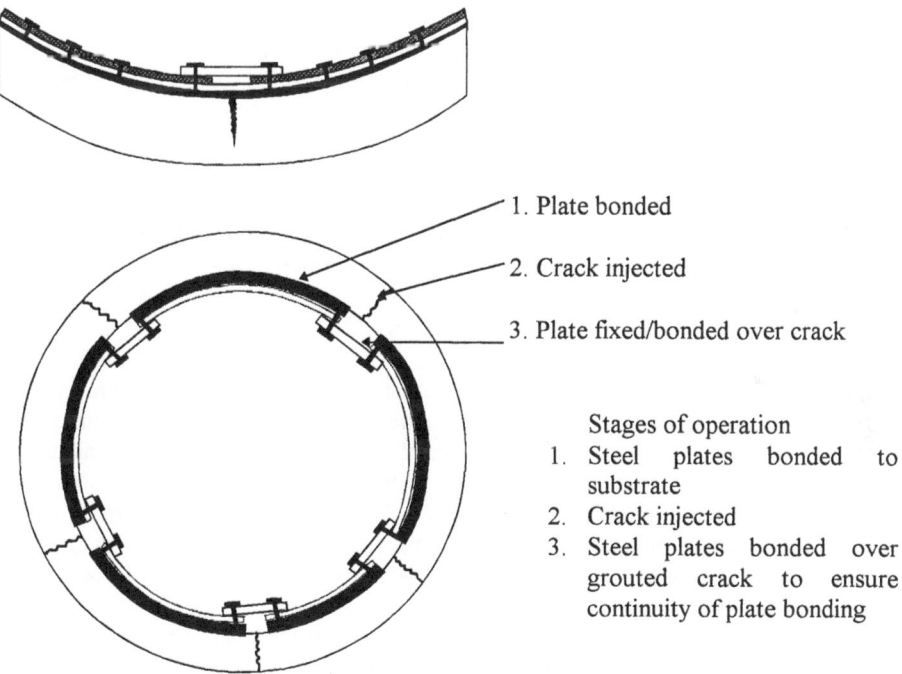

1. Plate bonded

2. Crack injected

3. Plate fixed/bonded over crack

Stages of operation
1. Steel plates bonded to substrate
2. Crack injected
3. Steel plates bonded over grouted crack to ensure continuity of plate bonding

Figure 9   Repairs and strengthening a concrete chimney

Case History III

Malaysian - Jetty

A sprayed mortar repair was carried out to the underside of a jetty in Malaysia. Due to excessive thickness and heat cracks were noted. A repair system was proposed based on crack injection and both the client and the engineer accepted this. As part of the jetty was below the high tide level i.e. it was submerged twice a day. conventional single entry injection would have been prohibitively expensive. To overcome these problems the following method was adopted. All crack were allowed to drain as the tide was filling. The surfaces were then grit blasted and cap sealed with injection ports at suitable centres. The injection ports were then capped and allowed to cure for up to five days. Then on a falling tide up to ten cracks were injected using the multi porting system. Cores were taken on instruction of the Engineer and in all cases the cracks were completely filled.

Case History IV
Sri Lanka - High Rise Building

Just prior to hand over of the twin storey Stock Exchange building in Colombo a near by bomb blast caused cracking in the architectural precast. Cracks were surveyed and measured and a repair method proposed. This involved coating all cracks less than 0.2mm with a triethyl butyl siloxane solution to prevent Chloride ingress and all cracks 0.5 mm or wider were to be injected with epoxy resin. the difficulty in this work was that while most of the cracks were only 20 -40 mm deep and did not go through the precast, some cracks terminated at the panel edge next to the curtain wall. The danger was therefore that the resin may either drain out of the crack or may damage the sealant. Contingency measures were allowed of drilling at the end of a crack and injecting a urethane foam and also having a thixotropic crack injection resin. In practice by carefully controlling injection pressures and when required regrouting. Careful grinding ensured the crack injection pressures were carefully controlled and regrouting was performed when required. After completion the cap seal was carefully ground off to restore the aestethic appearance

Conclusion

Over the last thirty years both materials technology and resin injection equipment have taken the guess work out of crack injection. However it is still a specialized field that requires training and an understanding of procedure. The advent of taking cores as the only means of ensuring the crack is filled has allowed Engineers to specify injection and have guaranteed results. The future will no doubt bring more portable twin line equipment with in built checks on the accuracy of the machine. Resins or more specifically the hardeners will become less toxic and no doubt will have a wider range of gel times to suit specific applications.

The writer would acknowledge support and guidance over the years from many companies and individuals. These include Taywood Engineering, Messrs Colebrand, Dr. R.D. Browne, L.H. McCurrich, John F. Trout Lily Corp, Robert W. Gaul and MBT.

HYDROPHILIC POLYURETHANE GROUTS AND THEIR APPLICATION ON ROLLER-COMPACTED CONCRETE DAMS

Y.H. Bao and D.N. Xu
East China Hydroelectric Power Investigation and Design Institute,
Hangzhou, China

Abstract
High strength and low strength hydrophilic polyurethane chemical grouting materials
(HW & LW) can be easily dispersed, emulsified and gelated when meeting water. The
solidified mass of HW & LW possess different strength and water-swell properties.
They have been used on stop seepage, strengthening crack and consolidating
foundation of concrete structures, particularly on the treatment of deformation crack,
contraction joint etc. This paper presents the gel time, swell data and strength when
mixing HW & LW in different proportion and the application on roller compacted
concrete (RCC) dam.
Keywords: anti-seepage, chemical grout, hydrophilic polyurethanes, RCC dam

1 Introduction

Roller compacted concrete technology is the significant innovation in hydraulic
structures, with the advantage of quick construction, and lower costs, which develops
very fast domestically and internationally. The seepage of RCC dam is the critical
problem which will increase uplift, decrease safety, consume large amount of water and
cause erosion to the internal concrete structure. Therefore, the design and construction
organization always make great efforts to take suitable measures to reduce or avoid
large seepage. Generally, the methods to solve the seepage problem are accepted by
using "Internal method" and "External method". "Internal method" indicates to take
the measures to control the separation between placement layers, such as adopting high
dosage of cementitious materials; adjust the exposed concrete time and place time; and
control the temperature. "External method" implies depending on the upstream
facilities to prevent seepage, such as adopting conventional concrete; set up the
independent anti-seepage on upstream face, like asphalt mortar, precast concrete facing;
and combine with the concrete facing and plastics. Although careful consideration is
made for the seepage during the design and construction of RCC dam, seepage will
occur with the large possibility due to the inferior combination around dam body and
foundation, crack of anti-seepage facing itself, and the crack in dambody caused by the

Polymers in Concrete, edited by Y. Ohama, M. Kawakami and K. Fukuzawa. Published in 1997
by E & FN Spon, 2–6 Boundary Row, London SE1 8HN, UK. ISBN: 0 419 22330 4.

rapid weather variation during construction or other structural problems when the dam is being built without longitudinal and transversal joints on the whole area, and the quality problems induced by the mechanical equipment, site condition, quality and responsibility of the construction operators. Therefore, a series of anti-seepage and reinforcement measures are needed to be taken for the RCC dam.

2   Selection of seepage prevention and reinforcement materials for roller compacted concrete dam (RCC dam)

The material firstly selected for seepage prevention and reinforcement of RCC dam naturally is the cement grouts, it has the lower price, non-toxic, wide resources and comparative familiar construction method. But some of its obvious disadvantages affect its application.   For example, on large volume leakage, a great quantity of the cement will be lost due to its long setting time; the new leaking passage will be produced due to the cemented block shrinkage caused by bleeding; the cemented block is sound solid without the capability to suit the deformation. In treating caulked joint, the mortar will cause cleavage and peel, usually because of the unfavorable adhesion and large shrinkage. Until now, the engineers have developed a series of materials for preventing and stopping seepage, reinforcing and consolidating. Among them, the polymer material are used widely in these application fields.

There are many kinds of chemical grouts, such as epoxy resin, polyurethanes, acrylamide, methylmethacrylate, acrylate, chrome-lignin, urea-aldehyde, etc., but only the polyurethane chemical material can be reacted rapidly with the water and gel. Let us introduce here of the polyurethane chemical grouts, combined with the seepage treatment for the RCC dam.

By the end of 1960's, the Tacss grouts was adopted by Japan initially in tunnel and sewer system for strengthening and anti-seepage. This is a kind of oil-soluble polyurethane chemical grouts. Later also in Japan, the Hycel-OH water soluble polyurethane material was applied widely in anti-seepage on subway, tunnel, dam, underground project and slope protection. In West German, the polyurethane material was used to consolidate the fissure of coal rock mass to protect against the falling down of broken coal roof, and the inflow of gas and water. They promoted the special mechanical equipment, formed the standard operating method and made the model products. In the middle of 1970's, De Neef Chemie Company promoted this kind of products to prevent seepage and consolidated soil in Europe and achieved great success. In 1980's, polyurethane grouts in North America was considered very useful, especially Tacss/Flex grouts has been applied in restoration of the dam. The materials with different property shall be selected depending on the actual condition. For example, Tacss 020NF can produce a tough foam, used in seepage treatment on non-active joint and under water structure; Flex 44 can produce the elastic foam, with 22% of elongation rate, used in anti-seepage for deformation and contraction joint; Tacss 025NF is a kind of low viscosity polyurethane grouts, mainly used for the curtain grouting and soil consolidation. These materials were used for treating on the seepage of vertical and horizontal joints by pipe-embedded or inclined hole drilling method. They also can be used for curtain grouting through the filter pipe or dual-pipe.

3 HW & LW hydrophilic polyurethane chemical grouts

The research on polyurethane grouts was started at the beginning of 1970's in China. Great achievements have been made in many organizations of hydroelectric, coal, tunnel and universities, and excellent results are obtained in these kinds of water-proof and consolidation. Among them, HW & LW have some special performances:

- HW and LW have fine hydrophilicity, they can be dispersed and emulsified uniformly and then be gelated. So water is not only their diluent but also their curing agent.
- HW and LW possess high strength and high elasticity respectively. They can be mixed in any proportion and get different properties to meet the construction requirement.
- HW and LW have adaptability to weather, they can be solidified in sea water, acid water and alkaline water with PH value from 3 to 11.
- The soaked liquid from solidified mass of HW and LW, show nonpoisonous type by animal test.
- LW is water-swollen elastomer. If the cracks open again after grouting, the solidified LW will restore the cracks by its swelling with water.
- The grout operation is very simple which can be adopted by one-component system without complex preparing. The handle pump is very convenient to carry and move in tunnel, gallery, etc.

The main properties are listed in the following Table 1:

Table 1. The main properties of HW & LW

Item	HW	LW
Viscosity (pa.s) 25°C	0.01	0.45
Specific gravity	~ 1.10	~ 1.08
Gel time	Adjustable	Adjustable
Adhesive strength (to wet face)	2.4Mpa	1.0Mpa
Tensile strength	7.8Mpa	2.1Mpa
Elongation	—	273%
Impermeability	S_{15}^{*}	1.8×10^{-9}cm/s
AIW (amount of involving water)	0.9 time	20-27 time
Rate of swelling volume	2-4%	150-300%

S_{15}^{*}: Impermeability Index; which means the solidified HW can resist water pressure of 15kg/cm^2.

3.1 The influence of catalyst and water on gel time
In order to make clear the gel time of different component of HW and LW, the influence of different catalyst and water content on gel time was tested as follows:

The determination of gel time was carried out by ball-falling method. When the grouts mixed, the time starts to reckon until the steel ball stops moving, this is the gel time. For those grouts gelated rapidly, the gel time starts from mixing to collapsing of the gelled grouts.

3.1.1   The influence of catalyst on gel time

The influence of content of catalyst (TEA) on gel time was indicated in the following Table 2 and Fig.1.

Table 2. The influence of content of catalyst (TEA) on gel time

Grouts	Water	TEA (%)					
		0	0.1	0.5	1	2	3
	3%	Gel time (minute´ or second˝)					
HW		13´	6´	2´	1´	48˝	36˝
LW		20´31˝	14´	4´	2´	1´7˝	42˝
HW:LW=80:20		16´13˝	6´6˝	2´19˝	1´4˝	33˝	
HW:LW=20:80		17´	16´	2´10˝	1´23˝		

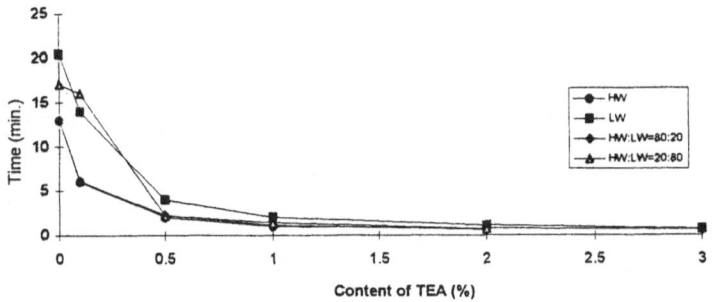

Fig. 1. The influence of TEA on gel time with different proportion of HW/LW

It can be known from the gel curve:

(1) If mixing 3% of water with grouts, and without catalyst, the gel time is 13 to 20 minutes according to HW/LW proportion, this shows the process of HW and LW construction is very simple without complex ingredient. It just needs to inject grouts to the leak position and gets gelated. As a matter of fact, HW and LW can be gelated as long as the concrete crack containing moisture and slight alkaline.

(2) HW gels quicker than LW when meeting water because of its more free NCO radicals compared with LW. Furthermore, HW:LW=80:20 or HW:LW=20:80 does have some HW composition, their gel time is shorter than the pure LW.

(3) The content of catalyst apparently affects gel time. Great acceleration can be achieved with more than 0.5% catalyst, the catalyst is controlled lower than 3% for sealing the large leak. Ordinarily, 2% is accepted. It is better to be as less as possible to use the catalyst.

3.1.2   The influence of water on gel time

Now as we wish to know the influence of water on gel time, we can put the catalyst in fixed dosage, e.g. 0.5%, then mix the water in 3%, 6%, 50%, 100%, 500%, 1000% and 2000% with the different component of grouts respectively, and then measure each of their gel time.

When analyzing the reaction of HW and LW with water, two points shall be noted. One is the reaction between NCO and H_2O, another is that hydrophilic polyurethane is dispersed, emulsified and gelled when meeting with water. The latter one is usually believed as hydrophilic effect, and indicated as "AIW"(amount of involving-water) which means the maximum water content when grouts meet water and gelate without bleeding. For instance, LW consists of enormous hydrophilic links, no bleeding happens when it is mixed with 20-25 times water, and at last results in the gel containing with water. But the water volume contained in HW is comparatively less, only within 1 time. The influence of water content on gel time is shown in the following Table 3 and Fig. 2.

Table 3. The influence of water content on gel time

Grouts	TEA (%) 0.5	Water (%) 3	6	50	100	500	1000	2000
		Gel time (minute or second")						
HW	2'	2'30"	2'26"	5'44"				
LW	4'	1'50"	39"	33"	38"		1'20"	8'7"
HW:LW=80:20	1'53"	1'46"	2'7"	2'13"	2'27"			
HW:LW=50:50	1'54"	1'25"	42"	1'23"	2'44"		40"	
HW:LW=20:80	1'41"	1'58"	59"	45"	1'14"	3'29"		

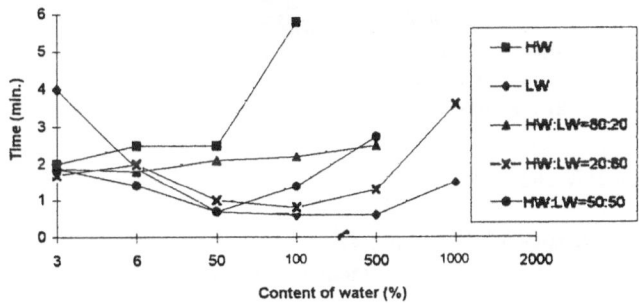

Fig.2. The influence of water on gel time with different HW/LW proportion

It can be seen from the gel curve:

(1) HW curve shows that the gel time usually is 2 minutes when water is between 3% to 50%, and delay to over 5 minutes if 100% water added in.
(2) LW curve indicates that the gel time usually is 1 minute with 6% to 1000% water (10 times). This phenomena sufficiently explain LW can be gelled in large range of water and emulsified, it is helpful in water stopping.
(3) When HW:LW=80:20, the gel time is mainly about 2 minutes with 3% to 500% water. This can illustrate the hydrophilic effect of LW promoting the hydrophilic capacity for HW.

3.2   Measurement of AIW on different proportion of HW/LW

Let us mix different proportion of HW/LW with different times of water respectively, then measure the maximum amount of involving water without bleeding phenomenon. The results are shown in the following Table 4.

Table 4.   The amount of involving water

	Water (%)										
	3	5	90	100	300	500	1000	2000	2200	2500	2700
HW	√	√	√								
HW:LW=80:20	√	√	√	√	√						
HW:LW=20:80	√	√	√	√	√	√	√	√	√		
LW	√	√	√	√	√	√	√	√	√	√	√

"√" represents it can involve water.

Table 4 shows that the LW has largest AIW because it contains many hydrophilic links. Therefore, the more LW, the more AIW.

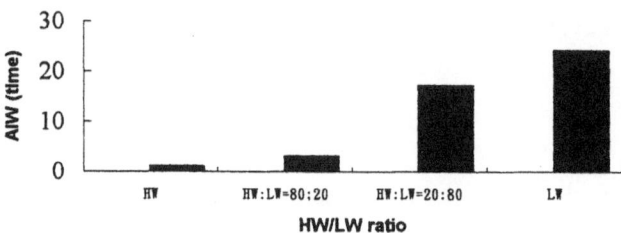

Fig.3   AIW of HW and LW

3.3   Measurement on the viscosity of HW and LW

The NDJ-1 revolving viscometer was used to measure the viscosity of different proportion of HW/LW.

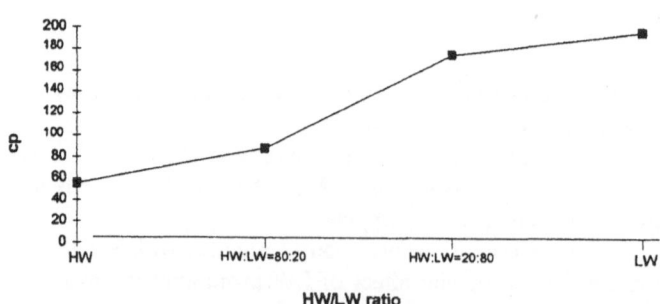

Fig. 4.   Viscosity of HW and LW

Fig.4 shows the viscosity of HW is greatly lower than that of LW. If certain content of LW is added into HW, the viscosity will raise to 55-195cp. Therefore, if the crack opening is not very small, mixing a proper proportion of LW with HW can increase the swell capability of gel. Of course, when grouting the contraction joint, LW is usually accepted for the reason of elasticity. A little bit of HW will be added to decrease the viscosity.

3.4   Measurement of water swell rate of HW-LW gel

After HW-LW is gelled, it also will swell when meeting with water. Therefore, if the crack grouted by LW-HW leaks again, LW-HW will apply the leaked water to swell its volume, and avoid seepage. It is so called as "stop water by water itself". In order to obtain the swell velocity of HW-LW in the presence of water, the measurement of water swell rate is conducted, the result is shown in Fig.5:

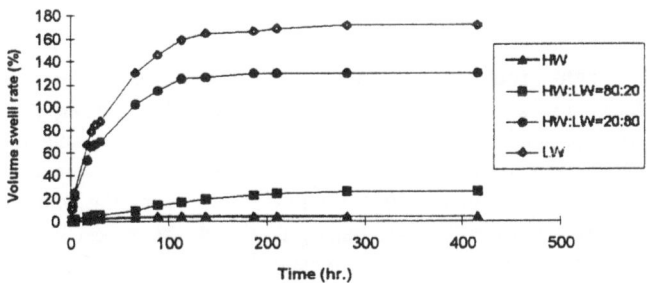

Fig.5.   Measurement of HW-LW water swell rate

The test shows LW swells quickest when meeting water, it expands 11.37% in 1 hour, 84.29% in 25 hours, the maximum can reach to 171%. HW expands slowest when meeting water, 0% in 1 hour, 2.04% in 25 hours, the maximum can reach to 4.04%.

4   Anti-seepage treatment for RCC dams-case study

Leakage is usually encountered on RCC dam. Willow Creek Dam of USA, one of the earliest constructed RCC dams, endured a significant leakage after completion because no effective measure for seepage control was taken. The total water leakage in the gallery was 6800 l/min. It was calculated that 5700 l/min. came from horizontal construction joints (compacted layers) and remaining 1100 l/min. came from the foundation. In China, various organizations in respect of engineering, construction, and operation have given emphasis to seepage problem. The method of chemical grouting using polyurethane is briefed as follows.

4.1   SD RCC dam

SD Dam is a RCC dam of continuous layered placement without any longitudinal and transverse joint. Mortared precast concrete section structure with acrylic latex mortar deep pointing was applied for anti-seepage control. Grooves were also provided with 2m intervals between the inner side of the precast sections and the RCC, and filled by woven plastic drainage belt with bagged fine stone. RCC placement for the dam section in the riverbed began from El. 90 m on November 23, 1992. The reservoir was impounded on November 19, 1993. When the water level rose to El. 121 m on November 24, 1993, a great amount of leaking water was found in the grouting gallery, which resulted in 3 drilling machines being inundated because of no time for evacuation. 2 sets of pumps with capacity of 30 m³/hr. each was required to pump water day and night. The causes of leakage were as follows:

(1) Many joint pointing overlaps between the precast sections form some leaking passages.
(2) The defects in the periphery joint between the foundation concrete and rock, and abundant honeycombs around the lateral foundation of the gallery due to inadequate vibration, in particular, were the key cause of water leakage.
(3) Three transverse cracks occur to the dam body.

For leakage treatment, cement grouting and cement-water glass grouting were ever applied. However, the leaky discharge was so big that cement grout was easily flushed away. Therefore, such grouting had only limited effect on locations where leakage was small, while the total leakage was not reduced obviously. In early 1994, it was decided to apply LW hydrophilic polyurethane grouts for leakage stopping. Then the water level was at El. 129 m. It was observed from the right bank gallery that the gallery below El. 107 m was still inundated and enormous spraying existed between El. 107 m and El. 119 m. A dozen of spraying spots were observed in both sides and the top of the gallery. It was obvious that the water sprayed from the lateral compacted layer and the top joints. Measurements were conducted on three points, the leaky quantity was 80, 72, and 36 l/min., respectively. Water sprayed from the one side of the gallery to the other side or poured down from the top of the gallery, which made it impossible to access or work in the gallery.

During the water-stop operation, initially, several pipes with diameter of 2 inch was embedded in location of large leakage and firmly fixed with cloth strip and wood bars. Due to the enormous leakage in the joint, normal method for joint sealing and treatment was not applicable, even though quick-setting cement was flushed away immediately. Therefore, grouting was operated under leaking condition. Automatic sealing and closure of the joint was achieved through quick consolidation of LW grout upon contacting water. For the purpose of speeding up consolidation and avoiding loss of grout diluted by excess water, 0.5 ~ 1% catalyst was added in LW.

Grouting operation was executed by handle pump of type HY-1 due to its simple application for steep slope gallery and easy movement. Since handle pump produces relatively small grout, for such large spraying water, four HY-1 pumps in parallel were applied to ensure adequate grout entering into the joint. As the grout penetrates in the joint and mixes with water, viscosity increases gradually and gel proceeds and leakage is finally stopped.

Pilot grouting of LW was conducted in the gallery between El. 107m and El. 119m, the test demonstrated that LW water-stop was quick and distinct. For example, through grouting operation for two half days, 400 kg grout was consumed, the dozen of spraying spots in the right bank gallery was basically blocked. During the process of LW grouting, the leaky quantities in the top plate and sidewall were gradually reduced until completely stopped once the grout penetrated into the cracks. Even the spraying spots 5 m away from the grouting pipe were automatically closed and sealed. This method has the effect on large area by one grouting hole. The estimation from the water quantities pumped indicated that a reduction of 40 m^3/hr. of leakage was achieved in two days from 110 m^3/hr. to 70 m^3/hr.

Because of the satisfactory testing result, the construction company applied the method in treating the left bank gallery from El. 107 m to El. 119 m and then El. 93 m to El. 107 m. After completing the grouting, holes of 1 ~ 1.5 m deep were drilled using pneumatic drills and then grouted with supplementary cement grout. In the meantime, drainage system was improved to obtain an integral combination of drainage and blockage.

4.2 SZ RCC Dam

The seepage control of the SZ RCC dam body depends mainly on upstream two-gradation aggregate RCC. In order to raise the density and anti-seepage capacity of the surface concrete, an amount of cement-flyash was added in the two-gradation aggregate RCC with 30cm away from formwork and compacted by vibrators. The surface concrete below dead water level is sprayed and coated with acrylic latex mortar. During construction, a crack of 2 ~ 4 mm width occurs to the dam at Station 0+117 (right) and Station 0+21.7 (left), respectively. The crack extends down to the baserock and up to the dam crest. It was treated before impoundment by gouging the crack on the upstream, bonding with BW water-seal, leveling the cracks with acrylic latex mortar, and finally strengthening with steel plates in the surface. However, since the crack then had not reached to the baserock datum while the reservoir water level rises continuously, a part of the crack above the baserock datum was not treated. After impoundment, the leaking quantity is measured at 380 l/min. when upstream water level is at El. 76 m. The leakage concentrates in two sections, from Station 0+12.7 ~ 0+27.7 (left) and from Station 0+110 ~ 0+140 (left). For leakage control, holes around the cracks were drilled for cement grouting. Grouting operation was executed in the contacting area between the right and left sidepiers with the RCC, and between the right and left conventional concrete and the RCC. After the treatment, leakage was reduced to 248 l/min.. The dynamic water flow has caused a heavy loss of cement grout and no pressure was possible to exert, therefore chemical grouting using LW grouts was employed to stop the enormous leakage. The chemical grout began in May 1995. In addition to the application of quick-setting method, the most recently developed product--PBM polymer mortar was applied for pipe embedding. The PBM polymer mortar has the characteristic of quick-setting in water and furthermore has certain expandability during quick-setting, it played a very positive role in pipe embedding and blockage under severe leakage condition. LW grouting to stop water exhibited evident effect.

During the first stage of test grouting (for three days), 400 kg grout was consumed, concentrated leakage in the gallery was reduced by 41.6% from 240 l/min. to 140 l/min.. The leakage in the whole gallery was further treated with LW grouting. The strengthening treatment by cement grouting along the cracks in the dam crest was also implemented. Leakage is significantly reduced by integral action combining the chemical and cement grouting.

5 Conclusions

(1) HW and LW can be mixed at any ratio to form a chemical grouting material that has the characteristic of solidification in water.
(2) When the water content is 3% and triethylamine is used over 0.5%, it takes less than 2 min. to gelate; When the water content is 3% and without catalyst, it takes 13 ~ 20 min. to solidify.
(3) When the grout is added with catalyst of 0.5%, and without contacting water, HW solidifies within about one day and LW about three days respectively. There are rarely foam in solidified grout.
(4) The grout will solidify within 10 ~ 12 hr. after contacting little amount water, such as 1%, without catalyst.
(5) The amount of involving water increases along with the add of LW, if HW is 0.9 times; HW:LW=80:20, 3 times; HW:LW=20:80, 22 times; LW is 27 times.
(6) The viscosity of HW is lower while that of LW is higher, mutual adjustment may be obtained. The viscosity ranges from 55 CP to 195 CP.
(7) HW with catalyst solidifies faster than LW does. So solidification process may be quickened by add of certain proportion of HW during stop water gush.
(8) The leakage of RCC dams can be quickly stopped by application of chemical grouting using HW & LW hydrophilic polyurethane grouts..

6 References

1. Yinhong Bao, "The Characteristics and Applications of Hydrophilic Polyurethane Chemical Grouting Material." *Water Resource and Hydropower Technology*, No. 10, p10-20, 1985.
2. Xiankang Zeng, "The Application and Development of Integral RCC Dam in Fujian Province of P.R.China." *Water Power of Fujian Province*, No. 1, 1992
3. Rudi Olinga, "Stop Leaks, Seal Cracks or Restore Strength with Polyurethane Grouts", *Concrete Repair Digest*, December/January 1992
4. James T. Joyce, "Polyurethane Grouts", *Concrete Construction*, July 1992
5. W. Glenn Smoak, "Polyurethane Injection Stops Water Tunnel Leaking", *Concrete Construction*, December 1988

PART FOUR
RECYCLING

EFFECTS OF *r*-PET AND PG ON THE MECHANICAL PROPERTIES OF POLYMER CONCRETE MADE FROM RECYCLED PET-BASED UNSATURATED POLYESTER RESIN

Y.S. Soh
Department of Architectural Engineering, Chonbuk National University, Chonju, Korea
Y.K. Jo and H.S. Park
Building Inspection Division, Korea Infrastructure Safety and Technology Corporation, Anyang, Korea
D.S. Lee
Dept. of Chemical Technology, Chonbuk National University, Chonju, Korea

Abstract
The purpose of this study is to evaluate the effects of r-PET(polyethylene terephalate) and PG(propylene glycol) on the mechanical properties of polymer concretes made from recycled PET-based unsaturated polyester resin(UP). The polymer concretes are prepared with various PET, PG and stylene monomer(SM) contents. It is found out that the degree of unsaturation of the resin was the most important property affecting the thermal and mechanical properties. The cured unsaturated polyester resin or polymer concretes made with resins of higher degree of unsaturation showed higher glass transition temperatures, and strengths, compared with those made with resins of lower degree of unsaturation. Except for some concretes, the effect of PET and PG contents on the compressive strength of polymer concretes is recognized, and the strengths of polymer concretes made from recycled PET-based are almost the same as those made from UP used commercially.
keywords: Polymer concrete, binder, polyethylene terephalate, unsaturated polyester resin

1 Introduction

PET is useful polymer used for fiber, film, and plastic containers such as carbonated beverage bottles. Recently, the recycling of polymers such as PET after use is attracting the attention of many researchers who are aware of environmental problems and attempts to find the ways to save earth resources. Previous studies showed that UP can be economically prepared from recycled PET and the resins may be useful for resin concrete(1-

Polymers in Concrete, edited by Y. Ohama, M. Kawakami and K. Fukuzawa. Published in 1997 by E & FN Spon, 2–6 Boundary Row, London SE1 8HN, UK. ISBN: 0 419 22330 4.

3). However, there is little information on the molecular features of the UP. Resins using recycled PET offer the possibility of a lower source cost of materials for making useful polymer concrete products. Also the recycling of PET in polymer concrete would help to solve some of the solid waste problems caused by plastics and to save energy. Excellent strengths were obtained with polymer concretes using resins based on recycled PET. Resins using a maximum amount of recycled PET are desirable because they do not adversely affect the materials while they help decrease the cost of polymer concrete products, thus making them more competitive. This experimental study attempts to synthesize various UP from PET. This experiment systematically varies PET and PG contents using for glycolyses of PET, chain flexibility and degree of unsaturation of the resins and finds out the effects of those variables on the mechanical properties of polymer concretes made from recycled PET.

2 Syntheses of unsaturated polyester resin

Fig.1 shows flow chart of products of unsaturated polyester based on recycled PET. Carbonated beverage bottles made of PET, high density polyethylene(HDPE)-based cup, and labels were collected, washed, and crushed into small fragments by using a crusher. The fragments of crushed PET bottles include PET, HDPE, and various labels from which PET fragments can be easily separated by density difference in water. The collected PET

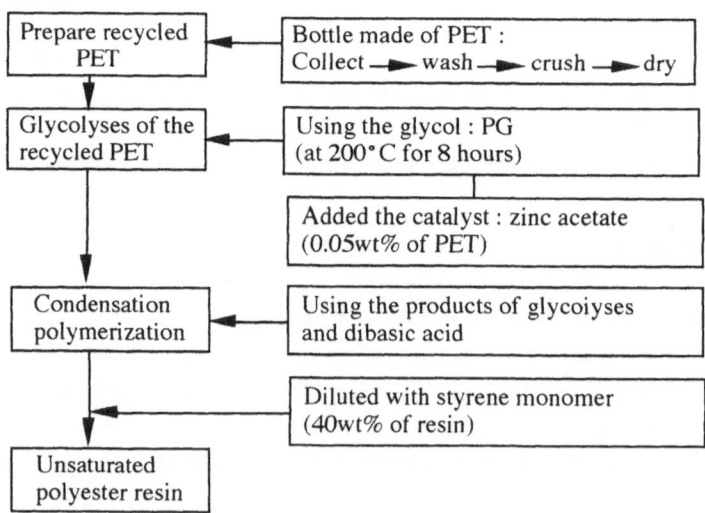

Fig.1 Flow chart of products of unsaturated polyester base on recycled PET.

fragments were dried in vacuum oven. Glycolyses of PET were carried out using propylene glycol(PG) at 200°C for 8 hours. For the glycolyses of PET, zinc acetate(0.05% by weight of the PET) was added as a catalyst. Unsaturated polyester resins were prepared by condensation polymerization at 200°C using the products of glycolyses and dibasic acids such as maleic anhydride and adipic acid. The resins were then diluted with styrene to make 44%(by weight) styrene solution after the polymerizations and hydroquinone(0.5% by weight of the resin) was added as an inhibitor. Variables in the syntheses of the UP were the type of glycol, the PET content, the molecular weight of the resin, and the degree of unsaturation of the UP. In Table 1, recipes of UP from recycled PET are given. Hydroxyl values of the resin were measured to check the number average molecular weight of the resin. Table 2 indicates properties of UP from recycled PET.

Table 1   Recipes(by molar ratios) for preparation of unsaturated polyester resin from recycled PET

Sample code	Recycled PET	Propylene glycol	Maleic anhydride
A1	1.2		
A2	1.0	1.4	
A3	0.8		
B1	1.2		
B2	1.0	1.2	1.0
B3	0.8		
C1	1.2		
C2	1.0	1.1	
C3	0.8		

Table 2   Properties of unsaturated polyester resin from recycled PET

Sample code	Glass transition temperature (°C)	Total heat of cure (cal / g)	Hydroxyl value (mgKOH / g)	Number average molecular weight	Degree of unsaturation	Viscosity (25°C, P)
A1	84.0	65.5	97.05	1154.10	29.4	8.6
A2	80.0	59.4	110.10	1017.26	31.3	5.8
A3	75.0	58.4	122.51	914.20	33.3	4.2
B1	107.0	71.9	76.08	1472.13	29.4	8.0
B2	103.5	66.4	75.58	1481.87	31.3	13.3
B3	91.5	55.3	75.43	1484.82	33.3	8.6
C1	110.0	73.2	55.09	2033.04	29.4	12.1
C2	100.0	64.2	50.05	2237.76	31.3	15.4
C3	90.0	64.0	72.07	1553.90	33.3	4.9

3 Materials for polymer concrete

3.1 Binder system

The binder system used for polymer concrete were based on unsaturated polyester resin from recycled PET together with styrene monomer(SM) as a diluent, methyl ethyl keton peroxide(MEKP) as an initiator and cobalt acetate(Co) as a promoter.

3.2 Filler and Aggregates

Commercially available ground calcium carbonate(Size ; 2.5 μm or finer) was used as a filler, and crused granite(Size ; 1.2-1.5mm), river sand(Size ; 0.3-1.2mm) and silica powder(Size ; 0.3mm or finer), as aggregates. Table 3 shows the physical properties of aggregates.

Table 3 Physical properties of aggregate

Type of Aggregate	Size(mm)	Specific Gravity	Content (wt.%)
Crused granite	2.5 - 5.0	2.43	21.7
River Sand	0.6 - 1.2	2.64	62.0
Silica Power	0.3	2.69	16.3
Unit Weight (kg / l)			1.67
Specific Gravity			2.60
Solid Volume Ratio (%)			64.3
Fineness Modulus			2.7

4 Test procedures

4.1 Preparation of specimen

According to KS F 2419(Method of making polyester concrete specimens), polymer concretes made from recycled PET-based UP were prepared with the mix proportions given in Table 4. Specimens 40x40x160mm were molded, then cured 20°C and 50% R.H. for 7 days.

Table 4 Mix proportions of PC using recycled PET-based UP

Resin (wt.%)	Filler (wt.%)	Aggregatre content (wt.%)
15.0	15.0	70

4.2 Strength test

The test for strengths of polymer concretes were carried out with KS F 2482(Method of test for flexural strength of polyester resin concrete) and KS F 2483(Method of test for compressive strength of polyester resin concrete using portions of beams broken in flexure).

5 Test results and discussion

Fig.2 and Fig.3 represent the effect of PET content on compressive and flexural strengths of polymer concretes made from recycled PET-based UP with PG contenst of 1.1, 1.2 and 1.4 mole%. In general, Tg's of the resins decrease with increase in PET content. Thus, the decrease of the Tg is attributable to the decreased crosslink density, as, PET content is increased in the resin. It is also observed that the higher are the molecular weight of the resin, the higher, Tg's of the cured resin. The compressive strength of polymer concretes with PG contents of 1.2 and 1.4 mole% reaches a minimum with PET content of 1.0 mole%, but that of polymer concretes with PG 1.1 mole% increases with increasing PET content except for SM content of 44%. The flexural strength of polymer concretes is not improved with PET content of 0.8 to 1.0 mole%, And on the other hand, that of polymer concretes reduces with PET contents of 1.2 mole%. The decrease of flexural strength with PET content may be attributable to lower unsaturation and lower crosslink density.

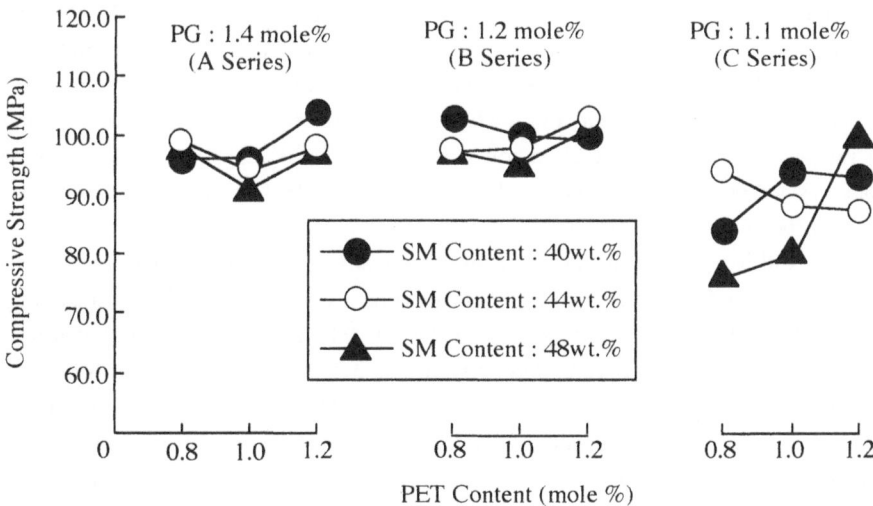

Fig.2 Effect of PET content on compressive strength of polymer concretes made from recycled PET-based unsaturated polyester resin.

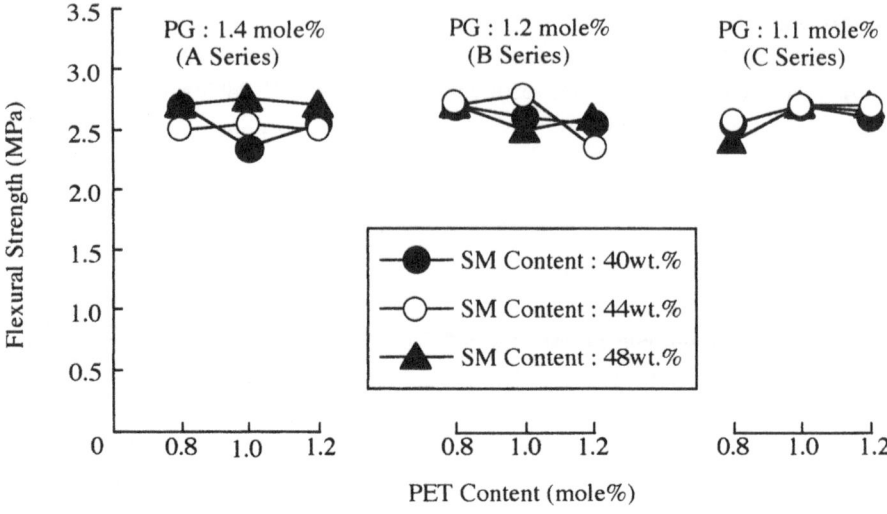

Fig.3 Effect of PET content on compressive strength of polymer concretes made from recycled PET-based unsaturated polyester resin.

Fig.4 and Fig.5 exhibit the effect of PG content on compressive and flexural strengths of polymer concretes made from recycled PET-based UP with PET contents of 0.8, 1.0 and 1.2 mole%, SM contents of 40, 44 and 48 wt.%. PG is usually used in order to regulate the molecular weight of the resin because it is mixed with styrene monomer easily. In this study, the molecular weights of the resin become about 1000, 1500 and 2000 according as PG contents mixed are 1.4, 1.2 and 1.1 mole% respectively. That is, the molecular weight increase with decreasing PG content. The compressive strength of polymer concretes made from recycled PET-based UP with PG content of 1.2 mole% is the highest in the three PG contents. However, the difference of the flexural strength of polymer concretes according to the increasing of the PG content is not recognized. That is, the polymer concretes made from recycled PET-based UP at the molecular weight of 1500 have the highest compressive strength, but the flexural strength of polymer concretes is not affected by the molecular weight. And it is observed that the compressive strength of the polymer concretes are affected by the SM contents so much, but this tendency can not be seen in flexural strength. That is, the polymer concretes made from recycled PET-based UP at the molecular weight of 1500 have the highest compressive strength, but the flexural strength of polymer concretes is not affected by the molecular weight. The compressive strength tends to increase with increase in PG content, and attains the maximum with PG content of 1.2 mole%. By contrast, except for PET content of 0.8 mole%, the flexural strength decreases with increaseing in PG content. It is supposed that the compressive strength rather than the flexural strength of polymer concretes made from recycled PET-based UP is affected by the PG content. Except some concretes, the strengths of polymer concretes at the same PG content tend to decrease with raising SM content. The SM plays an important role in the formation of the third dimentional network structure as well as a diluent. Therefore, workability is improved because the viscosity of the resin is made to reduce by SM. However, it is cosidered that the strengths of polymer concretes reduce due to the presence of a considerable amount of SM monomer which can not be hardened in the polymer concretes.

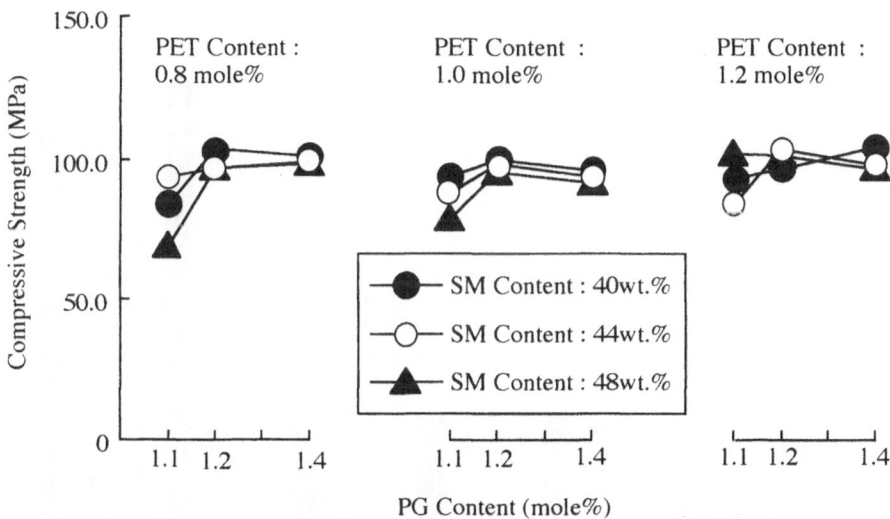

Fig.4 Effect of PG content on compressive strength of polymer concretes made fron recycled PET-based unsaturated polyester resin.

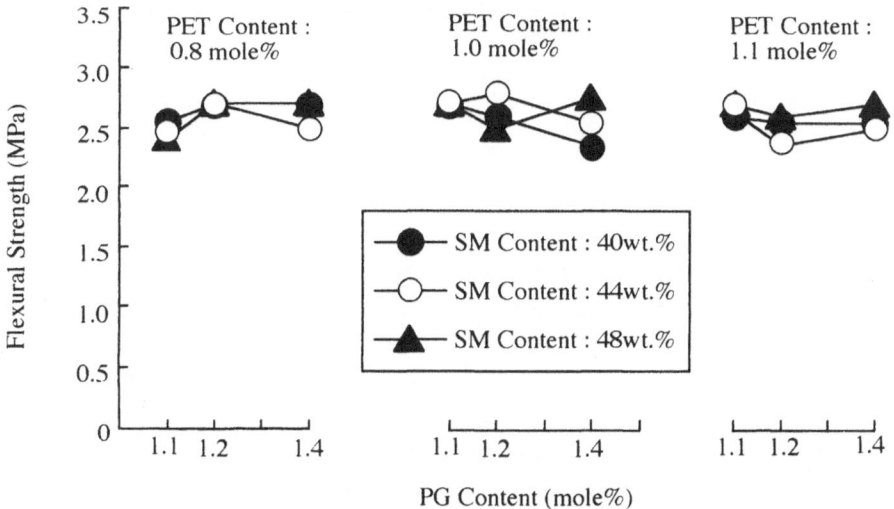

Fig.5 Effect of PG content on flexural strength of polymer concretes made from recycled PET-based unsaturated polyester resin.

Fig.6 indicates the comparision of strengths of polymer concretes made from recycled PET-based UP and ordinary polymer concretes using UP.   The flexural strength of polymer concretes made from recycled PET-based UP is about the same as that of ordinary polymer concretes. On the other hand, the compressive strength of polymer concretes made from recycled PET-based UP is somewhat smaller than that of ordinary polymer concretes using UP.

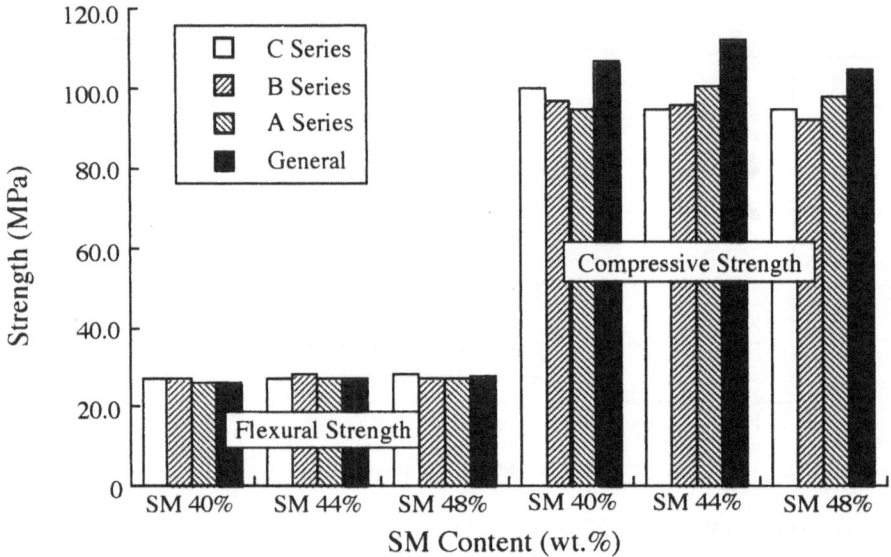

Fig.6 Comparision of strengths of polymer concretes made from recycled PET-based unsaturated polyester resin and unsaturated polyester resin used commercially.

6 Conclusions

The conclusions obtained from the test results can be summarized as follow :

(1) The compressive strength rather than flexural strength of polymer concrets made from recycled PET-based UP is affected by PET content.

(2) The compressive strength of polymer concretes made from recycled PET-based UP is the highest with PG content of 1.2 mole%.

(3) The strengths of polymer concretes made from recycled PET-based UP are about the same as those of polymer concretes made from UP used commercially.

(4) In the production of polymer concretes made from PET-based UP, the proper mix

proportions should be decided according to the use of the resin. And resins using a maximum amount of recycled PET are desirable because they do not adversely affect the materials while they help decrease the cost of polymer concrete products.

7 References

1. Pearson, W., Emerging Technologies in Plastics Recycling, Andrews, G.D. and Subramanian, P.M. (1992), Edt., Chapter 1, ACS Symposium Series 513, ACS, Washington D.C.
2. Rebeiz, K.S., Iyer, V.S., Fowler, D.W. and Paul, D.R. (1990), proceedings of 48th Annual Technical Conference(ANTEC'90).
3. Rebeiz, K.S., Fowler, D.W. and Paul, D.R., (1992), Emerging Technologies in Plastics Recycling, Andrews, G.D. and Subramanian, P.M.(1992), Edt., Chapter 1, ACS symposium Series 513, ACS, Washington D.C.

TENSILE CHARACTERISTICS OF POLYMER-MODIFIED ASPHALT CONCRETES

K.W. Kim, K.S. Yeon, Y.K. Choi and H.W. Joh
Dept. of Agricultural Engineering, Kangwon National University,
Chunchon, Korea

Abstract
This study was carried out to evaluate the effect of using polymer for modifying asphalt binder. Four polymer materials (SBS, SBR, LLDPE and LDPE) were selected and used for modification of AP-3 asphalt binder widely used in Korea. Tensile properties were measured from polymer modified asphalt concretes. The results showed that some of the polymer materials were more effective than others in improving tensile characteristics of asphalt concrete.
Keywords: Asphalt, Polymer, Polymer-modified asphalt, Tensile strength

1. Introduction

Since asphalt binder changes its characteristics significantly by temperature, the asphalt concrete pavement under service changes its behaviour significantly due to temperature variation. At low temperature, asphalt shows brittle characteristics which is known to be a major cause of thermal cracking in asphalt pavements[1]. At high temperature, asphalt shows easy deformation which is a major cause of rutting in the pavements. Characteristics of asphalt concrete pavement are highly dependent upon the asphalt binder property. Therefore, change or modification of asphalt binder property is considered important for improving performance of asphalt concrete pavement[2].

Lately, many polymers are proved to be good materials for asphalt modification purpose[3,4,5]. This study has been performed to develop a highly performing asphalt binder using polymer materials available in domestic markets in Korea. Four polymer materials were selected and polymer-modified asphalt (PMA) binders were made using the four polymers.

Polymers in Concrete, edited by Y. Ohama, M. Kawakami and K. Fukuzawa. Published in 1997 by E & FN Spon, 2–6 Boundary Row, London SE1 8HN, UK. ISBN: 0 419 22330 4.

The objective of this study is to evaluate tensile characteristics of the asphalt concretes which were made of the polymer modified asphalt binders. The purpose of this paper is to show study results about the effect of polymer modification on the asphalt concrete tensile property which is the most important characteristics for pavement performance under wheel load.

2. Experimental Program

2.1 Materials

AP-3 (penetration grade of 85-100) which is the most widely used asphalt cement as a binder for asphalt pavement construction in Korea was used as a base asphalt for all experiment in this study. Properties of this asphalt are shown in Table 1.

Four polymer materials were selected based on literature studies and availability in Korea. These are styrene-butadiene-styrene (SBS), styrene-butadiene-rubber (SBR), linear low-density polyethylene (LLDPE) and low-density polyethylene (LDPE). SBR is liquid, latex type, and the other three were powder type. Fig. 1 shows four polymer materials and Table 2 - 5 shows properties of the four polymer materials.

A fine and a coarse (maximum size 19mm) aggregate were used to make sand asphalt mixtures and asphalt concretes. The sand asphalt mixtures were used for direct tensile strength tests, and the asphalt concretes for Marshall stability and indirect tensile strength tests.

2.2 Method

A certain quantity of asphalt in a container was preheated in an oven set at 165°C. The preheated asphalt container was placed on a hot plate to keep the material at 163 ± 2°C and a polymer was slowly added into the base asphalt while stirring with a shear mixer. The content of each polymer was 5% by weight of base asphalt. The mixing process took approximately 30 minutes.

Penetration of each PMA binder was measured at temperatures of 5°C, 15°C, 25°C and 35°C by KS M 2252 to see temperature susceptibility of binder. The optimum asphalt content was determined as a 10% from job-mix design for sand asphalt mixes, and 6.1% for asphalt concrete.

A slender beam specimen was made as shown in Fig. 2. A hand Marshall hammer was used to compact the specimen. The temperatures for mixing and compaction were kept at the same level recommended for a standard Marshall specimen. The compaction effort (number of blow) was determined to keep the mixture density approximately at the same level that a 5% air void could be obtained from Marshall specimen with the same mixture.

Direct tensile strength test was carried out on the slender specimen by applying static tensile force with a rate of 50mm/min at 10°C, 25°C and 35°C. Two strain gages were placed on the two faces, opposite to each other, of the middle section of the specimen. Elastic modulus was measured from the stress-strain curve of the test.

Marshall stability and indirect tensile strength (ITS) were measured on asphalt concrete Marshall specimens with a loading rate of 50mm/min at 25°C. Mixing

Table 1. Physical properties of asphalt cement

Material	Penetration (25°C)	Ductility (cm)	Flash point (°C)	Pen. ratio after TFOT(%)	Asphaltene content(%)	Solubility in trichlorethylene (%)	Specific gravity
AP-3	91	150	45.0	60.4	8.0	99.4	1.03

Table 2. Physical properties of low density polyethylene

Material	Volatility (%)	Viscocity (cps)	Color	Tensile strength (kg/cm^2)	Type
LDPE	0.62	4,840	White	233	powder

Table 3. Fundamental properties of linear low density polyethylene

Material	Type	Specific Gravity	Color	Grain passing #50 sieve (%)	E at 20°C (kg/cm^2)	Type
LLDP	Powder	0.93	Blue	52.5	1,530	powder

Table 4. Physical properties of SBR Latex

Material	Specific gravity	Viscosity (ps)	Acid Value	Styrene content (%)	pH	Solid Content (%)	Type
SBR	0.98	2.25	20.0	40	10.5	50	Milky latex

Table 5. Physical properties of SBS

Material	Lime (%)	Density (g/cm^3)	Color	Foreign substance	Volatility (%)	Type
SBS	0.024	0.957	White	NIL	0.018	coarse powder

Table 6. Fundamental properties of sand asphalt mixture

Material	Density	Asphalt content (%)	Marshall stability (kg)	Flow (0.1mm)	Air void (%)	ITS (kg/cm^2)
Sand asphalt mixture	2.143	10.0	503	22	5.49	5.81

and compaction temperatures were approximately at 160°C and 140°C, respectively. These are the temperatures suggested for standard mixing and compaction of normal mixture. Three specimens were prepared for each mix type and condition for each test, and average value was reported.

3. Results and Analyses

3.1 Characteristics of PMA binders
SBR and SBS were well mixed (dissolved) in asphalt. Visual appearance of SBR-modified binder and SBS-modified binder was not different from normal asphalt at room temperature. LDPE and LLDPE were also mixed well with asphalt. However, it was not dissolved in asphalt because of its nature. But, the particles were evenly spread without aggregation or sedimentation when it was kept at room temperature. No phase separation was observed from the four PMA binders.

Penetration test result showed that SBS binder had the lowest penetration and was followed by LDPE binder, while normal asphalt (AP-3) showed highest penetration value at the high temperature, 35°C, as shown in Fig. 3. This means that PMA binders were less susceptible than normal asphalt to temperature variation.

Workability of mixtures using PMA was not much different from normal asphalt for mixing at 160°C. All the PMA binders were well mixed with aggregates at or above the mixing temperature. However, a relatively small temperature drop caused significant loss of workability in SBR binder mixture. This may be due to the nature of rubber, and therefore, keeping the mixing temperature above a certain level seemed to be very important for the particular mixture.

3.2 Tensile Characteristics of PMA mixtures

3.2.1 Sand PMA mixture
Fundamental properties of sand asphalt mixture using the base asphalt are shown in Table 6. Since asphalt content was relatively high (10%) due to exclusion of coarse aggregate particles, the density of the mixture was relatively low. However, Marshall stability and flow were in a generally acceptable level for surface asphalt mixture.

Specimens were made and aged for 1 day at the specified temperature to make sure the temperature throughout the specimen body is at the desired level before used in the test. During the tensile test, strain gage was cut off first and then crack appeared, but no specimen was completely separated at the end of test. Fig. 4 and Fig. 5 show direct tensile strength and elastic modulus of each mixtures by temperature. LDPE mixture showed the highest strength and the highest elastic modulus throughout the temperature ranges.

LDPE and SBS showed approximately 50% and 15% higher tensile strength than normal mixture (AP-3), respectively, at 10°C. Tensile strength of SBR was somewhat higher than AP-3 at 25°C and 35°C, but was lower than AP-3 at 10°C.

Table 7. Properties of asphalt concretes (at 25°C)

Property	Ap-3	LDPE	SBS
Marshall Stability (kg)	959	1072	1024
Indirect Tensile Strength (kg/cm^2)	5.2	11.6	9.0
Density (g/cm^3)	2.325	2.306	2.297
Air Void (%)	5.45	6.21	6.58

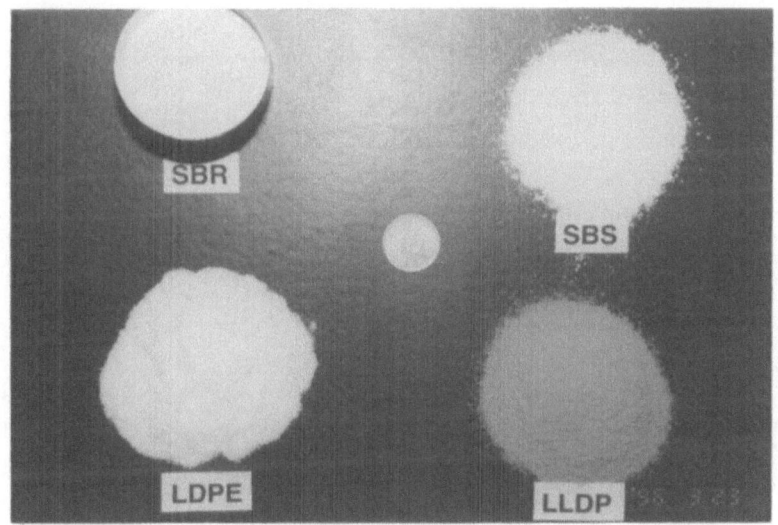

Fig. 1. Illustration of four polymer materials

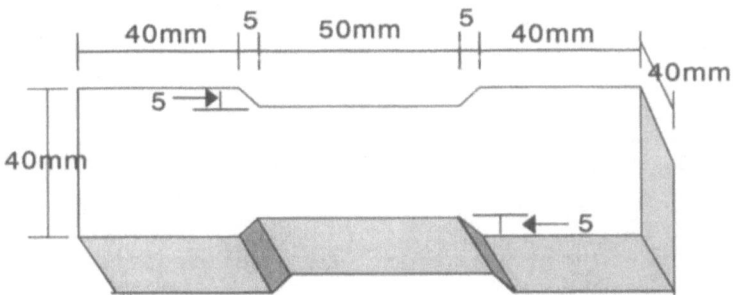

Fig. 2. Description of slender beam and its dimension

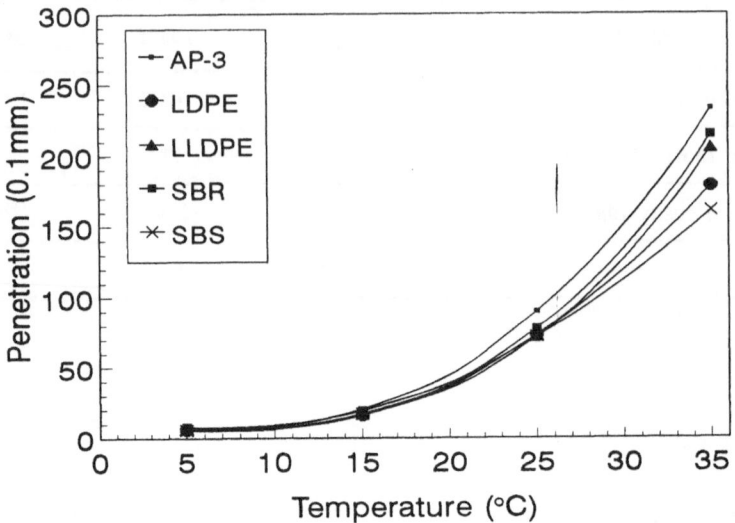

Fig. 3. Penetration of polymer modified asphalt binders

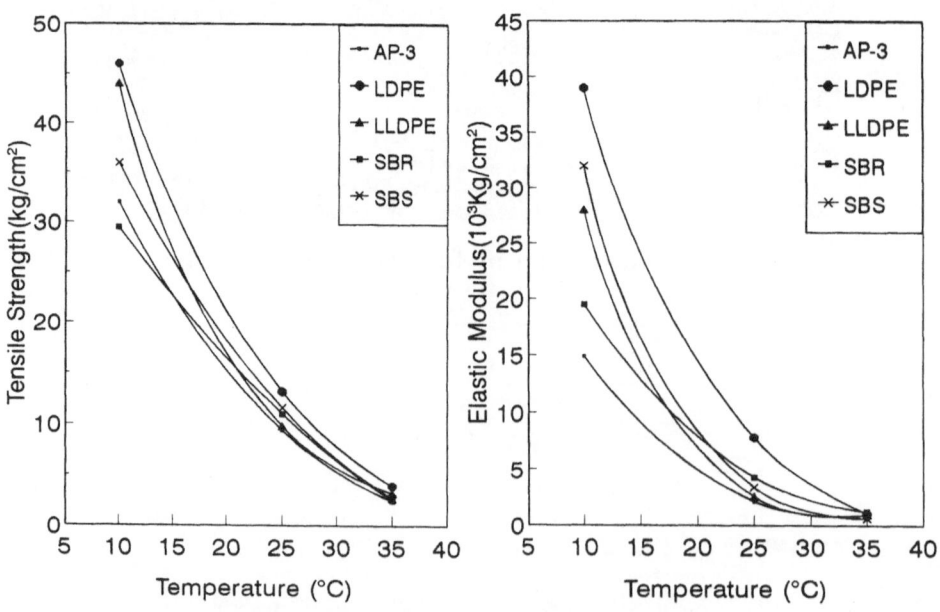

Fig. 4. Tensile Strength of sand mix   Fig. 5. Elastic Modulus of sand mix
        by Temperature.   by Temperature.

LDPE, SBS, SBR showed approximately 170%, 110% and 30% higher elastic modulus increment than AP-3, respectively, at 10°C. From these analyses, LDPE and SBS showed excellent improvement of asphalt concrete tensile properties at lower temperature, but SBR was not so good as the first two. LDPE also showed the most improvement at relatively high temperature (35°C) than other polymers. Therefore, LDPE and SBS were considered the best and the next for asphalt modification.

3.2.2 PMA concrete

In asphalt concrete evaluation, only LDPE and SBS binders were used because these two were found to be superior to the others in binder evaluation and sand-asphalt evaluation. Marshall stability and indirect tensile strength (ITS) test were carried out and results are shown in Table 7.

As revealed by density and air void, PMA asphalt concretes were less compacted than normal mixture with the same compaction effort. This means that PMA mixtures need either more compaction effort or higher temperature when they were compacted. Currently, compaction completed while the mixture was still at 140°C.

LDPE PMA concrete showed approximately 12% higher stability and 123% higher ITS than that of normal asphalt concrete respectively. While SBS PMA concrete showed approximately 7% higher stability and 73% higher ITS than normal asphalt concrete. Therefore, two polymer modified asphalt concretes did clearly improve tensile strength, but minorly improve Marshall stability.

4. Conclusions

In this study, four polymer-modified asphalt (PMA) binders were made and tensile strength characteristics of the PMA mixture were evaluated. The study results and conclusions are as follows.

1. The four polymers used in this study were well mixed in asphalt for making PMA binder using a shear mixer at approximately 165°C.
2. In general, PMA sand asphalt mixtures showed better tensile strength than normal asphalt mixture through the temperature range used in this study.
3. LDPE and SBS were the two best performing polymers, with LDPE being better, for improving tensile strength characteristics of sand asphalt mixture and asphalt concrete. These two also improved Marshall stability of asphalt concrete.
4. Since LDPE and SBS were found to have clear effect on improving asphalt concrete performance, further studies, such as, adjusting optimum polymer content and better adding method to be used in practice are suggested.

Acknowledgement:
This study was supported by the **Research Center for Advanced Mineral Aggregate Composite Products** designated by **KOSEF** at **Kangwon National University.**

5. References

1. Kim, K. W. and El Hussein, M. (1995) Effect of DTC on fracture toughness of asphalt materials at low temperatures. Journal of AAPT, Vol. 64, pp. 474-499.
2. Kim, K. W. (1996) Evaluation of polymer modified asphalt pavement materials. Proceeding, '96 Korean Symposium on Bituminous-Composite, Kangwon National University, Chun Chon, Korea.
3. Fleckenstein, L. J., Mahboub, K. and Allen, D. L. (1992) Performance of Polymer Modified Asphalt Mixes in Kentucky. ASTM STP1108.
4. Joseph, P., Dickson, J. H. and Kennepohl, G. (1992) Evaluation of polymer modified asphalts in Ontario. Proceedings, Canadian Technical Asphalt Assoc.
5. Srivastava, A., Hopman, P. and Molenaar, A. (1992) SBS polymer modified asphalt binder and its implications on overlay design. ASTM STP1108.

GPC CHARACTERIZATION OF POLYMER-MODIFIED ASPHALT BINDERS

K.W. Kim, K.D. Jeong, S.B. Lee
Dept. of Agricultural Engineering, Kangwon National University, Chunchon, Korea
K.A. Ahn
Dept. of Chemistry, Kangwon National University, Chunchon, Korea

Abstract
This study was performed to evaluate chromatographic characteristics of polymer-modified asphalt (PMA) binders in comparison with normal asphalt when they were aged in different conditions. Four polymer materials (SBS, SBR, LDPE and LLDP) were selected and used for modification of AP-3 asphalt widely used in Korea. The results show that aging caused increment of large molecular size in normal asphalt and most of PMA binders. However, LDPE polymer-modified asphalt was more effective than others in resisting against aging of asphalt material.
Keywords: Asphalt, Polymer, Polymer-modified asphalt, Chromatogram, Aging

1. Introduction

Characteristics of asphalt concrete pavement are highly dependent upon the asphalt binder property. Therefore, performance of asphalt concrete pavement can be improved by modification of asphalt binder property. Many studies have been performed to produce highly performing asphalt binders using various polymer materials[1,2,3]. This study was carried out for the same purpose using several polymers available in domestic markets in Korea. Four polymer materials (SBS, SBR, LDPE and LLDPE) were selected and polymer-modified asphalt (PMA) binders were made using the four polymers.

Asphalt binder changes its rheology due to aging. Aging is characterized as oxidization and hardening that are accompanied by both the heating during plant-mixing process and the weathering in the field. Material properties of asphalt are

Polymers in Concrete, edited by Y. Ohama, M. Kawakami and K. Fukuzawa. Published in 1997 by E & FN Spon, 2–6 Boundary Row, London SE1 8HN, UK. ISBN: 0 419 22330 4.

changed with oxidization by the time in the pavement and the oxidization affects the pavement performance. For example, aged asphalt pavement by oxidization becomes harder and easily cracked under wheel loading.

On the other hands, asphalt aging is known to be a chemical process that creates change of the molecular size distribution of the asphalt. Chromatographic analysis which shows a profile of molecular weight distribution has used to characterize aging of asphalt[4,5]. In this study, gel permeation chromatography (GPC) system was used to find molecular weight distribution of the PMA binders before and after aging in different conditions.

Modified asphalt binder performs better mechanically. However, a question has been arisen as to how much the PMA binder performs against aging compared with normal asphalt. If any polymer has better effect than any others on retarding aging process of the asphalt in addition to improving mechanical properties, use of the particular polymer will be a great advantage in extending service life of asphalt pavement. Therefore, the objective of this study is to characterize chromatogram of PMA binders in different aging conditions, among four PMA binders and the normal asphalt binder.

2. Experimental Program

2.1 Materials
Four polymer materials were selected based on literature studies and availability in Korea to make polymer-modified asphalt binders. Styrene-butadiene-styrene (SBS), styrene-butadiene-rubber (SBR), low-density polyethylene (LDPE), and linear low-density polyethylene (LLDPE) are the four selected polymers. SBR is latex type, and the other three were powder type. Properties of the four polymer materials are given elsewhere[6].

AP-3 (penetration grade of 85-100) which is the most widely used asphalt cement for pavement construction in Korea was used as a base asphalt. Properties of this asphalt are shown in Table 1.

Table 1. Properties of AP-3 asphalt cement

Material	Penetration (0.1mm)	Ductility (cm)	Flash point (°C)	Pen. ratio after TFOT(%)	Asphaltene content(%)	Solubility in trichlorethylene (%)	Specific gravity
AP-3	91	150	45.0	60.4	8.0	99.4	1.03

An HPLC grade tetrahydrofuran (THF) was used for a solvent and mobil phase in gel permeation chromatography analysis. AP-3 asphalt and the four PMA binders were used for the sample of GPC analysis.

2.2 Methods
A polymer-modified asphalt binder was made by adding 5% polymer by weight of asphalt into base asphalt which was preheated and being stirred using a shear

mixer at 165°C. Three artificial aging conditions were used to make differently-aged sample. A 20 grams of each PMA binder was placed in an aluminum dish ($\phi \approx$ 90mm) and aged in an oven at a specified temperature for a specified time. There were three aging conditions used, one day (24hours) at 80°C, three days (72 hours) at 80°C and one day at 163°C. After each aging, the weight of sample was measured and GPC test was carried out on each sample. Absolute viscosity of each PMA was measured by ASTM D2170.

An equipment with computerized control and data-acquisition software, Waters Millennium package, was used for GPC analysis. A differential refractive index meter (RI detector) was used for detection of elution. A series of two columns was used for separating constituencies of asphalt by molecular size. The mobile phase was THF flowing at a rate of 1ml/min. One test continued 30 min from injection to finish. A 50μl of a 0.5% (by weight) sample solution was used for GPC analysis. Elution started at approximately 10 min and ended at about 19 min from injection. A sample was tested three times and average value of the three was reported. More detailed process is given elsewhere[6].

A GPC profile represents a molecular size distribution of an asphalt sample. The area under a whole curve represents 100% of the molecules present in an asphalt. A GPC profile was divided into 12 equal-time slices and named as S1 to S12 from the beginning as shown in Fig. 1. Relative area of each slice was calculated by the software. Initial part of GPC profile represents larger molecular size group and rear part represents smaller molecular size group. Therefore, initial four slices (S1 - S4) are called large molecular size or LMS, middle four slices (S5 - S8) are called medium molecular size or MMS, and last four slices (S9 - S12) are called small molecular size or SMS. In general, LMS has the most significant correlation with aging and regarded as the most important molecular group in asphalt GPC analysis.

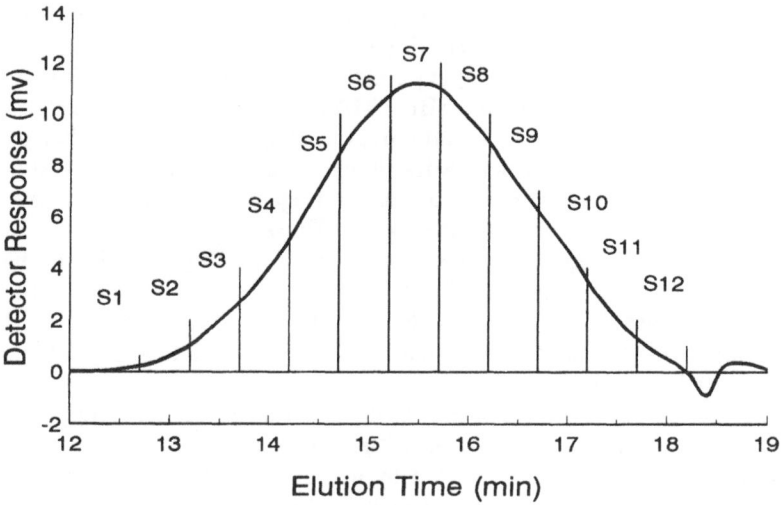

Fig. 1. Illustration of 12 slices in an asphalt chromatogram

3. Results and Discussion

Absolute viscosity of each PMA binder is shown in Fig. 2. SBS PMA showed the most significantly increased viscosity compared with other PMA.

Analyses of chromatogram of normal asphalt and PMA binders revealed that molecules of SBR and SBS were too big to be separated by the columns. They appeared at void volume area, front part of the asphalt molecular size distribution. Since pore size of the column used covers maximum molecular weight of 100,000, the molecular weight of SBS and SBR should be greater than 100,000. On the other hands, LDPE and LLDPE were not dissolved in THF and they were screened out by the 0.45μm syringe filter. Therefore, the profile of the two PMA chromatogram had no difference from normal asphalt.

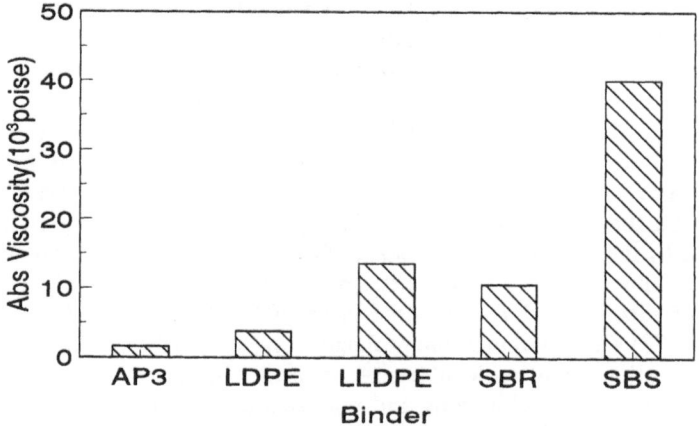

Fig. 2. Absolute viscosity of PMA binders

Aged asphalt showed increased LMS portion in chromatogram, although there were some differences in accordance with aging condition. Aging one day at 80°C was the least intensive artificial aging condition and one day at 163°C was the most intensive. Therefore, SBS and SBR PMAs aged for one day at 163°C showed extraordinary expansion of LMS, and corresponding reduction of other area.

This is an evidence of characteristics change of the polymer, chemical process between asphalt and polymer, and the asphalt aged at high temperature. The larger the LMS portion, the more the aging. The asphalt with greater LMS has the higher viscosity and the harder nature. The figures illustrating this phenomenon are given in Fig. 3 - Fig. 7. The PMA binder that showed the most LMS increment was SBR, SBS, LLDPE, and LDPE in descending order. LDPE was the only PMA that appeared to have strong resistance to aging. Others were aged more or less than normal asphalt.

As mentioned earlier, LMS is the most important molecular group, variation of which has a significant correlation with many physical properties of asphalt. Variation of S2, S3 and S4 are shown by aging condition in Figs. 8, 9, and 10, respectively. S1 is also one of the slices in LMS. But it is at very beginning of elution and its relative area is tiny. It is highly dependent upon determination of

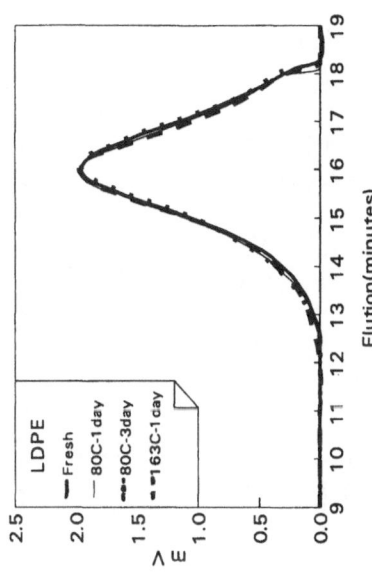

Fig. 3. Chromatograms of AP-3 in 4 aging Conds.

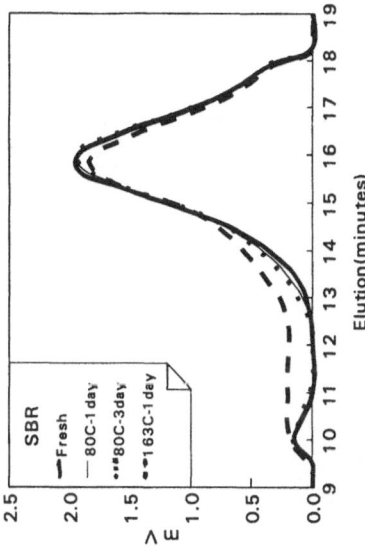

Fig. 4. Chromatograms of LDPE in 4 aging Conds.

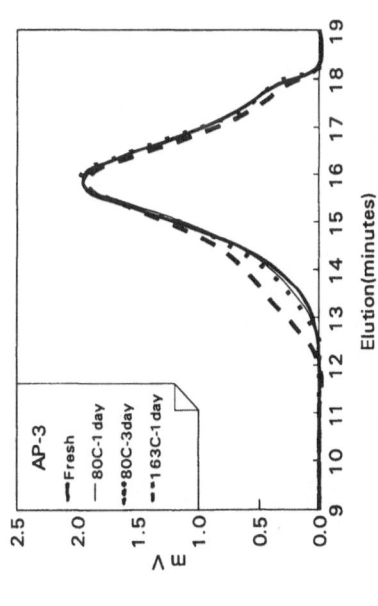

Fig. 5. Chromatograms of SBS in 4 aging Conds.

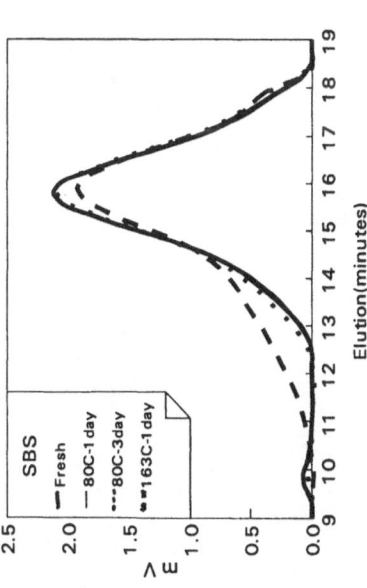

Fig. 6. Chromatograms of SBR in 4 aging Conds.

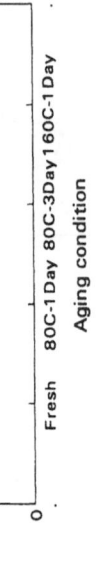

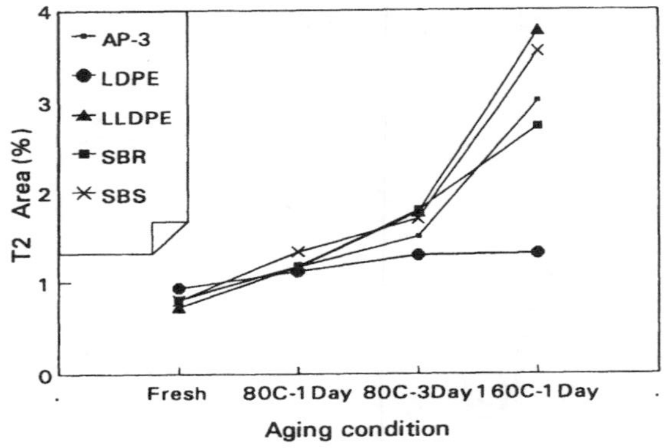

Fig. 8. S2 of 5 binders by aging conditions.

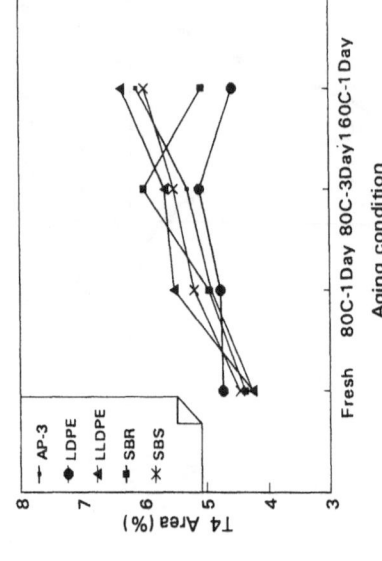

Fig. 10. S4 of 5 binders by aging conditions.

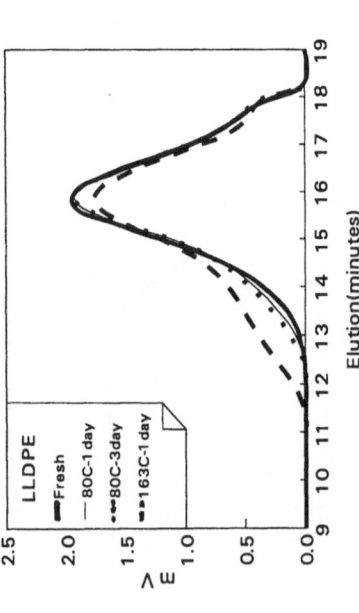

Fig. 7. Chromatograms of LLDPE in 4 aging conds.

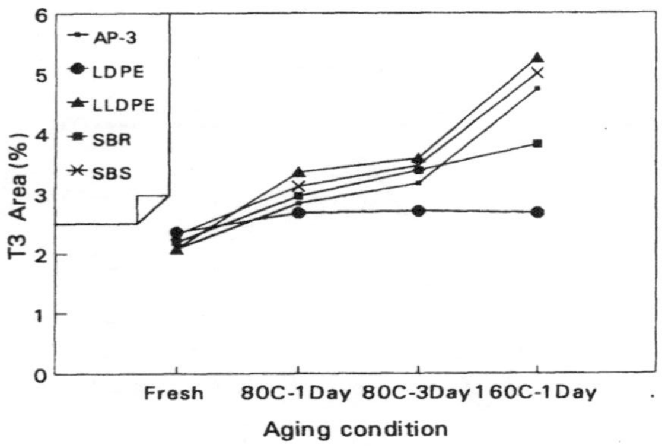

Fig. 9. S3 of 5 binders by aging conditions.

elusion starting point. Especially in case of SBS and SBR, relative area of S1 was very big due to melted polymer at 163°C. Therefore, it was discard in comparison of LMS slices. As shown in those figures, relative area of S2, S3, and S4 of LDPE were stable throughout the different aging conditions. But those of other PMA binders and AP-3 show increment with intensity of aging, except for S4 of SBR for one day at 163°C aging. From this, SBR seemed to be less aged for one day at 163°C than AP-3, SBS and LLDPE. Especially, by one day at 163°C aging, S2, the largest molecular size group, increased significantly except for LDPE binder. Therefore, LDPE seemed to have a strong effect of blocking (resistance) against aging progress in asphalt binder.

4. Conclusions

GPC characteristics of four polymer modified asphalt (PMA) binders were evaluated before and after three different artificial aging conditioning. Chromatographic analysis of PMA before artificial aging has no meaning because the molecular size of these polymers are much greater than asphalt molecules. They either appeared as a void volume in GPC profile or are screened out by micro filter. After artificial aging, PMA binders showed similar aging pattern to that of the normal asphalt (AP-3), except for LDPE PMA binder.

In general, large molecular size (LMS) increased by aging conditioning and this phenomenon was evidenced by chromatogram analysis. Relative areas of the larger molecules, S2, S3, and S4, increased significantly by artificial aging, except for LDPE PMA. This means that the PMA binders increase viscosity and become harder by aging, and LDPE polymer modified asphalt binder has a stronger resistance than any other PMA binders.

Acknowledgement:
This study was supported by the **Research Center for Advanced Mineral Aggregate Composite Products** designated by **KOSEF** at **Kangwon National University.**

5. References

1. Al Dhalaan, M. (1992) Field Trials with polymer modified asphalts in Saudi Arabia. ASTM STP 1108.
2. Joseph, P., Dickson, J. H. and Kennepohl, G. (1992) Evaluation of polymer modified asphalts in Ontario. Proceedings, Canadian Technical Asphalt Assoc.
3. Srivastava, A., Hopman, P. and Molenaar, A. (1992) SBS polymer modified asphalt binder and its implications on overlay design. ASTM STP1108.
4. Kim, K. W. and Burati, J. L. (1993) Use of GPC chromatogram to characterize aged asphalt cement," Journal of Materials in Civil Engineering, ASCE, Vol. 5. No. 1., pp. 41-52
5. Kim, K. W., Burati, J. L. and Park, J. S. (1995) Methodology for Defining

LMS Portion in Asphalt Chromatogram. Journal of Materials in Civil Engineering, ASCE, Vol. 7. No.1., pp. 31-40.

6. Kim, K. W. (1996) Evaluation of polymer modified asphalt pavement materials. Proceeding, '96 Korean Symposium on Bituminous-Composite, Kangwon National University, Chun Chon, Korea.

PHYSICAL AND MECHANICAL PROPERTIES OF POLYMER CONCRETE USING COAL MINE WASTE

K.S. Yeon, J.D. Choi, T.Y. Jang, M.K. Joo and D.S. Choi
Dept. of Agricultural Engineering, Kangwon National University, Chunchon, Korea

Abstract

Demand for high quality aggregate from construction and related industries have increased while supply of the aggregate lacks. This study was to explore a possibility of using coal mine waste as both coarse and fine aggregates in polymer concrete. Coarse and fine aggregates were produced from coal mine waste and tested. Different mixing proportions were tried to find an optimum mixing proportion of coal mine waste polymer concrete. Specimens were made to examine physical and mechanical properties of polymer concrete using coal mine aggregates. At 7 days of curing, compressive, flexural and splitting tensile strengths ranged between 79.6~93.2 MPa, 17.4~21.4 MPa and 9.2~11.1 MPa, respectively. It was concluded that coarse and fine aggregates from coal mine waste could be used in polymer concrete.
Keywords: Aggregate, coal mine waste, mixing proportion, polymer concrete, strength, stress-strain curve, unsaturated polyester resin.

1 Introduction

Shortage of concrete aggregate supply has been widening with the rapid growth of construction works such as infrastructures and buildings. Collection and supply of natural aggregates from river beds are not sufficient and moreover, environmental problems associated with the aggregate collection from river beds by dredging have caused strong protests from environmentalists and curbed

Polymers in Concrete, edited by Y. Ohama, M. Kawakami and K. Fukuzawa. Published in 1997 by E & FN Spon, 2–6 Boundary Row, London SE1 8HN, UK. ISBN: 0 419 22330 4.

collection of aggregates, resulting in a bitter aggregate shortage. Aggregate occupies more than 70% of concrete by volume. And with the growth of construction industry, supplying and securing aggregates in construction industry have been pending issues to solve in near future.

There have been numerous coal and other mines in Korea and the huge amount of mine wastes also have been produced as byproducts. As mines have closed because of the change of mine industry and government policy, mining areas have abandoned and the huge amount of mine wastes also have been left over and caused many environmental problems such as exudation of acid and soil contamination. It is very important to safely dispose or reuse the mine wastes to revitalize the devastated areas.

This study initiated to find a way to reuse the mine wastes as a precious resource. A way to reuse the mine waste is to use it as aggregate for construction industry. The goal of this study was to find a way to reuse the waste as a construction material so that it can greatly contribute to the ease of natural aggregate supply and also to the solution of environmental problems associated with it. The objectives of this study were (1) to find a way to reuse coal mine waste as aggregate in polymer concrete that has characteristics of high compressive strength, long durability, and corrosion resistivity and watertightness, (2) to test the polymer concrete that uses a coal mine waste as an aggregate with respect to physical and mechanical properties, and (3) to provide the test data for factory production of polymer concrete products. Coal mine wastes that have been abandoned in a coal mining area of southern Kangwon-do province, Korea, were used in this study.

2 Material and Methods

2.1 Material

2.1.1 Unsaturated polyester resin
An Ortho-type unsaturated polyester resin was used in this study and its properties are shown in Table 1.

2.1.2 Accelerator admixture
Mineral turpentine solution (CoOc) that is octanic acid cobalt 8% solution is mainly used as accelerator admixtures. Initiator used in this study is for normal hardening and its properties are shown in Table 2.

2.1.3 Shrinkage reducing agent
Shrinkage reducing agent is a solution that polystyrene that is a thermoplasticity is dissolved in styrene monomer and used in this study to reduce shrinkage of polymer concrete. Table 3 shows the properties of the shrinkage reducing agent.

Table 1. Properties of unsaturated polyester resin used

Specific gravity at 25 °C	Viscosity at 25 °C (poise)	Acid value	Styrene content (%)
1.138	3.0	20.0	40

Table 2. Properties of initiator used

Component	Specific gravity at 25 °C	Active oxygen
MEKPO 55% DMP 45%	1.12	10.0

Table 3. Properties of shrinkage reducing agent

Viscosity at 25 °C (poise)	Nonvolatile substance (%)	Appearance
31-41	34-38	Transparent

2.1.4 Filler
Common fillers are powders of calcium carbonate, alumina, silica, quarts or silicon carbide, oxide iron, cement and so on. Among the fillers, calcium carbonate powder was used in this study because it is relatively cheap and easy to buy.

2.1.5. Natural aggregates
Natural aggregates were used to mix with coal mine aggregates that was crushed coal mine wastes. Natural aggregates were to make control polymer concrete specimen and to adjust the particle size distribution of coal mine aggregates. Coarse aggregate was crushed granite that was produced at a quarry. Fine aggregate was river sand collected at a river bed. Both the quarry and the river were near Chuncheon, Kangwon-do province, Korea. Physical properties of the aggregates are shown in Tables 4 and 5.

2.1.6 Coal mine aggregate
Coal mine aggregate was crushed coal mine wastes collected at deserted waste dumps around Sabuk-eub, Cheongseon-kun, southern Kangwon-do province where there are many closed coal mines. A crusher was used to produce coal mine aggregates. The maximum particle size of coarse coal mine aggregate was 13 mm. And for fine aggregate, #4 sieve (4.75 mm) passed particles were used.

Table 4. Physical properties of natural coarse aggregate

Maximum size (mm)	Specific gravity	Absorption (%)	Unit weight (kg/m^3)	Abrasion (%)
13	2.62	0.65	1,701	26.5

Table 5. Physical properties of natural fine aggregate

Specific gravity	Absorption (%)	Unit weight (kg/m³)	#200 sieve passing (%)	F.M.
2.65	1.20	1,490	0.86	2.82

These aggregates were tested to use in polymer concrete and their properties are shown in Tables 6 and 7.

From Tables 4 and 6, it is shown that the properties of coarse coal mine aggregate are not different from those of natural coarse aggregate except that the unit weight of coal mine aggregate is a little smaller than that of natural one. And thus, it was thought that coarse coal mine aggregate would have no problems in using in polymer concrete. For fine aggregates, #200 sieve passing and absorption of coal mine aggregate were higher than those of natural aggregate but there were no significant differences between the other properties.

Table 6. Physical properties of coarse coal mine aggregate

Maximum size (mm)	Specific gravity	Absorption (%)	Unit weight (kg/m³)	Abrasion (%)
13	2.56	1.3	1,497	28.0

Table 7. Physical properties of fine coal mine aggregate

Specific gravity	Absorption (%)	Unit weight (kg/m³)	#200 sieve passing (%)	F.M.
2.50	6.0	1,380	3.9	3.17

2.2 Determination of mixing proportion

After many preliminary tests, 5 mixing proportions were tried to determine an optimum mixing proportion of polymer concrete using coal mine aggregate. Contents of unsaturated polyester resin, shrinkage reducing agent (SRA), filler and coarse coal mine aggregate (CW-CA) were fixed as in Table 8. No natural coarse aggregate was used because it had been proved that coarse coal mine aggregate could be used for polymer concrete production in this study. Total content of fine aggregate was 39% by weight and the 39% was shared by natural fine aggregate (N-FA) and fine coal mine aggregate (CW-FA) depending on the mix type as in Table 8.

2.3 Specimen

Specimens were prepared according to the Korea Standard Testing Methods, KS F 2419 (Specimen preparation methods for strength measure of polyester resin

Table 8. Summary of mixing proportions tried to determine the optimum mixing proportion of polymer concrete using coal mine aggregate

Mix type	Mix proportion (wt.%)						Total (wt.%)	CW (wt.%)
	Resin	SRA	Filler	CW-CA	CW-FA	N-FA		
A					0.0	39.0	100	50
B					11.7	27.3	100	65
C	9	3	10	39	19.5	19.5	100	75
D					27.3	11.7	100	85
E					39.0	0.0	100	100

Note: CW= percent of coal mine aggregate of the total aggregate

concrete). Polymer concrete was mixed by using a forced concrete mixer. Two types of specimen, i.e., cylindrical and cubic specimens, were made depending on the test described in the testing methods. A cylindrical specimen was formed by casting polymer concrete into a cylindrical mold in three layers. And a cubic specimen was formed by casting polymer concrete into a cubic mold in two layers. Each layer was compacted 25 times with a compaction hammer. After the compaction, the mold was put on a table vibrator and again compacted sufficiently by vibration. Compacted specimens were cured at 11 ∓ 3 °C that was the room temperature for one hour, moved into an incubator, and cured again at 28 ∓ 3 °C for upto 7 days.

2.4 Test methods

2.4.1 Physical properties
Unit weight was measured by weighing air-dried specimen and then by dividing the measured weight by specimen volume. Absorption test was performed according to KS F 2518 (Test methods for gravity and absorption of aggregate). The size of a cubic specimen was 5 cm x 5 cm x 5 cm.

Time of hardening was tested by using both penetration and touch methods among the three methods specified in KS F 2484 (Workable time measure methods of polyester resin concrete). Time of hardening was measured from the time when the initiator, MEKPO, was added to liquid resin. The room temperature was 12 ∓ 2 °C.

Slump test and flow test are usually used to test the workability of cement concrete. Flow test specified in ASTM C 24 was used to measure the workability of polymer concrete in this study. The room temperature was 12 ∓ 2 °C.

2.4.2 Mechanical properties
Compressive, flexural and splitting tensile strength tests were carried out

according to KS F 2481 (Compressive strength test method for polyester resin concrete), KSF 2482 (Flexural strength test method for polyester resin concrete), and KS F 2480 (Tensile strength test method for polyester resin concrete), respectively.

Tests for elastic modulus and poisson's ratio were performed by using the wire strain gauge method which is one of the two methods specified in KSF 2438 (Elastic modulus and poisson's ratio test method of cylindrical concrete specimen). Elastic modulus and poisson's ratio were obtained from the secant modulus that was computed from the tangent of stress-strain curve. A stress-strain curve was drawn by plotting the measures of stress and strain that was measured by repeatedly applying loads on a specimen. Loads were applied upto 40% of the failure load. The size of cylindrical specimen was ϕ7.5 cm x 15 cm and the length of strain gauges was 30 mm.

3 Results and Discussion

3.1 Physical properties

3.1.1 Unit weight and absorption
Table 9 shows the results of unit weight and absorption tests. Unit weight increased with the decrease of the content of coal mine aggregate. When the content of coal mine aggregate was more than 75% of the total aggregate weight, unit weight seemed to decrease rapidly. It was thought that the large amount of fine particles in fine coal mine aggregate absorbed more resin than natural aggregate did and thus, the workability and compactability of polymer concrete got lowered, resulting in more pores in polymer concrete and less unit weight. Measured unit weights of coal mine waste polymer concrete ranged from 2.2 to 2.3 t/m^3 and were similar to those of cement concrete.

Absorption ranged from 0.19 to 0.21% regardless of the content of coal mine aggregate. Absorption of coal mine waste polymer concrete was much less than that of cement concrete of 4 to 6% and was even less than that of natural aggregate of 0.5 to 4.0%. It proved that the watertightness of polymer concrete is superior.

3.1.2 Time of hardening
Measured time of hardening in this study is shown in Table 10. Time of hardening retarded with the increase of the content of coal mine aggregate. The retard was large when the content of coal mine aggregate was small. However, the differences between times of hardening on different contents of coal mine aggregate were at most 22 minutes. Setting times were between 65 and 67 minutes regardless of the content of coal mine aggregate.

Table 9. Results of unit weight and absorption tests

Mix type	Unit weight (t/m³)			Absorption (%)			
	Measured values		Mean	Measured values			Mean
A	2.30 2.31 2.30		2.30	0.17 0.18 0.21			0.19
B	2.29 2.28 2.30		2.29	0.21 0.19 0.19			0.20
C	2.29 2.28 2.27		2.28	0.22 0.20 0.21			0.21
D	2.27 2.26 2.26		2.26	0.19 0.19 0.20			0.19
E	2.25 2.22 2.23		2.23	0.21 0.23 0.20			0.21

Table 10. Measured time of setting and time of hardening of coal mine aggregate polymer concrete

Mix type	A	B	C	D	E
Time of setting (min)	67	66	67	65	65
Time of hardening (min)	191	200	207	210	213

Table 11. Results of flow tests

Mix type	A	B	C	D	E
Mean (%)	32.5	31.8	30.8	27.2	23.5

3.1.3 Workability

Flow test results are shown in Table 11. Flow decreased with the increase of the content of coal mine aggregate. Flow decrease was prominent when the content of coal mine aggregate exceeded 75%. Therefore, it is recommended that the content of coal mine aggregate had better not to exceed 75% of the total aggregate to maintain a good workability.

3.2 Mechanical properties

3.2.1 Properties of compressive, flexural and tensile strengths

Strength development of polymer concrete was assumed to be related to the time of curing and the content of coal mine aggregate. And thus, strength tests were performed with respect to the different time of curing and content of coal mine aggregate. Tables 12, 13 and 14 show the results of strength tests.

Measured strengths decreased with the increase of the content of coal mine aggregate. Strength decrease was prominent when the contents of coal mine aggregate were between 65 and 85%. The strength increment was large at the beginning of curing but the increment became small as the time of curing elapsed.

Table 12. Measured compressive strength (MPa) of coal mine waste polymer concrete with respect to time of curing and content of coal mine aggregate

Mix type	Time of curing (hours)				
	6	12	24	72	168
A	29.4	57.0	75.2	87.2	93.2
B	28.2	54.9	74.2	85.4	91.2
C	24.5	52.3	69.8	81.5	88.0
D	20.4	47.7	64.0	75.6	81.1
E	19.1	45.6	61.6	74.0	79.6

Table 13. Measured flexural strength (MPa) of coal mine waste polymer concrete with respect to time of curing and content of coal mine aggregate

Mix type	Time of curing (hours)				
	6	12	24	72	168
A	8.8	14.3	17.7	19.5	21.4
B	8.0	13.9	17.2	18.9	20.7
C	6.9	12.8	16.4	18.1	19.7
D	6.2	12.3	15.1	16.3	17.9
E	5.8	11.7	14.6	16.0	17.4

Table 14. Measured splitting tensile strength (MPa) of coal mine waste polymer concrete with respect to time of curing and content of coal mine aggregate

Mix type	Time of curing (hours)				
	6	12	24	72	168
A	4.1	7.0	8.8	9.8	11.1
B	3.9	6.8	8.3	9.4	10.7
C	3.4	6.7	8.0	9.1	10.2
D	3.1	6.0	7.5	8.5	9.5
E	2.6	5.7	7.2	8.2	9.2

3.2.2 Elastic modulus and poisson's ratio

Stress-strain curves of coal mine waste polymer concrete tested in this study is shown in Fig. 1, and elastic modulus and poisson's ratio obtained from the curves are summarized in Table 15.

Tangents of stress-strain curve increased with the decrease of the content of coal mine aggregate. However, by considering that the elastic behavior of coal mine waste polymer concrete ranged between 60-70% of the maximum strength, it was thought that the content of coal mine aggregate was closely related to the strength development but the content was not related to the elastic property.

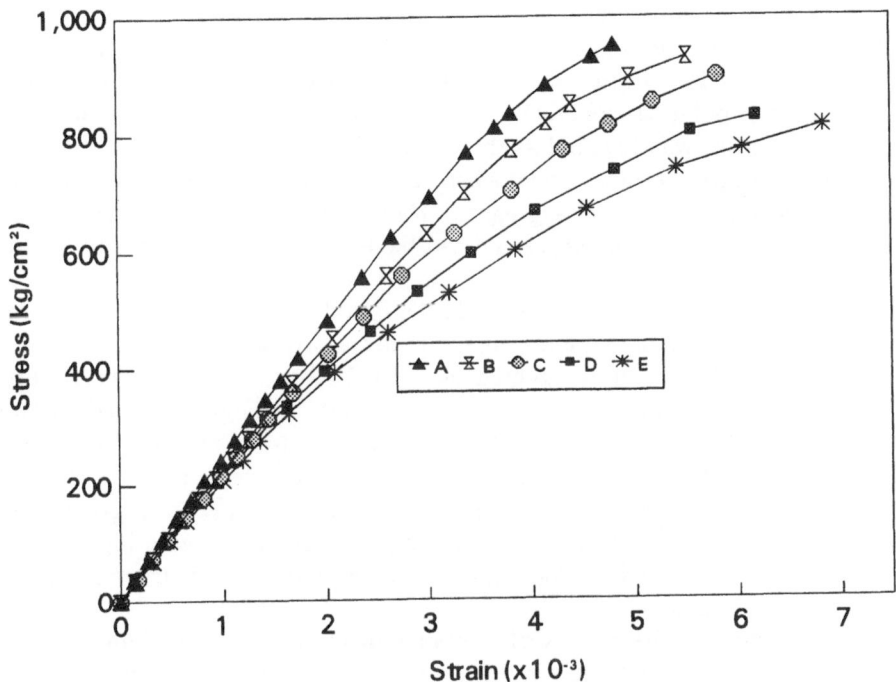

Fig. 1. Stress-strain curves of coal mine waste polymer concrete

Table 15. Obtained elastic modulus and poisson's ratio of coal mine waste polymer concrete

Mix type	Elastic modulus (E, $\times 10^4$ MPa)			Poisson's ratio (ν)		
	Measured values		Mean	Measured values		Mean
A	2.33 2.38 2.31		2.34	0.22 0.21 0.24		0.23
B	2.16 2.19 2.18		2.17	0.24 0.22 0.23		0.23
C	2.07 2.10 2.10		2.09	0.25 0.24 0.25		0.25
D	2.05 1.99 1.98		2.00	0.22 0.25 0.26		0.24
E	1.94 1.96 1.98		1.96	0.28 0.26 0.26		0.27

4 Conclusions

This study initiated to explore a possibility of using coal mine waste as both coarse and fine aggregates in polymer concrete. Specimens of polymer concrete using coal mine aggregate were made and tested for physical and mechanical properties. The followings were conclusions from the study.

1) Unit weight and absorption of coal mine waste polymer concrete ranged between 2.2~2.3 t/m^3 and 0.19~0.21%, respectively, and were similar to existing polymer concrete. However, time of hardening retarded and workability was lowered as the content of coal mine aggregate increased. Workability became poor when the content of coal mine aggregate exceeded 75% of total aggregate. Therefore, it was recommended that the content of coal mine aggregate had better to be less than 75% to maintain a proper workability.

2) Strength increment was higher for the first day of curing but the increment became small thereafter. Strength decreased with the increase of the content of coal mine aggregate. At 7 days (168 hours) of curing, compressive, flexural and splitting tensile strengths ranged between 79.6~93.2 MPa, 17.4~21.4 MPa and 9.2~11.2 MPa, respectively.

3) σ_t/σ_c and σ_b/σ_c of coal mine waste polymer concrete were 11.5~11.9% and 21.8~23.0%, respectively, and were larger than those strength ratios of cement concrete. It suggests that polymer concrete have a big advantage for the design and production of concrete products and structures that are not reinforced.

4) According to the results of stress-strain tests, elastic modulus tended to increase while poisson's ratio tended to decrease with the increase of the content of coal mine aggregate. Elastic modulus and poisson's ratio ranged between 1.96~2.34x10^4 MPa and 0.23~0.27, respectively. Elastic modulus and poisson's ratio of coal mine waste concrete were smaller and larger, respectively, than those of high strength cement concrete, which is a desirable property in that coal mine waste polymer concrete has larger tensile resistivity than high strength cement concrete.

5) Based on the above results, it was concluded that coarse and fine aggregates from coal mine waste could be used in polymer concrete. However, it is required to study on durability and long-term behavior of coal mine waste polymer concrete before industrial application.

Acknowledgement
This study was supported by the Research Center for Advanced Mineral Aggregate Composite Products designated by KOSEF at Kanwon National University.

5 References

1. Aguado, A., Martinez, A. and Salla, J.M. (1984) Effect of different factors in mixing and placing of polymer concrete, Proceedings of the 4th ICPIC, Darmstadt, Germany, pp.119-165.
2. Demura, K. (1983) Development of resin concrete for construction works, Ph.D. Thesis, Nihon University, Japan. (in Japanese)
3. Fowler, D.W. (1990) Status of concrete - polymer · materials, Proceedings of the 6th ICPIC, Shanghai, China, pp.10-27.
4. Korea Institute of Geology, Mining and Materials (1992) Studies on the practical use of coals and coal mine waste, Research report, pp.119-165. (in Korean)
5. Jung, S.J. (1995) Review of mineral waste utilization, Proceedings of the 1st international congress, Research center for advanced mineral aggregate composite products, Kangwon National University, Korea, pp.129-135. (in Korean)
6. Jo, Y.H. (1995) Prospect and counterplan on demand-supply of aggregate in Korea, Proceedings of the 1st international congress, Research center for advanced mineral aggregate composite products, Kangwon National University, Korea, pp.5-28. (in Korean)
7. Ohama, Y. Demura, K. (1980) Relation between curing conditions and compressive strength of polyester resin concrete, The international journal of cement composite and light weight concrete, Vol. 4, No. 318, pp.60-65.
8. Choi, M.S. (1991) Prospect and counterplan on demand-supply of aggregate in Korea, Journal of Korea concrete institute, Vol. 3, No. 2, pp.22-30. (in Korean)

BUILDING MATERIALS PRODUCTION BASED ON SULFUR ORE FROM THE KURIL ISLANDS

A.N. Volgushev
Research Institute of Concrete and Reinforced Concrete, Moscow, Russia
N.F. Shesterkina
"TEKOMA" Joint-Stock Company Limited, Moscow, Russia
A.V. Matyushkin
Institute of Organization and Technology Construction, Moscow, Russia
V.I. Mukhin
Research-Technical Center, "ROSVOSTOKSTROY" Joint-Stock Company, Moscow, Russia

Abstract
Preliminary results of studying physical and mechanical properties of sulfur concretes produced on the basis of sulphur ores from Kuril Islands are presented in the paper. It is shown that these materials have high strength and density as well as high waterproofness, cold and corrosion resistance. Recommendations for practical application of sulphur ores in the Kuril Islands Region are given.
Keywords: Building materials, concrete, granulation, sulphur, sulphur concrete, sulphur ore, granulation.

1 Introduction

In the Kuril Islands the construction is highly expensive. Basically building materials, products and constructions are mostly delivered from the continent during a short navigation period. This results in the rise of the construction cost and in the delay of putting industrial projects into operation.

Taking into account that the Kuril Islands Region is rich in natural sulphur, its use in the manufacture of sulphur astringent and, therefore, many elements and constructions on its basis, instead of delivered concrete and reinforced concrete, shall be considered as one of the possible ways to solve the problem of producing building materials.

Polymers in Concrete, edited by Y. Ohama, M. Kawakami and K. Fukuzawa. Published in 1997 by E & FN Spon, 2–6 Boundary Row, London SE1 8HN, UK. ISBN: 0 419 22330 4.

Manufacture of effective wall materials can be organized on the basis of slug pumice deposits combined with sulphur astringent. Sulphur concrete can be used in highway engineering.

2   Experiments

All the researches and studies have been carried out on the materials delivered from the Kuril Islands. The investigation program included analysis of properties and estimation of characteristics of initial raw materials. selection of compounds. determination of compound's. reological properties. working out techniques for its moulding. determination of basic physical and mechanical properties of the compounds received and setting primary products nomenclature. In future works on developing technologies. designing production processing equipment and organizing the manufacture shall be carried out. These activities will be aimed at establishing sulphur astringent manufacture on the basis of sulphur ores and building constructions for various use both in built up (sectional) and monolith variants.

2.1 Sulphur ores examination

Applicability assessment of sulphur ores for compositions to be blended on their basis was the main problem to be solved. Sulphur content in the ore should be the characteristic of primary importance.

Table 1. Ore types and results of their analysis

Name of the island	Ore type	Content (%)		Moisture content		Compression strength
		sulphur	mineral	Stand.	Max.	(MPa)
Kounashire	I	90	10	6	15.5	45.5
	II	70	30	–	–	37.0
	III	81	19	9	13.0	46.0
Ituroup	IY	7	93	4.5	8.0	–
	Y	21	79	–	–	46.0
	YI	10	90	–	–	–

The studies carried out proved that ores of types I.II.III.Y consisted of sulphur and mineral part. containing sand and powder. When heated they all become mastic paste that could be used to prepare sulphur compositions. Sample cubes moulded from this mastic paste proved rather high strength characteristics when tested for compression. Ore types IY.YI are ore rocks impregnated by sulphur and. therefore. ore pieces keep initial form and shape and geometric sizes after sulphur removal. In their initial condition these sulphur ores are not suitable for manufacture of sulphur compositions.

2.2 Proportioning of compositions

When proportioning compositions, ratios of the components under which both technological properties of the mixture mobility and physical and mechanical properties of the solidified compound have been obtained, were determined. The components ratio was determined by experimental and theoretical methods [1]. To obtain dense and rather strong concrete it was necessary that the mastic paste filled voids of the crushed stone-and-sand (ss) mixture. Hence, the quantity of the mastic paste (M mast) in the composition was determined by the following formula:

$$M \text{ mast} = \prod(ss) \cdot K \cdot V \rho \text{ (mast)},$$

where:
$\prod_{(ss)}$ – voidness of the stone-and-sand mixture:
K – grain uniformity coefficient:
V – form volume;
$\rho_{(mast)}$ – average mastic paste density.

2.3 Production of granulated sulphur astringent from sulphur ores

Granulated sulphur astringent can be produced from ore types I.II.III.Y without their special composition corrections. The studies carried out by Institute of Concrete and Reinforced Concrete in Moscow proved that sulphur astringent is characterized by the type of mineral powder and modifying addition [2].

Sulphur ore mastic melt is easy to produce technologically by heating at the temperature $150 + 5\,^{\circ}C$ and mixing the mastic melt to prevent segregation.

Mastic granulation can be performed at any existing granulator for sulphur (Fig.1).

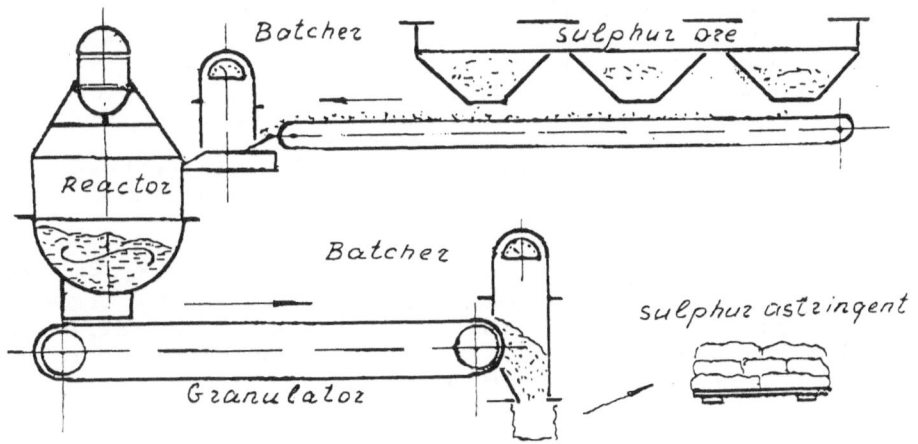

Fig. 1. Principal technological scheme for producing sulphur cement from ores.

2.4 Strength characteristics of sulphur concretes

Concrete samples of the size 4 x 4 x 16 cm have been produced on the basis of sulphur ore mastics. Compositions were mixed with heavyweight and lightweight aggregates.

Table 2. Characteristics of concretes based on sulphur ores

N	Ore type	Aggregate type	Density g/cm^3	Compression strength MPa
1	I	heavyweight	2.46	33.0
2	II	-"-	2.44	52.2
3	III	-"-	2.40	54.4
4	Y	-"-	2.40	54.7
5	I	lightweight	1.40	39.9

3 Results and conclusions

The fulfilled researches have shown that it is possible to use sulphur ores and local mineral aggregates in the technology for production of building material compounds including heavyweight, lightweight and porous concrete. In practice constructions produced from these compounds can be used because of their high strength, low water absorption, waterproofness, resistance to severe climatic conditions and attack by corrosion media [1,2].

Technological procedures for sulphur ores processing in the field of construction industry are not complicated and from the technical point of view are available at minimum capital costs. At the first stage of organizing production the processing equipment of asphaltic concrete factories could be used after certain up-dating.

There would be no difficulties in designing and manufacturing non-standard equipment for further improving and modernization of technological procedures.

Expenditures on organizing construction industry basis on the source of the Kuril Islands raw materials shall be, beyond any doubt, justified and repaid.

In addition to practical value of the above mentioned results which, in our opinion, are of great interest, the scientific aspect of studying problems of compound composition selection on the basis of sulphur astringent obtained from sulphur ores, that have been carried out for the first time, should be noted.

4 References

1.   Volgushev A.N., Shesterkina N.F. (1991) Production and application of sulphur concretes. *Survey Information*

TsNIITEIMS (in Russian)
2.   Volgushev A.N., Shesterkina N.F. (1993) Specific features of techniques and physical and mechanical properties of polymersulphur compositions.(in Russian).

SUPER-WORKABLE SLAG ALKALINE POLYMER CONCRETES

P.V. Krivenko, A.P. Semenyuk and T.A. Olbinskaya
Scientific Research Institute on Binders and Materials,
Kiev, Ukraine

Abstract
The present paper deals with a newly developed class of super-workable polymer-modified concretes, namely: slag alkaline polymer- modified concretes, in some details. The slag alkaline cements and concretes exhibiting a whole set of advantages as compared with conventional cements and concretes are known to be poorly sensitive to plasticizers. The reason for this is that the plasticizers themselves are alkalies. The formulations of slag alkaline polymer- modified concretes featuring an increased workability have been developed and investigated. The plasticizing effect was achieved by optimization of kind and density of alkaline component used, particle size distribution of aggregate as well as by introduction of an organic polymer- latex stable in alkaline environment. Mix proportions of the concretes intended for self-levelling floors have been developed. These concretes exhibit compressive strength of 10- 20 MPa at 1 day and 30- 40 MPa at 28 days after casting, linear shrinkage of 0.3 mm/ run.m in normal conditions and good corrosion resistance to aggressive media.
Keywords: Compressive strength, latex, polymer, slag alkaline cement and concrete, super- workable concrete, workability.

1 Introduction

High-workable concretes based on portland- and high- alumina cements in combination with superplasticizers are widely known [1]. Besides, it is already known that the introduction of different high- molecular compounds such as synthetic latexes into concrete compositions enhances rheological properties and physical and mechanical properties of the concretes [2, 3, 4].

Polymers in Concrete, edited by Y. Ohama, M. Kawakami and K. Fukuzawa. Published in 1997 by E & FN Spon, 2–6 Boundary Row, London SE1 8HN, UK. ISBN: 0 419 22330 4.

High- workable concretes should meet the stricter requirements in terms of strength, water- and corrosion resistance to aggresive media depending upon fields of application. The slag alkaline concretes which are known to possess the higher, as compared to ordinary portland cement concretes, physical and mechanical properties fully meet these requirements [6, 7].

However, the slag alkaline cements and concretes being greatly advantageous as compared to conventional cements and concretes are poorly sensitive to plasticizing action. The reason for this is that conventional plasticizers are not effective in alkaline cement systems because alkalies are plasticizers themselves. In the meantime, it is known that introduction of some organic additives into the slag alkaline concretes can improve their properties[8, 9].

The objective of this research was to develop and investigate the high- workable slag alkaline concretes, in which plasticizing effect would be attained by optimizing kind and density of alkaline component used, particle size distribution of aggregate, and by introducing organic polymer selected from a group of latexes stable in alkaline medium and which would provide their competitiveness with regard to portland cement- based concretes.

2 Materials and testing procedures

Used as basic component of high- workable concrete mixture made from slag alkaline cement was a blastfurnace granulated slag with modulus of basicity (M_b= CaO+ MgO/ SiO_2+Al_2O_3) 1.15 and modulus of activity (M_a=Al_2O_3/SiO_2) 0.148. The slag was preliminary ground in a ball mill to produce a specific surface 450 m^2/kg (by Blaine apparatus).

Used as alkaline component were: sodium silicate solution with characteristic of silicate modulus (M_s = SiO_2/Na_2O) 1 and density 1180- 1240 kg/m^3 and technical sodium metasilicate (Na_2SiO_3 $5H_2O$) in a solid state.

A quartz river sand with gradation factors of 1.38 and 2.64 was used as fine aggregate.

A styrene- butadiene latex (pH= 8.5, solid content= 53 % by mass) was used as a polymer additive. Mass share of the latex (calculated on dry matter) was 2- 15 % by mass of the slag. The slag to sand ratio was 1: 3 in the concrete mixtures. The mixing was carried out in a Hobart mixer. The workability was determined by a flow test using a Vicat cone. The flow diameter was found to be 200 mm.

Hardened concretes were evaluated for compressive strength using 4 cm- cube specimens. The hardening took place in normal conditions at 20 ± 2°C and 95- 100 % R.H.

Deformation characteristics of the high- workable concretes were measured by using a specially designed device with a dial indicator to an accuracy of 0.01 mm. A measurement at 1 day was taken as a basic one for reference. The specimens were stored under standard specified conditions of saturated sodium nitrite solution and in air at 65 % R.H.and at 95- 100 % R.H. The measurements of deformation characteristics of the slag alkaline polymer concretes were taken at 28 days of hardening.

3 Test results and discussion

Workability of concrete is known to be an important technological factor for determining simplicity and speed with which concretes may be placed.

The investigated concrete mixtures have been prepared from slag alkaline cement of two formulations and prepared in different manner. The first formulation of sodium silicate solution (M_s=1, density= 1180- 1240 kg/m^3) was added to preliminary mixed dry components, the second one was prepared with the use of solid sodium metasilicate (Na_2SiO_3 $5H_2O$) which was introduced into a mixture at a stage of preparation in a quantity of 12 % by mass of the slag and properly mixed until the components were homogeneously distributed, then the blend of dry components was mixed with water.

It was established by the researches that technological and physical and mechanical properties of the high- workable slag alkaline polymer concretes depend upon kind and state in which the alkaline component is taken (in a form of solution or solids), density of alkaline component, particle size distribution of aggregate and quantity of polymer additive introduced.

In the case of using sodium silicate solution the trend of increasing workability of the concretes with increasing density of the alkaline solution from 1180 to 1240 kg/m^3 was revealed. This may be attributed to a plasticizing action of alkalies. At a fixed workability of 200 mm, increasing density of the alkaline solution allowed to decrease the solution demand of the mixture by 15 % and more, and increase in compressive strength of the hardened concretes at 1 day- from 5 to 20 MPa or over 4 times thus surpassing greatly the requirements to these materials and accelerating the processes of placing concrete and hydroinsulation and corrosion resistant coatings.

Particle size distribution of the aggregate is known to affect strongly decline in solution/water demand of high- workable concretes. For this reason fine aggregate (quartz river sand) was properly graded and sands characterized by gradation factors 1.38 (fine-grained sand) and 2.64 (coarse- grained sand) were used.

It was established that the use of fine- grained sand having high specific surface and increased content of intergrain voids resulted in the increase in solution and water demand of the concrete mixtures as well as auxiliary cement consumption to add the required workability. Increasing the particle size (gradation factor) in aggregate from 1.38 to 2.64 allows to decrease the solution/water demand of the mixtures by 24 % and more with maintaining required workability (200 mm). In this case the solution to slag ratio may be lowered from 0.72 to 0.56 affecting positively the physical and mechanical properties of the concretes.

In light of the above facts, to further enhance the properties of the high- workable slag alkaline concretes it was found to be necessary to minimize the content of primary voids and water demand of the mixture at the expense of using plasticizing additives.

To reach this goal, along with the use of alkaline component of optimal concentration and aggregate with optimal particle size distribution, the introduction of the additives which would fill in the pores, lower void content in the aggregate, render plasticizing action and decrease water demand seemed to be of interest. All these requirements were found to be met at the expense of introduction of a polymer additive selected from a group of synthetic latexes, i.e. styrene- butadiene latex.

The additive (styrene- butadiene latex) was introduced at a stage of mixture preparation 1 minute before the end of mixing, which allowed to produce quick- hardening and high- strength concretes.

From the results of the study it can be concluded that technological properties of the slag alkaline polymer concrete mixtures depend upon quantity of polymer additive. The plasticizing action of latex additive on slag polymer concrete mixture is clearly seen in the cement formulation in which alkaline component was introduced in a form of solids. By using this cement formulation the mixtures are found to reach the required workability (200 mm) with water to slag ratio (w/s) of 0.61. The introduction of 5 % latex allowed to reduce water demand of the mixtures by 10 % and more (w/s = 0.54). With further increase in mass proportion of latex from 10 to 15 %, the water demand decreased by 24 %(w/s= 0,46) and by 30 % and more(w/s= 0.42) at a required workability of 200 mm.

Hence, the slag alkaline cement with the above formulation the workability of mixture tends to increase with increase in the mass proportion of latex from 5 to 15 % (Fig.1).

In the slag alkaline cement in which the alkaline component was introduced in a form of sodium silicate solution the plasticizing effect rendered on mixtures is found to be considerably greater as compared with that in which the alkaline component was introduced in a form of solids (Fig.1). The introduction of 5 % latex leads to the increase in workability and decrease in the solution demand by 7 %, the solution to slag ratio (s/s) decreasing from 0.55 to 0.52. With further increase in mass proportion of latex to 10 and 15 % by mass of the slag to maintain the required workability of 200 mm the solution to slag ratio should be increased up to 0.55 and 0.58, respectively.

Hence, the recommended optimal quantity of latex additive is 5 % by mass for the slag alkaline cement of this formulation(Fig.1).

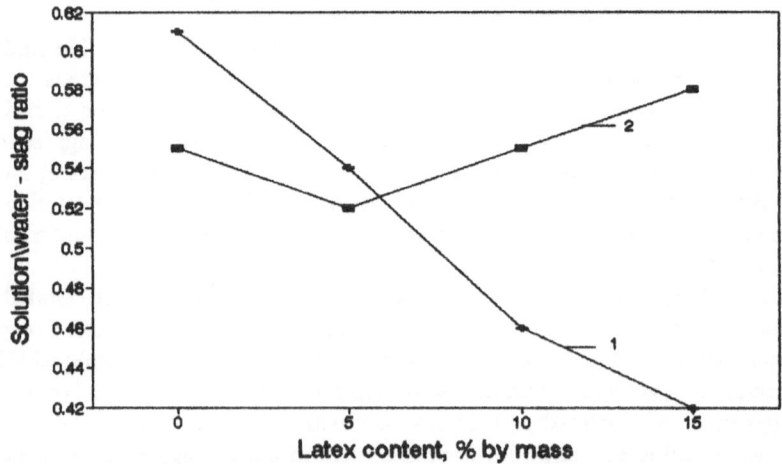

Fig. 1. Alkaline solution/ water demand of the mixtures vs latex content:
   1- slag alkaline cement formulation with technical sodium silicate solids;
   2- slag alkaline cement formulation with sodium silicate solution.

Thus, the latex additive was found to render a plasticizing action in a selective manner and is directly connected with cement composition, in particular, in which form the alkaline component is introduced.

The processes of hardening and performance properties of polymer concretes are known to be affected greatly by storage conditions [2]. In the specimens made with ordinary portland cement and stored in air- dry conditions (40- 50 % R.H.) water evaporates quickly resulting in the production of porous material with low strength characteristics, whereas when storing the polymer cement specimens under conditions of 100 % R.H. the polymers have no conditions for their full polymerization and remain in a weak state.

Taking this into account, while studying the high- workable slag alkaline polymer concretes the storage conditions that are known as the most unfavourable (95- 100 % R.H.and temperature of environment= 20± 2°C), have been selected.

The results of the study demonstrate that the introduction of 5 % latex enhances strength characteristics of high- workable slag alkaline polymer concretes by 10- 15 % as compared to those of additive-free ones and at 1 day of hardening they gain a compressive strength of 15- 20 MPa and at 28 days- 30- 40 MPa. With further increase in mass proportion of latex additive (10- 15 % by mass), the strength characteristics decline. Meanwhile, at maximum content of latex (15 %) the strength is at a level of 8- 10 MPa after 1 day of hardening and 20- 25 MPa after 28 days of hardening depending on slag alkaline cement formulation (Fig. 2).

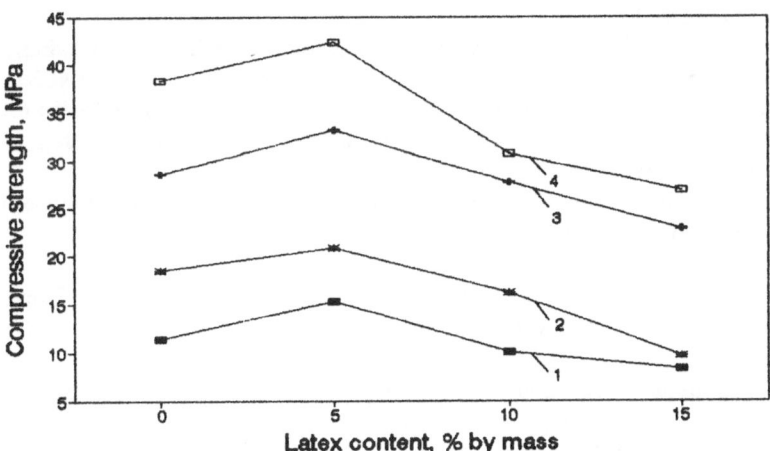

Fig. 2. Compressive strength vs latex content.
    1, 3 - slag alkaline cement formulation with technical sodium silicate solids
    at 1 day (1) and 28 days (3);
    2, 4 - slag alkaline cement formulation with sodium silicate solution
    at 1 day (2) and 28 days (4).

Strength development of the high- workable slag alkaline polymer concretes in the long- term was evaluated by change in compressive strength. The study's results demonstrate that the high- workable slag alkaline polymer concretes gain strength continuously and may be referred to as quick- hardening and high- strength super high-workable concretes. The characteristic compressive strength of concretes made with sodium silicate solutions at 1 day of hardening is 10- 20 MPa and at 28 days is 25- 40 MPa depending upon mass proportion of latex. The maximum strength of the concrete formulations containing 5 % latex is 20 MPa at 1 day age and 40 MPa at 28 days age (Table 1).

The concrete formulations with solid alkaline component are characterized by compressive strength of 8- 15 MPa at 1 day and 20- 30 MPa at 28 days depending upon mass proportion of latex. The maximum strength, as in previous case, is characteristic of the concrete formulations containing 5 % latex (15 MPa at 1 day age) and 30 MPa (at 28 days age) (Table 1).

Shelf life of the high- workable slag alkaline polymer concrete was defined as a time interval during which it maintained its working consistency. The effect of variations in latex content on shelf life of the developed concretes is quite obvious: with increase in the latex content the shelf life becomes longer.

The study's results demonstrate that shelf life of the additive- free concretes formulated with sodium silicate solution is 30 min. starting from the moment the alkaline solution was added.

Increasing the latex content to 5 - 10 % by mass results in longer shelf life (40 and 50 min., respectively). With further increase in latex content to 15 % by mass the shelf life is as long as 2 hrs.

The characteristic shelf life of latex- free concretes formulated with sodium metasilicate solids is as long as 50 min starting from the moment the water was added.

Increasing the latex content to 5 -10 % by mass results in longer shelf life (1 hr.-10 min. and 1 hr.-30 min., respectively). With further increase in latex content to 15 % by mass the shelf life is found to be 2 hr.- 40 min.

Thus, the developed formulations of high- workable slag alkaline polymer concretes meet the standard specified requirements for shelf life of these materials.

Table 1. Influence of kind of alkaline component and mass proportion of latex on strength development

Kind of alkaline component	Proportion of latex, % by mass of slag	S/S or W/S	Compressive strength (MPa) Age (days)			
			1	3	7	28
Na$_2$SiO$_3$ solution	0	0.55	18.5	22.8	24.7	38.3
	5	0.52	20.9	28.1	32.0	42.3
	10	0.55	16.2	18.4	21.2	30.8
	15	0.58	9.7	15.0	18.4	26.9
Na$_2$SiO$_3$ solids	0	0.61	11.4	18.7	21.4	28.6
	5	0.54	15.3	21.6	24.8	33.2
	10	0.46	10.1	17.3	20.5	27.7
	15	0.42	8.3	15.1	18.8	22.9

Deformation characteristics are known to be important characteristics of concretes. The results of the study demonstrate that at 95- 100 % R.H. the shrinkage deformations do not develop. Under standard conditions (65 % R.H.), the shrinkage deformations develop as soon as mass proportion of latex increases. However, they may be kept under control and be lowered to meet the required values at the expense of varying particle size distribution in the direction of increasing gradation factor.

The trend of exhibiting shrinkage deformations less than 0.3 mm/m by the concretes with increasing gradation factor to 2- 2.5 was revealed. This value of shrinkage deformations excludes cracking during the process of hardening.

A combination of high service properties of the slag alkaline cements and chemical resistance by styrene- butadiene latex determines high resistance of slag alkaline polymer concretes to different aggresive environments.

The slag alkaline latex- modified cements exhibit the increased corrosion resistance to concentrated solutions of salts of strong acids ($NaCl$, Na_2SO_4, Na_2NO_3), in sea water, in milk acid and other acidic media predicting the possibility of use for self- levelling floors and corrosion resistant coatings.

4 Conclusions

Novel slag alkaline polymer concretes with quick hardening, high strength and workability exhibiting excellent technological and service properties have been developed and investigated. The regularities allowing to govern the processing parameters of their manufacture have been determined. The formulations of slag alkaline polymer concretes for making self- levelling floors, hydroinsulation and corrosion resistant coatings have been developed. These concretes demonstrate the following characteristics: flow of 200 mm, shelf life varying from 30 min. to 2 hr.-40 min., compressive strength of 10- 20 MPa at 1 day and 30- 40 MPa at 28 days, enhanced corrosion resistance to different aggresive environments, including acid ones.

5 References

1. *Concrete Coatings for Industrial Floors* (ed.O.M.Ivanov) (1971), Moscow.
2. Cherkinsky, Yu.S. (1984) *Polymer- Cement Concrete*, Stroiizdat, Moscow.
3. Oye, B.A., Justnes, H.(1991) Microstructure and performance of polymer cement mortars based on latex. Proc. Int.Symp. "Concrete Polymer Composites", Germany.
4. Ohama, Y., Demura, K. (1991) Properties of polymer- modified mortars with expansive additives. Proc. Int.Symp."Concrete Polymer Composites", Germany.
5. Guk, V.G. (1990) *Polymer Cement Concrete in Road Construction*, Svit, Lvov.
6. Krivenko, P.V. (1992) *Special Slag Alkaline Cements*, Budivelnik, Kiev, Ukraine.
7. Krivenko, P.V. (1993) *Durability of Slag Alkaline Concrete*, Budivelnik, Kiev.
8. Krivenko, P.V., Raksha V.A., Raksha, L.V. (1996) Slag alkaline polymer cement concretes. Proc.Int.Cong. "Concrete in the Service of Mankind", Dundee, Scotland.
9. Krivenko, P.V., Raksha V.A., Raksha, L.V. (1996) Vibration- resistant slag alkaline polymer- cement concretes. Proc.Int.ICPIC Workshop on Polymers in Concrete for Central Europe, Bled, Slovenia.

PART FIVE
PRODUCTS

THE STUDY AND APPLICATION OF POLYMER CONCRETE GUTTER GRATES AND MANHOLE COVERS

H.Q. Zhu and D.W. Yang
Metals & Chemistry Research Institute, China Academy of Railway
Sciences, Beijing, China

Abstract
The thefts of cast iron manhole covers and gutter grates are becoming serious.
Therefore there has been an urgent need to develop an alternative material
for designing and manufacturing manhole covers and gutter grates. Choosing a
proper mix proportion and a cure system, steel bar reinforced polymer
concrete (SRPC) was used to manufacture gutter grates and manhole covers. The
first cracking load of a gutter grate was 90KN, the failure load was 150KN,
and the failure load of a manhole cover was 220KN. SRPC gutter grates and
manhole covers are now working well in the field and serving as an
alternative to conventional cast iron ones.
Keywords: polymer concrete, gutter grate, manhole cover.

1 Introduction

The conventional material used for manhole covers and gutter grates is cast
iron. But recently in China, the thefts of cast iron covers and gutter grates
are becoming very serious. This causes not only an economic loss but also a
menace to vehicles and pedestrians. More than 300 manhole covers and 6,000
gutter grates were stolen in Beijing in 1993. This caused direct economic
losses of 700 thousands RMB and accidences of pedestrians and bikeriders
dropping into open manholes. Accordingly, all kinds of measures were used to
protect the manhole covers and gutter grates from losing. But these measures
either increased the manufacture cost or brought inconveniences of using, and
the theft problem was not solved completely. Later, ordinary cement concrete
and steel fiber reinforced concrete (SFRC) covers and gutter grates were

Polymers in Concrete, edited by Y. Ohama, M. Kawakami and K. Fukuzawa. Published in 1997
by E & FN Spon, 2–6 Boundary Row, London SE1 8HN, UK. ISBN: 0 419 22330 4.

developed.But ordinary cement concrete is usually low in tensile strength and crack resistance, which easily affects its durability. Though the toughness of SFRC is improved, its compressive strength and other properties increase only a little, and the field performance is not satisfactory. In order to find a better way to solve this problem, research work has been undertaken in China Academy of Railway Sciences. Polymer concrete is a composite material consisting of aggregate particles held together by a polymeric binder. For its excellent physical and mechanical properties, SRPC is an alternative material for designing and manufacturing manhole covers and gutter grates.

2 Experiment

2.1 Materials
An unsaturated polyester resin (viscosity 34mPa.s, acid value 19.72, solid content 62.75%) is used as a binder. crushed stone (max. size 10mm) , river sand (fineness modulus 2.53) and quartz powder (80mesh+150mesh+250mesh) were used as aggregate and filler, and water content of these materials was controlled at less than 0.5 percent. 50% dibutylphthalate solution of cyclohexanone peroxide was used as an initiator, and 6% styrene solution of cobalt naphthenate was used as an accelerator. 20MnSi steel round bars of 10mm diameter were used as reinforcement. The design tensile strength of the bars is 340MPa.

2.2 Properties of polymer concrete
The polymer concrete mix proportion is given in table 1. The mixture was blended in a strong mixer and was moulded in required dimensions, then subjected to vibration for 2-3 minutes at a frequency of 50Hz prior to cure at 80°C for a period of 6 hours. The cured specimens were tested for physical and mechanical properties according to relative standard methods. The results are given in table 2.

Table 1   Mix proportion of PC

binder	aggregate		filler
UP(wt%)	fine(wt%)	coarse(wt%)	80mesh+150mesh+250mesh(wt%)
8.0-9.5%	30-40%	40-55%	10-15%

2.3 Gutter grate and manhole cover design and preparation
From the properties of polymer concrete given in table 2, we can see that they are much superior to that of ordinary cement concrete and SFRC. Since polymer concrete is a new structure material, there are still no design theories and specifications. So the design parameters for gutter grates and manhole covers were determined with reference to "Design Specifications of

Table 2　Phyical and mechanical properties of PC

Item	Test condition	Results
Density(kg/m^3)		2480
Compressive strength (MPa)	Size 10×10×10cm, 43 sets	$\bar{R}$=114.6, σ =6.9
Splitting tensile strength(MPa)	Size 10×10×10cm, 41 sets	$\bar{R}$=12.9, σ =1.0
Compressive modulus (MPa)	Size 10×10×20cm, 4 sets	$\bar{M}$=3.03×10^4
Fatigue intensity(MPa)	Size 10×10×40cm, ρ =0.2	45.0
Antifreeze	Slow freeze,100 cycles	Weight loss rate 0, Strength loss rate 2.6%
Creepage(10^6/MPa)	Constant 20℃, stabilized voltage load 24MPa	C(t,τ)=18.29 [1-e$^{-0.268(\tau}$,τ)$^{0.555}$], interrelation coef.0.9858
Anti-ageing	700W H.H lamp lighting 1470h	$\bar{R}$=123.5 MPa comparison groups $\bar{R}$=118.7
Anti-compact (Q=9.81×hammer weight ×impact height×times)	10×10×40cm, test piece arranged in sample supported beam,30cm biddle pivot points,impact on middle point	$\bar{Q}$=392.4J $\bar{Q}$=160J(steel fibre concrete)

Reinforced Concrete Structure" (TJ10-74) , "Design Specifications of Concrete Structure"(GBJ10-89)and some relative research results at home and abroad.

The determination of design parameters is as following: First the products of measured data and the relative size factors were converted to strength values of relative standard size samples, then the difference of the average strength and the mean square deviation was converted to standard strength of which the guarantee rate is more than 95%, and then the standard strength minus a mean square deviation, the difference times a necessary reduction factor, the product is the design strength, or get it by way of a experience formula with corresponding experience coefficient.

The determined design mechanical parameters are as following: Strength level R_b=98MPa, axial compressive strength Ra=51.2MPa, bending compressive strength Rw=64MPa, tensile strength R_1=6.2MPa, rupture strength R_r=6.8MPa, compressive static elastic modulus E_n=2.63×10⁴MPa.

Gutter grates and manhole covers have been designed on the base of 15 ton load vehicles and dynamic load coefficient 1.2. The dimensions of gutter grates are quite different from city to city in China, for example, 750× 450mm in Beijing, 670×450mm in Jinan, and 500×400mm in Harbin, the dimensions of manhole covers are quite different also. Since the bearing loads are different for gutter grates and covers of varied dimensions, different structure styles and reinforcement arrangements were used. Fig.1 and fig.2 are the illustration of the disign of gutter grates of 750×450mm and manhole covers of Φ 690mm respectively.

The manufacturing process is mainly divided into three steps: mixing, forming and curing. In the three steps, following should be paid attention to:
1. All the aggregates and fillers shall be dried and their water content shall be less than 0.5 percent. In order to decrease the viscosity of the mixture, in order to be convenient for mixing and forming, the temperature

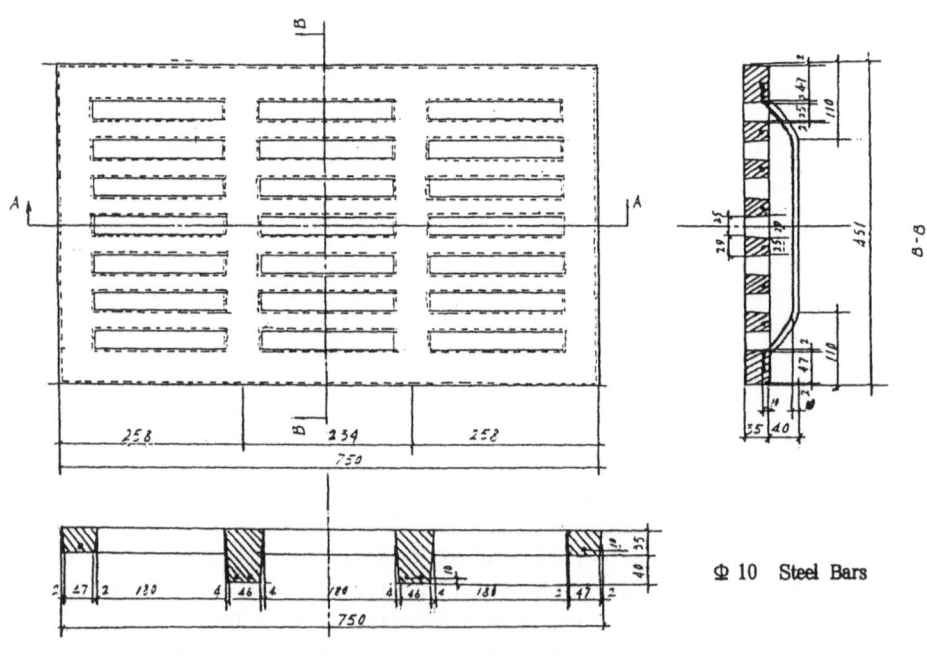

Fig.1 The disign of gutter grates of 750×450mm

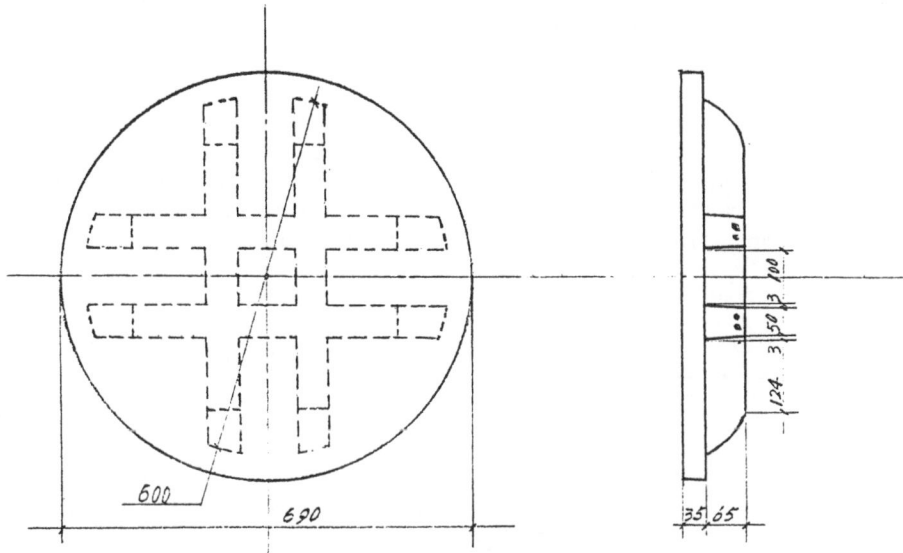

Fig. 2 The disign of manhole covers of Φ 690mm

of aggregates should be controlled between 40-50℃ before mixing.

2. Mixing must be carried out in a strong mixer.
3. Gutter grates and manhole covers are formed by vibration due to the presence of steel bar reinforcement, and built-up moulds are used.
4. In order to speed up moulds turnover, curing is carried out in two steps. First gutter grates and manhole covers are cured at room temperature for 1 -2 hours, and then demoulded and cured at about 80℃ for 5-6 hours.

2.4 Testing procedure

2.4.1 Load test

A gutter grate was put on a steel made four edges frame simply supported steadily. Three layers of rubber plates (200×350×5mm for each, the contacting square of a wheel on a gutter grate) were put on the middle of the gutter grate. Upon the rubber a steel plate (200×350×5mm) was set. A single concentrated load was supplied on the middle portion of the steel plate step by step and sustained for 10 minutes for each step. The first load of 10KN was increased by steps of 10KN. This test simulated the situation in which the load was supplied mainly on the short rib of the gutter grate. The load situation of test is illustrated in fig. 3. The test result shows that the first cracking load was 90KN, and failure load was 150KN. In fact the calculated load of gutter grates caused by 15 ton vehicles is only 42KN. Therefore the safety coefficient is sufficient.

The load test of manhole covers is almost the same as that of gutter grates. The load situation of test is illustrated in fig. 4. Because of the obstacle of the supporting structure, the first crack could not observed, the first crack load was not known, and the failure load was 220KN.

2.4.2 Impact test

This test simulated the situation in which objects on vehicles fall down and impact a gutter grate. A steel ball of 870g weight was held 1.2m above a gutter grate which was put on a steel made frame supported steadily, and dropped freely straight to a small rib of the grtter grate. No crack occured after 10 repetitions. If the steel ball impacted the edge of a small rib,

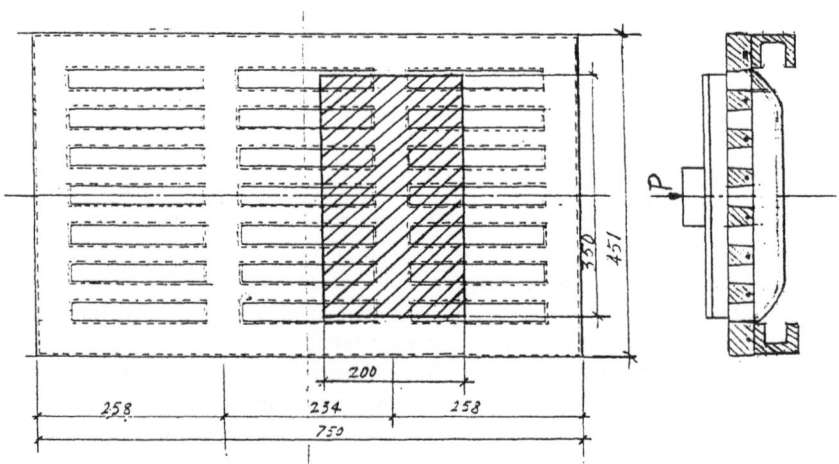

Fig.3 The load test of gutter grate

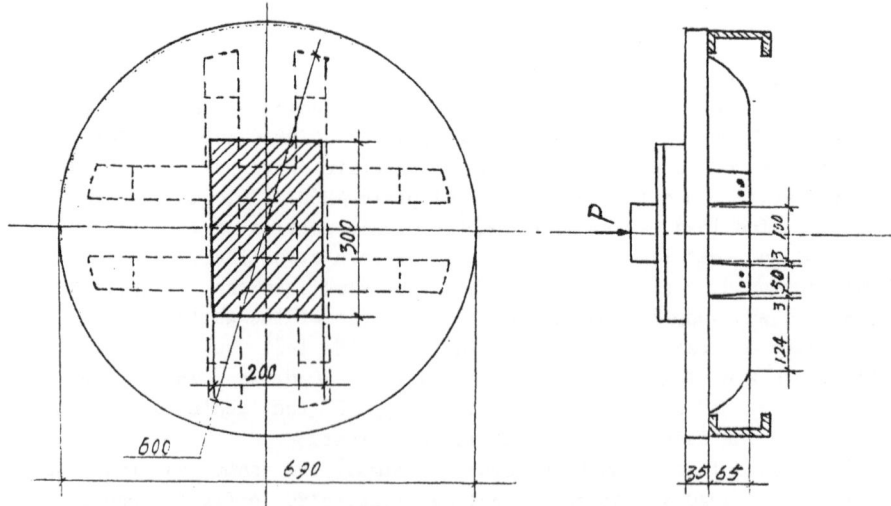

Fig.4 The load test of manhole cover

litter small broke off, but no crack occurred under the small rib. The test result shows that the impact resistance is sufficient to meet the needs of normal use of gutter grates.

There is no discharge opening on a manhole cover. As a whole, the impact resistance of a manhole cover is more supperior to that of a gutter grate. So the impact test is not necessay for covers.

2.4.3 Field test

After finishing preparation of the sample of SRPC gutter grates and manhole covers, iron cast gutter grates and manhole covers in factory road were replaced by SRPC ones. Varied heavy load trucks passed over them, the result of inspection afterwards was satisfactory and shows that SRPC gutter grates and manhole covers can be used in city road. Now they have been widely using in city road for more than two years in Jinan and Wuhan. Most of them are serving better, and well received by users.

3 Conclusions

1. Comparing with iron cast gutter grates and manhole covers, SRPC ones have many advantages, such as simple process, low cost and steel saving.
2. Because SRPC gutter grates and manhole covers have no resale value, thefts are avoided, and all troubles caused by them are solved. Therefore SRPC gutter grates and manhole covers are being used as a highly competitive alternative to conventional cast iron ones which are reported to attract " pilferage"because of their scrap resale values.
3. SRPC gutter grates and manhole covers of various dimensions can be designed and manufactured for 15-ton load of heavy duty vehicles.

4 References

1. Kimio Fukuzawa, and Tatsuya Numao. Flexural behavior of polymer concrete beams reinforced with fiber reinforced plastic rods. Proceedings of the First East Asia Symposium on Polymers in Concrete. Chuncheon, Korea, May 2 -3, 1994. pp. 325-331.
2. Yang Dianwen, and Zhu Hequan. Development of a polymer concrete railway bridge sleeper. Proceedings of the First East Asia Symposium on Polymers in Concrete. Chuncheon, Korea. May 2-3, 1994. pp. 337-344.
3. C. Vipulanandan. Mechanical behavior of polymer concrete system. Materials and Structures. 1988. 21. pp. 268-277.
4. O. P. Ratra. Durability of plastics fibres as reinforcement in cement concrete composites. Proceedings of the 6th International Congress on Polymers in Concrete. Sept. 24-27, 1990. Shanghai, China. pp. 434-441.

STUDY ON CENTRIFUGAL REINFORCED POLYMER CONCRETE PIPE

F. Omata
Technical Research Institute, Sho-Bond Corporation,
Ibaraki, Japan
H. Tokushige and M. Kawakami
Dept. of Civil Engineering, Akita University, Akita, Japan
O. Shinoe
Technical Research Center, Teihyu Corporation, Chiba, Japan
H. Okamoto
Resin Research Laboratory, Nippon Shokubai Co., Ltd.,
Osaka, Japan

Abstract
This paper describes the fundamental characteristics of centrifugally compacted reinforced polymer concrete pipe as a high quality sewer pipe. Firstly, mechanical and physical properties of the unsaturated polyester polymer concrete such as compressive and flexural strength, modulus of elasticity, surface roughness and abrasion were investigated. Secondly, reinforced polymer concrete pipes were manufactured by centrifugally compacted method as making Hume pipe. A new mixing and placing equipment for polymer concrete was proposed and developed. Thirdly, external load test and flexural test for the pipes were carried out and the results were discussed. Finally, the rational design methods were suggested based on the obtained test results.
Keywords:Centrifugal reinforced pipe, external load test, polymer concrete, roughness.

1 Introduction

It is urgent to spread and construct a sewer system in Japan. In the previous sewer system, polyvinyl chloride pipe (PVC pipe) has been used for relatively smaller size than diameter of 500 mm and its share is about half of total sewer pipes in Japan[1].
As this PVC pipe has a lot of advantage such as light weight, excellent durability to waste water, high waterproof quality and high tensile strength, it is widely used as flexible pipe. But this pipe has been applied for the restricted shallow cover earth depth because of non– reinforcing structure and lack of rigidity.

On the other hand, reinforced concrete pipe as an ideal pipe for transporting water has mainly used for the medium size of the diameter from 350 mm to 3000 mm. Ordinary reinforced concrete pipe has been made from cement concrete and

Polymers in Concrete, edited by Y. Ohama, M. Kawakami and K. Fukuzawa. Published in 1997 by E & FN Spon, 2–6 Boundary Row, London SE1 8HN, UK. ISBN: 0 419 22330 4.

reinforcing steel. Both concrete and steel have shortcomings to receive chemical attack, especially more acidic than pH 4. Therefore, recently the new technologies[2] have been developed the strong, durable and economical pipes such as the reinforced concrete coated at the interior surface by polymer concrete and the composite pipes[3][4] of cement concrete and PVC panel have gradually increased. In addition to the above composite pipes, centrifugally compacted reinforced polymer concrete pipe, which has the advantages of both conventional PVC pipe and reinforced concrete pipe, has developed.

This paper describes a study on the mechanical properties and characteristics of centrifugally compacted polymer concrete pipe. Mechanical and physical properties of polymer concrete used for polymer concrete pipe were firstly investigated. Subsequently mechanical properties of the pipe were clarified by loading tests and were compared with those of the conventional reinforced concrete pipe. Finally, rational design methods for polymer concrete pipe were discussed and proposed.

2 Materials and Test Procedures

2.1 Polymer concrete
1) Binder
Unsaturated polyester resin and low shrinkage polymer as binder were used. Properties of these resin are listed in Table 1. Furthermore, chemical formulation of the resin is shown in Table 2.

Table 1   Properties of unsaturated polyester and low shrinkage polymer

	Unsaturated polyester	Low shrinkage polymer
Specific gravity at 25 °C	1.05	0.95
Acid value	17	–
Viscosity at 25 °C (poise)	Smaller than 1	12
Gelation time at 25 °C (min)	5	–
Monomer content (%)	45	70

Table 2   Formulation of resin

Material	Weight (%)	Remarks
Unsaturated polyester resin	80	Medium reactive orthophtalic type resin
Low shrinkage polymer	17.56	Polystyrene type thermoplastic resin
Methyl ethyl keton peroxide	1.46	55 % solution in DMP
Cobalt naphtenate	0.49	Co contents 6 %
N,N'–Dimethyl aniline	0.49	10 % solution in SM
Total	100.0	–

2) Aggregate and filler
Silica sand as an aggregate and powder of calcium carbonate as a filler were used. Size of sand ranges from 0.15 mm to 2.5 mm and its specific gravity and fineness modulus were 2.63 and 2.76, respectively.

3) Mix proportion
In order to obtain higher compressive strength than 80 N/mm^2, the following mix proportion was designed. Weight ratio of resin : filler : sand was 1.00 : 0.72 : 6.43.

2.2 Dimension of pipe
Two types of pipes were manufactured as shown in Fig. 1. One is a bell type reinforced polymer concrete pipe and two kinds of spiral reinforcing steel of B–1 and B–2 were arranged. The other is a jacking pipe without bell socket and two kinds of thickness of J–1 and J–2 were made. The details of these pipes were also listed in Fig. 1. The effective length was 2000 mm for all pipes.

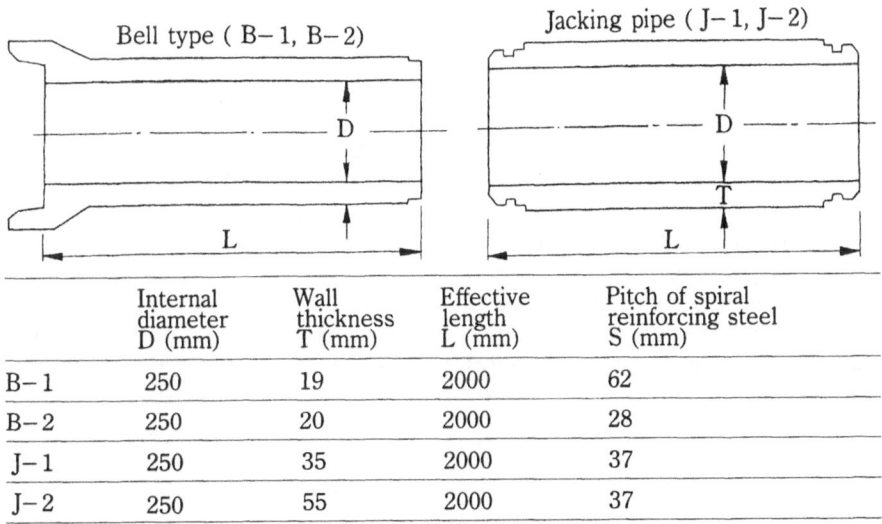

	Internal diameter D (mm)	Wall thickness T (mm)	Effective length L (mm)	Pitch of spiral reinforcing steel S (mm)
B–1	250	19	2000	62
B–2	250	20	2000	28
J–1	250	35	2000	37
J–2	250	55	2000	37

Fig. 1 Shape and dimension of tested pipes

2.3 Reinforcing steel
The diameter of reinforcing steel in spiral and longitudinal direction was 2.6 mm and 3.2 mm, respectively. The number of the reinforcing steel in longitudinal direction was ten. The tensile strength of reinforcing steel was 726 N/mm^2.

2.4 Mixing, placing, centrifugal compaction and curing
Mixing and placing were performed using the specially developed process equipment. The flow chart of this process is shown in Fig. 2. In this flow chart, aggregate and filler were pre–mixed at a storage tank and resin was storaged at another tank, and they were ejected through two separate nozzles from tanks and mixed together. The mixed polymer concrete was fed from the spray gun to the rotated cylindrical metal molds. Placing time was 2.5 minutes and duration and gravitational acceleration of

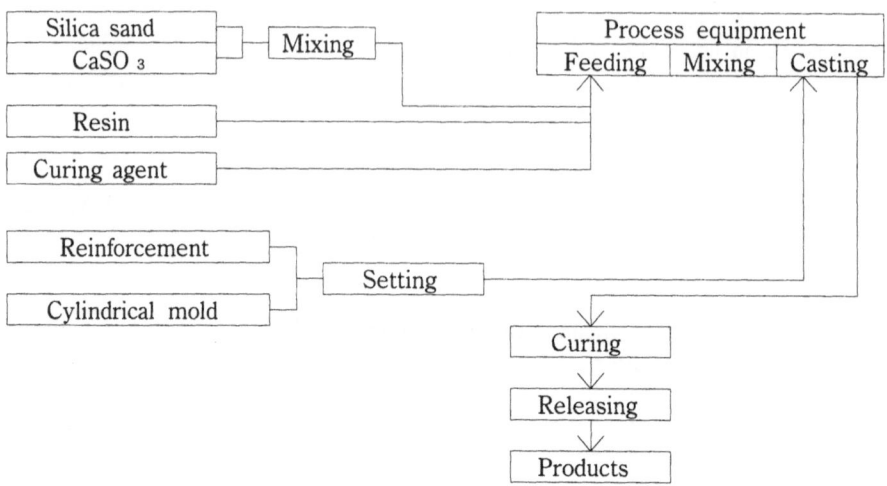

Fig. 2 Flow chart of manufacturing reinforced polymer concrete pipe

centrifugal compaction were 10 minutes and 10 G at the temperature of 25 °C . After compaction, curing was done for 20 minutes at the temperature of 80 °C . Then, the test pipe was released from cylindrical metal mold and was cured at room temperature of 20 °C and humidity of 65 %.

2.5 Test specimens and test method

According to JIS A 5303, external load test of polymer concrete pipe by two– edge bearing method as shown in Fig. 3 was carried out. Strains at inner and outer surface of the pipe and deflection of vertical direction due to deformation of diameter were measured. Flexural test by center– point loading method as shown in Fig. 4 was performed. Strain and deflection at span center were observed. Furthermore, failure test of the pipe under constant internal pressure as shown in Fig. 5 was done, according to JIS A 5303. It was observed that the failure of the test pipe is occurred by internal pressure and water is leaked out from the joints of two pieces of pipe for the straight alignment and deflected position.

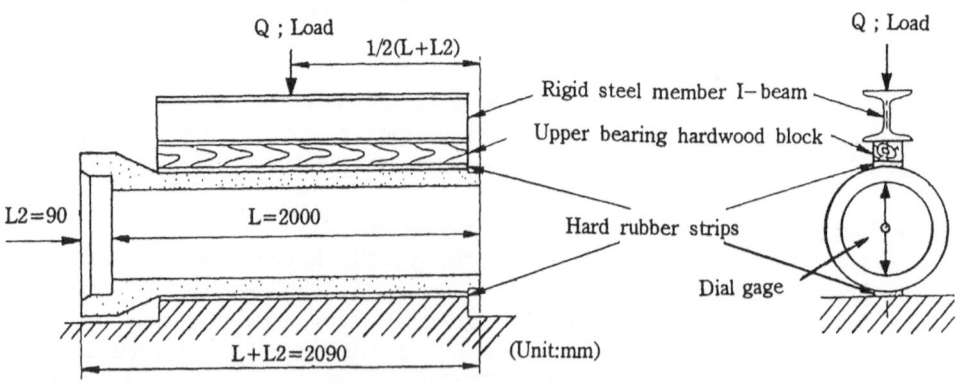

Fig. 3 External load test

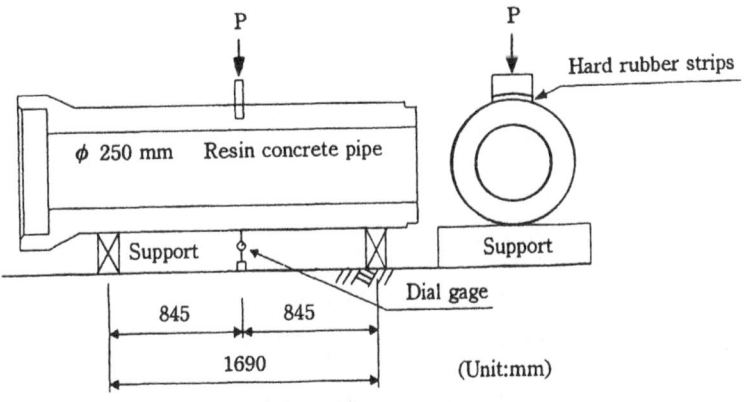

Fig. 4   Flexural test

Fig. 5   Internal pressure test

3 Results and Discussions

3.1 Mechanical and physical properties of polymer concrete

Flow value of polymer concrete tested according to JIS R 5201 was 150 mm. From the view point of sewer pipe structure, fundamental properties of polymer concrete were investigated and their results were listed in Table 3. It is worth notice that flexural strength of polymer concrete is about two to three times higher than that of cement concrete. Because flexural strength of the concrete is deeply related with cracking strength of pipe. Surface roughness of polymer concrete is very similar to that of PVC as shown in Fig. 6. Roughness coefficient of the polymer concrete pipe is estimated to be 0.010, however that of conventional cement concrete pipe is 0.013. Other physical properties of polymer concrete are generally superior to those of cement concrete.

3.2 External load test strength of polymer concrete pipe

Ultimate strengths of four pipes are shown in Table 4. It is noticed that cracking strength and ultimate strength should be almost same value. Elastic behavior had continued until just before cracking and the brittle failure mode was observed. At that time, most spiral reinforcing steels were torn and cracks ran at inner top and bottom surface and at outer lateral side along the longitudinal direction as shown in

Table 3   Properties of polymer concrete

	Test method	Specimen size	Value	Age
Compressive strength	JIS A 1181 [5] JIS A 1182 [6]	φ 50X100 mm	85.4 MPa	14 days
Flexural strength	JIS A 1181 [5] JIS R 5201 [7]	40X40X160 mm	24.8 MPa	7 days
Modulus of elasticity	JIS A 1181 ASTM C 469–65 [8]	φ 50X100 mm	24.7 GPa	14 days
Setting shrinkage	Length change	5X40X180 mm	0.1 %	14 days
Impact resistance	JIS K 5400 [9]	100X100X14 mm	Sound. greater than 110 cm	14 days
Abrasion	JIS K 5400	100X100X14 mm	0.467 g	14 days
Surface roughness	JIS B 0601 [10]	100X100X14 mm	9 μm	14 days

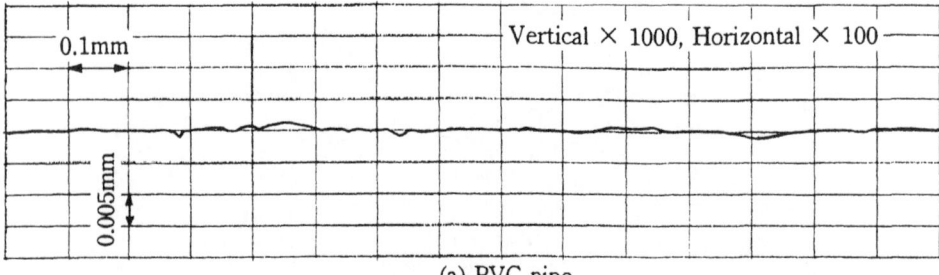

(a) PVC pipe

(b) Reinforced concrete pipe

(c) Polymer concrete pipe

Fig. 6   Surface roughness of the pipe

Table 4　Ultimate strength in external load test and flexural failure strength

	Ultimate strength (kN/m)	Flexural failure strength (kN)
B−1	36.45	52.14
B−2	39.44	unable
J−1	91.94	130.75
J−2	262.82	176.52

Fig. 7. Table 5 shows cracking and ultimate strengths of conventional bell type reinforced concrete pipe and jacking pipe with same size specified in JIS. Comparing these strengths and the specified strengths, strengths of polymer concrete pipes are very high. Ultimate strength of pipe J−2 is about five times larger than that of jacking pipe. Figures 8 and 9 show the relationships between external load and strain, and Figs. 10 and 11 show relationships between external load and vertical deflection. Compressive strains at points 2 and 3 increased lineally with the external

Fig. 7　Failure of polymer concrete pipe subjected external load

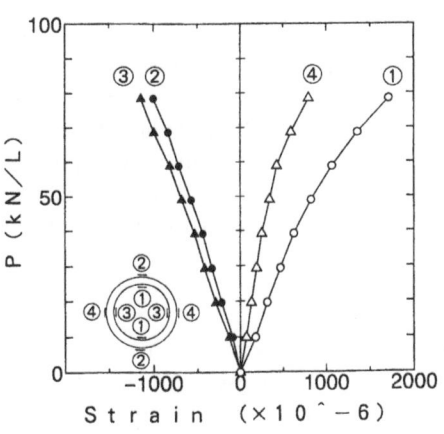

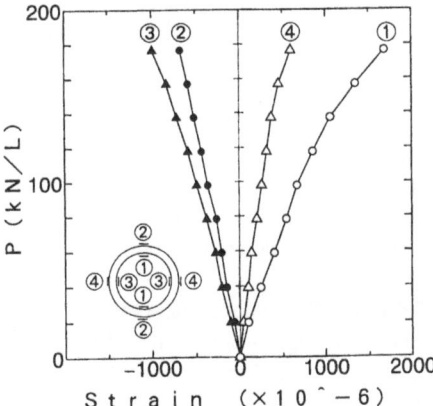

Fig. 8 Relationship between external load and strains of pipe B−2

Fig. 9　Relationship between external load and strains of pipe J−1 load and a little plastic tendency was observed for tensile strains at points 1 and 4. Indeed there were small differences between vertical deflection at socket and that at spigot, but the deflections were almost linear.

Table 5   Strength requirement for conventional pipes specified in JIS.

| | | Strength requirement (kN/m) | |
		Cracking	Ultimate
Ordinary RC pipe wall thickness 28 mm	Type 1	16.67	25.50
	Type 2	23.54	47.07
Jacking pipe wall thickness : 55 mm		32.36	49.03

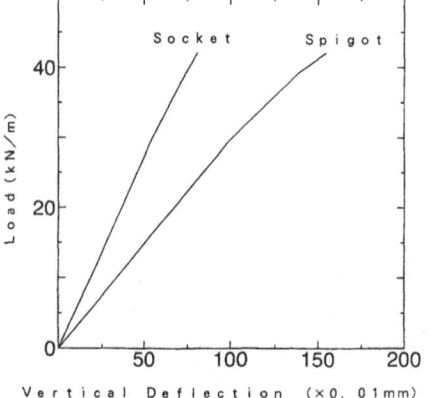

Fig. 10   Vertical deflection of pipe B−2

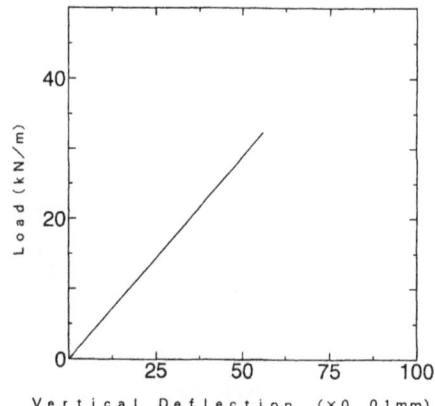

Fig. 11   Vertical deflection of pipe J−1

3.3 Flexural failure strength of polymer concrete pipe

Flexural failure mode of polymer concrete pipe was also quite different from that of cement concrete pipe.   One wide and deep crack at the bottom fiber closing to center of span initiated the failure.   Cracking and failure have occurred suddenly from the elastic state.   Flexural failure strengths are shown in Table 4.   As the flexural failure strength of the bell type cement concrete pipe with 250 mm diameter and 28 mm pipe thickness was 21.57 kN, polymer concrete pipe is about 2.5 times stronger than cement concrete pipe.   Figure 12 shows the relationships between load and deflection at center of span.   Deflection was lineally proportional to load for every pipe and the rate is mainly depend on pipe thickness.

3.4 Internal pressure resistance of polymer concrete pipe

The resisting capacities to internal pressure for bell type polymer concrete pipe and jacking one higher than 1 MPa and 0.3 MPa, respectively were proved by test. These values were 10 and 3 times larger than the internal pressure for conventional cement concrete pipes specified in JIS.   In the case of deflected joint, these polymer concrete pipes could resist the internal pressure of 0.3 MPa under the inclined degree of 3 ° 30'.   From these results of internal pressure test, it seems that polymer concrete pipes have sufficient capacities as sewer pipe.

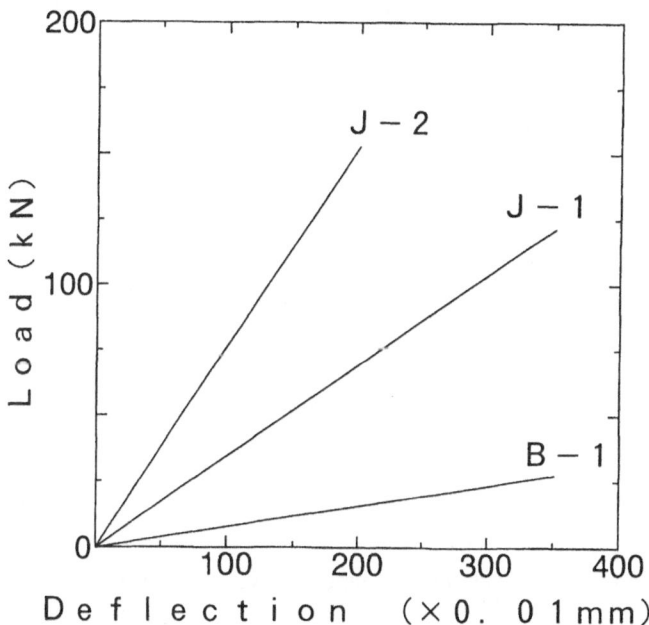

Fig. 12   Deflection of polymer concrete pipe in flexural test

4 Design of Centrifugal Reinforced Polymer Concrete Pipe

Applying polymer concrete pipe for sewer pipe, three conditions of high strength, excellent durability and good economy should be satisfied.   Indeed both conventional PVC pipe and reinforced concrete pipe have been mainly used for sewer pipe system, but these pipes should be used in the restricted conditions mentioned above. Therefore a lot of composite pipes have been tested and proposed.   Then, it is expected that progressive research for the composite pipe will be further advanced. Strength and durability of polymer concrete pipe were excellent, because polymer concrete itself is of high strength and high durability.   In this study cracking strength and ultimate strength of polymer concrete pipe were almost same value.   But it is convenient for practical application that the ultimate strength is greater than cracking strength.   For these results, it seems that increase of reinforcing steel and use of new reinforcing materials such as carbon fiber and aramid fiber will give a good solution.

Furthermore, initial cost of polymer concrete pipe is about 1.3 times more expensive   than that of conventional cement concrete pipe.   But comparing total costs including initial cost and maintenance one of polymer concrete pipe with other pipes, polymer concrete pipe expecting excellent strength and durability is not always expensive.

5 Conclusions

Strength of centrifugal reinforced polymer concrete pipe was investigated and design method was discussed.   The following remarkable conclusions were obtained.

1. Mechanical and physical properties of polymer concrete used for the pipe are clarified.   Comparing polymer concrete with cement concrete, flexural strength and surface roughness of polymer concrete are specially superior.

2. A newly developed process equipment for mixing and placing of polymer concrete for centrifugal compaction was used and introduced.

3. Cracking strength and ultimate strength obtained by external load test were almost same value, which was much higher than the specified strength of JIS for reinforced cement concrete pipe of same diameter. Therefore, the brittle failure of the polymer concrete was observed.

4. Flexural failure strength and internal pressure resistance were high and polymer concrete pipe can be sufficiently applied for sewer pipe.

5. Design of polymer concrete pipe was discussed, being based on the obtained test results.

Development and use of high quality pipe are very important to accumulate the infrastructures.

6 References

1.   Kawakami, M. et al. (1990) Design of Composite pipe Using Polymer Mortar and Expansive Concrete. *Proceedings of the 6th International Congress on ICPIC*, Shanghai, pp. 548– 555.
2.   Kawakami, M. et al. (1989) Coating Technologies for Sewer Pipe. *Concrete International of ACI*, Vol. 11, No. 11, pp. 86– 88.
3.   Kawakami, M. et al. (1984) Study on Sewer Pipes Using Plastics. *Proceedings of the International Symposium on Future for Plastics in New Construction and in Maintenance, Rehabilitation Repair and Reinforcement of Existing Structures*, Liege, pp. 1B. 18. 1– 6.
4.   Kawakami, M. et al. (1984) Strength of Polymer Composite Pipe. *Proceedings of the 4th International Congress on ICPIC*, Darmstadt, pp. 169– 174.
5.   JIS A 1181 "Method of Making Polyester Resin Concrete Specimens"
6.   JIS A 1182 "Method of Test for Compressive Strength of Polyester Resin Concrete"
7.   JIS R 5201 "Physical Testing Methods of Cement"
8.   ASTM C469– 65 "Test for Static Modulus of Elasticity and Poisson's Ratio of Concrete in Compression"
9.   JIS K 5400 "Testing Method for Organic Coatings"
10.   JIS B 0601 "Surface Roughness"

PROPERTIES OF POLYMER-IMPREGNATED SILICA FUME MORTAR FOR PRECISION SURFACE PLATE

N. Maeda
Maeta Concrete Industry Ltd., Sakata, Japan
T. Kobayashi, M. Maita and S. Miura
Maeta Techno-Research, Inc., Sakata, Japan

Abstract
This paper deals with a possibility of polymer-impregnated silica fume mortar for use as precision surface plate instead of granite or cast iron. Mechanical properties and length change of silica fume mortars impregnated with methyl methacrylate monomer (MMA) or alkyl alkoxy silane are examined. To measure the length change, the mortars are subjected to the cyclic conditions of dry and moist. The effect of specimen size on the length change of partial MMA-impregnated silica fume mortar are also studied. As a result, the impregnation of MMA monomer gives a significant improvement in strength and modulus of elasticity, and no noticeable change in the length of silica fume mortar, and can successfully be used as precision surface plate applications.
Keywords : Alkyl alkoxy silane, length change, methyl methacrylate, partial polymer-impregnation, precision surface plate, silica fume mortar, specimen size, strength.

1 Introduction

The dimensional stability is an essential requirement of the surface plates, when they are used for size examination of workpieces. Generally the surface plates are made by granite or cast iron. However, it is difficult to produce the granite surface plates with large scale or complicated shapes because of a little existence of large scale granite without internal defects and their high processing cost. The aim of this study is to investigate a

Polymers in Concrete, edited by Y. Ohama, M. Kawakami and K. Fukuzawa. Published in 1997 by E & FN Spon, 2–6 Boundary Row, London SE1 8HN, UK. ISBN: 0 419 22330 4.

possibility of using high-strength concrete or mortar for the precision surface plates.

Silica fume concrete or mortar has very high strength, low permeability, low creep, etc. [1, 2] in contrast with ordinary cement concrete or mortar. However, the possibility of length change, namely, shrinkage or expansion can take place during its period of service, and it may exert a bad influence on flatness of the precision surface plates. To discuss above problem, silica fume mortar specimens were impregnated by methyl methacrylate monomer (MMA) or alkyl alkoxy silane. In this paper, their mechanical properties and length change under the cyclic conditions of dry and moist were examined. The effect of specimen size on the length change of partial MMA-impregnated silica fume mortar were also studied.

2 Materials

2.1 Silica fume mortar

Ordinary portland cement, commercially available Norwegian silica fume, crushed sand (specific gravity : 2.8) and water-reducing agent were used for the preparation of silica fume mortar. To reduce the shrinkage of silica fume mortar due to hydration, an expansive additive was used in one mixture. The properties of silica fume are listed in Table 1.

Table 1. Chemical and physical properties of silica fume.

Chemical composition (%)								Particle size (μm)	Specific surface area (m^2/g)
SiO_2	C	Fe_2O_3	Al_2O_3	Na_2O	K_2O	MgO	SO_3		
89.6	1.92	1.30	0.87	0.48	2.15	2.22	0.62	0.13-0.16	15-20

2.2 Impregnant

Monomer system for impregnation was prepared by mixing of methyl methacrylate monomer (MMA), 2,2'-azobisisobutyronitrile (AIBN) as a catalyst and γ-methacryloxypropyltrimethoxy silane (Silane) as a coupling agent. The formulations of impregnant were MMA : AIBN : Silane = 100 : 1 : 1 (by weight). Alkyl alkoxy silane (AAS) was also used as a waterproofing agent for another impregnation.

3 Testing procedures

3.1 Preparation of base mortars

Silica fume mortars listed in Table 2 were molded and cured as shown in Table 3. The flow of the mortar was 200±10. All the specimens were subjected to both steam and autoclave curing in order to minimize drying

shrinkage of the mortar and prevent crack initiation during 150 °C drying for
polymer-impregnation.

Table 2. Mix proportions of silica fume mortar.

Mix No.	$\dfrac{W}{(C+E)}$ (%)	$\dfrac{SF}{C}$ (%)	Unit weight (kg/m^3)					
			Water W	Cement C	Silica fume SF	Sand	Water-reducing agent	Expansive additive E
A10	30	10	237	791	79.1	1251*	8.99	-
A20	30	20	220	733	147	1251*	15.4	-
B10	30	10	202	672	67.2	1529**	8.48	-
B20	30	20	188	626	125	1529**	11.8	-
B20-E	30	20	188	570	125	1529**	11.8	56.0

Notes;   *   1.2mm or finer, sand volume of 45%
         ** 5.0mm or finer, sand volume of 55%

Table 3. Outline of the test.

Series of test	Mix No.	Specimen size (mm)	Curing method	Impregnant	Test
A	A10 A20	40x140x160 φ 50x100	80℃-4h steam + 180℃-4h autoclave	MMA AAS	Polymer loading Compressive strength Length change (50%R.H. and 85%R.H.)
B	B10 B20 B20-E	40x40x160 60x60x240 100x100x400	80℃-4h steam + 180℃-4h autoclave	MMA	Polymer loading Polymer depth Flexural Strength Length change (50%R.H.)

3.2 Polymer-impregnation

Cured silica fume mortars were dried in an oven at 150°C for 72 hours and
cooled to room temperature. For impregnation of MMA monomer system,
the cooled specimens were evacuated at 10mmHg or lower for one hour.
Then, the specimens were soaked in MMA monomer system under an
atmospheric pressure for 18 hours. After MMA-impregnation, the
specimens were placed in hot water at 80°C for 4 hours for thermal
polymerization and allowed to cool at 20°C, 50% R.H. Alkyl alkoxy silane
was impregnated by another method; five coats of the alkyl alkoxy silane
were applied by a brush on the cooled specimens. Polymer loading of the
mortar specimens was determined by calculating the ratio of the change of
weight to the value before the impregnation.

   In this study, the specimens were not subjected to overpressure during
impregnation process, hence complete impregnation is not possible. The
effect of specimen size on the length change were evaluated by using
specimens in series B.

3.3 Strengths test

Cylindrical specimens were tested for compressive strengths in accordance with JIS A 1108 (Method of Test for Compressive Strength of Concrete). Beam specimens were tested for flexural strengths in accordance with JIS R 5201 (Physical Testing Methods of Cement) and JIS A 1106 (Method of Test for Flexural Strength of Concrete).

3.4 Length change test

Polymer-impregnated and unimpregnated specimens were tested for length change according to JIS A 1129 (Methods of Test for Length Change of Mortar and Concrete). In series A test, the specimens were subjected to cyclic conditions of 20°C-50%R.H. for 12 days and 20°C-85%R.H. for 2 days. In series B test, the specimens were maintained at 20°C-50%R.H.

4 Results and discussion

4.1 Effect of cyclic conditions of dry and moist

Figs. 1 and 2 show the compressive strength and modulus of elasticity of the polymer-impregnated(MMA and AAS) and unimpregnated (Normal) silica fume mortars. Generally, the compressive strength and modulus of elasticity of the mortars are increased by MMA-impregnation regardless of silica fume content. The compressive strength of AAS-impregnated mortars is slightly higher and the modulus of elasticity of the mortars is lower than those of

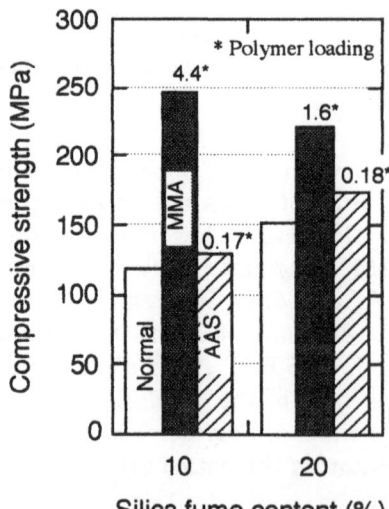

Fig.1. Compressive strength of impregnated and unimpregnated silica fume mortars.

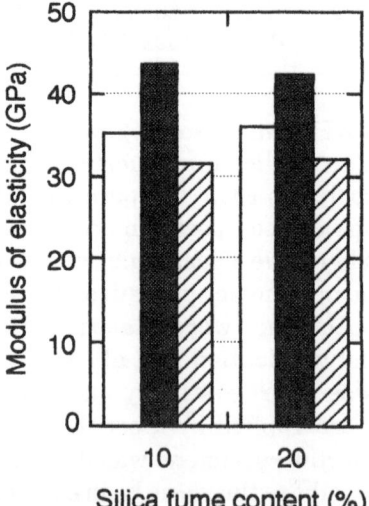

Fig.2. Modulus of elasticity of impregnated and unimpregnated silica fume mortars.

unimpregnated mortars. AAS-impregnation does not influenced on mechanical properties of silica fume mortars and the slight improvement in compressive strength may be attributed to the effect of drying of the mortars before impregnation. The difference in polymer loading in MMA-impregnation between mortars containing silica fume content of 10% and 20% may be due to less porosity of mortars.

Figs. 3 and 4 represent the length change of polymer-impregnated and unimpregnated silica fume mortars under cyclic conditions of dry and moist. The behavior of length change varies with methods of treating silica fume mortar. Generally, AAS-impregnated silica fume mortar tends to expand, although unimpregnated silica fume mortar tends to wholly shrink, regardless of the silica fume content. It is considered that the expansion of AAS-impregnated silica fume mortar is caused by moisture absorption during exposure in 50%R.H. condition because the specimens were absolutely dried and applied alkyl alkoxy silane which has moisture vapor transmission before length change test. In case of the unimpregnated silica fume mortar, alternating drying shrinkage and expansion were observed. This length change behavior seems to be not suitable for precision surface plates which require accuracy of flatness in the order of micro meters.

On the other hand, MMA-impregnated silica fume mortar hardly shows any expansion and shrinkage even under cyclic conditions of dry and moist. Generally speaking, the length change of polymer-impregnated mortars tend to expand because of the stress relaxation caused by setting shrinkage during polymerization of MMA monomer [3]. However, such expansion of

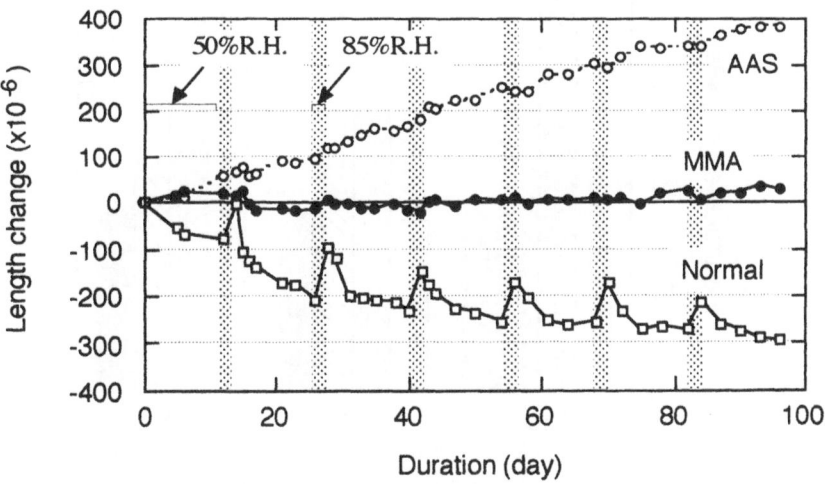

Fig.3. Length change of impregnated and unimpregnated mortars with silica fume content of 10%.

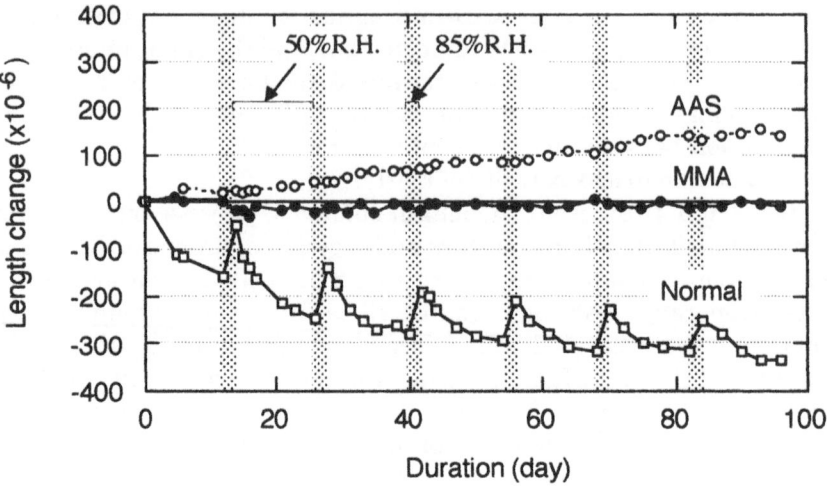

Fig.4. Length change of impregnated and unimpregnated mortars
with silica fume content of 20%.

MMA-impregnated silica fume mortar is hardly recognized within 96-day
exposure. This may be due to the low polymer loading depending on low
porosity of silica fume mortar.

4.2 Effect of specimen size
Fig. 5 exhibits the effect of specimen size on flexural strengths of MMA-
impregnated silica fume mortars.

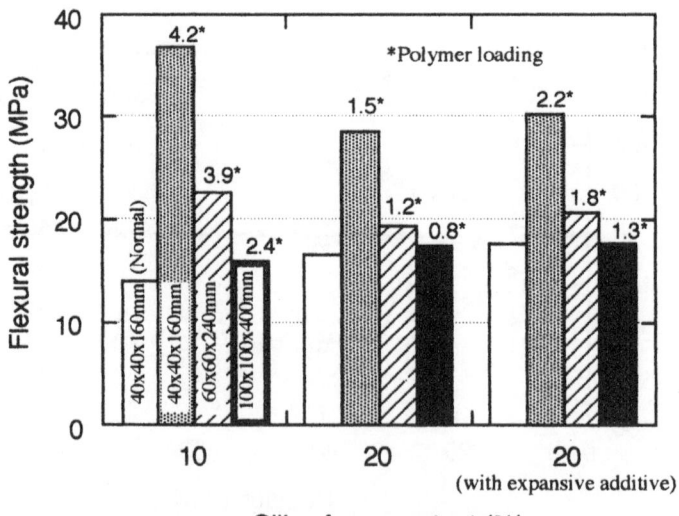

Fig.5. Flexural strengths of MMA-impregnated mortars with variation
of specimen sizes.

Depending upon polymer loading and depth, the flexural strength of the MMA-impregnated mortars is decreased with an increase in the specimen size regardless of silica fume content. The flexural strength of MMA-impregnated silica fume mortars containing expansive additive is slightly higher because of high polymer loading due to more porosity in hydration process. Except for the 40x40x160mm beam mortar with silica fume content of 10%, the mortars were partially impregnated. Therefore, the polymer loading is strongly influenced by the size of specimens. The polymer depth lies in the range of 20 to 25mm for silica fume content of 10%, and 6 to 10mm for silica fume content of 20%.

Figs. 6, 7 and 8 represent the length change of MMA-impregnated and unimpregnated silica fume mortars with different specimen sizes. As a matter of course, the shrinkage of unimpregnated mortars rises with an increase in additional dry curing period. The slight expansion of MMA-impregnated mortars are recognized. The effect of specimen size on length change of MMA-impregnated mortars are also recognized, and the length change of 100x100x400mm specimens are larger than those of 40x40x160mm and 60x60x240mm specimens, regardless of silica fume content. However, the value of length change of MMA-impregnated mortars are quite smaller than that of unimpregnated mortar. The expansion of MMA-impregnated mortars may caused by moisture absorption. The effects of silica fume content and the expansive additive on length change of MMA-impregnated silica fume mortars are hardly recognized.

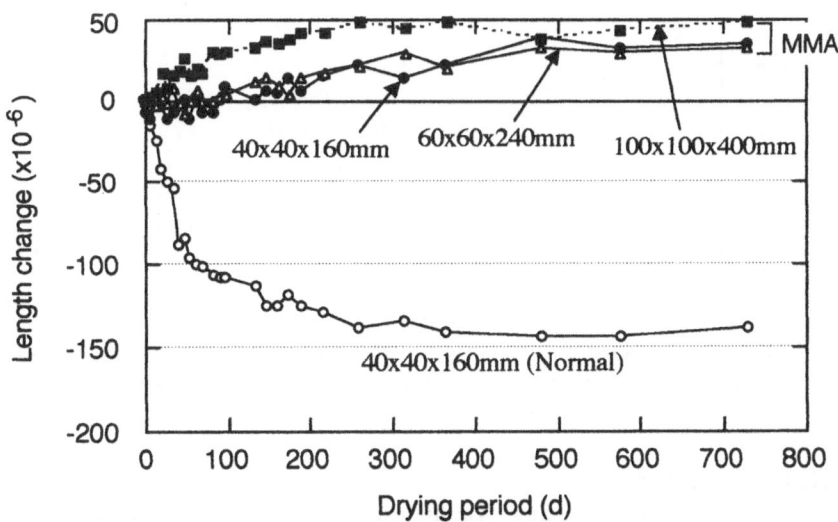

Fig.6. Effect of specimen sizes on length change of MMA-impregnated mortars with silica fume content of 10%.

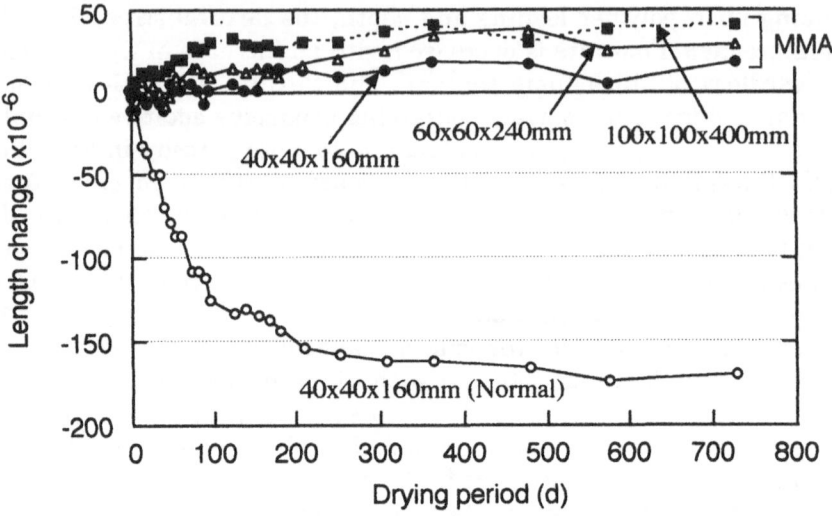

Fig.7. Effect of specimen sizes on length change of MMA-impregnated mortars with silica fume content of 20%.

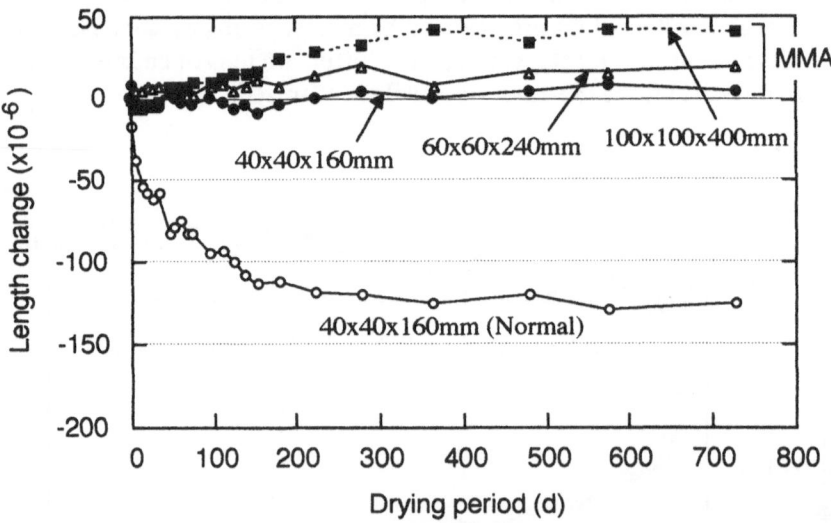

Fig.8. Effect of specimen sizes on length change of MMA-impregnated mortars with silica fume content of 20% and expansive additive.

4.3 Trial production

A precision surface plate using MMA-impregnated silica fume mortar was made on an experimental basis as shown in Fig. 9. This precision surface plate was made to use in a machine (i.e. thickness measuring instrument of floppy discs) as an integral part of it. The plate consists with holes and some iron inserts. These inserts were connected each other by a steel wire mesh in order to allow the electricity to pass through them. Such work seems to be impracticable for granite or cast iron plates. It is also advantageous that silica fume mortar plate is not needed drilling or trenching because it can be easily molded into complex shapes. MMA-impregnation prevented the formation of dust from silica fume mortar surfaces. The dimensional stability, namely flatness of the plate are being measured.

Fig.9. Precision surface plate of MMA-impregnated silica fume mortar.

5 Conclusions

It is concluded from the above test results that the impregnation of MMA monomer shows a significant improvement in strength and modulus of elasticity, and restrains length change of silica fume mortars. MMA-impregnated silica fume mortars can successfully be used as precision surface plate applications instead of granite or cast iron plates.

6 References

1. Houde, J., Prezeau, A. and Roux, R. (1987) Creep of Conrete Containing Fibers and Silica Fume, American Concrete Institute SP-105, pp.101-118.
2. Gjorv, O.E. (1983) Durability of Concrete Containing Condensed Silica Fume, American Concrete Institute SP-79, pp.695-708.
3. Kobayashi, K. and Tazawa, E. (1980) Fiber-Reinforced Concrete/ Polymer Concrete (in Japanese), Sankaido, Tokyo.

EVALUATION OF ADHESION BETWEEN FRESH POLYMER MORTARS AND FRESH CEMENT CONCRETE

M.A.R. Bhutta, T. Kobayashi and T. Kawano
Maeta Techno-Research, Inc., Sakata, Japan
N. Maeda
Maeta Concrete Industry Ltd., Sakata, Japan
Y. Ohama and K. Demura
Dept. of Architecture, Nihon University, Koriyama, Japan

Abstract

To prepare specimens for adhesion and flexural strength tests, fresh cement concrete is placed on fresh polymer mortars at several time intervals after fresh polymer mortar placings with different thicknesses, and subjected to various curing conditions. After that, the specimens are tested for adhesion in tension and flexural strength. The effects of the thickness of polymer mortar, time interval after fresh polymer mortar placing and curing conditions on the adhesion in tension and flexural strength are discussed. As a result, the production of precast concrete panels with polymer mortar surface layers is found to be possible.
Keywords: Adhesion test in tension, curing conditions, flexural strength, fresh polymer mortars, fresh cement concrete, time interval after polymer mortar placing, thickness.

1 Introduction

In recent years, a strong interest has been shown in the applications of polymer mortars and concretes for various construction works such as the repair and restoration of reinforced concrete structures because of their versatile properties such as good workability, low- temperature curability, high early strength development, watertightness, chemical resistance and abrasion resistance. Recently, polymer mortars have successfully been developed for underwater placement in Japan (1, 2). In this study, polymethyl methacrylate (PMMA) mortar and unsaturated polyester resin (UP) mortar are used, and considered for the protection of the rapid deterioration of reinforced concrete panels for concrete structures, caused by the penetration of

Polymers in Concrete, edited by Y. Ohama, M. Kawakami and K. Fukuzawa. Published in 1997 by E & FN Spon, 2–6 Boundary Row, London SE1 8HN, UK. ISBN: 0 419 22330 4.

carbon dioxide, oxygen, chloride ions and water in various environmental conditions. To prepare specimens for adhesion and flexural strength tests, fresh cement concrete is placed on fresh polymer mortars with different thicknesses at several time intervals after fresh polymer mortar placing, and subjected to various curing conditions. The effects of thickness of polymer mortar, time interval after fresh polymer mortar placing and curing conditions on the adhesion in tension between the fresh polymer mortar and fresh cement concrete are discussed.

The objective of this study is to evaluate the adhesion between fresh polymer mortar and fresh cement concrete to consider the production of precast concrete panels with polymer mortar surface layers.

2 Materials

2.1 Materials for binder systems

Binder system for polymethyl methacrylate (PMMA) mortar is based on methyl methacrylate (MMA) monomer, together with trimethylolpropane trimethacrylate (TMPTMA) as a crosslinking agent, unsaturated polyester resin (UP) and polyisobutyl methacrylate (PIBMA) as shrinkage-reducing agents, benzoyl peroxide (BPO) as an initiator, and N,N-dimethyl-p-toluidine (DMT) as a promoter. Hydrophilic unsaturated polyester resin (UP) mortar was employed as a liquid resin for UP mortar, together with methyl ethyl ketone peroxide (MEKPO) as an initiator and cobalt octoate (CoOC) as an accelerator.

2.2 Filler and fine aggregates

Commercially available ground calcium carbonate (size; 2.5μm or finer) was used as a filler, and silica sands (sizes; 0.04-0.30mm and 0.21-1.19mm) were done as fine aggregates.

2.3 Materials for cement concrete

Ordinary portland cement, river sand (size; 2.5 mm or finer) and river gravel (size; 5-25 mm) were used with an admixture.

3 Testing procedures

3.1 Preparation of cement concrete

Ordinary cement concrete was prepared with the mix proportions (kg/m^3) of cement to water to sand to gravel 565 : 170 : 622 : 1134, a water-cement ratio of 30%, an admixture content of 0.01% (of cement), an air content of 2.0% and a slump of 19.0 cm.

3.2 Preparation of polymer mortars

According to JIS A 1181 (Method of Making Polyester Resin Concrete Specimens), polymer mortars were mixed with the binder formulations and mix proportions as shown in Tables 1 to 3. The working lives of the polymer mortars with the binder formulations and mix proportions as shown in Tables 1 and 3 were controlled to be 30±5 minutes.

Table 1 Formulations of binder for PMMA mortar

Formulations by mass					
(%)				(phr*)	
MMA	TMPTMA	UP	PIBMA	DMT	BPO
67.40	1.80	23.10	7.70	0.50	2.00

Note; * : Parts per hundred parts of resin.

Table 2 Formulations of binder for UP mortar

Formulations by mass		
(%)	(phr*)	
UP	DMA	BPO
100	1.00	1.00

Note; * : Parts per hundred parts of resin.

Table 3 Mix proportions of polymer mortars

Type of mortar	Mix proportions by mass (%)				Binder-Filler ratio (B/F)
	Binder	Filler	Silica sand		
			No.4	No.7	
PMMA	15.00	15.00	35.00	35.00	1.00
UP	15.00	15.00	35.00	35.00	1.00

3.3 Preparation of specimens

As seen in Figs. 1 to 3, firstly, polymer mortars were placed with thicknesses of 0, 5, 10 and 15 mm in molds 50x100x100 mm and 400x400x100 mm for adhesion test in tension and flexural strength respectively, and then fresh cement concrete was placed on the fresh polymer mortars at time intervals of 0, 5, 10, 15, 20 and 30 minutes after fresh polymer mortar placings at 20°C and 50% R.H. in accordance with JIS A 1138 (Method of Making Test Samples of Concrete in the Laboratory) and JIS A 6909 (Coating Materials for Textured Finishes of Buildings). Specimens were subjected to different curing conditions: (1) 1-d-20°C-80% R.H. moist plus 4-h-60°C steam plus 28-d-20°C-80% R.H. moist cure (steam cure), and (2) 1-d-20°C-80% R.H. moist plus 28-d-20°C water cure (water cure).

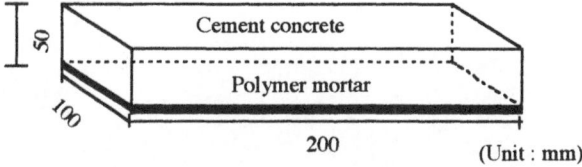

Fig. 1 Specimen for adhesion test in tension.

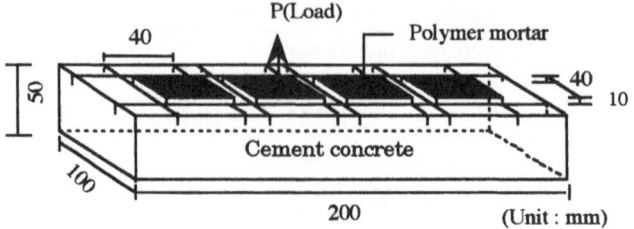

Fig. 2 Specimen for adhesion test in tension.

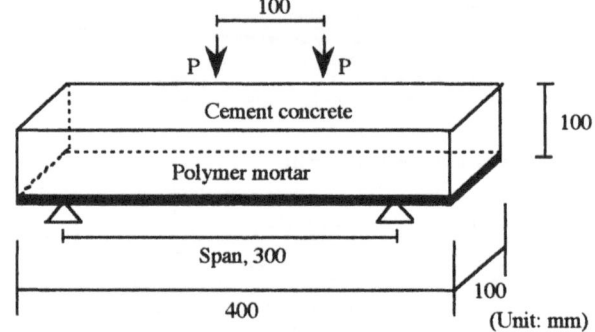

Fig. 3 Specimen for flexural strength test.

3.4   Adhesion test in tension and flexural strength

According to JIS A 6909, the cured specimens were tested for adhesion in tension by using a manually operated pull-gage as illustrated in Fig. 2. After curing, specimens were tested for flexural strength according to JIS A 1106 (Method of Test for Flexural Strength of Concrete) by use of the Amsler-type universal testing machine as shown in Fig. 3. After adhesion test in tension, the specimens were observed for failure modes, which are classified in Fig. 4. The total area of the bonded surfaces of each specimen was supposed to be 10, and the respective approximate ratios of A, M and S areas on the failed crosssection were expressed as suffixes for A, M and S.

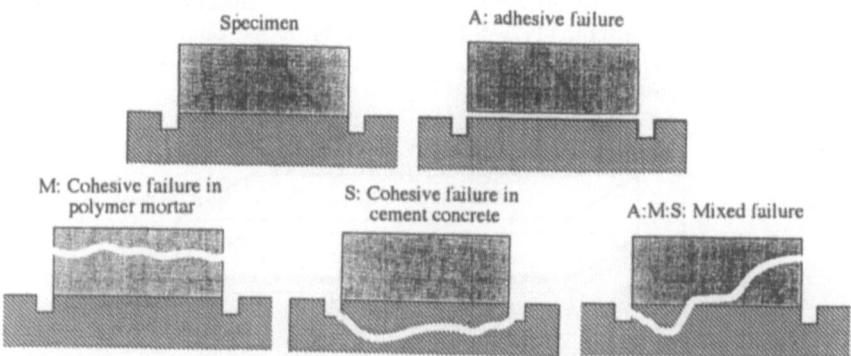

Fig. 4 Types of failure modes in adhesion test in tension.

4 Test results and discussion

4.1 Effects of time interval and thickness of polymer mortars on adhesion in tension

Figs. 5 and 6 show the effects of time interval after the placings of polymer mortars (PMMA mortar and UP mortar) and their thickness on the adhesion in tension of fresh polymer mortars to fresh cement concrete in water cure or steam cure. In general, the adhesion in tension of the polymer mortars to the cement concrete in water cure is much higher than in steam cure. This causes the sufficient hydration of the cement concrete, and provides the strong adhesion between the polymer mortars and cement concrete. The adhesion in tension of the polymer mortars to the cement concrete at time intervals of 0 to 15 minutes after polymer mortar placings, is much larger than that at prolonged time intervals of 15 to 30 minutes. The adhesion in tension between the polymer mortars and cement concrete is gradually decreased with a prolongation in the time interval after polymer mortar placings, and is increased with an increase in the thickness of the polymer mortars regardless of curing conditions. A prolongation in the time interval after polymer mortar placings provides the poor adhesion of the polymer mortars. Irrespective of the curing conditions and thickness of the polymer mortars, the adhesion in tension of PMMA mortar to the cement concrete is higher than that of UP mortar. Especially, a significant improvement in the adhesion in tension of PMMA mortar to the cement concrete is achieved in water cure at a time interval of 10 minutes after PMMA mortar placing at a thickness of 15 mm, and the highest adhesion in tension is 5.8 MPa. The adhesion in tension of UP mortar to the cement concrete at a time interval of 0 minutes after UP mortar placing at a thickness of 15 mm in water cure is 4.8 MPa. Such a remarkable improvement in the adhesion in tension of the polymer mortars to the cement concrete is achieved by a good timing due to the suitable working life of the polymer mortars.

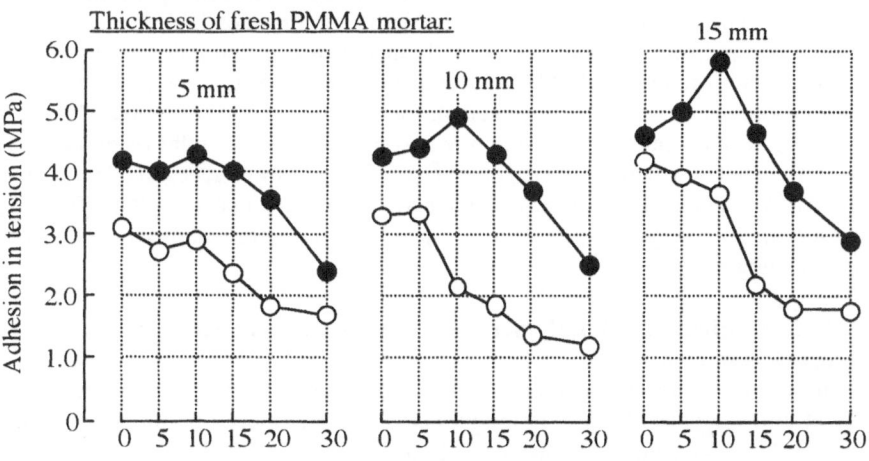

Fig. 5 Adhesion in tension between fresh polymer mortars and fresh cement concrete.

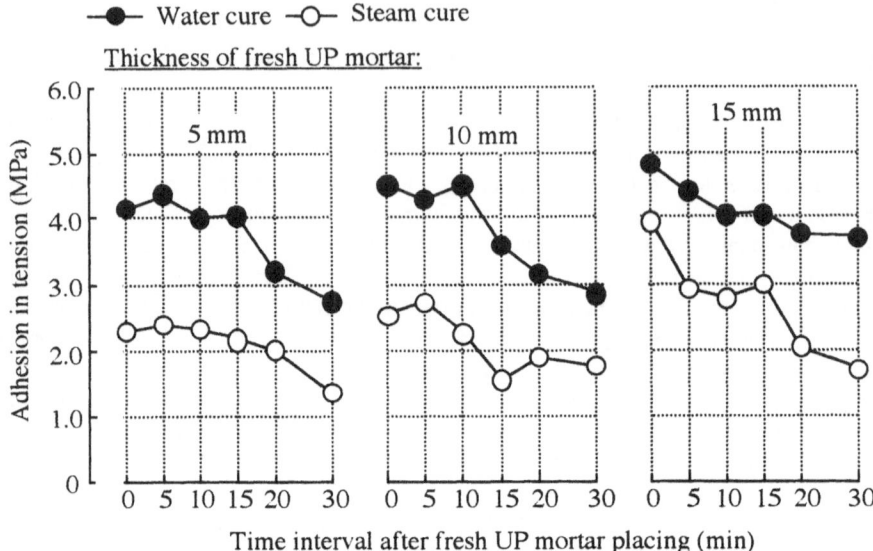

Fig. 6 Adhesion in tension between fresh polymer mortars and
fresh cement concrete.

Fig. 7 demonstrates the failure mode distribution and adhesion in tension of fresh polymer mortars to fresh cement concrete in watre cure or steam cure. The adhesion in tension of the polymer mortars to the cement concrete is increased with an increase in the ratio of cohesive failures. Regardless of the thickness of the polymer mortars, the ratio of adhesive failure in adhesion test in tension in steam cure is larger than that in water cure. The effect of the thickness of PMMA mortar on the failure modes is hardly recognized in steam cure. The ratios of cohesive failure in PMMA mortar and in cement concrete tend to decrease with a prolongation in time interval after PMMA mortar placing, and to increase with an increase in the adhesion in tension regardless of the thickness of PMMA mortar and the curing conditions. PMMA mortar bonded to the cement concrete at time intervals of 0 to 20 after PMMA mortar placing in water cure, indicates almost cohesive failure in cement concrete. On the other hand, the effects of the thickness and the curing conditions on the failure modes in adhesion test in tension of UP mortar to the cement concrete at differnt time intervals after UP mortar placing are hardly recoginzed. The ratio of the adhesive failure is larger in UP mortar than PMMA mortar.

4.2 Effect of thickness of polymer mortars on flexural strength of cement concrete

Fig. 8 shows the flexural strength of the cement concrete placed on the fresh polymer mortars at time intervals of 0 and 30 mintues after polymer mortar placings at different thicknesses in water cure or steam cure. Like the adhesion in tension, the flexural strength of the cement concrete in water cure is also higher than that in steam cure irrespective of the thickness of the polymer mortars. The flexural strength of the cement concrete is increased with an increase in the thickness of the polymer mortars regardless of the curing conditions. The flexural strength of the cement concrete placed on PMMA

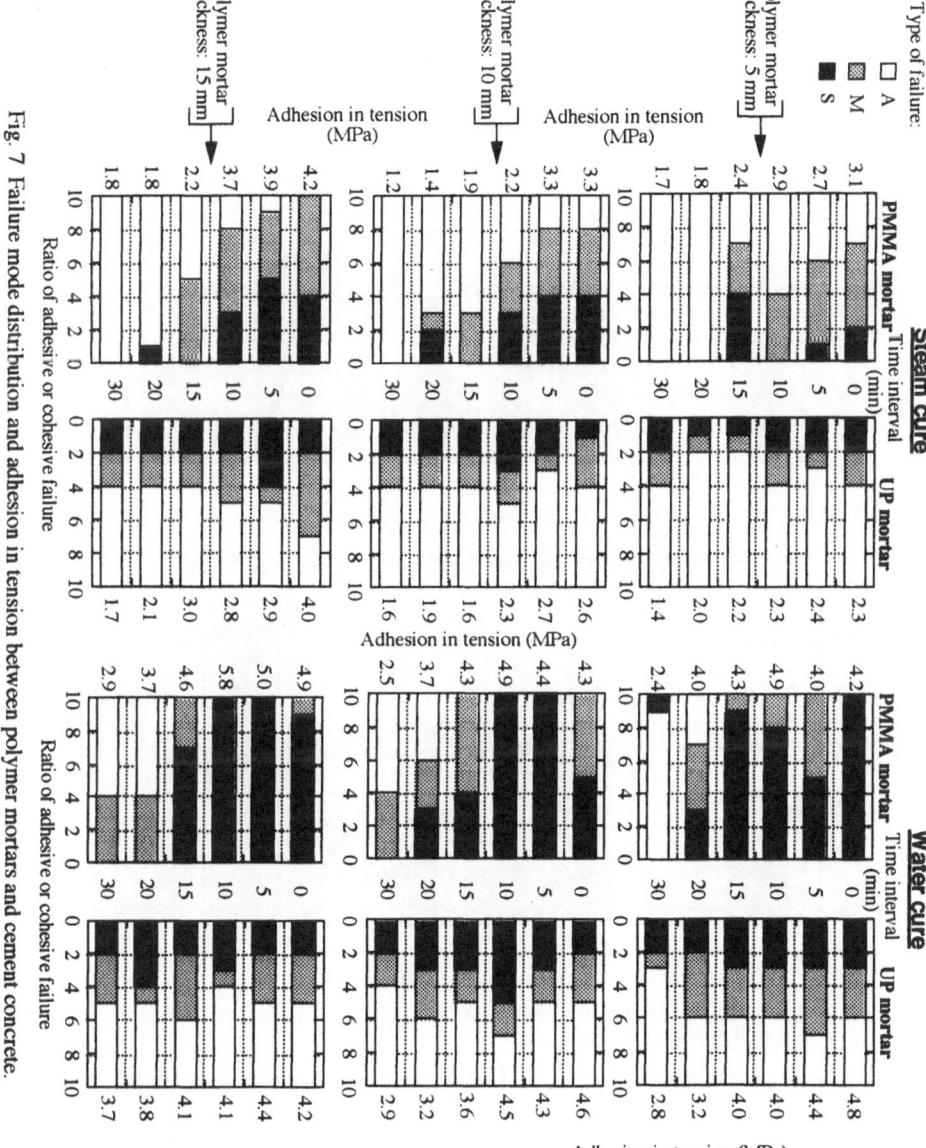

Fig. 7 Failure mode distribution and adhesion in tension between polymer mortars and cement concrete.

mortar is higher than that on UP mortar in both water cure and steam cure. The reason for such improvement in the flexural strength of the cement concrete placed on PMMA mortar at a time interval of 30 minutes with increasing thickness of PMMA mortar is due to the high early strength deveopment of PMMA mortar and the sufficient hydration of the cement concrete.

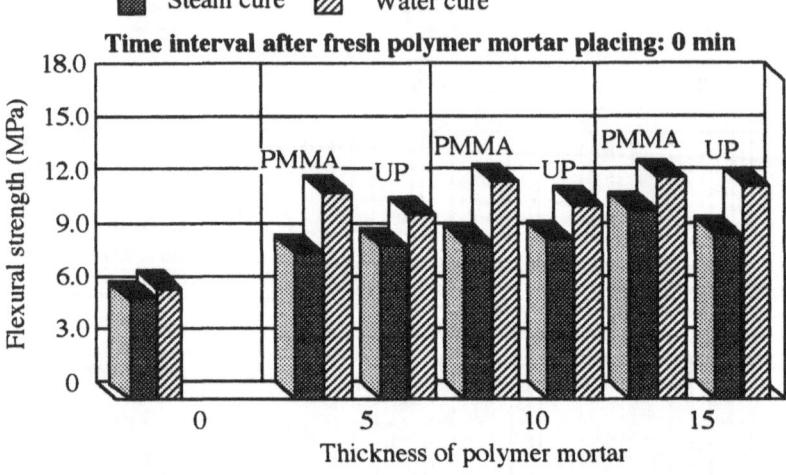

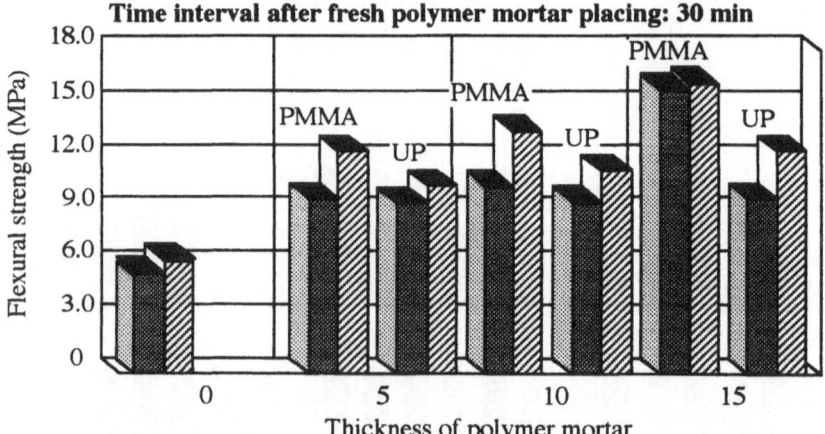

Fig. 8 Flexural strength of cement concrete placed on polymer mortars with different thicknesses.

4.3 Adhesion mechansim between fresh polymer mortar and fresh cement concrete

Fig. 9 exhibit the adhesion mechanism and conditions of the adhesive interfaces between fresh polymer mortars and fresh cement concrete placed on the polymer mortars at time intervals of 0 and 30 minutes. As seen in Fig. 9, the working life of the polymer mortars was controlled to be 30±5 minutes. At time intervals of 0 to 15 mintues, the conditions at the interfaces between the polymer mortars and cement concrete are A, which causes the high adhesion in tension of the polymer mortars to the cement concrete due to strong mechanical interlockings between them. PMMA mortar shows A at time intervals of 0 to 15 like UP mortar, but the adhesion in tension of PMMA mortar to the cement concrete is higher than that of UP mortar because of the good workability of PMMA mortar compared to UP mortar. Because of the low viscosity of PMMA binder, it may easily penetrate into the cement concrete compared to UP binder, and develop strong mechnical interlockings and van der Waals bonds due to good timing and the proper binder formulations of PMMA mortar. A prolongation in the time interval provides almost C, and may cause the formation of a thin film of the binders on the surfaces of the polymer mortars which forms a barrier to the fresh cement concrete and prevents the penetration of the binders to the fresh cement concrete. This leads to obstruct the mechanical interlockings or van der Waals bonds between the polymer mortars and cement concrete.

Condition of adhesive interface:

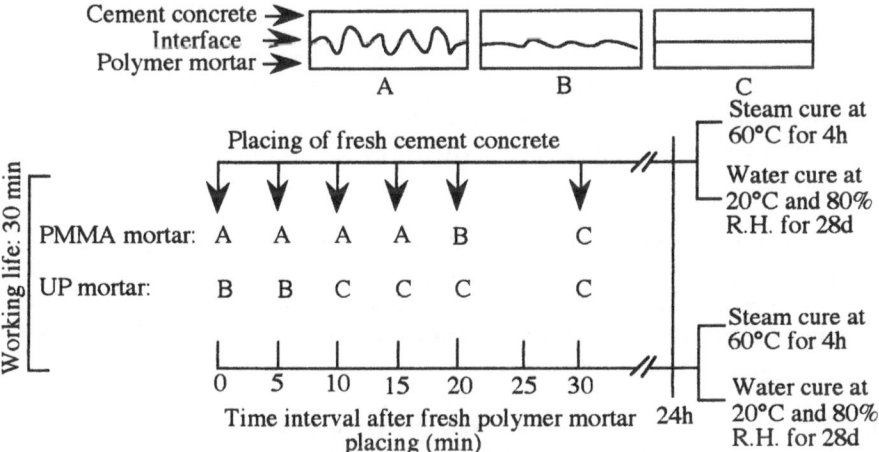

Fig. 9   Conditions of interfaces between polymer mortars and cement concrete placed after different time intervals.

Figs. 10 and 11 demonstrate the interfaces between polymer mortars and cement concrete after hardening at time intervals of 0 and 30 minutes.

Fig. 10   Interfaces between polymer mortars and cement concrete after hardening at time interval of 0 mintues.

Cement concrete ➡

Interface ➡
Polymer mortar ➡

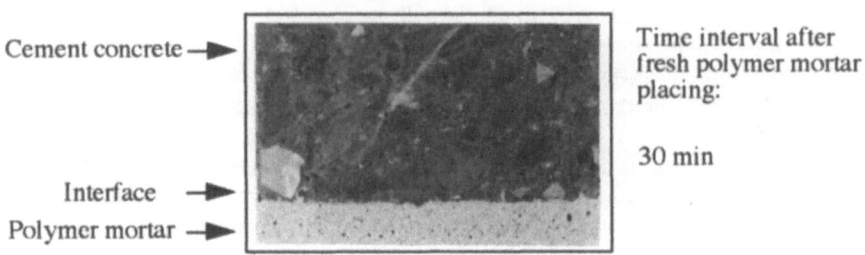

Time interval after
fresh polymer mortar
placing:

30 min

Fig. 11 Interfaces between polymer mortars and cement concrete after
hardening at time interval of 30 mintues.

5   Conclusions

(1) The placing of fresh cement concrete at time intervals of 0 to 15 mintues after the placings of fresh polymer mortars on the polymer mortars with a thickness of 15 mm provides a good adhesion between both polymer mortars and cement concrete in water cure, causing almost cohesive failure in the cement concrete and polymer mortars.

(2) A significant improvement in the adhesion in tension of fresh polymer mortars (PMMA and UP mortars) to fresh cement concrete is achieved in water cure at a time interval of 10 minutes after polymer mortar placing with a thickness of 15 mm, and the highest adhesions in tension are 5.8 MPa and 4.8 MPa for PMMA mortar and UP mortar respectively.

(3) The highest flexural strength of about 16 MPa is obtained at a time interval of 30 minutes after PMMA mortar placing with a thickness of 15 mm, and is more than twice the flexural strength of the cement concrete with a thickness of 0 mm.

(4) The production of precast concrete panels with polymer mortar surface layers is found to be possible.

6   References

1.   Ohama, Y., Demura, K. and Bhutta, M.A.R., (1992) Polymer mortars for underwater construction, *Proceedings of the 7th International Congress on Polymers in Concrete*, Berecom, Moscow, pp. 86-97.

2.   Bhutta, M.A.R., Ohama, Y. and Demura, K., (1995) Cement paste primers for underwater adhesion of polymethyl methacrylate mortars to cement mortar substrates, *Proceedings of the 8th International Congress on Polymers in Concrete, Technological Institute of the Royal Flemish Society of Engineers, Antwerp, Belgium, pp. 521-526.*

3.   Sasse, H.R. and Fiebrich, M., (1983) Bonding of Polymer Materials to Concrete, *Materials and Structures, Research and Testing,* Vol.16, No.94, pp. 293-301.

DEVELOPMENT OF A LOW-TEMPERATURE STOREHOUSE USING SANDWICH PANELS WITH POLYMER MORTAR FACINGS

K.S. Yeon, J.D. Choi, K.H. Jung, Y.I. Park,
M.K. Joo and D.S. Choi
Dept. of Agricultural Engineering, Kangwon National University,
Chunchon, Korea

Abstract
A low temperature storehouse was developed by using polymer concrete sandwich panels which were consisted of expanded polystyrene form as a core and polymer mortar as facings. Glass fiber was used to reinforce the panel. The panel's strength, thermal insulation and other physical characteristics were proved that the polymer mortar panel was a good material to build the low temperature storehouses in rural agricultural areas. A small storehouse was built with the panel and tested.
Keywords: Construction technique, core, facing, polymer mortar, sandwich panel, storehouse, strength, thermal insulation.

1 Introduction

Agricultural products storage facilities play a very important role in agriculture because they can reduce losses and adjust the shipment of the products. Therefore, the demand of the facilities have been increased and the Korea government also invests a considerable mount of the budget to supply the facilities.

Existing small and low temperature storehouses used in rural agricultural areas are mainly constructed with sandwich panels which are made of expanded polystyrene foam as a core and galvanized sheet metal as facings. These panels have many problems caused by poor insulation between the panel joints and corrosion of the sheet metal. However, newly developed polymer mortar

Polymers in Concrete, edited by Y. Ohama, M. Kawakami and K. Fukuzawa. Published in 1997 by E & FN Spon, 2–6 Boundary Row, London SE1 8HN, UK. ISBN: 0 419 22330 4.

sandwich panels can minimize the energy loss at the joints and have no corrosion problem while they are reasonably light in weight and strong in strengths and have a long durability.   And the panels are considered as an excellent construction material for low temperature storehouses, mushroom growing facilities and so on for small farms in rural agricultural area.

The objectives of this study were to develop a technique of building a small and efficient low temperature storehouse with polymer mortar sandwich panels for small farms in Korea and to test the storehouse with respect to temperature.

2 Production of Polymer Mortar Sandwich Panels

2.1 Material

Polymer mortar sandwich panels were composed of facings, core and reinforced material.   Facings of the sandwich panels made with polymer mortar that was a mixture of unsaturated polyester resin, calcium carbonate ($CaCO_3$) and fine aggregates.   Physical and mechanical properties of the polymer mixture measured at 7 days of curing age were summarized in Table 1.   Elastic modules, compressive strength, splitting tensile strength and flexural strength were measured according to the respective Korea Standard (KS) testing methods.

Expanded polystyrene which is a porous foam plastic was used as the core of the panels.   For the panel reinforcement, a woven glass fiber (E-glass) was used and its properties were summarized in Table 2.

Table 1. Physical and mechanical properties of the polymer mortar used for facings

Specific gravity	Elastic modules (MPa)	Compressive strength (MPa)	Splitting tensile strength (MPa)	Flexural strength (MPa)
2.1	13,720	63.7	8.9	17.1

Table 2. Physical and mechanical properties of the E-glass fiber

Specific gravity	Tensile strength (MPa)	Tensile elastic modules (MPa)	Remarks
2.55	3,430	72,520	

2.2 Production procedures

The two different sizes and 18 different shapes of the sandwich panels were precisely predetermined based on the storehouse to be built. The two sizes were 90 cm x 180 cm and 90 cm x 300 cm, respectively, and the thickness of core styrofoam was 10 cm for both sizes. The thickness of a polymer mortar facing was 5 mm for both sides. Schematic production procedures of the sandwich panels were as follow:

1) Prepare the core (expanded polystyrene) on a plain surface.
2) Bond the woven glass fiber on both sides of the core.
3) Prepare polymer mortar, place the mortar over the glass fiber, and make the surface smooth while maintaining the facing thickness of 5 mm.
4) Cure the panel about 1 hour at room temperature.
5) Bond the woven glass fiber on both sides of facings.

2.3 Engineering properties of the polymer mortar sandwich panel

Thermal insulation tests and local compression resistance tests were performed according to the KS L 9016, KS F 2273, and the test results are shown in Tables 3 and 4, respectively. Simple flexural tests were carried out according to the KSF 2273. Loads were applied at three equally spaced points (three point loading) and the test result is shown in Table 5. Axial compression tests with an eccentricity of upto $t/6$, where t is the thickness of the panel, also were performed according to the KS F 2273 and the test result is shown in Table 6.

Table 3. Thermal insulation test results

Part thickness (mm)		Thermal conductivity $(W/m^2 h \, °K)$	Heat transmission coefficient $(kcal/m^2 h \, °C)$
Facing	Core		
5	100	0.0410	0.531

Table 4. Local compression resistance test results

Jacked depth (mm)	Local compressive strength (MPa)	Residual deformation (mm)	Local compressive failure load (KN)
8.4	11.2	4.2	5.4

Table 5. Simple flexural test results

Failure load (KN)	Flexural strength (MPa)	Maximum deflection (mm)	Max. moment strength (MPa)	Maximum strain (x10⁻⁶)		Max. shear force (KN)	Max. flexural shear strain (x10⁻⁵)
				Tensile	Comp.		
15.7	102.5	28	0.3	465	440	7.8	572

Note: Specimen size 90 cm (B) x 180 cm (H)

Table 6. Axial compression test results

Buckling load (KN)	Maximum horizontal displacement (mm)	Buckling moment (N•m)	Strain (x10⁻⁵)	
			Comp. side	Tensile side
165	6	3,368	250	220

Note: Specimen size 90 cm (B) x 180 cm (H)

3 Construction Method

The 18 different shapes and 2 different sizes of polymer mortar sandwich panels were precisely designed before the production of the panels so that a box-shaped storehouse body can be built by fabricating the panels.   Therefore, in the panel design stage, the exact size of storehouse was determined and the shapes, sizes and corresponding numbers of each panel were decided. The basic sizes of wall panel and bottom and ceiling panels were 90 cm x 300 cm and 90 cm x 180 cm, respectively. However, there were 5 to 10 cm differences in panel size depending upon the fabrication location. The total of 136 polymer sandwich panels were produced and fabricated to build the storehouse.   The outside dimensions of the storehouse were 370 cm wide, 950 cm long and 310 cm high. After the box-shaped storehouse was built, roof and refrigerating systems were installed.   The followings were the procedures to build a small and low temperature storehouse in this study.

1) Determine the size of storehouse and make a reduced scale model (Fig. 1).
2) Precisely design the shapes, sizes and corresponding numbers of polymer mortar sandwich panels.
3) Produce the panels while reclaim the construction site and cast a cement concrete footing (Fig. 2). In this study, all polymer mortar sandwich panels were manufactured by hand. And a 410 cm wide, 950 cm long and 10 cm deep cement concrete footing reinforced with wiremeshes was made. The surface of the base must be precisely leveled to evenly place the panels.

Fig. 1. Scaled model of the storehouse

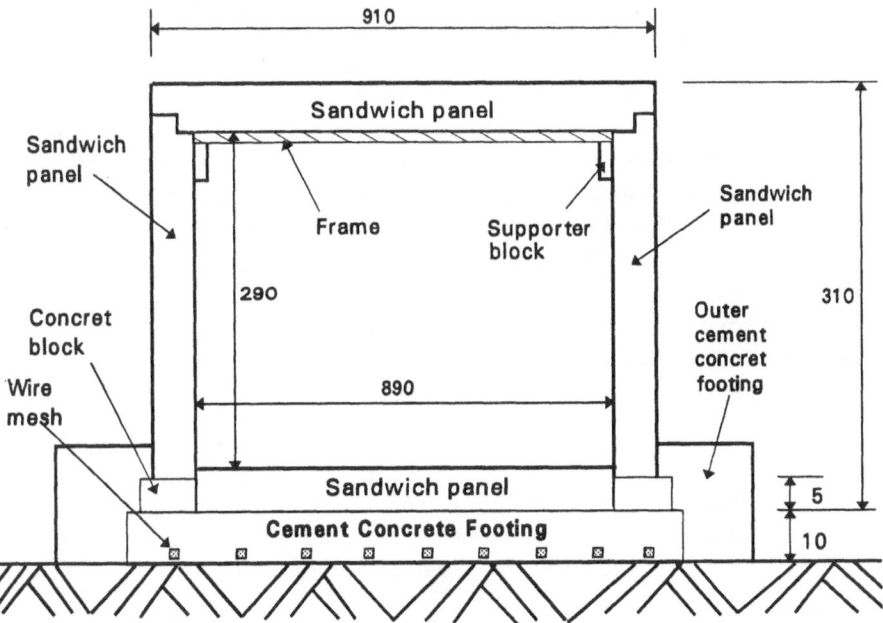

Fig. 2. A cross-section of the storehouse (unit: cm)

4) The bottom of the storehouse was assembled and bonded first according to the predesigned order.
5) Concrete support of 15 cm wide and 5 cm high was made along the bottom panel to place wall panels.
6) Wall panels were assembled and bonded with a polymer resin. The panels were fixed with bolts and nuts while the bonded surfaces were cured. After the surfaces were cured, most of the bolts and nuts were removed and the bolt holes were filled with a polyurathane foam insulator. But, least numbers of bolts and nuts were not removed to strengthen the structural safety of the storehouse.
7) Steel frame to support the ceiling panels was installed. About 40 cm x 40 cm x 10 cm polymer mortar blocks were made and bonded to the wall and steel frame was placed and bolted on the top of the blocks.
8) Ceiling panels were fabricated.
9) Small crevices between the panels were filled with a polyurathane foam insulator.
10) Outer cement concrete footing was casted and roof was installed.
11) Refrigeration system was installed.

4 Preliminary Temperature Monitoring Results

Twenty seven thermocouples were installed inside the storehouse and three thermocouples were placed outside to monitor the temperature changes . Temperature changes are simultaneously monitored by using a Campbell Scientific's CR10 data logger with an AM416 Relay Multiplexer. The refrigeration system was specially developed to maximize the use of small and private agricultural storages. The system has both cooling and heating capabilities. Therefore, the storehouse can be used to store fresh agricultural products at low temperature to keep their freshness and quality, and to dry various agricultural products at high temperature. However, only preliminary results can be presented in Table 7 because the temperature monitoring is not completed.

Many small agricultural low temperature storages are built and used to store various agricultural products before shipping to the market. The storages are not designed to dry agricultural product by increasing temperature but

Table 7. Preliminary temperature monitoring results

	Start temp. t1 (°C)	Reached temp. t2 (°C)	Elapsed time(t1→t2) (hours)	Elapsed time(t2→t1) (hours)	Outside temp. (°C)
Heating	23	48	2	33	day 28, night 15
Cooling	26	-22	9	34	day 28, night 15

designed only for low temperature storage. Moreover, temperatures of the most of the storages that are currently being used can be lowered to -10 °C at most. Therefore, it is thought that the trial storehouse built with polymer mortar sandwich panels works satisfactorily.

5 Conclusions

This study initiated to develop a small but multi-purpose storehouse for small farms in agricultural area by using newly developed polymer mortar sandwich panels. Design and production of polymer mortar sandwich panels, tests of engineering properties of the panels, construction methods, and cooling and heating system development were carried out and temperature monitoring of the storehouse that was built with the panels is currently made. The following conclusions were drawn from the study:

1) Polymer mortar sandwich panels had relatively high resistivity against flexural and axial compression loads compared to the existing sandwich panels in the market. Also, the panels showed relatively good resistivity against local compression and impact loads.
2) Water resistivity of polymer mortar sandwich panels was excellent and air-tight interfaces of the panel fabrication could be reached. Therefore, polymer mortar sandwich panels are thought to be very suitable to build low temperature storehouses for small farms in agricultural areas.
3) It is thought that polymer mortar sandwich panels are suitable material for structures that need a durability because the polymer mortar facings of the panels are anti-corrosive material. Existing sandwich panels using galvanized sheet metals as facing material have problems with chemical corrosion such as rust.
4) Polymer sandwich panels showed the possibility that they can be used to build various agricultural structures requiring thermal insulation such as mushroom growing chambers.
5) Preliminary temperature monitoring showed that temperature monitoring both in heating and cooling in the storehouse built in this study was satisfactory.

Acknowledgement
This study was supported by R&D Promotion Center for Agriculture, Forestry, and Fishery, Ministry of Agriculture and Forestry, Korea.

6 References

1. ACI. (1985) Building Code Requirements for Reinforced Concrete (ACI 318-83).
2. Benjamin, B.S. (1969) Structural Design with Plastics, Van Nostrand Reinhold Company, pp.118-133.
3. Fowler, D.W. (1991) Structural Design of Polymer Concrete, ICPIC Working Papers, North American Workshop, San Francisco.
4. Huson, J.A. (1965) Precast Concrete Wall Panels: Flexural Stiffness of Sandwich Panels, ACI8 SP-11.
5. Kuenzi, E.W. (1970) Sandwich Panel Design, Symposium on Panelized Building Systems, Sir George Williams University, Montreal.
6. Yeon, K.S., Kim, K.W. and Hwang, J.Y. (1992) Structural Behavior of Sandwich Panels with Polymer Mortar Facings, Proceedings of the 7th International Congress on Polymers in Concrete, Moscow, Russia, pp.550-557.

DEVELOPMENT OF POLYMER MORTAR ONDOL PANEL

K.S. Yeon, S.S. Kim, M.K. Joo and D.S. Choi
Dept. of Agricultural Engineering, Kangwon National University,
Chunchon, Korea
T.B. Jeon
Dept. of Industrial Engineering, Kangwon National University,
Chunchon, Korea

Abstract
An Ondol panel using thermal-hardening polyester resin for hot press has been developed in this research. Test results not only indicate that the Ondol panel has a good performance, high strength, high impact resistance, good heat storage property and reasonable workability, but also has a good possibility to put into practical usage.
Keywords : Ondol panel, high strength, impact resistance, heat storage

1 Introduction

Ondol system which is a traditional heating system in Korea is a key component of Korean housing structure. Ondol system is composed of fire place, heat and smoke paths, chimney, Ondol and finishing material for smooth floor. Hot gas and smoke generated in the fire place heat Ondol while passing through the horizontal heat and smoke paths and emit through the chimney. An Ondol is a kind of thin and flat panel that can bear floor loads and keep heat long once heated. Thin and flat stones and cement concrete Ondol panels (KS F 4024) were mostly used when direct heat sources from burning wood or coal were used. As the heating sources have changed from direct heating to various boiler systems, Ondol system also have changed to fit to modern housing system while the concept of traditional Ondol system maintains.

Polymers in Concrete, edited by Y. Ohama, M. Kawakami and K. Fukuzawa. Published in 1997 by E & FN Spon, 2–6 Boundary Row, London SE1 8HN, UK. ISBN: 0 419 22330 4.

Many prefabricated Ondol systems have been developed and used. An Ondol system is a prefabricated panel that contains copper pipes in the winding grooves of an insulator panel for hot water circulation as heating source and uses aluminum sheet metal as radiating surface cover for floor. This system radiates heat fast but lacks in heat storage capacity and floor stability. Other system is an Ondol system that pipes as heat source and materials with high specific heat are arranged in the frame of a panel and then cover the surface with Ondol panels. This Ondol system is relatively light, easy to install and has a high heat efficiency. The Ondol panel must be thin, easy to handle, has a strength to bear floor loads and provides stability. However, no such Ondol panel have been developed yet and thus, the Ondol system has not been successfully marketed.

The objectives of this study were to develop an Ondol panel that has higher strength, better thermal property, easier joint treatment than existing Ondol panels by using polymer mortar and to experimentally test to describe the physical and mechanical properties of the panel.

2 Materials and methods

2.1 Materials

2.1.1 Unsaturated polyester resin
The polymer resin used in this study was a thermal-hardening unsaturated polyester resin from a company in Korea. And its properties are shown in Table 1.

Table 1.   Properties of unsaturated polyester resin

Specific gravity (25°C)	Viscosity (25°C, ps)	Acid value	Response time (82.2°C, min)		Peak exotherm temp. (°C)
			Gel time	Min. hardening time	
2.3	6~9	20 or less	4.5±1	6.5±1	220±15

2.1.2 Initiator
The initiator used in the experiment was also provided by the company and Table 2 displays its properties.

Table 2.   Properties of initiator

Components	Specific gravity	Viscosity (25°C, ps)	Active oxygen (%)
t-Butyl Peroxybenzoate 98 % or above	1.04	6.22	8.03 or above

2.1.3 Filler
The filler for the experiment was heavy calcium carbonate having the properties shown in Tables 3 and 4.

Table 3. Physical properties of heavy calcium carbonate

Unit weight (gr/cc)	Absorption (cc/gr)	Moisture content (%)	pH	Avg. gradation (μm)	Retained ratio by #325 sieve (%)
1.1	0.20	0.3 or less	8.8	13	0.03

Table 4. Chemical composition of heavy calcium carbonate   (unit: %)

CaCO	Al_2O_3	Fe_2O_3	SiO_2	MgO	Ig. Loss
53.7	0.25	0.09	2.23	0.66	42.4

2.1.4 Aggregate
The aggregate used in this study was river sand and sieve analysis results are given in Fig. 1.

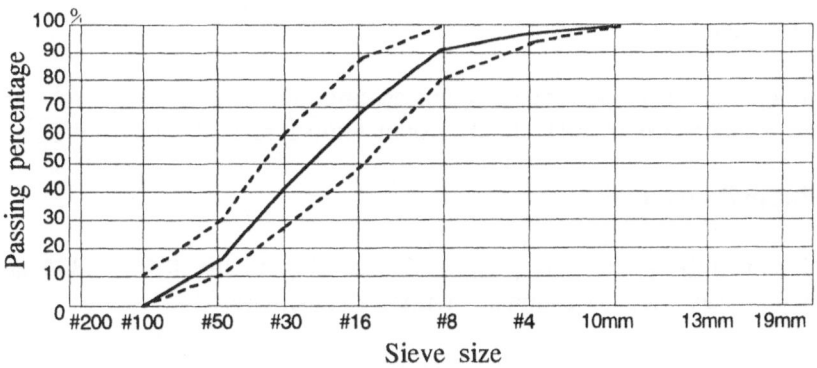

Fig. 1. Gradation curve of sand

2.1.5 Reinforcement
A polyester fiber fabrics that has mesh size of 5×5 *mm* was used as the reinforcement material for the panel.

2.2 Test specimens

2.2.1 Mixture of polymer mortar
The mix ratio of polymer mortar is given in Table 5. When a bio-ceramic powder was added, it replaced 2wt.% of calcium carbonate.

Table 5. Mix ratio of polymer mortar   (unit : wt.%)

Unsaturated polyester resin	Accelerator	Initiator	CaCO₃	Sand
13.5	2 phr	2 phr	14.5	72

* phr : parts per hundred parts of resin

2.2.2 Manufacture of Ondol panel
The size of an Ondol panel was decided as 60 x 80 x 1 cm and steel molds to produce the panel were prepared. For one-side reinforced Ondol panel, the reinforcement fiber was placed first on the bottom of the mold and polymer mortar was casted. And then, the mold was put in a chamber and treated with a pressure of 9.8 MPa at 140-150°C for 90 seconds. Two-side reinforced Ondol panels also were produced with the similar manner.

2.3 Test methods

2.3.1 Heat storage

Heat storage test was performed by measuring the cooling time from the preheated temperature of 50°C to the room temperature of 14.4-16.9°C.

2.3.2 Thermal conductivity

Thermal conductivity test was performed according to the KS L 3306 and Equation (1) was used to compute thermal conductivity.

$$\lambda = 0.24 \times \frac{I^2 R}{4\pi} \times \frac{\log t_2/t_1}{\theta_2 - \theta_1} \times 3.6 \tag{1}$$

where, λ is thermal conductivity ($Kcal/mh°C$), I electric current (A), R electric resistance (Ω/m), t_1 and t_2 are elapsed times (min), and θ_1 and θ_2 wire temperatures (°C) at t_1 and t_2, respectively.

2.3.3 Bearing stress

Bearing stress of the Ondol panel was measured under the two different bases and two different load conditions, i.e., 12 mm thick plywood base and 50 mm thick polystyrene foam base, and $\varnothing 35$ mm steel round bar and $\varnothing 25$ mm steel ball, respectively.

Ondol panel was first placed on one base and bearing stress was measured by using a $\varnothing 35$ mm steel bar, and $\varnothing 25$ mm steel ball and $\varnothing 25$ mm bar as shown in Fig. 2.

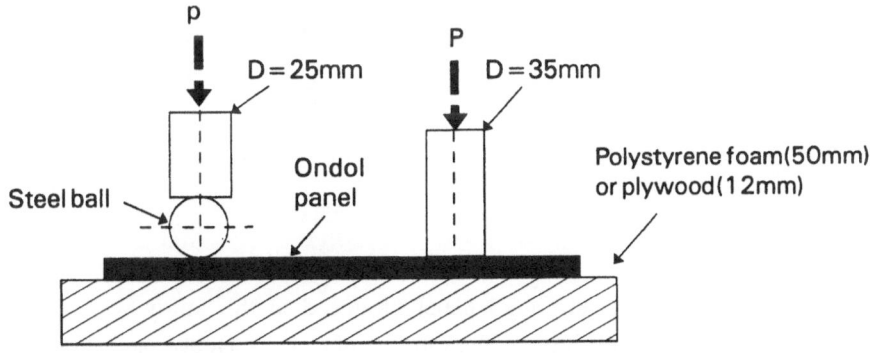

Fig. 2.   A schematic sketch of bearing strength test set-up

2.3.4 Bending strength

The Ondol panel size for bending test was 40 x 10 x 1 cm. Pure span length was 30 cm and test load was applied according to the three-point-loading method.   LVDTs and 67 mm long strain gauges were used to measure displacement and strain during the test.

2.3.5 Impact strength

An Ondol panel was placed on a 50 mm sand base as shown in Fig. 3 and the impact strength of the Ondol panel was measured by dropping an 1 kg pendulum from 30 - 300 cm high.

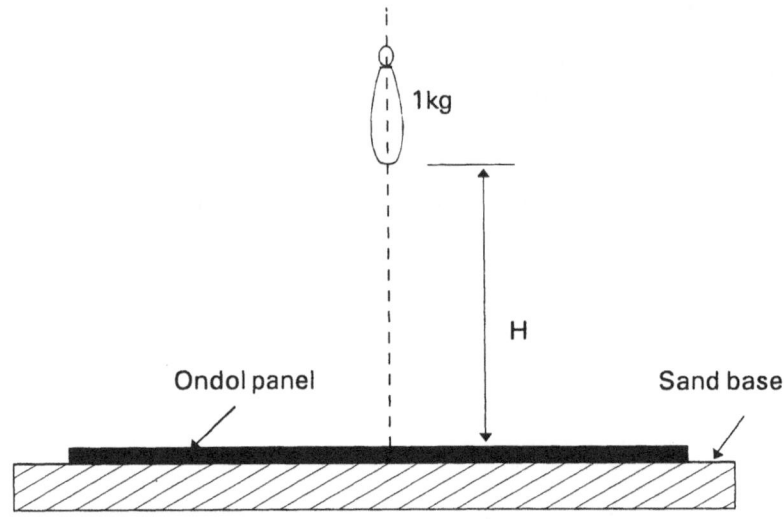

Fig. 3. A schematic sketch of impact strength test set-up

3 Results and Discussion

The size and weight of an Ondol panel produced in this study were 60 x 80 x 1 cm and 10 kg, respectively. The unit weight of the Ondol panel was 2,083 kg/m^3 which was about 10% less than the unit weight of 2,300 kg/m^3 of regular Portland cement concrete.

3.1 Heat storage

The results of heat storage test are shown in Fig.s 4 and 5. From Fig. 4, it is seen that the cooling time of the Ondol panel from preheated temperature of 50°C to the room temperature of 14.4°C was about 29-34 minutes when the both sides of the panel was exposed to the air. For the Ondol panels that bioceramic powder was added, the cooling time was about 8 minutes longer than the regular Ondol panels.

Fig. 5 shows that when the preheated Ondol panel was placed on polystyrene foam, i.e., only one side of the panel was exposed to the air and the other side insulated, the cooling time was about 100-125 minutes, which is about 3 times longer than that of 2-side test. For the bioceramic powder added Ondol panels, the cooling time extended to about 25 minutes compared to the regular ones.

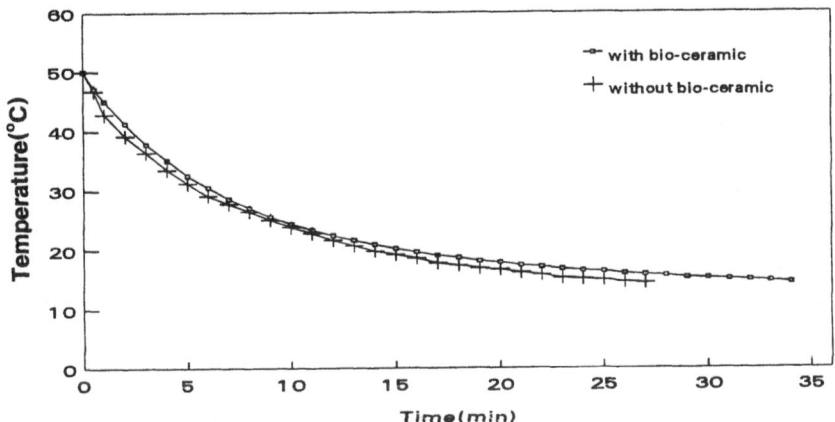

Fig. 4. Cooling time of the Ondol panel when both sides of the panel were exposed to the room temperature.

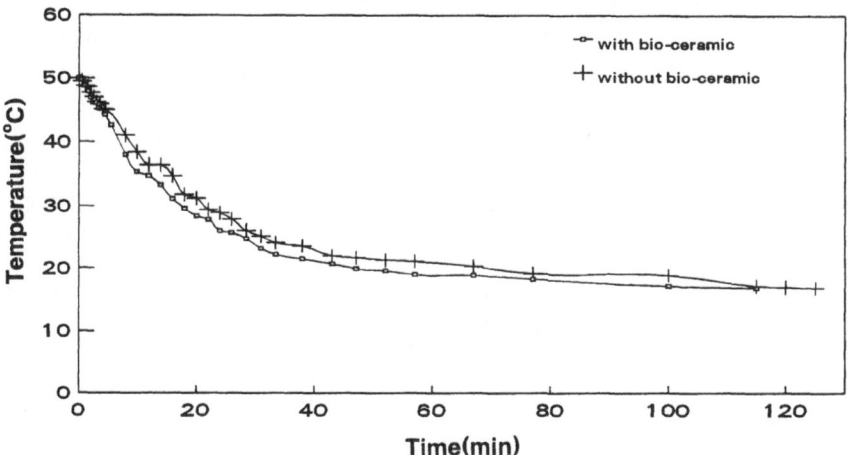

Fig. 5. Cooling time of the Ondol panel when one side was insulated and the other side exposed to the room temperature.

3.2 Thermal conductivity

Measured thermal conductivity of the Ondol panel was about 1.21 kcal/mh℃. By considering that the thermal conductivity of common polymer concrete ranges between 1.1-1.2 kcal/mh℃, it seemed that the conductivity of the Ondol panel was a little larger than that of common polymer concrete. However, the thermal conductivity of the Ondol panel was a little smaller than that of 1.3 kcal/mh℃ of 15 mm thick cement concrete.

3.3 Bearing stress

3.3.1 With 12 mm-thick plywood base

Table 6 represents the results of bearing stress test with 35 mm round bar. For one-side reinforced Ondol panels, penetration began at the load of 19.6 KN and when the load increased to 24.5 KN, the penetration depth reached 1.4 mm. For two-side reinforced Ondol panels, penetration began at the load of 27.4 KN, and when the load increased to 39.2 KN, penetration depth was 1.6 mm. Bearing stresses at the beginning of penetration were 20.4 MPa and 28.5 MPa for one-side and two-side reinforced Ondol panels, respectively.

Table 7 shows the results of bearing stress test with 25 mm steel ball. For one-side reinforced Ondol panels, penetration began at the load of 6.9 KN, and when the load increased to 9.8 KN, penetration depth was 3.9 mm. For two-side reinforced panels, penetration began at the load of 6.9 KN, and when the load increased to 19.6 KN, penetration depth was 4.4 mm.

Table 6. Bearing stress under the loading of 35 mm round rod with 12 mm-thick plywood base

One-side reinforcement			Two-side reinforcement		
Load (kN)	Stress (MPa)	Deformation (mm)	Load (kN)	Stress (MPa)	Deformation (mm)
14.7	16.3	0	19.6	20.4	0
19.6	20.4	0	26.9	28.5	0
22.5	23.4	0.5	30.4	31.6	0.4
24.5	25.4	1.4	39.2	40.7	1.6

Table 7. Bearing stress under the loading of 25 mm steel ball with 12 mm-thick plywood base

One-side reinforcement		Two-side reinforcement	
Load (kN)	Deformation (mm)	Load(kN)	Deformation (mm)
4.9	0	4.9	0
6.9	0	6.9	0.2
7.8	0.4	9.8	1.5
9.8	3.9	19.6	4.4

3.3.2 With 50mm thick polystyrene foam base

Table 8 represents the results of bearing stress test with 35 mm round rod. For one-side reinforced panels, deformation and cracks occurred at the load of 2.9 KN and the rod penetrated through the panel at the load of 3.0 KN. For two-side reinforced panels, deformation and cracks began at the load of 3.3 KN and the rod penetrated through the panel at the load of 3.5 KN. Bearing stresses at the maximum loads were 3.0 MPa and 3.4 MPa for one-side and two-side reinforced Ondol panels, respectively. These bearing stresses were much smaller than those obtained from the plywood base tests.

Table 9 shows the results of bearing stress test with 25 mm steel ball. For one-side reinforced panels, cracks and deformation began at the load of 2.4 KN and the ball penetrated through the panel at the load of 2.9 KN. For two-side reinforced panels, cracks and deformation began at the load of 2.6 KN and when the load increased to 3.5 KN, penetration depth was 3.3 mm but the ball did not penetrate through the panel.

From these results, it was thought that the bearing stress of Ondol panels was dependent on reinforcement, load condition, and base support condition at the bottom. However, it was thought that bearing stress of Ondol panel were high enough to be used for the Ondol system of Korean housing.

Table 8. Bearing stress under the loading of 35 mm round rod with 50 mm polystyrene foam base

One-side reinforcement			Two-side reinforcement		
Load (kN)	Stress (MPa)	Deformation (mm)	Load (kN)	Stress (MPa)	Deformation (mm)
2.4	2.5	0	2.4	2.5	0
2.6	2.7	0	2.9	3.0	0
2.9	3.0	2.5	3.3	3.4	2.4
3.0	3.1	Fail	3.5	3.6	Fail

Table 9. Bearing stress under the loading of 25 mm steel ball with 50 mm polystyrene foam base

One-side reinforcement		Two-side reinforcement	
Load (kN)	Deformation (mm)	Load (kN)	Deformation (mm)
0.98	0	0.98	0
1.96	1.0	2.06	0.3
2.45	4.0	2.65	0.7
2.94	Fail	3.23	3.3

3.4 Bending stress

3.4.1 Load-deflection relationship

Fig. 6 shows load-deflection relationship obtained from the bending test. Failure load of two-side reinforced Ondol panel was higher than that of one-side reinforced panel. While deformation of one-side reinforced Ondol panel was larger than that of two-side reinforced panel. Failure loads were 637 N and 735 N for one-side and two-side reinforced panels, respectively. And the ultimate deflections were 16.6 mm and 15.9 mm for one-side and two-side reinforced panels, respectively.

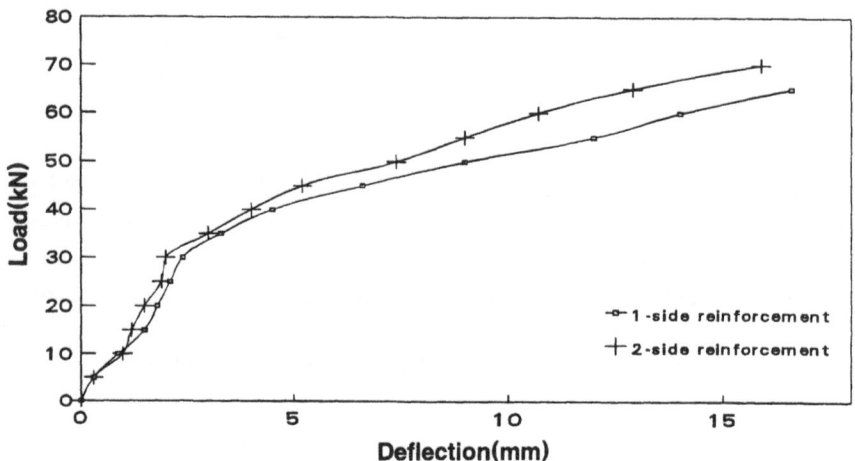

Fig. 6. Load-deflection relationship of polymer mortar Ondol panel

3.4.2 Bending stress-tensile side strain relationship

Mean bending stresses of one-side and two-side reinforced Ondol panels were 28.7 MPa and 33.1 MPa, respectively (Table 11). Mean bending stress of two-side reinforced Ondol panel was about 15% higher than that of one-side reinforced panel.

Fig. 7 shows the relationship between tensile side strain and bending stress. As can be seen in the table, bending properties were improved by reinforcing the both sides of Ondol panel.

Table 11. Bending stress of Ondol panel   (unit : Mpa)

Specimen	Mean
One-side reinforced panel	28.7
Two-side reinforced panel	33.1

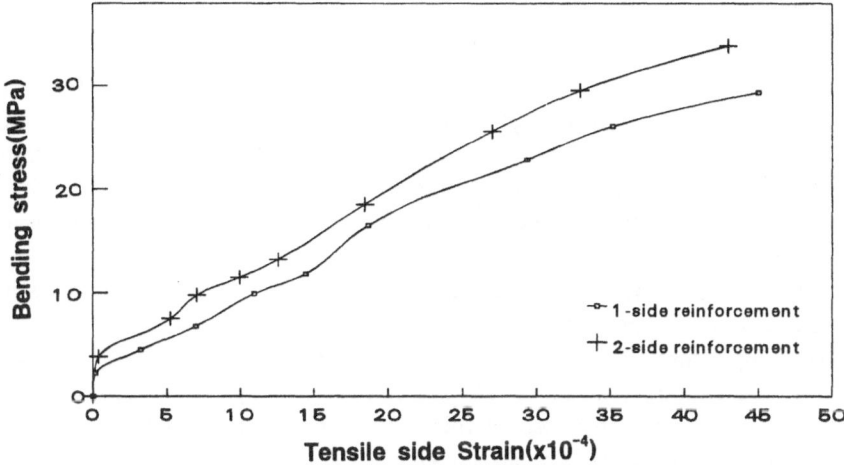

Fig. 7. Relationships between tensile-side strain and bending stress of Ondol panel

3.5 Impact test

The results of impact test are shown in shown in Table 12 and impact force was computed with Equation 2.

$$F_{avg} = \frac{m}{\Delta t}(V_2 - V_1)$$

where, F is impact force (N), m the mass of pendulum, Δt falling time (second) and V_1 and V_2 are falling velocities (m/sec) of pendulum, respectivly. Δt is very small and negligible, V_1 is 0 and V_2 is $\sqrt{2gh}$. For one-side reinforced Ondol panels, cracks appeared at the drop height of 150 cm while for two-side reinforced panels, blow marks were observed but no clear cracks appeared even at the drop height of 300 cm. Impact resistance of Ondol panel was higher with two-side reinforcement than that with one-side reinforcement.

Table 12. Impact test results on the center part of Ondol panel

One-side reinforced panel			Two-side reinforced panel		
Drop height(cm)	Impact force(N)	Surface state	Drop height(cm)	Impact force(N)	Surface state
80	3.96	no change	150	5.42	no change
100	4.43	no change	180	5.94	no change
150	5.42	1cm length crack	200	6.26	∅0.3cm blow mark
200	6.26	3cm length crack	250	7.00	∅0.5cm blow mark
250	7.00	5cm length crack	300	7.67	∅1cm blow mark

4 Conclusions

A hot press-type Ondol panel using polymer mortar was developed and evaluated its performance in this study. And the following results were obtained.

1) Cooling times of Ondol panel from the preheated temperature of 50℃ to the room temperature of 14.4-16.9℃ were 29-34 minutes and 100-125 minutes for two and one side exposures, respectively. When bioceramic powder was added to Ondol panel, the cooling took 8-25 minutes longer to reach the room temperature.
2) Thermal conductivity of Ondol panel was 1.21 kcal/mh℃ and slightly smaller than that of 1.3 kcal/mh℃ of cement concrete.
3) Bearing Stress was largely dependent on reinforcement, load condition and base support condition. However, it was thought that bearing stress of the Ondol panel developed in this study were high enough to be used for the Ondol system of Korean housing.
4) Mean bending stresses were 28.7 MPa and 33.1 MPa for one-side and two-side reinforced Ondol panels, respectively. And the maximum deflections were 16.6 mm for the former and 15.9 mm for the latter. It implied that bending property of two-side reinforced panels improved over one-side reinforced panels.
5) Ondol panel resisted well to the impact force applied.
6) It was concluded that the newly developed Ondol panel in this study could be used for the Ondol system of Korean housing without causing problems.

Acknowledgement
This study was supported by the Research Center for Advanced Mineral Aggregate Composite Products designated by KOSEF at Kangwon National University.

5 References

1. ACI Committee 548 (1977) Polymers in Concrete. American Concrete Institute, Detroit, p 92.
2. Agaewal, B. A. and Broutman, L. (1980) Analysis and Performance of Fiber Composite. John Wiley & Sons, New York.
3. Chuo Kouzai Co., Ltd. (1987) Resin Concrete-Teachnical Data. p 20.
4. Fowler, D. W. (1990) Status of Concrete-Polymer Materials. Proceedings of the Sixth International Congress on Polymers in Concrete, Shanghai, China, pp. 10~27.
5. Lin, L. and Cochran, J. K. (1987) Optimization of a Complex Flow Line for Printed Circuit Board Fabrication by Computer Simulation. Journal of Manufacturing Systems, Vol. 6, No. 1, pp. 47~57.

REINFORCEMENT OF A LOAD-BEARING WALL USING EPOXY RESIN MORTAR

K.S. Yeon, K.W. Kim, S.S. Kim, D.S. Choi and K.H. Kim
Dept. of Agricultural Engineering, Kangwon National University, Chunchon, Korea

Abstract
This study was carried out to evaluate the effect of thin-layer polymer reinforcement on the surface of a deteriorated load bearing wall using polymer mortar. The results showed that load bearing capacity of the wall was highly improved in bending and compression tests. The moment at failure was improved $2.5 \sim 2.9$ times in bending test, ultimate load was improved approximately 2.5 times in axial compression tests. Therefore, it was found that thin layer in both sides of the wall using epoxy resin mortar revealed an excellent reinforcement effect.
Keywords: Polymer, epoxy resin, load bearing wall, polymer mortar, reinforcement.

1 Introduction

Recent years, many precast items are produced for building and civil engineering structural members using concrete materials. Precast products have some advantages in economical and time saving point of view for construction of fabricating structures.

Among those is the load bearing wall. For some reasons, the wall panel is often produced with an inferior strength and used in the fabrication without close inspection for the quality. In that case, the wall panel must be replaced with a new one. However, there are various difficulties involved in replacing

Polymers in Concrete, edited by Y. Ohama, M. Kawakami and K. Fukuzawa. Published in 1997 by E & FN Spon, 2–6 Boundary Row, London SE1 8HN, UK. ISBN: 0 419 22330 4.

because replacing the load bearing wall in already-made structure usually accompanies demolition. Therefore, reinforcement is an alternative to make the wall perform satisfactorily without demolition.

The objective of this study is to evaluate the performance of the load bearing wall that was reinforced with epoxy resin mortar and welded wire-mesh. The purpose of this paper is to present the results of the experimental study.

2 Materials and Method

2.1 Materials
Materials for thin layer reinforcement were an epoxy mortar and welded wire mesh. The epoxy mortar was produced with an epoxy resin, a filler, and a silica sand. Mix proportion of epoxy mortar are given in Table 1. Two types of the wire mesh were used and the mesh diameters were 3mm and 5mm with mesh size of 50x50mm and 75x75mm, respectively. Mechanical properties of epoxy mortar were measured on the cylinder specimen (ϕ70x140mm) and beam specimen(60x 60x240mm) which were made during working for reinforcement on the wall. The specimens and reinforced wall panel were cured at $26 \pm 3°C$ for 7 days before testing, and the results are in Table 2.

Table 1. Mix proportion of epoxy mortar (unit: wt.%)

Epoxy resin	Filler	Silica sand	Coarse aggregate
15	15	45	25

Table 2. Mechanical properties of epoxy mortar (unit: MPa)

Elastic modulus	Compressive strength	Flexural strength	Split tensile strength
25,578	82.8	27.3	13.1

2.2 Test specimens
Test specimen is the existing load bearing wall which was reinforced using epoxy mortar. The wall panel was originally made using cement concrete in which a wire mesh of ϕ8mm with mesh size 100x100mm was embedded.

Since thin layers of epoxy mortar are placed on both sides of the wall panel, the existing concrete load bearing wall becomes a core. Properties of the cement concrete of the wall are shown in Table 3. In this table, unreinforced-1 represents the normal wall panel that performs properly as designed. unreinforced-2 is the wall panel that produced with inferior strengths and needed a reinforcement.

Table 3. Strengths of cement concretes as original wall materials   (unit: MPa)

Specimen	Compressive strength	Flexural strength	Split tensile strength
Unreinforced-1	27.9	3.4	2.7
Unreinforced-2	18.9	2.9	2.4

Detailed description of the wall panel specimens are given in Table 4, and dimension and shape are shown in Figs. 1 and 2.

Table 4. Description of wall panel specimens

Specimen	Type	Dimension(mm)	Note
Unreinforced-1	No reinforcement	Thickness: 150 Width: 400 & 500 Length: 2,450	Comp. strength: 27.9 Mpa Standard load bearing wall
Unreinforced-2	″	″	Comp. strength: 18.9 Mpa Deteriorated wall, need a reinforcement
Reinforced-I	Epoxy mortar reinforcement	Thickness: 180 Width: 400 & 500 Length: 2,450	Core: unreinforced-2 Reinforced with 15mm of epoxy mortar in both sides Wire mesh: $\phi 5 \times 75 \times 75$ mm
Reinforced-II	″	″	Core: unreinforced-2 Reinforced with 15mm of epoxy mortar in both sides Wire mesh: ϕ $3 \times 50 \times 50$ mm

(Unit : mm)

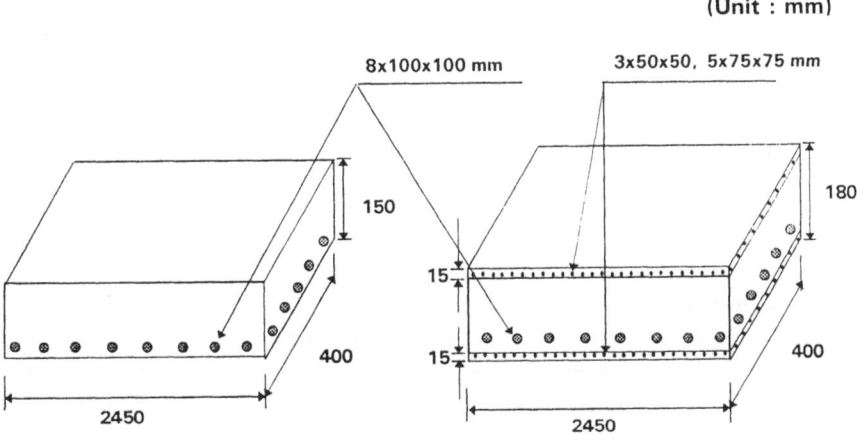

Fig. 1.   Shape and dimension of specimens for bending test

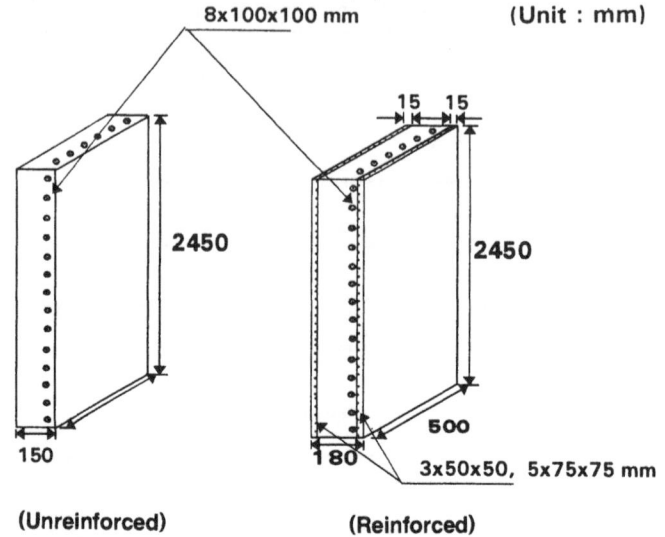

Fig. 2. Shape and dimension of specimens for axial compression test

2.3 Test Methods

2.3.1 Bending test
Flexural strength test was performed by three-point loading according to KS
F 2408 (Method of test for flexural strength of concrete). Fig. 3 shows points
at which strain gages were placed on the specimen.

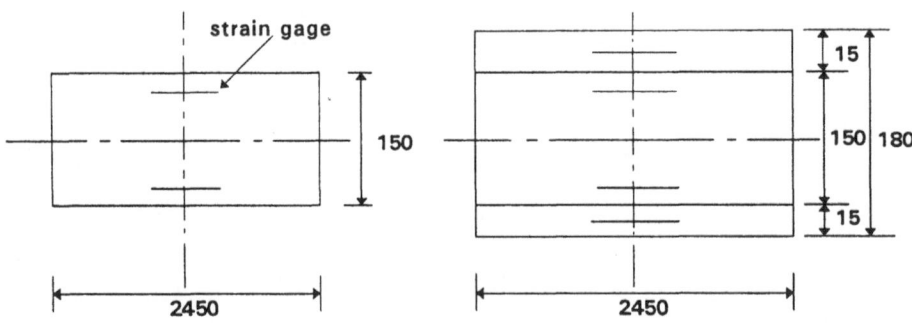

Fig. 3. Description of strain gage placement.   (Unit : mm)

2.3.2 Axial compression test
Uniaxial compression test was carried out according to KS F 2273 (Methods
of performance test of panels for building construction). A 30mm thickness
steel plate was used for application of an eccentric load at 1/3 of wall thickness.
To measure longitudinal strain during compression, strain gages were placed
at 1/4, 1/2 and 3/4 of the wall height along the center line of both sides.

3 Results and Analysis

3.1 Bending test

Fig. 4 shows load and deflection relationship of 4 different specimens for flexural strength test. The reinforced walls showed lower deflections under the same load and higher ultimate strengths. The linear portion of the reinforced walls was approximately 60% of ultimate load, but that of the unreinforced walls were approximately 30% of ultimate load.

Fig. 5 shows load-compressive strain relationship of 4 different specimens for flexural strength test. The increment rate of compressive strain of the reinforced walls under unit load was smaller than that of the unreinforced walls.

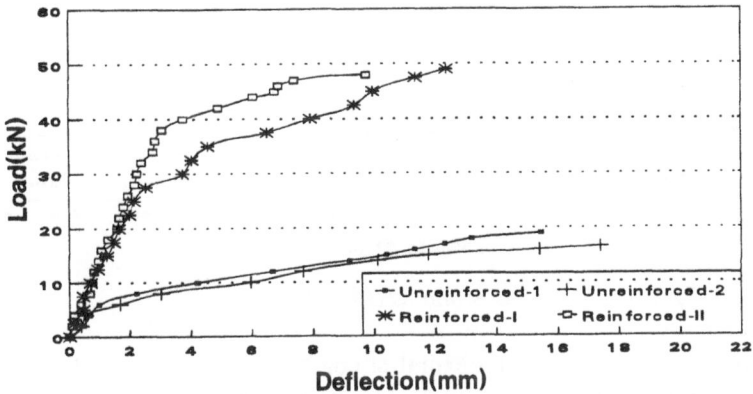

Fig. 4. Load-deflection curves for bending test

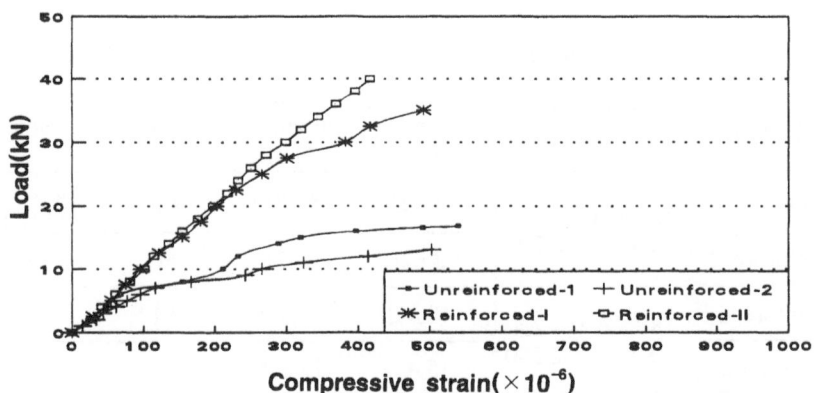

Fig. 5. Load and compressive strain relationship

Fig. 6 shows load and tensile strain relationship. It can be seen from the figure that the increment rate of tensile strain of the reinforced walls under unit load was smaller than that of the unreinforced walls.

Cracks appeared at the site where maximum bending moment was developed, and more number of cracks appeared in reinforced specimens than unreinforced specimens. This is explained by the stress distributing effect of epoxy mortar layer on the surfaces.

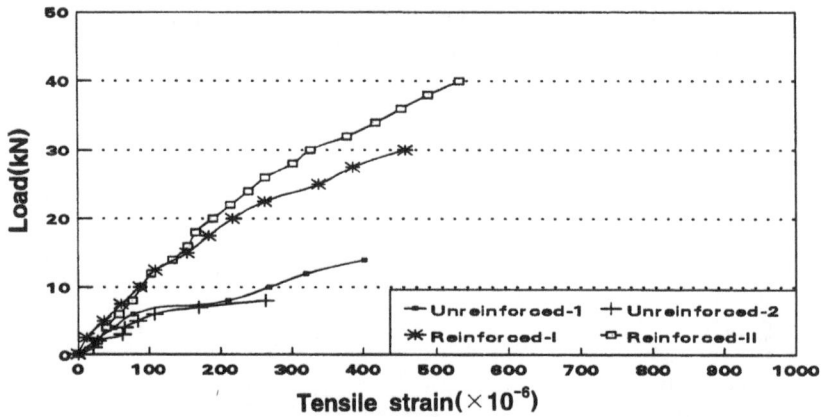

Fig. 6. Load and tensile strain relationship

Table 5 shows results of flexural strength test. The moment at failure of unreinforced-1, normal wall panel, was larger than that of unreinforced-2, deteriorated wall panel. The moment at failure of reinforced-II was larger than reinforced-I. In comparison of the reinforced wall with the unreinforced wall, the reinforced wall showed 2 or 3 times higher failure moment than the unreinforced. The reinforcement of the wall with a polymer mortar did increase resistance capacity of the wall against bending.

Table 5. Results of the flexural strength test

Specimen	Dimension (mm)	Cracking moment (N·m)	Failure moment (N·m)	Flexural strength (MPa)	Note
Unreinforced-1	150×400×2,450 Span = 2,200	5,027	6,938	4.6	Failed at central part
Unreinforced-2		3,626	6,036	4.0	"
Reinforced-I	180×400×2,450 Span = 2,200	13,475	17,395	8.1	"
Reinforced-II		15,454	17,640	8.2	"

3.2 Axial compression test

Fig. 7 shows axial compression load and compressive strain relationship of 4 different specimens. Since the curves of the reinforced walls showed steeper slope, stiffness of the reinforced walls should be larger. Fig. 8 shows axial compression load and tensile strain relationship. In the other way in this result, tensile strain of the reinforced wall was larger than the unreinforced wall. It is assumed that initially tension and compression sizes are under pure compression, but by increasing axial load, the tension side is subjected to tensile-like compression.

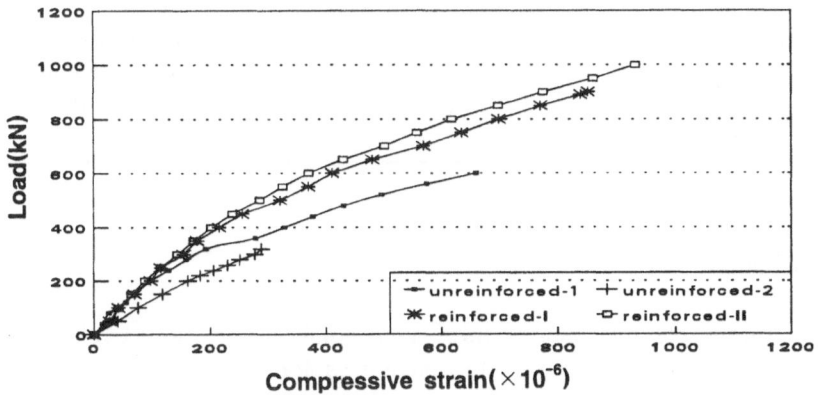

Fig. 7. Axial compression load and compressive strain relationship

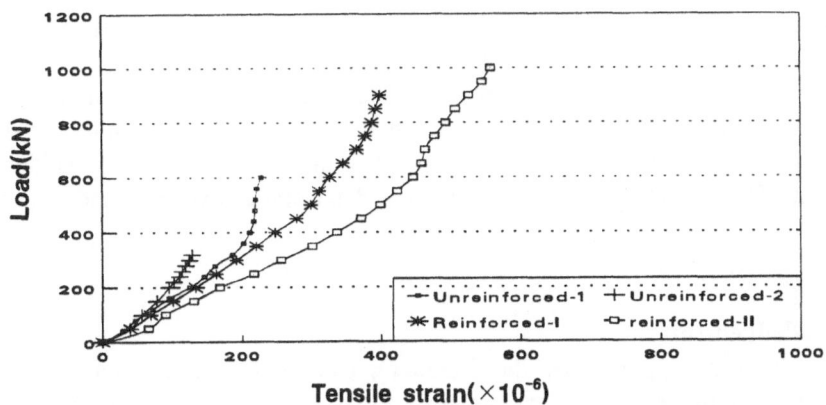

Fig. 8. Axial compression load and tensile strain relationship

Axial compression test results are shown in Table 6. Compressive strength of the normal load bearing wall panel, unreinforced-1, was 1.6 times greater than the deteriorated wall panel, unreinforced-2. The strength of reinforced-I was a little larger than reinforced-II. In general, the reinforced walls were 1.5 times stronger than unreinforced-1, and 2.5 times stronger than unreinforced-2. Therefore, application of epoxy mortar thin layers on both sides of the wall was considered very effective for reinforcement of the deteriorated load bearing wall.

Table 6. Results of axial compression test on wall panel

Specimen	Dimension(mm)	Ultimate load (KN)	Note
Unreinforced-1	150 × 400 × 2,400	598	Local bearing failure after 200mm length crack appearance
Unreinforced-2		363	
Reinforced-I	180 × 400 × 2,450	960	
Reinforced-II		872	

4 Conclusions

In this study, polymer mortar thin layer reinforcement was evaluated and conclusions drawn from the evaluation are as follows.

1) Moments at failure in bending test of reinforced walls were 2.5 or 2.9 times higher than unreinforced walls. Reinforcement of epoxy mortar layers was found to be very much effective for bending resistance.

2) In bending test, crack appeared at the site where maximum bending moment was developed. The more number of cracks showed up form reinforced wall specimens than unreinforced specimens. This is due to stress distributing effect of epoxy mortar layer on the surfaces.

3) The reinforced walls were 1.5 times stronger than unreinforced standard load bearing wall, and 2.5 times stronger than unreinforced deteriorated wall.

4) Application of an epoxy mortar thin layer on both sides was considered as an effective reinforcing method for a deteriorated existing load bearing wall.

Acknowledgement
This study was supported by the **Research Center for Advanced Mineral Aggregate Composite Products** designated by **KOSEF** at Kangwon National University.

5 References

1. Abdel-Halim, M. A. and McClure, R. M. (1985) Flexural Behavior of Reinforced Polymer-Portland Cement Concrete Beams. *Polymer Concrete,* SP-89, American Concrete Institute, Detroit, pp. 105-206
2. Huson J. A. (1965) Precast Concrete Wall Panels: Flexural Stiffness of Sandwich Panels. *ACI SP-11.*
3. Hsu, H. T. (1984) Flexural Behavior of Polymer Concrete Beams. *Ph. D Dissertation,* The University of Texas at Austin, p. 250
4. Yeon, K. S., Kim, K. W. and Hwang, J. Y. (1992) Structural Behavior of Sandwich Panels with Polymer Mortar Facings. *Proceedings of the 7th International Congress on Polymers in Concrete,* Moscow, Russia, pp. 550-557
5. Yeon, K. S., Flower, D. W. and Wheat, D. L. (1987) Static Flexural Behavior of Various Polymer Concrete Beams. *Proceedings of the 4th ICPIC.*
6. Allen, Howard G. (1969) Analysis and Design of Structural Sandwich Panels. Pergameon Press.
7. ACI Committee 533 (1992) Precast Concrete Wall Panels. Concrete International, Vol. 14, No, 11, pp. 33-35
8. Erki, M. A. and Rizkalla, S. H. (1993) FRP Reinforcement for Concrete Structure. Concrete International, Vol. 15, No, 7, pp. 48-53

PART SIX

THERMAL EFFECTS

THE THERMAL INFLUENCE ON THE BOND OF POLYMER-MODIFIED CONCRETE OVERLAYS

H. Abeyruwan
Dept. of Civil Engineering, University of Peradeniya,
Peradeniya, Sri Lanka

Abstract
Polymer Modified Concrete (PMC) is a material commonly used to repair damaged structural concrete members. The strong bond between PMC repair concrete and ordinary concrete substrate is one of the key factors which make PMC the choice. Based on the performance of repaired members at room temperature, it is generally considered that they are almost equal in structural performance to monolithically cast members, under all conditions. However, the dependence of bond on the temperature does not seem to have been sufficiently investigated.

This paper reports the results of an investigation launched primarily to assess the influence of elevated temperature on the residual bond between PMC containing styrene butadiene, and ordinary concrete substrate. The test specimens were subjected to varying temperature levels up to 500 °C, and the bond, compressive and tensile strengths at each level were measured. Results show how the efficacy of PMC as structural repair compound is changed when it is subjected to elevated temperature.
Keywords: Bond, overlay, polymer modified concrete, strength, fire endurance.

1 Introduction

It is the current practice to repair and use moderately damaged concrete structures. A successfully repaired concrete structure should behave equally or better than the original structure in every sense. Polymer modified concrete (PMC) is a material often used in repairing damaged structural concrete members [1]. The strong bond between PMC repair concrete and the ordinary concrete substrate is a key factor which makes PMC the choice. Based on the performance of repaired members at room temperature, it is generally considered that repair with PMC will result in structural members which are equal in structural performance to monolithically cast members. However, the fire

Polymers in Concrete, edited by Y. Ohama, M. Kawakami and K. Fukuzawa. Published in 1997 by E & FN Spon, 2–6 Boundary Row, London SE1 8HN, UK. ISBN: 0 419 22330 4.

endurance of such members may not be the same, as the mechanical properties including the bond are affected by the high temperature. Even though the effect of high temperature on the properties of ordinary concrete has been investigated [3] the dependence of the properties of PMC on high temperature does not seem to have been sufficiently investigated.

The investigation described herein examined the effects of elevated temperature on the bond between PMC overlay and ordinary concrete substrate. It also examined the temperature effects on the compressive and indirect tensile strengths of PMC. The testing programme covers a temperature range from 26 °C to 500 °C.

2 Experimental work

2.1 Materials

Ordinary Portland cement, river sand and crushed gneiss coarse aggregate (12-5mm) were used as ingredients in ordinary concrete substrate. A commercially available styrene-butadiene emulsion (20% polymer solids) was used additionally in polymer modified concrete. Following the guideline that the optimum degree of polymer modification is usually achieved at 10-20 percent dry polymer solids by weight of cement in the mix [1], 15 percent polymer solids by weight of cement were included. The mix details are given in Table 1.

Table 1. Mix details

Ingredient	Quantities (kg/m³)	
	Ordinary concrete	Polymer modified concrete
Cement	500	264
Fine aggregate	887	1128
Coarse aggregate	752	752
Water	215	158
Styrene - butadiene solids in emulsion	--	40

2.2 Test specimens

The types of test specimen are shown in Fig. 1, and their use, in Table 2.

Type A specimens [2] consist of ordinary Portland cement substrate and PMC overlay. The substrate of the specimen was cast against a dummy section placed in a steel cylindrical mould, fitted with a base plate. These half-cylinders were kept undisturbed for 1 day at 26 °C and 95% relative humidity and transferred to a water bath at 70 °C for curing for another day before allowing to dry in the laboratory air for another 5 days. Each of these half-cylinders was soaked in water for 1 hour, wiped off of excess water and applied with a stiff brush, a prime coat consisting of 1:1 cement:polymer emulsion to the diagonal face which forms the interface between the substrate and overlay. These substrate portions were set in the cylindrical moulds and filled the moulds with PMC, and hand-tamped, while the prime coat remained wet. The moulds were kept undisturbed for 1 day at 26 °C and 95% relative humidity. Then, the specimens were demoulded and kept submerged in water at 20°C for another day, and were removed from water to allow to dry at 26 °C and 70% relative humidity for another 26 days until tested.

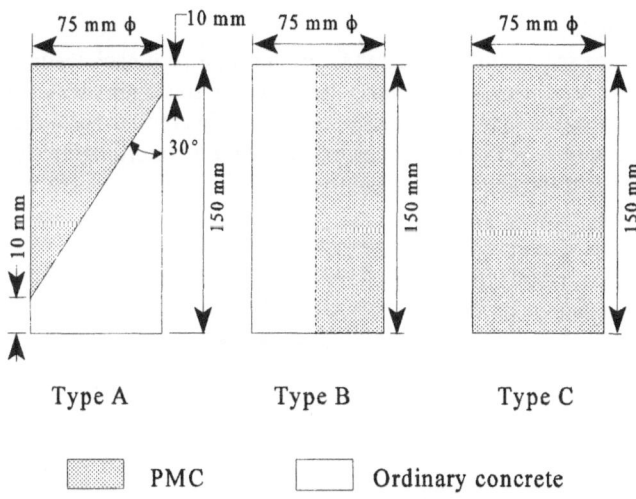

Fig. 1. Test specimens

Table 2. Testing programme

Specimen type	Test performed	Properties tested	Temperature range (°C)
A	Slant shear bond test	Shear bond strength	26-500
B	Split cylinder bond test	Indirect tensile bond strength	26-500
C	Cylinder compression test	Compressive strength	26-500
	Split cylinder test	Indirect tensile strength	

Type B specimens consist of ordinary Portland cement substrate and PMC overlay. First, 75 mm diameter 150 mm long ordinary concrete cylinders were cast in steel cylindrical moulds, fitted with base plates. The curing of these were identical to that of the substrate sections of Type A specimens. These cylinders were split into two equal halves along the longitudinal axis by applying a diametrically opposite line loads parallel to the axis. Each of the substrate sections so obtained was soaked in water, wiped off surface water, and primed with cement-polymer slurry, in a similar manner as for Type A specimens, and placed in a cylindrical mould to cast the remaining half with PMC. These were cured as same as Type A composite specimens.

Type C specimens were cast in steel cylindrical moulds. The moulds were filled with PMC in two layers and compacted by vibration. These were then cured as same as the PMC overlays in Type A and Type B specimens.

2.3 Testing procedure

From each type three specimens were tested for each property at each testing temperature; each of the three was heated separately. One set of specimens was tested at room temperature which was 26 °C. All the other specimens were first heated to the required temperature at a rate 5 °C/min in an electronically temperature controlled oven. The specimens were kept at the final temperature for one and a half hours before being allowed to cool in the outside air.

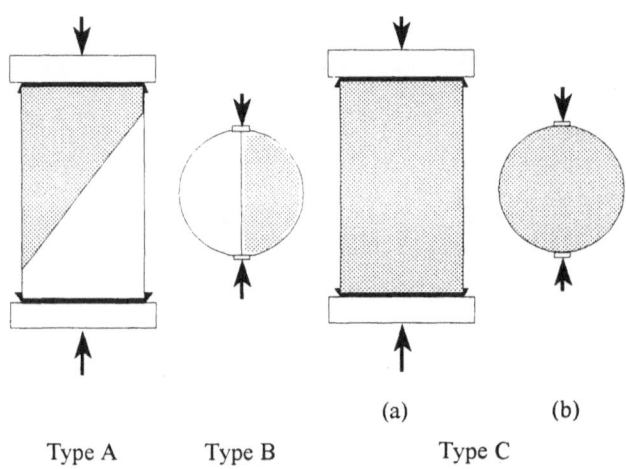

Type A   Type B   Type C

Fig. 2. Test loading

The testing programme covered a temperature range from 26 °C to 500 °C. The thermal influence on the mechanical properties of PMC was studied by testing specimens for residual strength of shear bond, indirect tensile bond, cylinder compression and indirect tension, after returning to initial temperature. The type of loading to which the specimens of each type were subjected is shown in Fig. 2. The failure loads were recorded and the modes were observed.

3 Test results

The values of the four parameters, viz. shear bond strength, indirect tensile bond strength, compressive strength and indirect tensile strength, of unheated specimens are shown in Table 3.

Table 3. Strength of unheated specimens

Property	Strength (MPa)
Shear bond strength	6.45
Indirect tensile bond strength	2.05
Compressive strength	10.40
Indirect tensile strength	1.53

In Table 4 and Fig. 3, the residual strengths are presented as percentages of the corresponding strength of the unheated specimen. The results show the influence of high temperature on the residual shear bond, indirect tensile bond, compressive and indirect tensile strengths of PMC.

Table 4. Residual strength after heating to high temperature

Temperature (°C)	Residual strength as a percentage of initial value				
	PMC overlay on ordinary concrete		PMC		Ordinary concrete [3]
	Shear bond	Indirect tensile bond	Compressive	Indirect tensile	Compressive
26	100	100	100	100	100
100	75	78	106	101	96
150	93	95	122	121	91
200	76	81	114	104	86
250	66	67	111	96	81
300	58	48	99	82	76
350	45	34	82	68	70
400	37	23	67	56	65
450	27	22	53	42	57
500	12	17	34	32	50

Both the shear bond and indirect bond specimens showed complete bond failure upon loading. The residual bond strength, whether it was shear bond or indirect tensile bond, showed a reduction after the specimens were heated. The residual shear bond strength and the indirect tensile bond strength, respectively, at 500 °C was about 12% and 17% of the

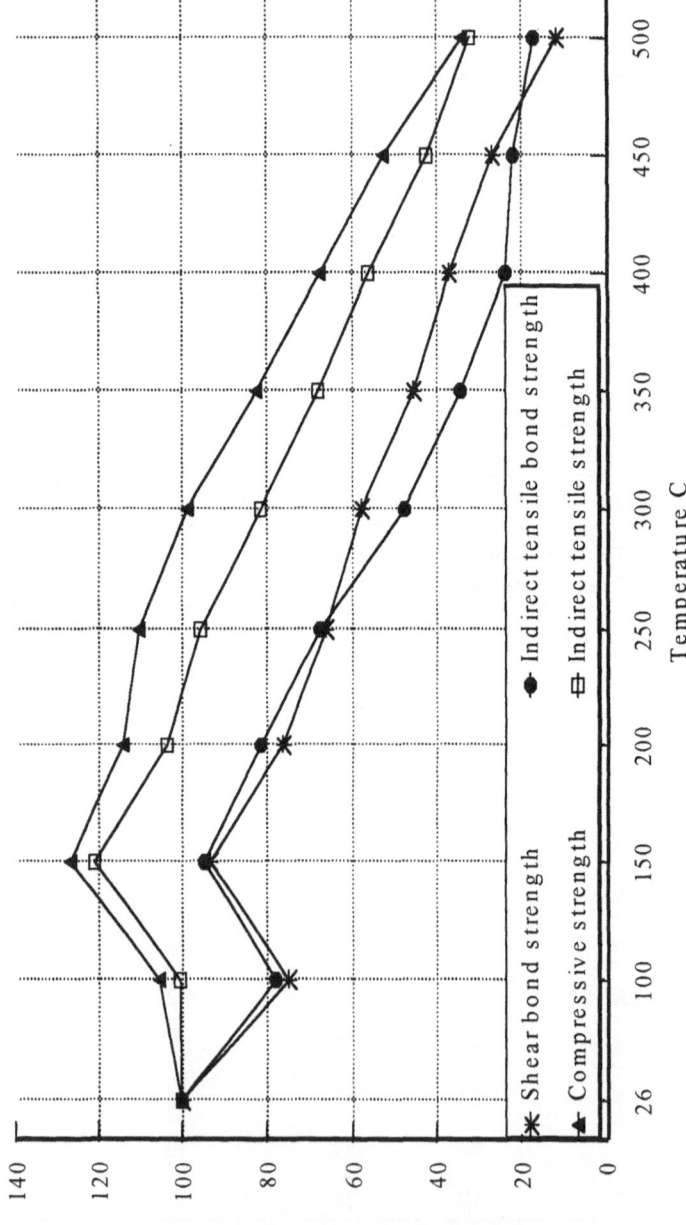

Fig. 3. Residual strengths of PMC after heating to high temperature

initial strength. However, there was an improvement in bond strength for temperature rise from 100 °C to 150 °C. The compressive and indirect tensile strengths of PMC improved when the temperature was increased, with their maximums occurring around 150 °C. Further increase in temperature caused a gradual reduction in strength. This behaviour of PMC is different to that of ordinary Portland cement concrete, which shows a gradual drop in the residual strength with the temperature rise [3]. Below 450 °C the thermal endurance of PMC appeared to be better than that of ordinary concrete, and above 450 °C the converse is true. At 500 °C temperature elevation, the residual compressive strength of ordinary Portland cement concrete was around 50% of the initial value, whereas that of PMC was 34%.

4 Conclusions

The investigation described herein examined the effects of elevated temperature on four important strength parameters governing the performance of structural overlay.

Based on the limited observations made on the thermal behaviour of Portland cement concrete modified with styrene butadiene polymer, the following conclusions are drawn:

1. The residual values of shear bond strength and indirect tensile bond strength of PMC laid on hardened ordinary concrete, and the compressive and indirect tensile strengths of PMC are significantly affected by the temperature to which the hardened materials are heated.
2. Except within the range of 100 °C-150 °C, the residual bond strength and indirect tensile bond strength decreases with increase in temperature.. The residual bond strength after heating to 500 °C drops to about 12% of that of the unheated material. However, within the temperature range of 100 °C to 150 °C, the larger the temperature the greater is the residual bond strength.
3. The residual compressive and indirect tensile strengths of PMC increase by about 20% when the material is heated to 150 °C. Further heating reduces the compressive and indirect tensile strengths at a similar rate.
4. PMC sheds its compressive strength more than the ordinary Portland cement concrete does at temperatures greater than 450 °C. The residual compressive strength of PMC after heating to 500 °C is about 34% of the initial, whereas that of ordinary Portland cement concrete is about 50%.

5 Acknowledgements

The author wishes to acknowledge the permission to use the laboratory facilities and financial support given by the Department of Civil Engineering, University of Peradeniya, Sri Lanka. Thanks are due in general to the laboratory staff for their assistance, and in particular to Messrs. Manjula Jayasinghe and Noel Laksiri for co-ordinating the experimental work.

6 References

1. American Concrete Institute. (1991) *State-of-the-Art Report on Polymer-Modified Concrete.* ACI,Detroit. ACI548.3R-91.
2. American Society for Testing and Materials. (1985) *Standard Test Method for Bond Strength of Latex Systems Used With Concrete.* ASTM, PA. C1042-85.
3. American Concrete Institute. (1989) *Guide for Determining the Fire Endurance of Concrete Elements.* ACI, Detroit. ACI216R-89.

SHRINKAGE STRESSES IN POLYMER CONCRETE OVERLAYS

Y.K. Jo
Building Inspection Division, Korea Infrastructure Safety & Technology
Corporation, Anyang, Korea
D.W. Fowler
Dept. of Civil Engineering, The University of Texas at Austin,
Austin, USA
Y. Ohama
Dept. of Architecture, Nihon University, Koriyama, Japan

Abstract

Polymer concrete overlays are subject to shrinkage and thermal stresses which can lead
to delamination. Thermal stress can be predicted; shrinkage can be measured using a
new, innovative test. Research is carried out to determine the shrinkage stresses as a
function of polymer type (epoxy, polyester, acrylic), overlay thickness, time after curing
and temperature. The test involves the placement of a PC overlay on a concrete beam 15
cm x 15 cm x 1 m. The middle 50 cm is unbonded. A measuring device with a 25-cm
gage length is placed in the unbonded portion of the overlay. At a given time after the PC
has cured, one end of the unbonded section is cut. Shrinkage stresses in the PC cause it
to shorten and the shortening is measured; with the modulus of the PC and strain
known, the stresses can be calculated. It has been found that, depending on the type of
polymer, the shrinkage stresses are eliminated by relaxation in time ranging from a few
hours to a few days.
Keywords : Polymer concrete, shrinkage stresses, delamination, epoxy, polyester,
polymethyl methacrylate.

1 Introduction

In recent years, polymer concrete has been used as overlays to correct many bridge deck
problems such as deterioration of the concrete, delamination caused by corrosion of the
reinforcement, and low skid resistance. Polymer concrete has also been used
successfully for highway and airport pavement repairs and overlays [1 - 3].

Polymers in Concrete, edited by Y. Ohama, M. Kawakami and K. Fukuzawa. Published in 1997
by E & FN Spon, 2–6 Boundary Row, London SE1 8HN, UK. ISBN: 0 419 22330 4.

Generally, polymer concrete made with epoxy resin (EP) has a low shrinkage strain, but shrinkage strains for those made with unsaturated polyester resin (UP) and poly methyl methacrylate (PMMA) are generally high. The shrinkage of polymer concrete overlays causes interface shear and normal stresses and axial stress in the overlay (4). These stresses were found to be a function of the coefficient of thermal expansion of the PC and portland cement concrete(PCC), the temperature change, relative thicknesses of the PC and PCC, and relative modulus of the PC and PCC. In general, the interface and axial stresses are found to be present for a distance from the edge or boundary equal to the thickness of the PC and PCC (4). The purpose of this study is to find the effect of the polymer types, overlay thickness, time after curing and temperature on the shrinkage stresses of polymer concrete overlays. Polymer concrete overlays were prepared with various polymer types, the shrinkage strains were measured by the new method, and the shrinkage stress were calculated. The effect of various factors on the shrinkage stresses of polymer concrete overlays is discussed.

2 Materials

2.1 Polymer concrete overlays
2.1.1 Binder system
Unsaturated polyester resin (UP) was employed as a liquid resin for UP concrete overlays, together with styrene monomer (St) as a diluent, methyl ethyl keton peroxide (MEKP) as an initiator, and cobalt octoate (CoOc) as an accelerator. Commercially available epoxy resin and hardener were used for epoxy resin (EP) concrete overlays. A binder system for PMMA concrete overlays was based on methyl methacrylate (MMA) monomer, with trimethylolpropane trimethacrylate (TMPTMA) as a cross-linking agent, N-N-dimethyl-P-toluidine (DMT) as a promoter, and benzoyl peroxide (BPO) as an initiator.
2.1.2 Filler and aggregates
Commercially available white portland cement (Blaine's specific surface of 3350 cm^2 /g) was used as a fine filler. Crushed granite (5 mm or finer) and Colorado river sand (2.5 mm or finer) were used for polymer concrete overlays. A premixed filler and silica sand system actually used in construction was used for epoxy concrete overlays.

2.2 Latex-modified concrete overlays
Ordinary portland cement, crushed granite (5 mm or finer) and Colorado River sand (2.5 mm or finer) were used for latex-modified concrete overlays. Commercially available latex was a styrene-butadiene rubber (SBR) latex. The latex-modified concrete overlays were prepared for comparison with polymer concrete overlays.

2.3 Ordinary portland cement concrete for substrate
Ordinary portland cement and Colorado River sand (5 mm or finer) and gravel (5 to 20 mm) were used for ordinary portland concrete substrates.

3 Testing procedures

3.1 Preparation of ordinary portland cement concrete substrate

Ordinary portland cement concrete beams, 15 cm x 15 cm x 1 m, were cast for the substrate concrete with the following mix proportions: cement content of 390 kg/m^3, sand-aggregate ratio of 44%, water-cement ratio of 35% and slump of 16 cm. Superplasticizer was added for workability. This concrete was designed to develop a compressive strength of 45 MPa. The cement concrete substrate was cured at 23.9℃ and 50%R.H. for 28 days. The portland cement concrete substrates were cured at 23.9℃ and 50% R.H. for 28 days.

3.2 Preparation of polymer concrete and latex-modified concrete overlays

According to ACI 548.5R (Guide for Polymer Concrete Overlays) and ACI 548,4 (Standard Specification for Latex-Modified Concrete (LMC) Overlays) (5), polymer concrete and latex-modified concrete overlays were mixed with the proportions as shown in Tables 1 and 2. The thickness of the polymer concrete overlays ranged from 6 mm and 12 mm, and for latex-modified concrete overlays, 25 mm.

Table 1 Mix proportions of polymer concretes.

1) UP concrete

Constituent	Percent by weight
Trap rock (6 mm or finer)	39
Sand	39
White potland cement	11
Resin	11

2) PMMA concrete

Constituent	Percent by weight
Trap rock (6 mm or finer)	35
Sand	45
White potland cement	10
Resin	10

3) EP concrete

Constituent	Percent by weight
Resin	21
Filler (Filler and Sand)	79
Trap rock (6 mm or finer)	38

Table 2 Mix proportions of latex-modified and portland cement concretes.

1) Latex-modified concrete

Constituent	Percent by weight
Trap rock (6 mm or finer)	29.7
Sand	44.6
Ordinary portland cement	17.2
Styrene-butadiene rubber latex	3.4
Water	5.1

2) Portland cement concrete

Constituent	Percent by weight
Trap rock (19 mm or finer)	44.2
Sand	33.0
Ordinary portland cement	17.0
Water	5.9
Superplasticizer	0.09

3.3 Curing shrinkage test

According to ACI 548.4, curing shrinkage of polymer concrete and latex-modified concrete overlays can be measured by the DuPont device as shown in Fig.1.

The polymer concrete and latex-modified concrete overlays were allowed to bond to the substrate in the two end portions. The DuPont device (electronic transducer) was placed within the limits of the unbonded central portion. Restraint provided by bond to the substrate through the end portions was then removed by cutting the overlay transversely near one end of the unbonded central portion at various times after curing

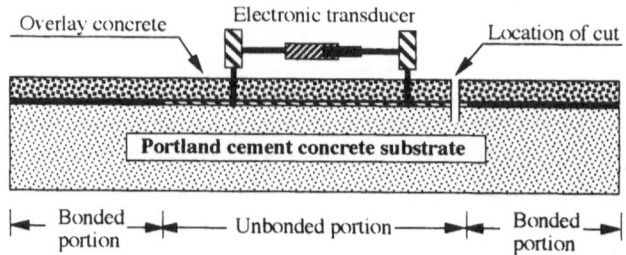

Fig.1 Curing shrinkage test for in overlays.

was completed.Any residual shrinkage strain was measured by the DuPont device. Concrete overlays were cured at 23.9°C and 50% R.H. and 32.2°C and 50% R.H. The shrinkage-induced axial stress was calculated as follow;

$$\sigma = \varepsilon \times E$$

where σ : Shrinkage stress (MPa)

                  ε : Shrinkage strain (μm/m)

                  E : Modulus of elasticity (MPa)

Unrestrained shrinkage was also measured using PC and LMC placed in the same mold as restrained shrinkage without bonded portion between overlay concrete and substrate.

4 Test results and discussion

4.1 Basic properties of polymer concrete, latex-modified concrete overlays and ordinary portland cement concrete substrate

Table 3 shows the basic properties of polymer concrete overlays, latex-modified concrete overlays and ordinary portland cement concrete used in this study. It is evident that the

Table 3 Basic properties of overlay concretes and portland cement concrete substrate.

Type of concrete	Compressive strength (MPa)	Modulus of elasticity ($\times 10^4$MPa)	Flexural strength (MPa)	Set time (min.)	Cure time (min.)
	23.9 ℃	23.9 ℃	23.9 ℃	23.9 ℃	23.9 ℃
	32.2 ℃	32.2 ℃	32.2 ℃	32.2 ℃	32.2 ℃
UP	92 / 104	2.34 / 2.79	19.6 / 19.7	45 / 35	246 / 220
PMMA	89 / 71	2.30 / 2.04	17.2 / 12.5	42 / 30	93 / 80
EP	39 / 38	0.10 / 0.16	32.7 / 26.7	55 / 30	84 / 60
Latex-modified	34 / -	2.33 / -	-	-	-
Portland cement concrete	47 / -	3.60 / -	-	-	-

basic properties of polymer concrete overlays are affected by curing temperature. The modulus of elasticity is an important factor influencing overlay stresses. The modulus of portland cement concrete was much higher than that of polymer concrete and latex-modified concrete in this study. From Choi[4] and Zalatimo [6], the interface stresses were found to be influenced by the modular ratio between polymer concrete overlays and portland cement concrete. The bigger the modular ratio, the higher the interface stresses.

4.2 Shrinkage strain
4.2.1 UP concrete overlays

Figures 2 and 3 represent the restrained and unrestrained shrinkage strain for UP concrete overlays having thicknesses of 6 mm and 12 mm at curing temperatures of 23.9°C and 32.2°C. Generally, the properties of UP concrete also vary with temperature, and the shrinkage of polymer concrete overlays take place during the curing period. The curing shrinkage strain of UP concretes is generally large compared with PC made from thermosetting resins. In this experiment, the restrained UP concrete overlays of 6-mm thickness cured at 23.9°C and 32.2°C were cracked at two portions within the 25-cm gage length after about 90 minutes from the time of mixing due to concentration of shrinkage stress within unbonded portion before the UP concrete was hardened. The restrained UP concrete overlays with a 12-mm thickness and cured at 32.2°C cracked at the boundary between the unbonded and bonded portions after about 2 hours. Unrestrained shrinkage strain for UP concrete showed no change for nearly 2-1/2 hours and then increased suddenly with temperature change in the UP concrete. Unrestrained shrinkage strain for UP concrete overlays with a 6-mm thickness cured at 23.9°C was about one-half of that for a curing temperature of 32.2°C. It is obvious that unrestrained shrinkage strain for UP concrete is affected by curing temperature. The restrained shrinkage strain of UP concrete overlays with a 6-mm thickness of 6 hours was much smaller than that for a thickness of 12 mm, irrespective of curing temperatures. From these results, it is apparent that the restrained and unrestrained shrinkage strains of UP concrete overlays are affected by curing temperature and overlay thickness. For UP concrete overlays, about 2000 μm/m of residual shrinkage strain was detected when restraint was removed after 6 hours at 23.9°C, and about 100 μm/m when removed after 72 hours at 23.9°C. Obviously, some relaxation occurred during the 6 hours, but the stresses were not completely relieved.

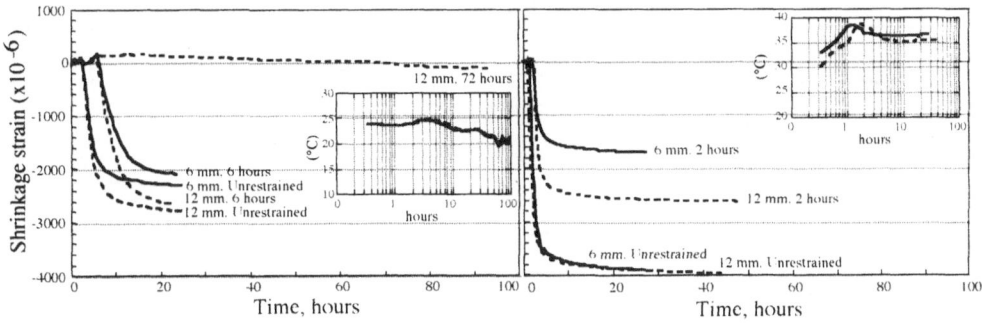

Fig.2. Shrinkage strain for UP concretes having thickness of 6 mm and 12 mm and cured at 23.9 °C.

Fig.3. Shrinkage strain for UP concretes having thickness of 6 mm and 12 mm and cured at 32.2 °C.

4.2.2   PMMA concrete overlays

Figures 4 and 5 show the restrained and unrestrained shrinkage strains for PMMA concrete overlays having thicknesses of 6 mm and 12 mm and cured at 23.9℃ and 32.2℃. Generally, PMMA concrete has several advantages such as rapid development of high compressive, flexural, and tensile strength; good bond to dry surfaces; good resistance to acids; and excellent abrasion resistance, but curing shrinkage is high. The restrained PMMA concrete overlays of 6-mm and 12-mm thicknesses cracked after about 24 hours from the time of mixing. The unrestrained and restrained shrinkage strains for PMMA concrete overlays with a thickness of 12 mm are much higher than for those of 6-mm thickness. Unrestrained shrinkage strain for PMMA of 6-mm thickness was negligible. It is evident that the shrinkage strain of PMMA concrete overlays is affected by curing temperature. The restrained shrinkage strain of PMMA concrete overlays increased at an early age, and then, became smaller and smaller with time due to relaxation. The maximum unrestrained shrinkage strain of PMMA concrete overlays for a thickness of 6 mm is about one-fifth of that for a 12-mm thickness, and the maximum restrained shrinkage strain for a 6-mm thickness after 2 hours is about one-third of that for 12 mm. This tendency of unrestrained and restrained shrinkage strains of PMMA concrete overlays is considerably different from that of UP concrete overlays. It seems that this reason is due to different properties between thermosetting and thermoplastic resin. Although the PMMA concrete overlays were restrained at the two end portions, the shrinkage strain was markedly increased at an early age. The shrinkage strain of PMMA concrete overlays cured at 32.2℃ was smaller than that of PMMA concrete overlays cured at 23.9℃. This tendency of PMMA concrete overlays is the opposite of UP concrete overlays.

4.2.3   EP concrete overlays

Figures 6 and 7 illustrate the restrained and unrestrained shrinkage strain for EP concrete overlays having thicknesses of 6 mm and 12 mm cured at 23.9℃ and 32.2℃. Generally polymer concrete made with epoxy binders is less brittle than portland cement concrete

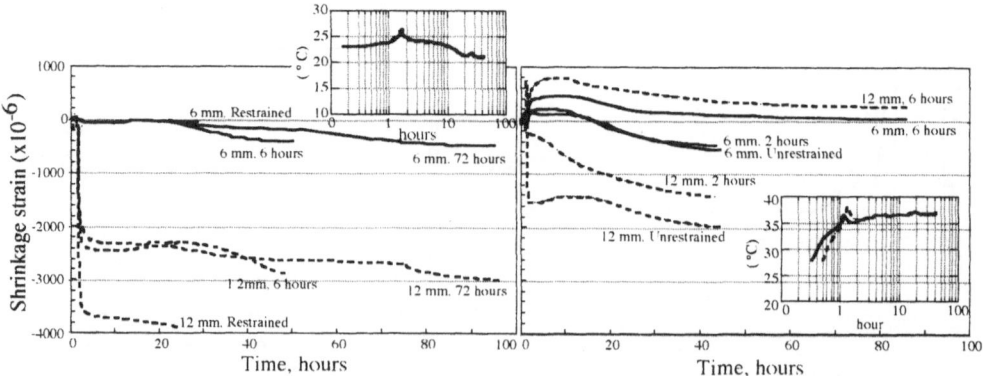

Fig.4. Shrinkage strain for PMMA concretes having thickness of 6 mm and 12 mm and cured at 23.9 °C.

Fig.5. Shrinkage strain for PMMA concretes having thickness of 6 mm and 12 mm and cured at 32.2 °C.

and has a greater coefficient of thermal expansion. Epoxies are known to be good adhesives due to their tenacious bond, are capable of developing high compressive and tensile strengths, and are resistant to chemical attack and wear. Polymer concrete made with epoxy resin chosen in this study has a very low modulus of elasticity and compressive strength, but the flexural strength is very high compared with those of other polymer concretes. This means the EP polymer concrete has a high toughness. The unrestrained and restrained shrinkage strain was the smallest of the three polymer concrete overlays. The shrinkage strains of EP concrete overlays cured at 23.9℃ ranged from 10 μm/m to 100 μm/m, and the difference of shrinkage strain between thickness of 6 mm and 12 mm was negligible. At the curing temperature of 32.2℃, the effect of overlay thickness on the shrinkage strain is clear.

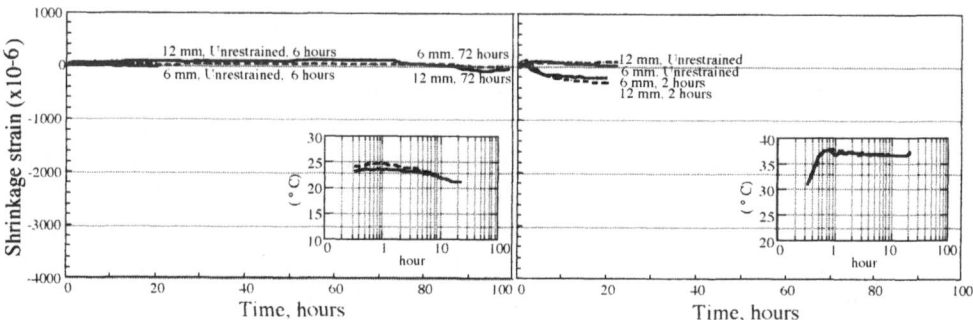

Fig.6. Shrinkage strain for EP concretes having thickness of 6 mm and 12 mm and cured at 23.9°C.

Fig.7. Shrinkage strain for EP concretes having thickness of 6 mm and 12 mm and cured at 32.2°C.

4.2.4 Latex-modified concrete overlays

Figure 8 exhibits the restrained and unrestrained shrinkage strain for latex-modified concrete overlays having thicknesses of 25mm at 23.9℃. The unrestrained shrinkage strain of latex-modified concrete overlays increases with increasing curing time and

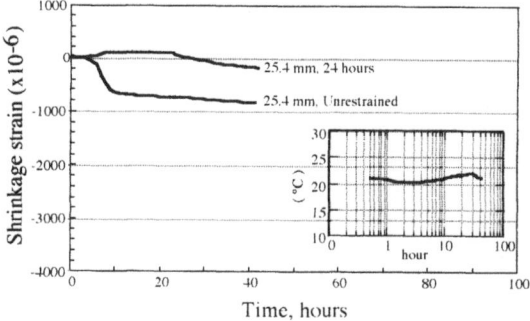

Fig.8. Shrinkage strain for Latex-modified concretes having thickness of 25.4 mm and cured at 23.9°C.

becomes constant at about 850μm/m after about 10 hours. On the other hand, the restrained latex-modified concrete overlay slightly expanded to about 70 μm/m, and then decreased with increasing curing time. About 250 μm/m residual shrinkage strain was recorded for this latex-modified concrete when restraint was released after 9 hours. The shrinkage strain of latex-modified concrete overlay is smaller than for the polymer concrete overlays.

4.3 Shrinkage stress

Figures 9 to 12 indicate the shrinkage-induced axial stress of polymer concrete and latex-modified concrete overlays. Residual shrinkage stresses of UP concrete overlays decrease with increasing restraint time, and the shrinkage stresses were nearly eliminated by relaxation when restraint was released after 72 hours. The shrinkage stress of UP concrete overlays was influenced by curing temperature and overlay thickness. In case of an overlay thickness of 6 mm cured at 23.9℃ and 32.2℃, the restrained shrinkage stress is apt to relax in UP concrete overlays but not for a 12-mm thickness. The shrinkage stresses of PMMA concrete overlays cured at 23.9℃ are much higher than for those cured at 32.2℃. The effect of overlay thickness on the residual shrinkage stress relaxation of PMMA concrete overlay with a 6-mm thickness is larger than that of 12mm. The shrinkage-induced axial stress of EP concrete overlays is much smaller than for other polymer concrete overlay and ranged from 0.043 MPa to 0.49 MPa. The shrinkage stress of latex-modified concrete overlays when restraint was removed after 24 hours was one-third of the unrestrained shrinkage stress, about same as that for UP concrete overlays having an overlay thickness of 12 mm, for restraint removed after 72 hours, and cured at 23.9℃.

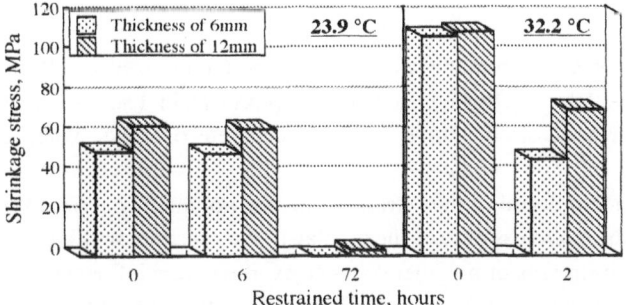

Fig.9 Shrinkage-induced axial stress of UP concrete overlays.

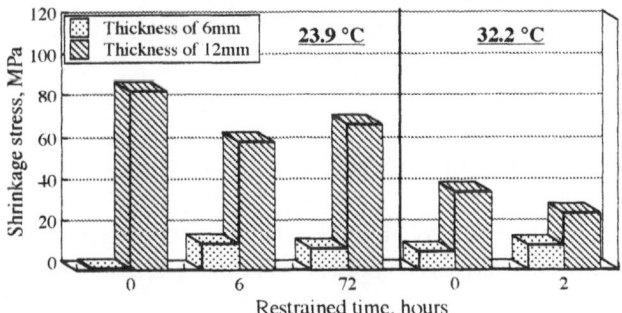

Fig.10 Shrinkage-induced axial stress of PMMA concrete overlays.

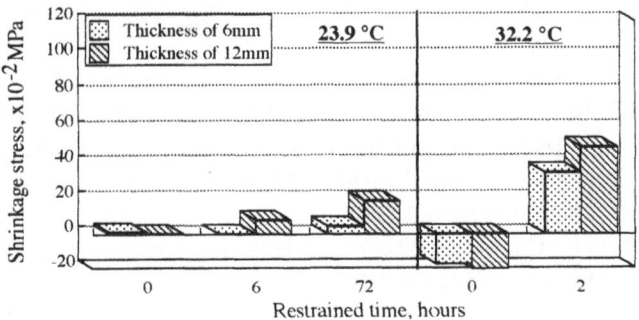

Fig.11 Shrinkage-induced axial stress of EP concrete overlays.

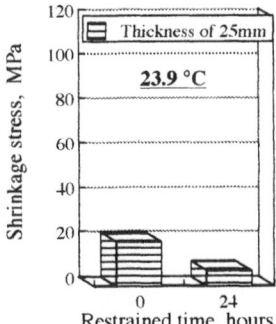

Fig.12 Shrinkage-induced axial stress of
latex-modified concrete overlays.

5 Conclusions

The conclusions obtained from the test results can be summarized as follows:
(1) Polymer concrete overlays made from epoxy resin had the smallest residual shrinkage stresses of the three polymer concrete overlays when restraint was removed.
(2) The shrinkage-induced axial stresses in the polymer concrete overlays become large when the thickness ratio of the overlay to the substrate was increased.
(3) The shrinkage stresses of polymer concrete overlays were influenced by the restraint time after curing. Generally, shrinkage stresses were reduced by relaxation, particularly for 6-mm thickness.
(4) The shrinkage-induced axial stresses of polymer concrete overlays were considerably influenced by the curing temperature.

6 References

1. Sprinkel, M. M. (1991) Polymer Concrete Bridge overlays, ICPIC Working Papers, International Congress on Polymers in Concrete, San Francisco, California, p.13.
2. Fowler, D. W. and James T. D. (1992), Polymer Concrete for Overlays and Precast Components, Conference on Advance in Concrete Technology, Athens, Greece.
3. Cater, P. D. (1990), Thin Polymer Wearing Surfaces for Preventive maintenance of Bridge Decks, Prepared for ACI Fall Convention, Philadelphia.
4. Choi, Donguk. (1992), An Analytical Investigation of Thermally-Induced Stresses in Polymer Concrete-Portland Cement Concrete Composite Beams, Master's Thesis, The University of Texas at Austin, Austin, Texas, May 1992.
5. ACI Maunal of Concrete Practice, Masonry, Precast Concrete, Special Processes, American Concrete Institute.
6. Zalatimo, J. A. (1993), Analysis, Design, Construction, and Durability of Polymer Concrete Overlays, Ph.D.Dissertation, The University of Texas at Austin, Austin, Texas.

ESTIMATION OF THE CREEP COEFFICIENT OF RESIN CONCRETE AT HIGH TEMPERATURE

T. Yamasaki and T. Idemitsu
Dept. of Civil Engineering, Kyushu Institute of Technology,
Kitakyushu, Japan

Abstract
Cement concrete material has some problems such as chemical attack and water proofing from the view point of durability. In these days,resin concrete (written as REC in following) became to be used as a material to substitute a cement concrete such as a members of under ground structures, a block manholes,a panel for parmanent form, a pipe and a shield segments, etc. As REC is composed of liquid resin, filler and aggregates, its mechanical properties at various temperature depend on the nature of resin used as binder. Ordinarily, thermosetting resin is superior at mechanical and chemical properties under normal tempe-rature, but some types of resin change remarkably the mechanical prop-erty according to the change of temperature. In this study, the test results of mechanical properties depending on temperature, especially on coefficient of creep aere described. From which, the extent of the temperature that REC is possible to be used as structural marerials practically is discussed. Furthermore, the equation to calculate a creep coefficient is derived considering the age, elapsed time, stress level and temperature.
Keywords: Creep, Resin concrete, Strength, Stress level, Temperature, Unsaturated Polyester.

1 Introduction

An amount of creep strain of REC varies depending on types and volumes of resin mixed, elapsed time from loading, age of REC at loading, loading stress level and temperature, etc. Unsaturated polyester resin is commonly used in Japan according to economical reason and the mix proportion of resin is selected to be necessarily minimum. In this

Polymers in Concrete, edited by Y. Ohama, M. Kawakami and K. Fukuzawa. Published in 1997 by E & FN Spon, 2–6 Boundary Row, London SE1 8HN, UK. ISBN: 0 419 22330 4.

creep test, unsaturated polyester resin concrete which mix proportion is shown in Table 1 is used.

Creep coefficient of REC has a tendency to be smaller for the elder aged specimen at the loading time and be smallest for the sufficient heat cured specimen. So, the authors tried to investigate the relations between creep and age at the loading time and between creep and loading stress level. Considering these test results, the equation from which the final creep coefficient could be calculated was derived. Furthermore, the effect of temperature to the creep of REC was investigated and defined the correctable coefficient of equation which estimate the final creep coefficient.

2. Outline of test

2.1 Specimen

Cylindrical specimen $\phi 7.5 \times 15$cm was used to the uniaxial compressive creep teste. The compressive strength of REC reached more than 98 Mpa, so the size of specimen was decided to be $\phi 7.5$ cm by the capacity of test apparatus. The mixture of ortho-phthalic acid type unsaturated polyester 60% and styrene monomer 40% was used as a material of liquid resin. Each weight of the material composed of REC that is liquid resin, calcium carbonate, fine aggregate and coarse aggregate is shown in Table 1. In which, the maximum particle diameter of the coarse aggregate was 10 mm. As an additives of liquid resin, styrene monomer

Table 1 Mix of resin concrete, kg/m³

Resin	CaCO₃	Fine aggregate	Coarse aggregate	s/a %
241	289	612	1267	36

Table 2 Mix of additives

Additives	Weight % to Resin
Diluent	6
Hardener	0.5
Accelerator	0.7

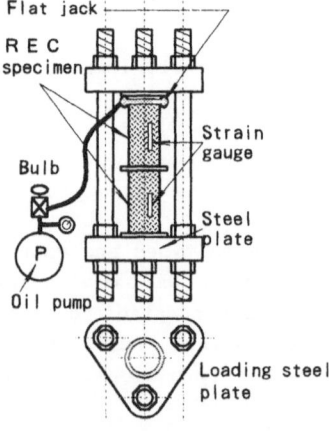

Fig. 1 Creep test apparatus

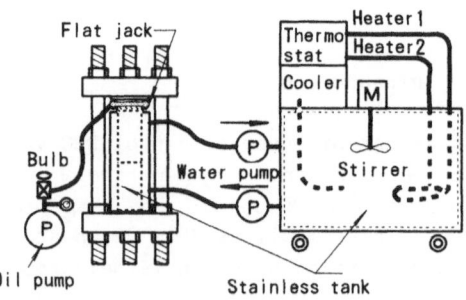

Fig. 2 Thermo control creep test system

for diluent, MEKPO.(55%) plus dimethyle phthalate (45%) for hardener and cobalt naphthenic acid 6% solution for accelerator were used like as Table 2. The aggregates were used after drying with temperature of 110℃ for 24 hours. An ordinary curing temperature after casting was 20 ℃ and a high curing temperature was 80℃ for 12 hours.

2.2 Creep Test apparatus

The creep Test apparatus is shown in Fig.1. Two cylindrical specimens were put one upon another in each thermostatic stainless tank as shown in Fig.2. Creep strain of REC specimen was measured by strain gauge under the state that specimens were loaded constant uniaxial force by the flat type oil jack on the top of specimen. Loaded force on speci- men was confirmed constant by the value of oil pressure gauge of jack. Temperature of specimen was controlled by heated circulating water, the wire strain gauge was waterproofed by overlapping three times with the polyester resin, butyl rubber and silicone rubber. A lead wire of strain gauge was special water proofing type for long term.

2.3 Thermostatic water circulation system

Thermostatic water circulation system is shown in Fig.2. Test tempe- rature was 20, 30, 40, 50, 60℃. Each temperature in cylindrical stain- less tank which size was ϕ 20 × 30 cm was controlled within ±1.5℃ .

3. Test results and consideration

3.1 Compressive strength due to the age and curing temperature

If REC is cured in the normal temperature 20℃ , the compressive strength will be increasing for some months. It corresponds to the field casting condition of under ground structure. To estimate the loading stress level of creep test, compressive strength of REC at each age of 1, 3, 7, 14, 28 days were measured. As the result is shown in Fig.3, the compre- ssive strength after curing for 28 days at temperature 20℃ was 86% of the final strength attained by the curing temperature of 80℃ and tested

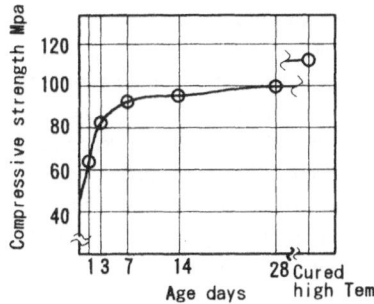

Fig. 3 Relation between compressive strength and age of REC cured 20℃.

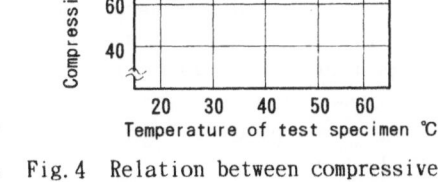

Fig. 4 Relation between compressive strength and temperature of REC.

at 20℃. Supposing to the circumstance in high temperature, compressive strength test was carried out keeping each temperature of specimen constant at 30, 40, 50, 60℃. To eliminate the curing effect of the test temperature, all kind of specimens were cured with temperature 80℃ for 24 hours before the strength test. From the test results, the relation between compressive strength and temperature of specimen was obtained as shown in Fig.4. The strength reduction due to increment of temperature from 20 to 60℃ was about 25%.

3.2 The elastic strain and loading stress level of the creep test.

To make clear the relation between creep property and the age of REC at loading, creep test was carried out loading at the age of 3, 7, 14, 28 days and after high temperature curing. The strength and elastic strain of specimen at loading time were listed in Table 3. In which, the strength ratio means the proportion of compressive strength of REC cured with 20℃ to that cured 80℃. Objective stress of creep test was confirmed by both of elastic strain measured at the loading time and pressure gauge of oil pump.

Table 3　Elastic strain measured and calculated.

Age at loading		Objective stress level. %				Compressive strength Mpa	Strength ratio %
		10	20	30	40		
1 day	———	—-	—	—	—	61.9	54
3 days	Measured strain	327	618	1147	1180	80.1	70
	Stress level	9.9	18.8	34.8	35.8		
7 days	Measured strain	381	642	1052	1335	89.5	79
	Stress level	11.3	19.1	31.3	39.8		
14 days	Measured strain	343	660	1046	1406	93.0	82
	Stress level	10.2	19.6	31.1	41.8		
28 days	Measured strain	307	626	899	1120	97.8	86
	Stress level	8.8	18.0	25.8	32.2		
Cured at Temp. 80℃	Measured strain	—	787	1073	1215	110.2	100
	Stress level	—	24.4	33.2	37.6		

3.3 The relation between creep coefficient and age at the loading

To make clear the relation between creep property and the age at the loading time, creep test was carried out from each age of 3, 7, 14, 28 days and after high temperature curing. Test was operated in room condition which temperature was 20℃ and humidity 70%.

Creep coefficients calculated from dividing the measured strain by the elastic one in Table 3 were plotted on normal-logarithmic coordinate. In which, the results of the test loaded at 28 days and after curing in high temperature were illustrated in Fig.5 and Fig.6 respectively. From these figures, the relation curve of creep coefficient and loading time could be assumed as straight line over the elapsed time after 10 days.

To investigate the details of straight relation between creep coeffi-

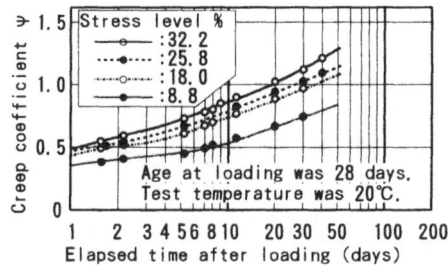

Fig. 5 Change of creep coefficient due to elapsed time (loading age was 28 days.)

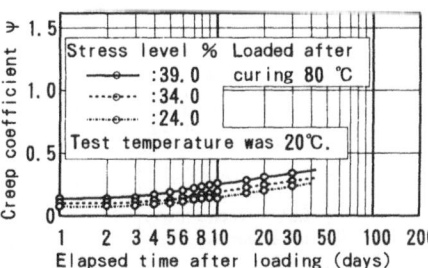

Fig. 6 Change of creep coefficient due to elapsed time (loading age was 28 days.)

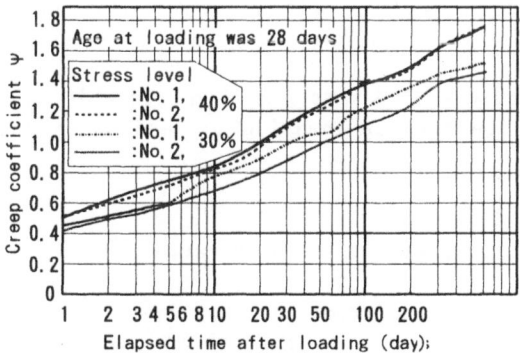

Fig. 7 Change of the creep coefficient for 2 years after loading at the age of 28 days.

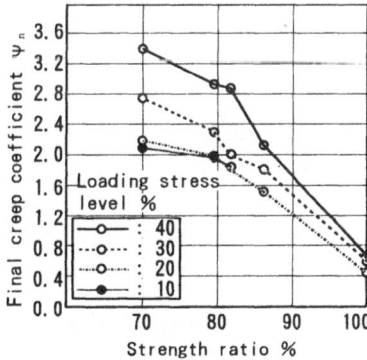

Fig. 8 Relation between final creep coefficient and compressive strength ratio at the loading time.

cient and elapsed time on normal-logalithmic coordinate, the elapsed time extended for 600 days as shown in Fig.7. From the test results, the increment of creep coefficient found to be small after 200 days and to be possible to estimate a final value extending the linear relation up to 10000 days which corresponded to 30 years. The final creep coefficients estimated by this way were 2.0~2.2 from the data measured for 600 days and ones shown in Fig.8 were estimated 1.8~ 2.2 from the test data for 30 days. From these test results, The final creep coefficient estimated by the test data for 30 days and that for 600 days were found to be almost same values. Then, the final creep coefficient estimated by the data measured for 30 days could be used for the practical design of REC members of structures. From these results, the final creep coefficient of REC was thought to be about 0.5 if REC would be cured sufficiently. The strength ratio in Fig.8 was thought to express a curing degree of REC as well. So, strength ratio was selected as an index to estimate the creep coefficient loaded at various age of REC. The coefficient α_r from which the curing effect on final creep coefficient could be estimated was derived from the values shown in Fig.9. The coefficient α_r means a ratio of final creep coefficient loaded at insufficient curing ages versus that load-

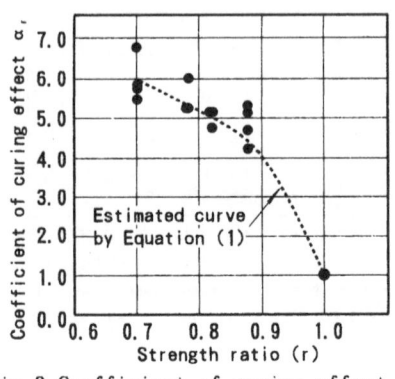

Fig. 9 Coefficient of curing effect α_r due to strength ratio r.

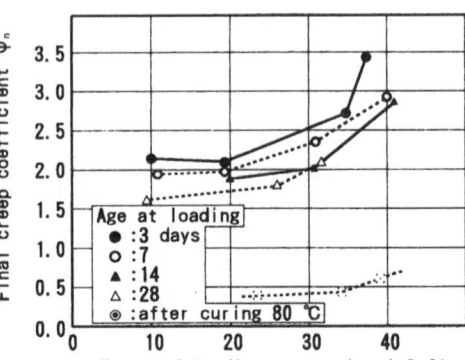

Fig. 10 Relation between final creep coefficient ϕ_n and stress level.

ed after curing of high temperature. The function of α_r is able to be expressed 3 dimensional function like as equation (1) and shown such as dotted line in Fig.9.

$$\alpha_r = -283\,r^3 + 665\,r^2 - 526\,r + 146 \tag{1}$$

In which, $0.7 \leqq r \leqq 1.0$, r: compressive strength ratio.

3.4 Estimation of the effect of stress level on creep coefficient

The creep coefficient of REC changes due to the age at loading (it means a curing effect or reaction degree) and loading stress level as well. To estimate the effect of loading stress level on final creep coefficient ϕ_n, the relation between ϕ_n and the loading stress level measured from creep test are expressed in Fig.10. From the relation in Fig.10, final creep coefficient ϕ_n is getting larger in the case that the age at loading is earlier and that the loading stress level is higher. These correlation curves loaded at each age of specimen were seemed to be similar shape, therefore to standardize each correlation curves, measured final creep coefficient on same curve were divided by the value of stress level of 30 %. Divided values are expressed in Fig.11 as α_s and named coefficient of stress level. The relation between α_s and stress level S was derived as functions of two types expressed by equation (2) or (3).

$$\begin{array}{ll} \alpha_s = 8.5\,S^2 - 2.2\,S + 1.10 & (\,0.13 \leqq S \leqq 0.4\,) \\ \alpha_s = 0.38\,S + 0.91 & (\,0 \leqq S < 0.13\,) \end{array} \tag{2}$$

$$\alpha_s = 25\,S^3 - 9.0\,S^2 + 0.91 \quad (\,0 \leqq S \leqq 0.4\,) \tag{3}$$

in which, S means the loading stress level based on the compressive strength of specimen at the loading age.
As equation (2) had a minimum extream value of α_s at the S of 0.13, α_s was expressed by another function for S under 0.13.

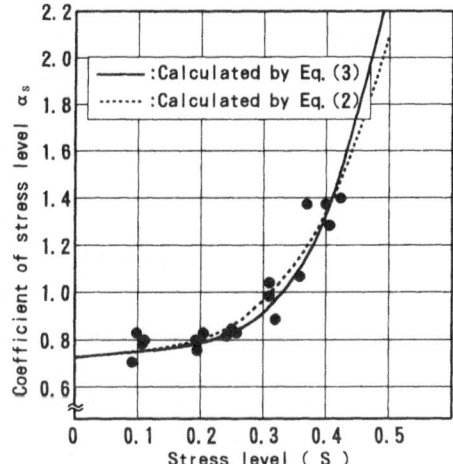

Fig. 11 Relation between coefficient of stress level and tested stress level

Tab. 4 Mechanical properties of REC at various temperature

	Temperature ℃				
	20	30	40	50	60
Compressive strength Mpa	111	112	98	88	83
Strength ratio %	100	102	89	79	75
Elastic Modulus Gpa	35	34	30	25	24

Table 5 Measured elastic strain when loading $\times 10^{-6}$

		Objective Stress level		
		20%	30%	40%
Temperature ℃	20	787	1073	1215
	30	709	1017	1410
	40	581	994	1342
	50	777	1049	1437
	60	633	1231	1423

The coefficient of curing effect α_r and coefficient of stress level α_s has been known as above, the final creep coefficient of REC which strength ratio was r and loading stress level S is able to be estimated as $\phi_{r,s,n}$ like as equation (4).

$$\phi_{r,s,n} = \alpha_r \cdot \alpha_s \cdot \phi_0 \qquad (4)$$

Here, $\phi_0 = \phi_{1\cdot 0,20,n}$

3.5 Creep properties of REC in high temperature

A creep tests at the temperature of 20, 30, 40, 50, 60℃ were carried out with the stress level from 20 to 40 % using the specimens cured by high temperature before the tests. The modulus of elasticity and compressive strength at each temperature were measured like as the results shown in Table 4. The total strain measured by the creep test with temperature 40 and 50℃ are shown in Fig. 12 and Fig. 13 respectively. The creep strain curves shown in Fig. 12 predict to be stable with the loading stress level under 40 %. However, the strain which loading stress level is 40 % in Fig. 13 is found to increase rapidly and to lead up to creep failure of specimen. In this connection, at the temperature of 60℃, the creep failure occured even at the loading stress level of 30 and 40 % as well. On the specimen occuring creep failure, the strains at the original point of accelerative creep were observed to be 3200 $\times 10^{-6}$ for 50℃ -40 %, 3300 and 2800 $\times 10^{-6}$ for 60℃ - 30 and 40 % respectively, and the creep coefficient were calculated to be 1.23 (3200/1437-1), 1.68 (3300/1231-1) and 0.97 (2800/1423-1) using the elastic strain shown in Table 5.

The example of creep coefficients corresponding to Fig. 12 and 13 are expressed as Fig. 14 and 15 respectively and the creep coefficient in

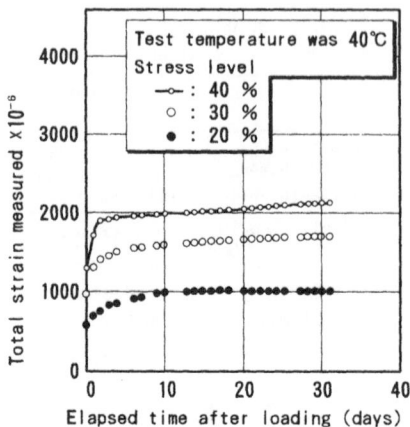

Fig. 12   Relation between total strain strain and elapsed time after loading with temperature 40℃

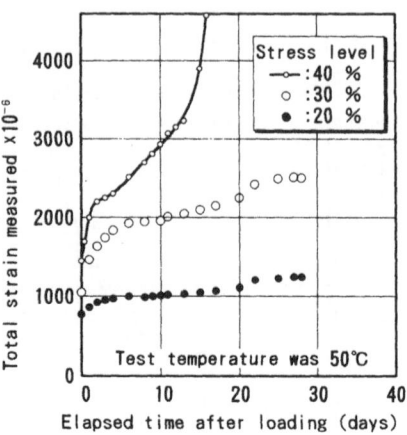

Fig. 13   Relation between total strain and elapsed time after loading with temperature 50℃

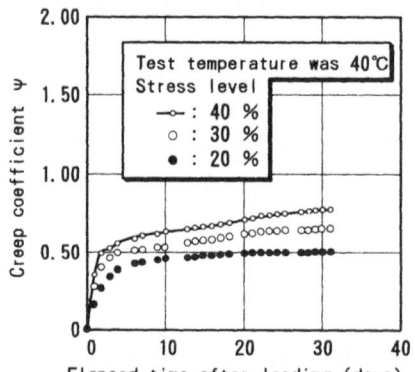

Fig. 14   Relation between creep coefficient and elapsed time after loading with temperature 40℃

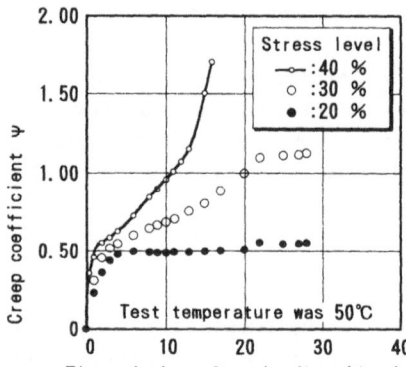

Fig. 15   Relation between creep coefficient and elapsed time after loading with temperature 50℃

stable state was recognized to be under 0.9.

Here, the test results with the temperature of 20℃ were already shown in Fig.5 and 6. Selecting a creep coefficients at the elapsed time of 28 days, it was made clear that the creep coefficient increased according to the increment of temperature of specimen and loading stress level. The equation to estimate the effect of loading stress level had been derived as equation (2) or (3) and creep coefficient at each stress level was possible to calculate by $a_s \times \phi_0$. The effect of the temperature of material were discussed basing on the comparison between the rate of each creep coefficient at the elapsed time of 28 days and the value of $a_s \times \phi_0$ as shown in Table 6. Here, this ratio of coefficient due to the temperature was defined a_t as equation (5) and named coefficient of temperature.

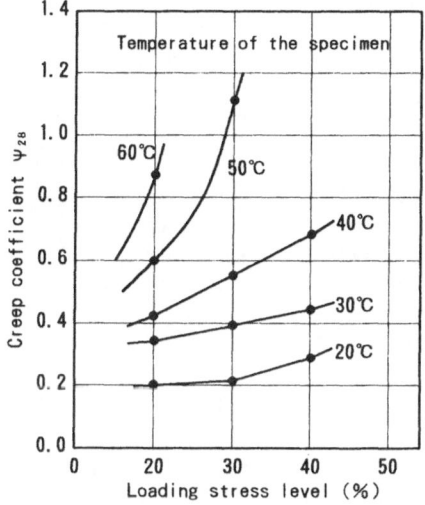

Fig. 16 Relation between creep coefficient
at the elapsed time of 28 days and loading
stress level.

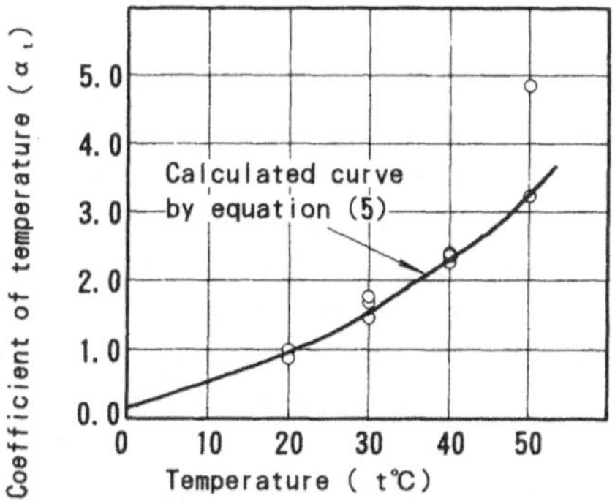

Fig. 17 Relation between temperature
and coefficient of temperature α_t
on creep.

The relation between temperature of material and coefficient of temperature on creep is expressed in Fig. 17 and approximated by function
such as equation (6). Therefore, the final creep coefficient derived
as equation (4) was rewrited like as equation (7) added the coefficient of temperature α_t.

$$\alpha_t = \phi_{28} / (\alpha_s \cdot \phi_0) \tag{5}$$

$$\alpha_t = (0.05 \ t^2 + 3.5 \ t + 10)/100 \tag{6}$$

$$\phi_{t,s,r,n} = \alpha_t \cdot \alpha_s \cdot \alpha_r \cdot \phi_0 \tag{7}$$

In which, $\phi_{t,s,r,n}$ means final creep coefficient of REC loaded the
stress level S at the curing age which compressive strength ratio was
r% under the condition of temperature to be t℃.

4. Conclusion

The authors tried to estimate the creep properties of REC and derived
the equation as mentioned above from which the creep coefficient under
various loading condition was calculated. These properties discussed
in this paper should be taken care to be limited within polyester
resin concrete.

The results obtained in this study on creep of REC are as follows.

(1) The compressive strength of REC cured by room temperature increase even after the age of 28 days.

(2) The reduction of compressive strength of REC according to the incre-ment of temperature from 20℃ to 60℃ was 25%.

(3) The final creep coefficient of REC cured by room temperature tended to being smaller according to the age at the loading and could be calcu-lated by the equation (4) using the coefficient of strength ratio α_r and coefficient of stress level α_s.

Table 6 Creep coefficient and coefficient of temperature α_t

		Stress level %		
		20	30	40
	$\alpha_s \times \Psi_0$	0.19	0.23	0.29
Temperature ℃	20	0.19	0.21	0.29
	α_t	1.00	0.91	1.00
	30	0.32	0.40	0.43
	α_t	1.68	1.73	1.43
	40	0.43	0.55	0.69
	α_t	2.26	2.39	2.38
	50	0.61	1.12	—
	α_t	3.21	4.86	—

(5) The final creep coefficient of REC loaded after curing of high temperature is estimated to be very small value such as 0.3 at the elapsed time of 45 days and about 0.5 at that of 30 years after loading.

(6) The creep coefficient of ortho-phthalic acid type REC was larger for the higher temperature and could be calculated by equation (7) using the coefficient of temperature auch as α_t.

(7) For the usage of structural members, REC is desireble to be cured sufficiently for creep property as well as for mechanical property.

References

1) YAMASAKI,T., IDEMITSU,T., WATANABE,A. and Miyakawa,K.(1991) A study on Characteristic of Unsaturated Polyester Resin Concrete. Journal of Material Science Japan, Vol.40, No.456, pp.1178-1184.

2) Ayyar,R.S.,Joshi,S.N. and Despande,S.N.(1987) The Production Performance & Potential of Polymers in Concrete, The 5th Internatio-nal Congress on Polymers in Concrete, pp.103.

3) YAMASAKI,T.,MIYAKAWA,K.,WATANABE,A.(1982) Study on The Calculation of Hardening Shrinkage Stress in Reinforced Resin Concrete, Proceed-ings of J.S.C.E.,pp.127-137.

THERMAL STRESSES OF POLYMER MORTAR COMPOSITE HOLLOW CYLINDERS STRENGTHENED BY STEEL PLATE

F. Omata
Technical Research Institute, Sho-Bond Corporation,
Ibaraki, Japan
M. Kawakami, H. Tokushige and M. Kagaya
Dept. of Civil Engineering, Akita University, Akita, Japan

Abstract

As a reinforcing method of polymer mortar hollow cylinders, a method that a steel plate is adhered around the periphery of a polymer mortar hollow cylinder is considered.

In this study, the experimental and analytical investigation on effectiveness in reinforcement of the cylindrical specimen and thermal stresses of hollow cylinders were conducted[1].

Firstly, the improvement of proof stress was investigated in compression test and splitting test. Secondarily, the thermal stress analysis of curved concrete member was performed. Finally, the conclusion that analysis enables the calculation of thermal stresses of reinforced hollow cylinders was obtained.

Keywords : Hollow cylinders, steel—plate—bonding, thermal stress analysis, epoxy resin, effectiveness in reinforcement

1 Introduction

Reinforced concrete piers are conventionally strengthened with the purpose to improve a flexural strength and a shear strength by such method that halved or quartered steel plates are firstly prepared, secondarily they are welded, and finally epoxy resin or shrinkage compensating mortar is injected in the gap existing between the steel plates and concrete, thereby achieving composite construction. If necessity to reinforce the concrete hollow cylinders arises such as pipelines, it is presumed that they can be reinforced by placing curved steel plates around their peripheries as with piers.

In this study, unconfined compression tests and splitting tests were carried

Polymers in Concrete, edited by Y. Ohama, M. Kawakami and K. Fukuzawa. Published in 1997
by E & FN Spon, 2–6 Boundary Row, London SE1 8HN, UK. ISBN: 0 419 22330 4.

out with the aim to understand the basic effectiveness in reinforcement by using a reinforcing method that steel plates are adhered around the periphery of polymer mortar hollow cylinders.

In addition, experimental and analytical consideration was made on thermal strains developed by heating a steel–plate–bonded hollow cylinder from outside, on the assumption that there would be a difference in temperature between inside and outside of a hollow cylinder.

2 Experiment

2.1 Materials used

Polymer mortar composing of unsaturated polyester resin, silica sand and calcium carbonate was used. Weight ratio of constituents was " 110 : 1921 : 215 (kg/m^3) ". It should be remembered that polymer mortar is widely and practically used for centrifugal reinforced resin concrete pipes, etc.

In addition to polymer mortar, hot rolled mild steel plates (JIS G 3131 : SPHC) of 1.2 mm in thickness were used. As an adhesive, liquid type epoxy resin (3 mm in thickness) was used. Physical properties of the materials used are shown in Table 1.

Table 1 Physical properties of materials used

Material	Strength (N/mm^2)	Young's modulus ($\times 10^3$ N/mm^2)	Coefficient of thermal expansion ($\times 10^{-6}$/°C)
Polymer mortar	83.2 (Comp.) 27.0 (Flex.)	18.6 (at 20 °C) 12.1 (at 60 °C)	26.2
Steel plate	286 (Yield)	202	10.8
Epoxy resin	48.3	2.81	79.6

2.2 Specimens and test methods in compression test and splitting test

Two kinds of specimens were used in compression test and splitting test, i.e., an original cylindrical specimen of which dimensions were 10 cm in diameter and 20 cm in height and a reinforced cylindrical specimen, of the same dimensions, adhered with steel plate on its periphery with the purpose of restraining radial expansion affected by Poisson's ratio of the core concrete. Note that the height of the steel plate was 18 cm by reducing 1 cm at both ends avoiding a direct loading on the steel plate in compression test.

A strain gage was adhered on the specimen used for compression test so as to measure circumferentical and longitudinal strains of the specimen in such way that the gage intersected the corresponding side face of the polymer mortar. A strain gage was adhered on each circular end face of the specimen intended for splitting test so as to measure vertical and horizontal strains. Note that compression test was conducted in conformance with JIS A 1108 and splitting test was in conformance with JIS A 1113.

2.3 Details of the specimen and testing procedure on the thermal strain test

A hollow cylinder with dimensions of 200 mm in inside diameter, 27 mm in wall thickness and 460 mm in length was used on the thermal strain test. It was same shape to the centrifugal reinforced concrete pipe with nominal diameter of 200 mm referred to in JIS A 5303 but was cut to a predetermined length of 460 mm. Note that the specimen used in this study was unreinforced.

As with the case described in 2.2, two kinds of specimens were used, i.e. an original cylindrical specimen and a reinforced cylindrical specimen adhered with steel plate on its periphery. Schematic diagram of the reinforced specimen is shown in Fig. 1. As shown in the figure, a strain gage for measuring circumferentical and longitudinal strains was adhered on the outer surface and the inner surface of the specimen. In addition to the strain gages, a copper–constantan thermocouple was placed at the location of each strain gage.

The specimen was heated by a flexible sheet–form heating medium closely fitted against the periphery of the specimen that was placed in a laboratory where the temperature was 20 °C . A piece of expanded rubber polyethylene foam was attached onto the top and bottom surfaces of the specimen for the purpose of thermal insulation.

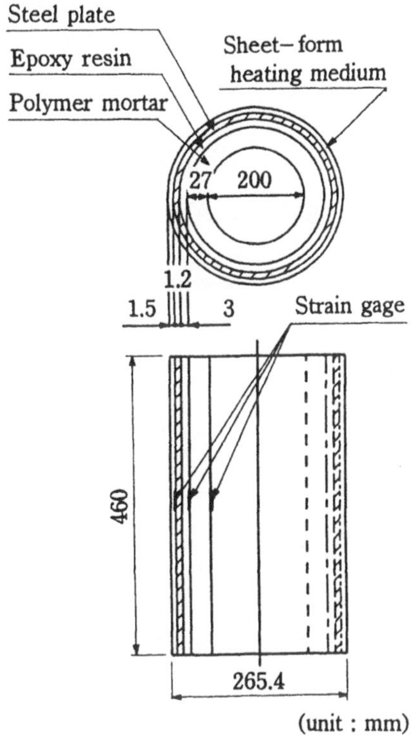

Fig. 1   Hollow cylinder specimen

(unit : mm)

The sheet–form heating medium was a 120 V type silicon rubber heater and was adjusted so that its outer surface was as hot as approximately 60 °C . The specimen was heated until its outside and inside temperatures presented a stationary state. Heating was terminated at that time. Age and strains and temperature at a predetermined location were measured. Measuring was started simultaneously with the start of heating and terminated when the temperature of the specimen lowered after the termination of heating almost to the room temperature.

Strains measured were calibrated using quarts glass that was small in coefficient of thermal expansion due to changes in temperature (coefficient of thermal expansion of quarts glass : $0.5 \times 10^{-6} / °C$).

3 Thermal stress analysis of curved concrete members

3.1 Basic equations

Analysis of stress and strain will based on the assumptions that plane cross sections remain plane after deformation and that stress–strain relations for concrete and steel are linear.

Consider a curved member cross section (Fig. 2) subjected to a normal force, N at a reference point O combined with a bending moment M. The reference point O is arbitrarily chosen, not necessarily at the centroid. A segment ABCD is shown in elevation before deformation and after deformation (A'B'C'D'). The length of the segment is assumed unity at the level of the reference point O; at any fiber y, the length before deformation is $[(r-y) / r]$; where r is radius of the reference axis O.

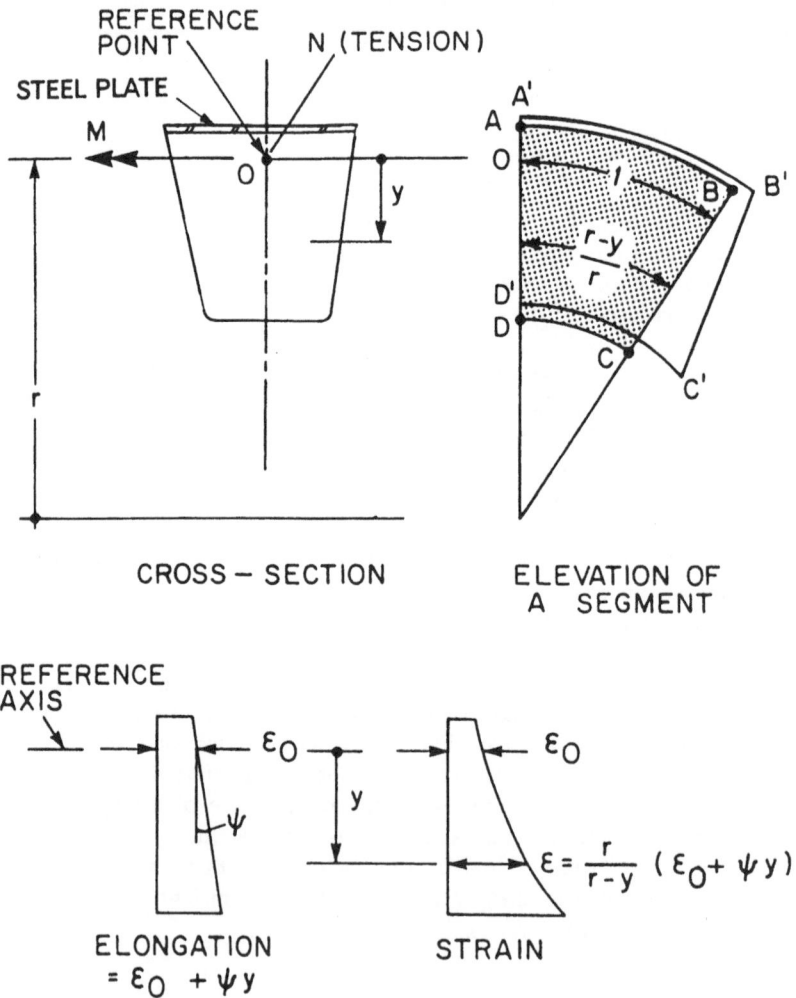

Fig. 2   Strain distribution in cross section of curved concrete member

The strain at any fiber is equal to elongation of the fiber divided by its original length; this gives:

$$\varepsilon_y = \frac{r}{r-y}(\varepsilon_0 + \psi y) \quad \cdots\cdots\cdots\cdots\cdots\cdots\cdots\cdots\cdots\cdots\cdots\cdots\cdots (1)$$

Although ε varies nonlinearly with y, the strain distribution is defined by the parameters ε_0 and ψ, representing the axial strain at O and the change in curvature of the reference axis through O.

The stress in concrete or steel is:

$$\sigma = E \varepsilon \quad \cdots\cdots\cdots\cdots\cdots\cdots\cdots\cdots\cdots\cdots\cdots\cdots\cdots\cdots\cdots\cdots (2)$$

where $E \equiv E_c$ or E_{ns}, modulus of elasticity of concrete or steel.

The stress resultants N and M are related to stress:

$$N = \int \sigma dA \quad \cdots\cdots\cdots\cdots\cdots\cdots\cdots\cdots\cdots\cdots\cdots\cdots\cdots\cdots\cdots (3)$$

$$M = \int \sigma y dA \quad \cdots\cdots\cdots\cdots\cdots\cdots\cdots\cdots\cdots\cdots\cdots\cdots\cdots\cdots (4)$$

where dA = elemental area of concrete or steel. Substitution of (1) and (2) into (3) and (4) gives:

$$\begin{Bmatrix} N \\ M \end{Bmatrix} = E_c \begin{bmatrix} A_r & B_r \\ B_r & I_r \end{bmatrix} \begin{Bmatrix} \varepsilon_0 \\ \psi \end{Bmatrix} \quad \cdots\cdots\cdots\cdots\cdots\cdots\cdots\cdots\cdots\cdots\cdots (5)$$

where A_r, B_r, and I_r are transformed area properties defined as:

$$A_r = r\int \frac{dA}{r-y}; \quad B_r = r\int \frac{y}{r-y}dA; \quad I_r = r\int \frac{y^2}{r-y}dA \quad \cdots\cdots\cdots (6)$$

The quantity $[r\, dA / (r-y)]$ in (6) represents an elemental transformed area; it is equal to an elemental concrete area multiplied by a weighting factor $r / (r-y)$; it can also be equal to cross−section area of a steel layer multiplied by weighting factor $[\alpha_{ns}\, r / (r-y)]$; where $\alpha_{ns} = E_{ns} / E_c$. When the member is straight, $r \rightarrow \infty$, $A_r = A$, $B_r = B$ and $I_r = I$ where A, B, and I are the area of conventional transformed section and its first and second moments about an axis through O.

Furthermore, if O is chosen at the centroid of the transformed section, $B = 0$ and (5) takes the more familiar forms: $N = \varepsilon_0 EA$ and $M = \psi EI$.

When N and M are given at any section of a curved member, the strain parameters can be determined by solution of (5), thus giving the two parameters defining strain distribution.

$$\begin{Bmatrix} \varepsilon_0 \\ \psi \end{Bmatrix} = \frac{1}{E_c(A_r I_r - B_r{}^2)} \begin{bmatrix} I_r & -B_r \\ -B_r & A_r \end{bmatrix} \begin{Bmatrix} N \\ M \end{Bmatrix} \quad \cdots\cdots\cdots\cdots\cdots\cdots (7)$$

3.2 Thermal stress of curved member

Consider the effect of a temperature increase $T(y)$, nonlinearly varying with y (Fig. 3). It is necessary to find the strain and stress variation over the cross section.

The analysis can be in three steps:

- Step 1: Determine the hypothetical strain that would occur at any fiber were free to expand

$$\varepsilon_{free} = \alpha_t T \quad \cdots\cdots\cdots\cdots\cdots\cdots\cdots (8)$$

   where α_t =coefficient of thermal expansion (degree $^{-1}$).
- Step 2: The stress that can prevent thermal expansion

$$\sigma_{restraint} = - E \varepsilon_{free} \quad \cdots\cdots\cdots\cdots\cdots\cdots (9)$$

   where $E = E_c$ or E_{ns}, with E_c being the modulus of elasticity of concrete at the time of introduction of temperature variation.
   (The effects of creep, during the time in which the temperature change occurs, on the thermal stress or strain are beyond the scope of this paper).
- Step 3: Determine the resultants of $\sigma_{restraint}$

$$\Delta N = \int \sigma_{restraint} \, dA \quad \cdots\cdots\cdots\cdots\cdots\cdots (10)$$

$$\Delta M = \int \sigma_{restraint} \, y dA \quad \cdots\cdots\cdots\cdots\cdots\cdots (11)$$

   The integrations in (9) and (10) may have to be evaluated numerically. Eliminate the artificial restraint by application of forces $- \Delta N$ and $- \Delta M$ on transformed section; the resulting strain and stress can be calculated by (7), (1) and (2).

$$\begin{Bmatrix} \Delta \varepsilon_0 \\ \Delta \psi \end{Bmatrix} = \frac{1}{E(A_r I_r - B_r{}^2)} \begin{bmatrix} I_r & -B_r \\ -B_r & A_r \end{bmatrix} \begin{Bmatrix} -\Delta N \\ -\Delta M \end{Bmatrix} \quad \cdots\cdots\cdots (12)$$

And the thermal stress at any point y are calculated by following equation.

$$\sigma = \sigma_{restraint} + \Delta\sigma = - E\alpha_t T + E\frac{r}{r-y}(\Delta \varepsilon_0 + \Delta \psi y) \quad \cdots\cdots\cdots (13)$$

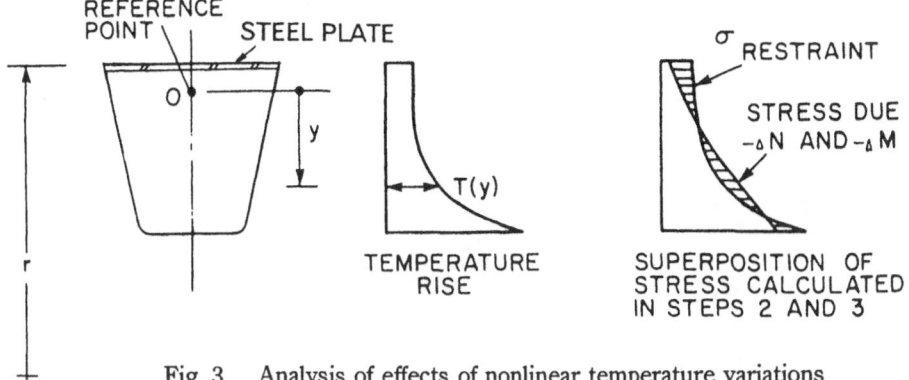

Fig. 3   Analysis of effects of nonlinear temperature variations

4   Results and discussions

4.1   Mechanical properties of the polymer mortar hollow cylinders strengthened by steel plate

Load–strain relation of each specimen is shown in Fig. 4.   Fig. 4 indicates that the polymer mortar provided higher proof stress and developed larger maximum strain in compression test as compared with general cement concrete.   When comparing the strain developed at an equal load between the original specimen and the reinforced specimen, the latter only developed a strain of approximately 80 % of that developed by the former.   This proves that the reinforced specimen has an effectiveness in reinforcement.

In splitting test, the maximum horizontal strain of the polymer mortar was approximately four times as high as that of general cement concrete.   This indicates that the polymer mortar provides a higher elongation ability.   When comparing the strain developed at an equal load, the reinforced specimen developed a slightly smaller strain than that of the original specimen.   This indicates that the clear effectiveness in reinforcement is not recognized.

Table 2 shows the maximum load (maximum stress) in compression test and the cracking load (the load with which core concrete cracks) in splitting test.   The indicated values are mean values of the three specimens under the same conditions. When comparing the maximum stress (compressive strength) of the original specimen and the reinforced specimen in compression test against Table 2, the latter gives a value that is 1.15 times as large as that of the former.   This proves that the reinforced specimen has a certain effectiveness in reinforcement.

In the case of splitting test, there was a scares difference between the original specimen and the reinforced specimen in terms of the cracking load.   This means that steel–plate–reinforcement proved a scares effectiveness in reinforcement.

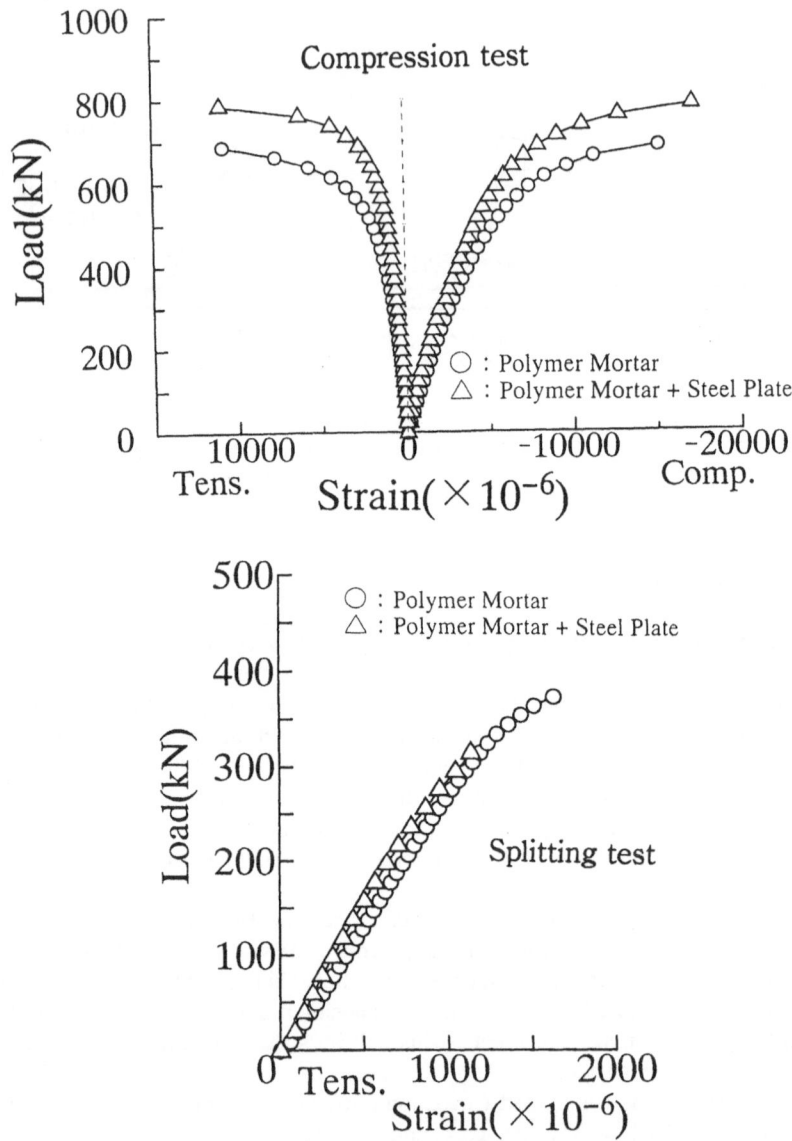

Fig. 4   Load–strain relation in compression test and splitting test

Table 2   Results of compression test and splitting test

Specimen	Compression test Max. load (kN)	Max. stress (N/mm²)	Splitting test Cracking load (kN)
Original	681.1	86.7	36.90
Reinforced	780.6	99.4	35.00

4.2 Strains of hollow cylinders with the temperature changed

For the original specimen, circumferentical inner strain measured in the experiment was 683×10^{-6} , against 1150×10^{-6} obtained by analysis, when the temperature raised by 28.3 ℃ . When the temperature raised by 43.1 ℃ , circumferentical external strain measured in the experiment was 1069×10^{-6} , against 1150×10^{-6} obtained by analysis.

The relation between the external temperature and strain, for the reinforced specimen, is shown in Fig. 5 as a representative example (temperature of outer surface of steel plate raised by 35.4 ℃ (20 ℃ → 55.4 ℃) , temperature of inner surface of hollow cylinder raised by 30.6 ℃ (20 ℃ → 50.6 ℃).

Through the experiment with the temperature changed, comparatively good responses were obtained.

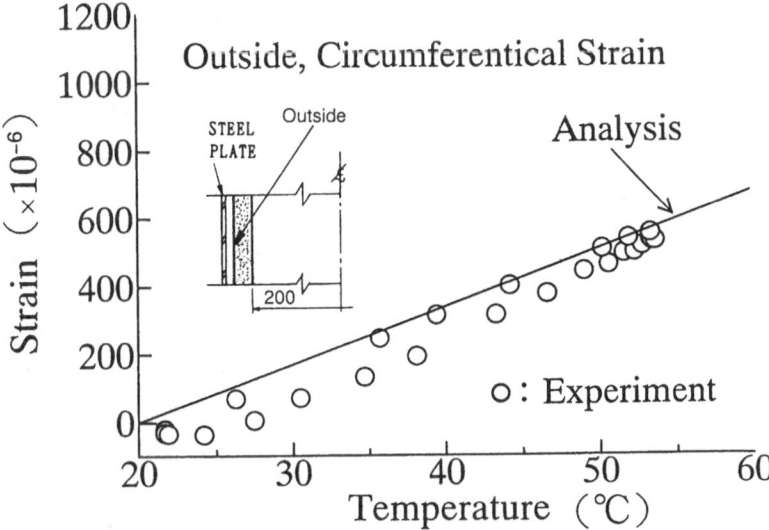

Fig. 5   Temperature–strain relation

4.3 Parametric analysis of thermal stress

As described above, thermal strain and thermal stress of hollow cylinders can be calculated by analysis.

Fig. 6 shows the calculated thermal stresses of steel–plate–bonded polymer mortar in terms of circumferentical direction. Fig. 6 shows the results of thermal stresses of two pieces of the same material, only different in thickness of epoxy resin (adhesive) as 3 mm and 10 mm. In general, thickness of epoxy resin used as adhesive for reinforced piers is specified to approximately 4 mm as standard. However, in this experiment, thickness of epoxy resin was specified to 3 mm considering the case that the epoxy resin can be slightly thinner than 4 mm in some practical applications and 10 mm considering the case that the epoxy resin can be actually thicker than 4 mm due to unevenness of concrete surface in the field.

For the temperature in the experiments, 55.4 ℃ (raised by 35.4 ℃ from the room temperature) measured on the steel plate surface and 50.6 ℃ (raised by 30.6 ℃) measured on the inner surface of the hollow cylinder were employed.

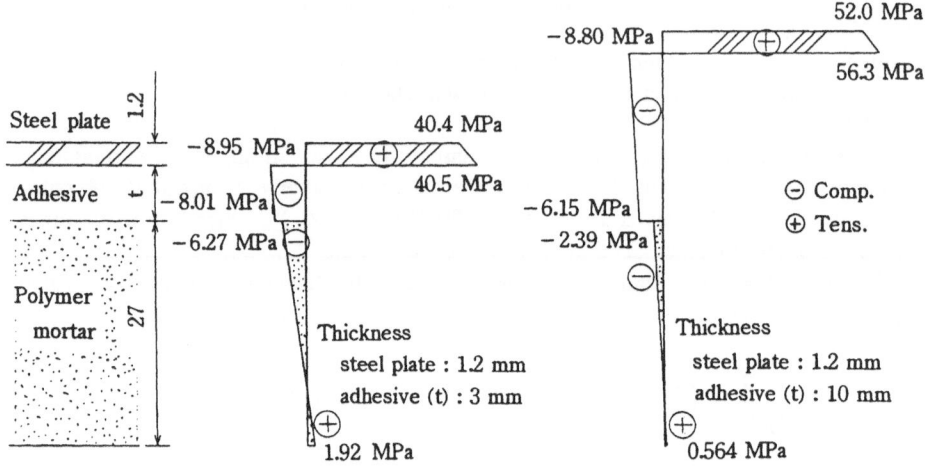

Fig. 6 Calculation results of thermal stresses

From this figure, if the thickness of adhesive is 10 mm, the stress of steel plate is larger and that of polymer mortar is smaller than 3 mm. In short, the thickness of the adhesive is not a serious problem in terms of the thermal stress.

5 Conclusions

Through the investigation on effectiveness in reinforcement and thermal stress with respect to a method where a steel plate is adhered on the periphery of a polymer mortar hollow cylinder, the following conclusions were obtained.

1. In the case where a steel plate was adhered onto the periphery of a polymer mortar hollow cylinder, proof stress was slightly improved in compression test but no other outstanding effectiveness in reinforcement was observed.
2. Analyzing method of thermal stress for steel–plate–bonded polymer mortal hollow cylinders has been demonstrated.
3. Experimental value and analytical value of strains developed when a steel–plate–bonded polymer mortar hollow cylinder is heated from outside correspond to each other. This means that such strains can be calculated through analysis.

6 Reference

1. Omata,F., Kawakami,M., Tokuda,H. and Kagaya,M. (1995)
A study on Temperature Variations of Concrete Members Strengthened by Bonding Plates. *Procedings of the 8th international congress of ICPIC*, pp.591–596.

STRESSES AND STRAINS OF RESIN MORTAR DURING SETTING

A. Moriyoshi
Dept. of Civil Engineering, Hokkaido University, Sapporo, Japan
F. Omata
Technical Research Institute, Sho-Bond Corporation,
Tsukuba, Japan
M. Kawakami and H. Tokushige
Dept. of Civil Engineering, Akita University, Akita, Japan
Y. Ohama
Dept. of Architecture, Nihon University, Koriyama, Japan

Abstract
This paper describes changes of stresses and strains of prismatic specimen of resin mortar during setting. The new apparatus measuring setting shrinkage stress and strains at both ends of resin mortar specimens at the constant temperature was developed. This apparatus is simple in structure, while consisting of a noncontact–type displacement device and a load measuring device respective which are purpose–specific. The setting shrinkage stresses and strains, and conversion Young's modulus of three kinds of resin mortar were discussed[1–3]. Furthermore, the mechanical properties of three kinds of resin mortar were investigated.
Keywords: Setting, stress, strain, laser beam displacement meter,
conversion Young's modulus

1 Introduction

Because of its high early strength, high strength and resistance to chemical attack, resin mortar is widely used as repairing material for concrete structures. If resin mortar should provide significant shrinkage or expansion during setting, it is possible to presume that a concrete structure repaired with this material could be adversely affected resulting in troubles such as exfoliation and cracks. This means that it is extremely important to understand the behavior of resin mortar during setting.

At present, various methods for measuring the behavior of resin mortar during setting have been proposed such as the use of many kinds of contact–type displacement meters and the use of mold gages. For the method that uses a contact–type displacement meter, resiliency of its built–in spring can adversely affect the measurements. On the other hand, the mold gage is unable to measure the strain

Polymers in Concrete, edited by Y. Ohama, M. Kawakami and K. Fukuzawa. Published in 1997
by E & FN Spon, 2–6 Boundary Row, London SE1 8HN, UK. ISBN: 0 419 22330 4.

of resin mortar in its early stage of setting. From these disadvantages, neither methods are considered to be satisfactory ones.

The authors have developed a new apparatus that is capable of simultaneously measuring a displacement of resin mortar during setting by means of a laser beam displacement meter without contacting the specimen and measuring a load developed during setting by load cell.

Firstly, this paper introduces the newly–developed apparatus. Secondly, it refers to the development of compressive strength and flexural strength, investigated through the research of three different kinds of resin mortar that are used as repair materials. Thirdly, it reports the strains and stresses of the aforementioned three kinds of resin mortar measured using the newly–developed apparatus.

2 Apparatus

Fig. 1 shows the measuring apparatus that has been newly developed. This apparatus consists of two devices, i.e., a purpose–built noncontact–type displacement device and a purpose–built load measuring device. It features the capability of simultaneously measuring a displacement and a load during setting by placing of resin mortar simultaneously into each of the two devices.

The purpose–built noncontact–type displacement device consists of a bottom plate and side plates adhered with a Teflon sheet, an aluminium L–shaped contact adhered with a Teflon sheet for measuring displacement and a laser beam displacement meter. The bottom plate and side plates are assembled in advance and L–shaped contact for measuring displacement are installed at both ends of a specimen to allow the contact to move freely (take care not to allow the resin mortar to leak from between the side plates and the L–shaped contact). The resin mortar is placed in the space formed by the bottom plate, the side plates and the L–shaped contact for measuring displacement. Immediately after the placing of the resin mortar, displacement is measured with the laser beam displacement meter at given intervals. It should be remembered that screws are provided for inside of the L–shaped contact so as to improve bond strength. The dimensions of resin mortar are $25 \times 25 \times 220$ mm.

The purpose–built load measuring device also uses a bottom plate and side plates that are adhered with a Teflon sheet as with the purpose–built displacement device. Both ends of the load measuring device are fixed jigs that are rigid–jointed to clamped frame made of Invar metal (Young's modulus=1.77×10^5 N/mm^2, coefficient of thermal expansion=1×10^{-6}/°C). A load cell is connected to one fixed jig, this enables the measuring of a load during setting (responds to both a development of tension by restraining shrinkage and a development of compression by restraining expansion). As with the L–shaped contact for measuring displacement, the fixed jigs are provided with screws. The resin mortar is placed in the load measuring device at the same time when one is placed in the purpose–built displacement device. Immediately after the placing of resin mortar, a load is measured at given intervals. Dimensions of the resin mortar are $25 \times 25 \times 220$ mm as with the purpose–built displacement device.

With these two devices, the following is made possible. Firstly, a stress can be found by dividing the load measured by the purpose–built load measuring device by the cross sectional area (2.5 cm × 2.5 cm = 6.25 cm ^{2}). Secondly, a strain can be found by dividing the displacement measured by the purpose–built displacement device by the initial length (220 mm) of the resin mortar. Furthermore, a Young's modulus is calculated backward using the stress and strain obtained. This enables continuous calculation of a Young's modulus which has not conventionally been found unless a loading test is performed. In particular, the newly–developed apparatus has made it possible to find a Young's modulus of resin mortar in its very early age.

Noncontact–type displacement measuring device

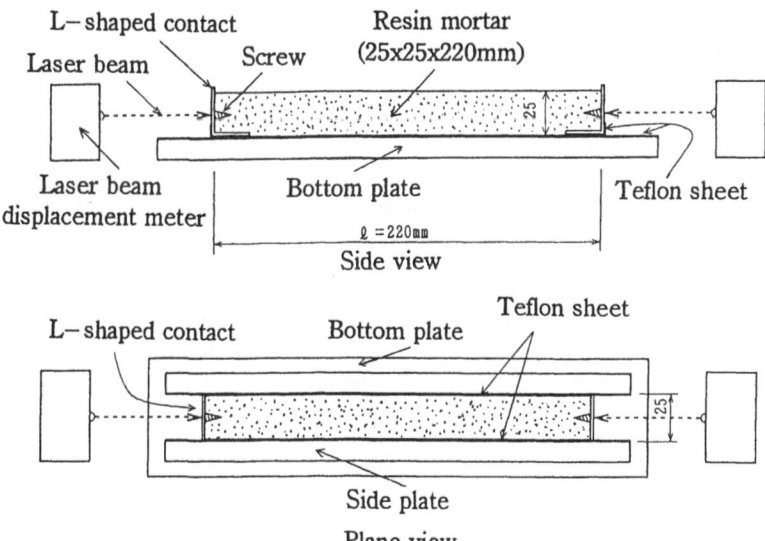

Load measuring device

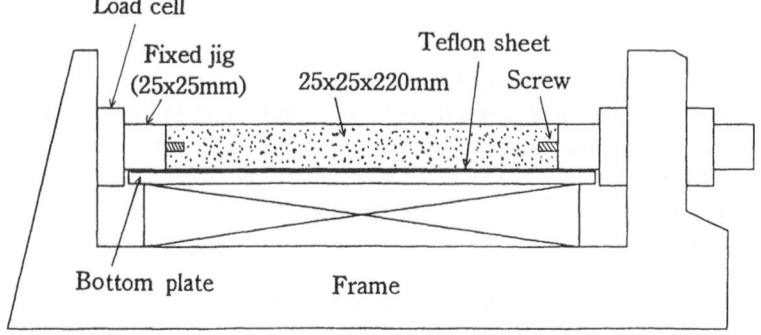

Fig. 1   The newly–developed apparatus

3   Experiments

3.1   Material used
The following three different kinds of resin mortar generally used as repair materials were investigated.
 (1) Glycerol methacrylate/Styrene resin mortar (GM/St)
 (2) Methyl methacrylate resin mortar (MMA)
 (3) Unsaturated polyester resin mortar (UP)
    Physical properties of the three kinds of resin are shown in Table 1. Converter slag and silica sand were used as aggregate and calcium carbonate was used as filler, all of which were used in dry state.   Physical properties of converter slag, silica sand and calcium carbonate are shown in Table 2.   Weight ratio of resin mortar was resin : converter slag : silica sand : calcium carbonate = 1 : 4.74 : 2.58 : 1.88

Table 1   Properties of three kinds of resin

Material	Specific gravity (at 20 ˚C)	Viscosity (at 20 ˚C)
GM/St resin	0.99	35 mPa/s
MMA resin	0.99	200 mPa/s
UP resin	1.05	330 mPa/s

Table 2   Properties of aggregate and filler

Material	Specific gravity	Grading
Converter slag	Larger than 3.00	5mm– 2.5mm
Silica sand	Larger than 2.50	5mm– 0.3mm
Calcium carbonate	Larger than 2.60	Smaller than 0.15mm

3.2   Compression test and flexural test
The compressive test and flexural test were performed to obtain the relationship between age after placing and strength of three kinds of resin mortar.   Ages employed were 1.5 hours, 3 hours, 6 hours, 1 day and seven days after placing.   The temperature was set to 20 ˚C .   Specimens and test methods were in compliance with JIS R 5201 (40 × 40 × 160 mm).

3.3   Measuring displacement and load during setting
For three kinds of resin mortar, ages, displacements and loads after placing were measured using aforementioned newly– developed apparatus.   The temperature was set to 20 ˚C .

4 Results and discussions

4.1 Mechanical properties

Result of the compression test is shown in Fig. 2. This Figure indicates that approximately 60 % of the 7 day's strength of UP was developed at 1.5 hours after the placing. This proves that an adequate compressive strength was developed in a shorter period of time after the placing. For MMA, an adequate compressive strength was developed in a shorter period time as with the case UP. On the other hand, GM/St developed a lower compressive strength as compared with the other two kinds of materials at both seven days and 1.5 hours. However, it provided a compressive strength of 34.5 N/mm ^{2} at three hours. This strength is similar comparison with the compressive strength of 20 to 40 N/mm ^{2} that is developed by general cement concrete. All of the three kinds of resin mortar developed a compressive strength of 50 to 108 N/mm ^{2} at seven days. This proves that the materials provide a higher degree of compressive strength.

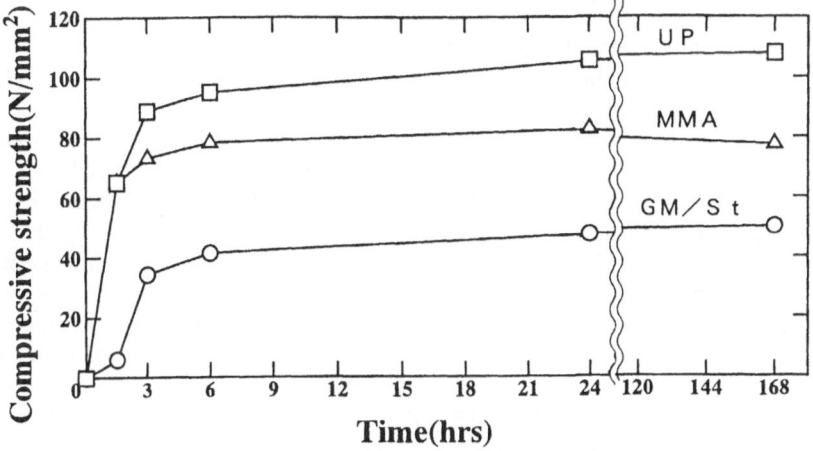

Fig. 2   Result of the compression test

Result of the flexural test is shown in Fig. 3. This Figure indicates that MMA developed approximately 80 % of its 7 day's flexural strength (25.6 N/mm ^{2}) at 1.5 hours. This proves that MMA provides an adequate flexural strength in a short period of time. For Up, its flexural strength at three hours was almost equivalent to the 7 day's strength. This means that UP also provides an adequate strength in a comparatively short period of time. On the other hand, flexural strength of GM/St at 1.5 hours and that at three hours were approximately 5 % and 41 % of its 7 day's strength. GM/St developed flexural strength that was 83 % of its 7 day's strength at six hours. It was found that GM/St produced flexural strength, at three hours, around 3 to 6 N/mm ^{2} that was the flexural strength achieved by general cement concrete. It can be, therefore, determined that there would be no practical problems to use GM/St. All of the three types of resin mortar produced 7 day's flexural strength of 13 to 27 N/mm ^{2} . It is possible to conclude that the tested materials provide remarkably high flexural strength.

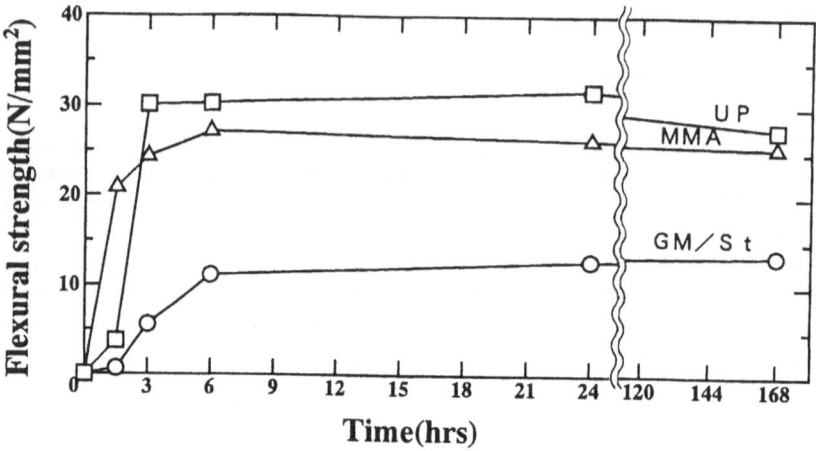

Fig. 3   Result of the flexural test

4.2 Displacement and load during setting

The measurement results are shown in Figs. 4–6 as the time–strain relation (displacement / 220 mm) and the time–stress relation.

For GM/St, strain was scarcely developed until approximately 20 minutes passed after the placing, then a slight degree of shrinkage was observed. Instead, expansion was developed after 40 minutes and beyond after the placing. At the time when 100 minutes passed after the placing, GM/St produced the largest expansion (approximately 3000×10^{-6}). At the final stage, the expansion of approximately 2800×10^{-6} remained. With respect to the time stress relation, no stress was observed by 50 minutes after the placing. Thereafter, tensile stress started to be developed by restraining expansion. The tensile stress tended to increase in proportion with the lapse of time until approximately 10 minutes after the placing. The tensile stress almost reached the highest value, 1.07 N/mm^2 at 166 minutes after the placing. At the final stage, 0.95 N/mm^2 tensile stress remained. This proves that strains were firstly developed and the strains then caused stresses to be developed. Strains before the development of stresses were produced while mortar was still in the state of fluid. It is presumed that these strains therefore do not have adverse affects, but strains after the development of stresses have some.

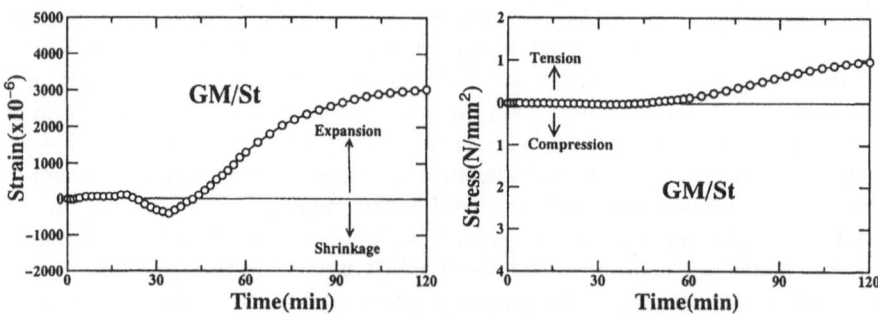

Fig. 4   The time–strain and time–stress relation (GM/St)

For MMA, strains were scarcely developed, as with GM/St, up to approximately 20 minutes after the placing. Then, some degree of expansion was observed. When approximately 50 minutes passed, the material suddenly started shrink. The highest degree of its shrinkage, i.e., 5500×10^{-6} was reached 70 minutes after the placing. After this moment, expansion and shrinkage were nearly in equilibrium. From this result, it is presumed that , an extremely large shrinkage remaining could give rise to adverse affects on cement mortar repaired with MMA such as cracks. With respect to the time—stress relation, stresses were scarcely produced up to 60 minutes after the placing. Thereafter, compressive stress due to restraining of shrinkage started to be developed. Compressive stress increased steeply up to approximately 90 minutes after the placing. Thereafter, it increased moderately until the highest tensile stress of 2.67 N/mm^2 was reached when 200 minutes passed. After this moment, compressive stress was nearly in equilibrium.

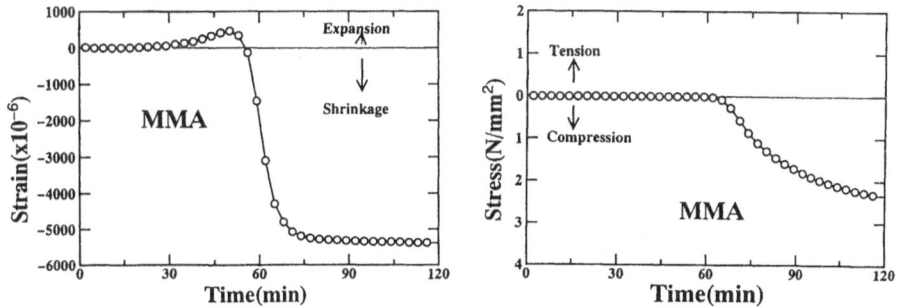

Fig. 5　The time—strain and time—stress relation (MMA)

In the case of UP, some degree of expansion was observed immediately after the placing. Shrinkage suddenly replaced expansion when 12 minutes passed. Thereafter, the degree of shrinkage increased in proportion with the lapse of time. This tendency continued up to approximately 70 minutes after the placing. The highest degree of shrinkage of 5700×10^{-6} was almost reached when 200 minutes passed. This leads to a presumption that UP, as with MMA, could exert adverse affects on the repaired cement concrete. With respect to the time—stress relation, no stress was developed approximately 20 minutes after the placing. Thereafter, compressive stress due to restraining of shrinkage started to be developed. Compressive stress increased continuously until 200 minutes passed after the placing through the rate of increase reduced around 70 minutes. Compressive stress of 3.13 N/mm^2 was developed 200 minutes after the placing.

Now, the relation between the age after the placing and the conversion Young's modulus for three kinds of resin mortar is shown in Figs. 7—9. Conversion Young's modulus was found through the following procedure. A conversion strain was determined by assuming the strain at the time when a load started to occur to be "0" while ignoring then strain developed. The conversion Young's modulus to be sought can be obtained with the following equation.

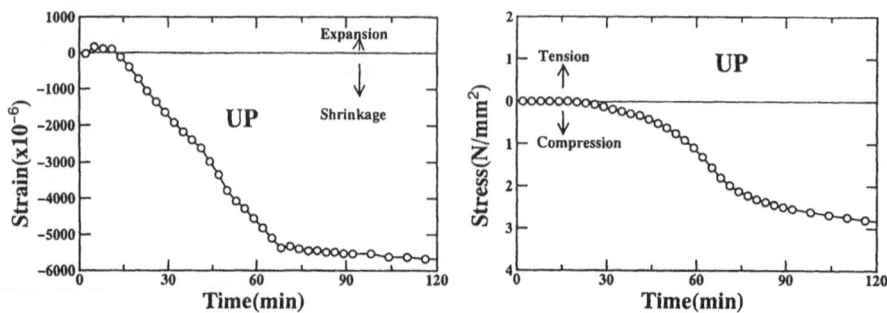

Fig. 6   The time–strain and time–stress relation (UP)

$$E_n = (P_n - P_{n-1}) / (A \times \varepsilon_n - A \times \varepsilon_{n-1})$$

where

t_0 : Time when stress started to be developed

$t_1,\ t_1,\ -----,\ t_{n-1},\ t_n$: Given time

$P_1,\ P_2,\ ----,\ P_{n-1},\ P_n$: Load corresponding to a given time

$\varepsilon_1,\ \varepsilon_2,\ ----,\ \varepsilon_{n-1},\ \varepsilon_n$: Conversion strain = $(\triangle L_n - \triangle L_{n-1}) / L$

$\triangle L_1,\ \triangle L_2,\ ---,\ \triangle L_{n-1},\ \triangle L_n$: Displacement corresponding to a given time

$E_1,\ E_2,\ ----,\ E_{n-1},\ E_n$: Conversion Young's modulus

For GM/St, conversion Young's modulus started to be produced when 50 minutes passed after the placing.   Thereafter, the conversion Young's modulus increased moderately with some variations until the highest modulus of 1060 N/mm^2 in tension was reached around 105 minutes after the placing.   After this moment, the conversion Young's modulus was nearly in equilibrium.

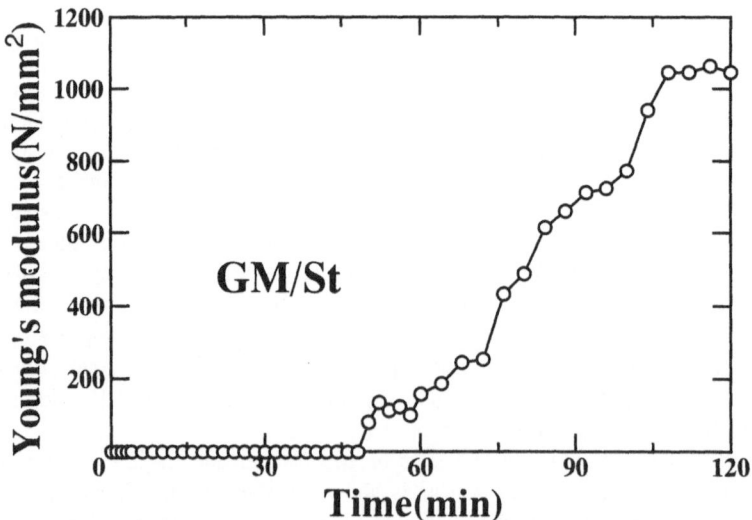

Fig. 7   The time–conversion Young's modulus relation (GM/St)

For MMA, conversion Young's modulus started to be developed approximately 57 minutes after the placing. The conversion Young's modulus increased suddenly in its starting stage. When approximately 90 minutes passed, the conversion Young's modulus of 11000 N/mm^2 in compression was reached. After this moment, the conversion Young's modulus was nearly in equilibrium.

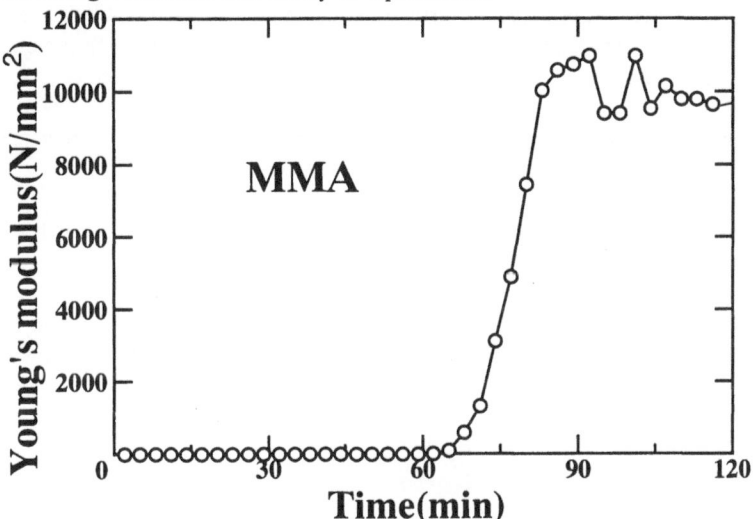

Fig. 8 The time–conversion Young's modulus relation (MMA)

For UP, conversion Young's modulus started to be developed when approximately 15 minutes passed after the placing. Up to around 70 minutes, the conversion Young's modulus increased. Thereafter, the conversion Young's modulus was nearly in equilibrium at the highest value of 1050 N/mm^2 in compression.

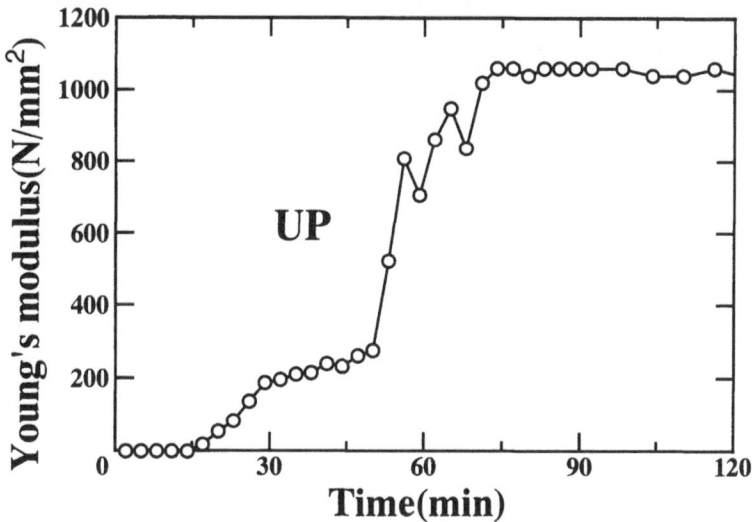

Fig. 9 The time–conversion Young's modulus relation (UP)

5 Conclusions

The conclusions obtained in this study are as follows:

1. The three kinds of resin mortar provide sufficient compressive strength and flexural strength in a shorter period of time after placing as compared with generally cement concrete.

2. The newly–developed apparatus consisted of a noncontact–type displacement device and a load measuring device was introduced.

3. The setting shrinkage stresses and strains were measured using newly–developed apparatus and discussed.

4. The conversion Young's modulus can be calculated (it can be calculated continuously in a shorter period of time after placing) using newly–developed apparatus and discussed.

6 Reference

1. Moriyoshi,A., Hirano,T., Ogasawara,A. and Nagata,S. (1995)
   Thermal Properties of Polymer Concrete Using Glycerol Methacrylate/Styrene System at Low Temperature. *Proceedings of the 8th international congress of ICPIC*, pp.509–514.
2. Koyanagi,W., Uchida,Y., Hayashi,F. and Ohshima,M. (1995)
   Internal Stresses due to Setting Shrinkage in Polyester Resin Concrete. *Proceedings of the 8th international congress of ICPIC*, pp.435–440.
3. Omata,F., Kawakami,M., Wakayama,S. and Yamamura,H. (1995)
   Thermal Stress and Setting Shrinkage stress of Concrete Members Repaired by Polymer Mortar. *Proceedings of the 8th international congress of ICPIC*, pp.113–118.

COMPRESSIVE STRENGTH AND ADHESION STRENGTH OF PCC UNDER SUSTAINED LOAD AT ELEVATED TEMPERATURE

H.R. Sasse and J. Hannawald
Institute for Building Research, RWTH Aachen, Aachen, Germany

Abstract

This paper reports results from compressive and tensile adhesion tests under sustained load with two PCC and one CC for comparison at an elevated temperature of 40 °C. Styrene acrylicester (SAE) and vinylacetate-ethylen (VAE) were used as cement matrix modifiers. The experience from the tests will be used to propose suitable test methods and requirements for repair mortars in future European Standards for the protection and repair of concrete structures.

Keywords: PCC, compressive test, tensile test, sustained load, static fatigue, adhesion strength

1 Introduction

In assessing the strength of concrete, usually short term tests on small specimens at an age of 28 days are conducted, with a typical time of load application of only a few minutes. From short-term testing it is known that the strength of fully hardened concrete decreases with increasing loading time [1,2]. In real structures, loading occurs to a great extent even more unfavourably, i. e. loads are being applied during short time and held constant during the service life. Compressive tests under sustained load showed that the strength of concrete is reduced to about 80% of the short-term strength [3].

From the author's knowledge there is a lack of information concerning the strength reduction under sustained compression for PCC. The same is true for the reduction of the adhesion strength of the repair mortar to the concrete substrate under sustained tensile stresses. Both properties are very important for concrete repair measures, where the structural stability is affected. The German „Guidelines for the Protection and Repair of Concrete Components" [4] introduces the aspect of static fatigue for repair mortars in the load-bearing field, but definite test methods and requirements have not yet been fixed.

This paper reports results from compressive tests and tensile adhesion tests on PCC under sustained load at a moderately elevated temperature, carried out in order to arrive

Polymers in Concrete, edited by Y. Ohama, M. Kawakami and K. Fukuzawa. Published in 1997 by E & FN Spon, 2–6 Boundary Row, London SE1 8HN, UK. ISBN: 0 419 22330 4.

at an proposal for a suitable test method and strength requirements for PCC in the load-bearing field.

2 Testing procedures

2.1 Materials
Ordinary Portland cement according to DIN 1164 and silica sand were used for the mortar mixes. A styrene acrylicester (SAE) latex (solid content: 50 %) and a vinylacetate-ethylen (VAE) redispersible powder were used as cement matrix modifiers. The glass transition temperature of the SAE and the VAE is 298 K and 243/263 K (double transition) respectively.

2.2 Preparation of specimens
Mortars were prepared with the mix proportions given in Table 1.

Table 1. Mix proportions of polymer-modified mortars

Type of mortar	cement:sand c/g	water:cement w/c	polymer:cement p/c
		wt-%	
unmodified	38,5	40	0
VAE-modified	27,8	55	12
SAE-modified	33,0	40	9

The fresh mortar properties are listed in Table 2.

Table 2. Fresh mortar properties of polymer-modified mortars

Type of mortar	air content %	flow table cm	bulk density kg/dm^3
unmodified	4,7	12,4	2,264
VAE-modified	6,2	13,0	2,141
SAE-modified	6,3	19,1	2,188

For the compressive tests, cylindrical mortar specimens $\varnothing$=50 mm, h=150 mm were moulded and moist cured at 23 °C for one day and then demoulded. After demoulding, the specimens were sealed with aluminium foil in order to prevent evaporation and stored at 40 °C for another 27 days. One series of specimen made from the SAE-modified mortar was stored at 23 °C for comparison purposes.

For the tensile adhesion tests, cores were drilled out of concrete slabs coated with the mortar to be tested. Coating thickness was 2 cm for all mortars. The coated concrete slab was moist cured at 23 °C for one day, then sealed in total with aluminium foil and stored at 40 °C. The cores were taken 20 days after coating. Steel plates were glued on the top and bottom surfaces of the cylinders in order to introduce the tensile load. The cores were sealed individually with aluminium foil and kept for another 7 days at 40 °C.

In order to examine the effect of the warm/moist-storage on the short term strength two series of mortar prisms 40x40x160 mm^3 were moulded. One series was given the storage as described above for the cylindrical specimens and the other series was stored at 23 °C/50 % R.H. after one day of moist curing at 23 °C. The storage conditions are summarised in Table 3.

Table 3. Storage imposed on specimens and kind of tests performed

Type of storage	Type of specimen and details of storage	Type of test
40 °C/moist	cylinders and prisms sealed[1], T = 40 °C	short term (prisms[2]), tensile[3] and compressive fatigue (cylinders)
23 °C/moist	cylinders sealed[1], T = 23 °C	compressive fatigue (cylinders, SAE-modified mortar only)
dry	prisms, T = 23 °C, 50 % R.H.	short term (prisms[2])

1) sealed with aluminium foil in order to prevent evaporation,   2) flexural and compressive strength
3) tensile adhesion on cores drilled out of coated concrete slabs

2.3 Testing climate

The static fatigue tests are performed under climatic conditions, which are unfavourable for polymer modified mortars but still realistic for structural members. In this sense, a test temperature of 40 °C has been selected, which is considered to be representative of the maximum temperature in a weathered structural member in central europe. The high humidity of the specimen due to the sealing, is expected to aggravate the test conditions in addition.

2.4 Compressive Loading

Loading commenced 28 days after the preparation of the cylinders. Load levels for the fatigue tests were chosen at ratios $f/f_{c,max}$ of 0.8, 0.75, 0.7, 0.65 and 0.6, where $f_{c,max}$ designates the short-term compressive strength of each batch, determined in the usual short term test with a duration of approximately 1 min. For the fatigue test the load was applied by means of a hydraulic loading arrangement within approximately 1 min. Once the desired load has been reached, the hydraulic piston was connected to a constant nitrogen pressure source in order to assure the constancy of the load despite the creep deformation of the specimen.

The loading device was placed in a climate chamber tempered to the test temperature.

2.5 Tensile loading

The compound cylinders were loaded 28 days after the coating of the concrete slabs had taken place. Load levels for the fatigue tests were planned at ratios $f/f_{a,max}$ of 0.8, 0.7 and 0.6, where $f_{a,max}$ designates the short-term adhesion strength of each batch, determined in a tensile test on the compound cylinders with a duration of approximately 1 min. The sustained tensile load was applied by means of a dead weight cantilever apparatus. The loading device was placed in a climate chamber tempered to the test temperature.

3 Results

3.1 Effect of warm/moist storage

The effect of the storage conditions on the short term strength for the unmodified and the polymer modified mortars is illustrated in Fig. 1 in terms of the flexural (Fig. 1 a) and compressive (Fig. 1 b) strength, determined on standard mortar prisms. For both polymer modified mortars, the flexural strength is reduced in comparison to the standard dry storage, whilst the compressive strength is not altered significantly. The reduction of the flexural strength is more severe for the SAE-modified mortar as compared to the VAE-modified mortar.

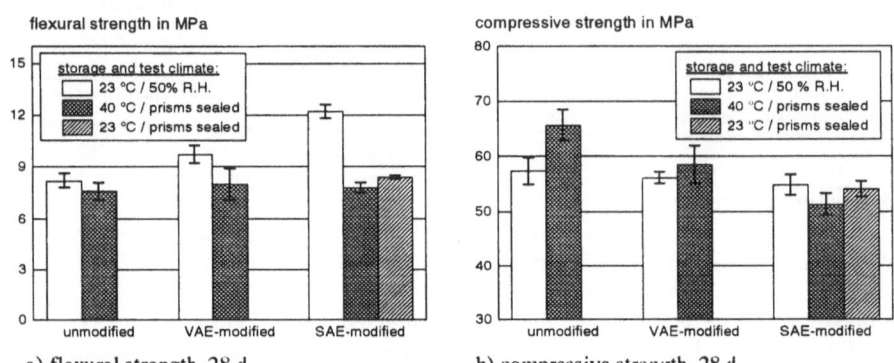

a) flexural strength, 28 d   b) compressive strength, 28 d

Fig. 1. Effect of storage and test climate on the strength of the unmodified and modified polymer mortars

For both polymer modified mortars, the flexural strength drops down to the level of the unmodified mortar, i. e. the polymer film has lost its crack stopping capacity, to which the increase in flexural or tensile strength of polymer modified mortars is usually attributed [5], completely. Since an increased temperature is expected to favour polymeric film formation [6], this effect has to be attributed to the high humidity of the specimen due to sealing. This interpretation is supported by the results obtained with the SAE-modified mortar at 23 °C.

3.2 Static fatigue tests in compression

The present status of the investigations is depicted in Fig. 2. It is seen that the static fatigue strength of both polymer modified mortars amounts to only 60 % of the respective short term strength, Figs. 2 a and 2 b. Thus, the static fatigue strength for the two polymer modified mortars is considerably reduced as compared to the 80 % limit, which is usually accepted for cement concrete. In contrary, the unmodified mortar only tightly fails to reach this limit.

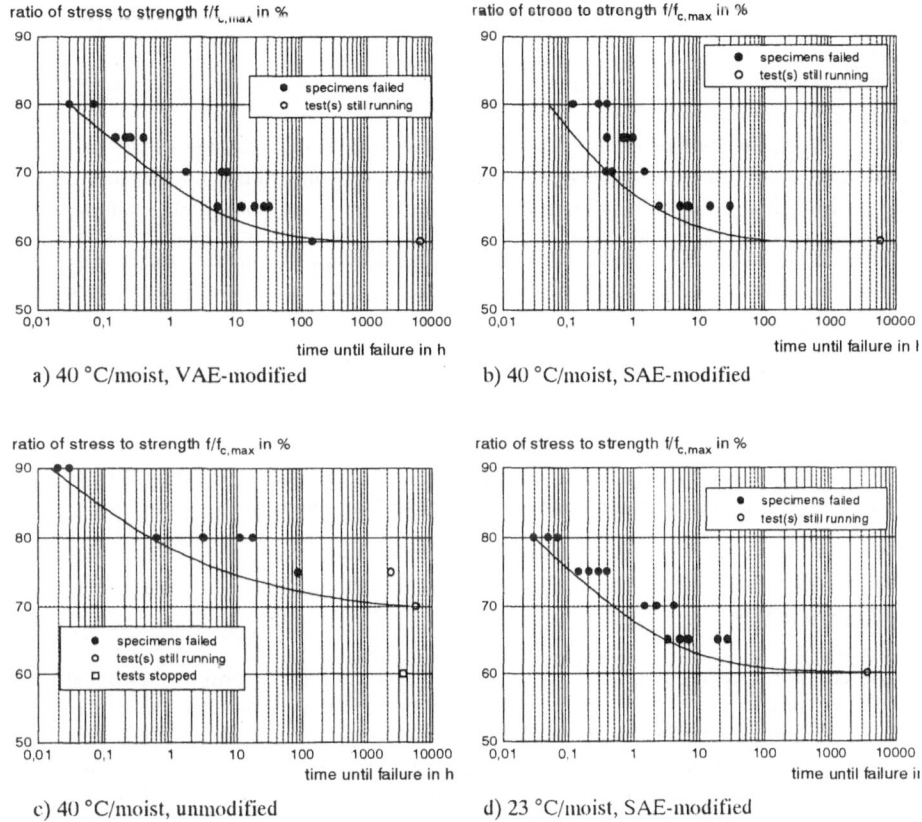

Fig. 2. Results of static fatigue tests in compression for the unmodified and polymer modified mortars.

It is interesting to note that the severe reduction of the static fatigue strength observed for the polymer modified mortars cannot be attributed to the elevated test temperature, as the results obtained for the SAE-modified mortar at 23 °C clearly show, Fig. 2 d.

The effect of the glass transition temperature on the fatigue strength cannot be assessed from the test results.

Since the compressive strength of the polymer modified mortar is unaffected by the warm/moist condition and furthermore the static fatigue strength is not affected by the test temperature, it seems that, for simplicity, the fatigue test in compression under moist condition can be performed at 23 °C.

3.3 Static fatigue of compound specimens in tension

On the 80% load level no failure could be observed until now in the tests on the unmodified and SAE-modified mortars (test duration: 8 months). With the VAE-modified mortar failure was detected on the 80 % and 70 % load level after 2 weeks and 6 months respectively. In all cases, the type of failure was cohesive failure in the mortar or the substrate concrete. The same holds for the short term tests on all mortars, which had been carried out in advance in order to obtain the reference strength for the selection of the sustained load.

Since no adhesive failure occured, neither in the tests under sustained load nor in the short term tests, the conclusions from the test results concerning the adhesion fatigue are limited. But it can be stated that the adhesion strength under sustained load continues to exceed the cohesive strength, if this had been the case in the short term test. (Nothing can be stated for the case of adhesive failure in the short term test.) This is contrary to our expectations that the adesive strength should become the limiting parameter in the compound system with increasing test duration.

Assuming for the coating mortars that the reduction of strength under sustained load in tension is similar to the one in compression, the fatigue tensile strength of the compound system can be estimated to at least 60 % of the short term tensile strength.

It seems that the adhesion fatigue cannot be assessed quantitatively with this particular type of tensile test on compound cylinders. The design of a test, in which the specimen is forced to fail in the adhesion zone, is left for the future work.

4 Conclusions

1) As compared to the standard dry storage, the warm/moist storage causes a drop of the flexural strength of both polymer modified mortars down to the value of the unmodified mortar. The compressive strength is unaffected.
2) The static fatigue strength of both polymer modified mortars amounts to only 60 % of the respective short term strength and is thus considerably reduced as compared to what is usually expected for cement concrete.
3) The severe reduction of the static fatigue strength is independent of the type of polymer.
4) The reduction of the static fatigue strength for the SAE-modified mortar is independent of the test temperature. (The effect of the humidity has yet not been studied separately.)
5) The adhesion fatigue cannot be assessed with the particular type of tensile test on compound cylinders. The tensile fatigue strength of the compound system is estimated to exceed 60 % of the short term tensile strength.

5 References

1. Sparks, P.R. and Menzies, J.B.: The Effect of Rate of Loading upon the Static and Fatigue Strengths of Plain Concrete in Compression. in: Magazine of Concrete Research 25 (1973), Nr. 83, p. 73-80.
2. Wesche, K. ,Weber, J.W. and Kunze, W.: On the Fatigue Behaviour of Plain Concrete, State of Knowledge. Düsseldorf : Verein Deutscher Eisenhüttenleute, 1973. - In: Dritte Internationale Tagung über den Bruch, München, 8. bis 13. April 1973, Teil VIII Bruchvorgänge in Verbundwerkstoffen, S IX.

3. Rüsch, H.: Researches Toward a General Flexural Theory for Structural Concrete. In: Journal of the ACI Proceedings 57 (1960), Nr. 1, S. 1-28.

4. Deutscher Ausschuß für Stahlbeton ; (DAfStb): Richtlinie für Schutz und Instandsetzung von Betonbauteilen ; Teil 1: Allgemeine Regelungen und Planungsgrundsätze ; Teil 2: Bauplanung und Bauausführung (August 1990) ; Teil 3: Qualitätssicherung der Bauausführung. (Februar 1991) ; Teil 4: Qualitätssicherung der Bauprodukte (November 1992). Berlin : Beuth Verlag, 1991.

5. Henschel, R. und Schorn, H.: Über die Eignung langsam erhärtender Zemente für die Verwendung in kunststoffhaltigen Zementmörteln und Betonen. In: Beton-Informationen 33 (1993), Nr. 5-6, S. 51-55.

6. Sasse, H.R. , Schießl, P. , Fiebrich, M. , Kwasny-Echterhagen, R. , et al: Schutz und Instandsetzung von Betonbauteilen unter Verwendung von Kunststoffen. Sach-standsbericht. Berlin : Beuth, 1994. - In: Schriftenreihe des Deutschen Ausschusses für Stahlbeton (1994), Nr. 443.

THERMAL RESISTANCE OF POLYMER-MODIFIED MORTARS USING REDISPERSIBLE POLYMER POWDERS

W.K. Kim, Y. Ohama and K. Demura
Dept. of Architecture, Nihon University, Koriyama, Japan

Abstract
Redispersible polymer powders are usually produced from aqueous polymer dispersions by spray drying process. In this paper, polymer-modified mortars using various redispersible polymer powders are prepared with various polymer-cement ratios, and tested for flexural and compressive strengths before and after 28-day heating at 150 °C to evaluate their thermal resistance. It is concluded from the test results that the polymer-modified mortars using the redispersible polymer powders with higher glass transition point have a higher thermal resistance.
Keywords : Thermal resistance, redispersible polymer powders, polymer dispersions, polymer-modified mortars, polymer-cement ratio

1 Introduction

The mechanical properties of polymer-modified mortars vary at higher temperature because of the temperature dependence of those of polymers used. The strengths of the polymer-modified mortars are sharply decreased with a raise in heating temperature. This tendency is generally marked at heating temperatures of 100 to 150 °C over the glass transition point of the polymers, depending on the type of polymers used and the polymer-cement ratio [1]. The thermal resistance of the polymer-modified mortars is governed by the type of polymers used, the polymer-cement ratio, the heating temperature, and finally the thermal degradation of the polymers. From the standpoint of their practical applications, the maximum allowable temperature for the strength retention of most polymer-modified mortars is

Polymers in Concrete, edited by Y. Ohama, M. Kawakami and K. Fukuzawa. Published in 1997 by E & FN Spon, 2–6 Boundary Row, London SE1 8HN, UK. ISBN: 0 419 22330 4.

considered to be approximately   150°C [2].

In this paper, polymer-modified mortars using commercial redispersible polymer powders as cement modifiers are prepared with various polymer-cement ratios, and tested for flexural and compressive strengths before and after 28-day heating at 150°C to evaluate their thermal resistance. The properties of the mortars modified with a styrene-butadine rubber (SBR) latex and a poly (ethylene-vinyl acetate) (EVA) emulsion as controls for the redispersible polymer powders are examined in the same manner as the redispersible polymer powder-modified mortars. From the test results, the thermal resistance of the redispersible polymer powder-modified mortars is discussed.

2 Materials

2.1 Cement and fine aggregate
Ordinary portland cement and Toyoura standard sand specified in JIS (Japanese Industrial Standard) were used for all the mortar mixes.

2.2 Cement modifiers
Commercial cement modifiers used were three [poly (ethylene-vinyl acetate) (EVA)], two [poly (vinyl acetate-vinyl carboxylate) (VA/VeoVa)], one [poly (methyl methacrylate-butyl acrylate) (MMA/BA)], one [poly (styrene-butyl acrylate) (St/BA)] and one styrene-butadiene rubber (SBR) redispersible polymer powders, and one [poly (ethylene-vinyl acetate) (EVA)] emulsion and one styrene-butadiene rubber (SBR) latex as controls. Their basic properties are listed in Tables 1 and 2. Before

Table 1. Properties of redispersible polymer powders

Type of polymer	Appearance	Average particle size (μ m)	Glass transition point, Tg (°C)	pH [10% Water Dispersion] (20°C)
EVA-1	White powder	85	-5	5. 4
EVA-2	White powder	72	14	9. 1
EVA-3	White powder	60	0	5. 0
VA/VeoVa-1	White powder	75	-3	7. 5
VA/VeoVa-2	White powder	64	24	5. 3
MMA/BA	White powder	70	26	11. 7
St/BA	Light-brown powder	68	9	7. 0
SBR	White powder	<45	17	8. 7

mixing, a synthetic ester-type antifoamer was added to the redispersible polymer powders in a ratio of 1. 0% of them. Also, a silicone emulsion-type antifoamer was added to EVA emulsion and SBR latex in a ratio of 0. 7% of the silicone solids of the antifoamer to the total solids of the emulsion or latex.

Table 2. Properties of polymer dispersions

Type of polymer	Specific gravity (20℃)	pH (20℃)	Viscosity (20℃, mPa·s)	Total solids (%)
EVA	1. 060	5. 70	1588	44. 2
SBR	1. 020	9. 70	64	44. 6

3 Testing procedures

3.1 Preparation of specimens

According to JIS A 1171 (Method of Making Test Sample of Polymer-Modified Mortar in the Laboratory), polymer-modified mortars were mixed with the mix proportions given in Tables 3 and 4, and their flow was adjusted to be constant

Table 3. Mix proportions of polymer-modified mortars using redispersible polymer powders

Type of mortar	Cement: sand (by mass)	Polymer-cement ratio (%)	Water-cement ratio (%)	Air content (%)	Flow
Unmodified	1 : 3	0	75. 0	7. 0	170
EVA-1	1 : 3	5	66. 0	7. 4	171
powder-		10	66. 0	7. 2	168
modified		15	66. 0	7. 0	172
		20	65. 0	7. 8	169
EVA-2	1 : 3	5	64. 0	9. 6	173
powder-		10	63. 0	8. 8	168
modified		15	64. 0	8. 6	168
		20	64. 0	8. 8	169
EVA-3	1 : 3	5	71. 0	9. 0	167
powder-		10	67. 0	7. 8	175
modified		15	65. 0	6. 8	172
		20	63. 0	7. 6	170
VA/VeoVa-1	1 : 3	5	71. 0	7. 8	170
powder-		10	69. 0	8. 0	168
modified		15	70. 0	7. 8	173
		20	71. 0	8. 4	168
VA/VeoVa-2	1 : 3	5	66. 0	8. 2	170
powder-		10	63. 0	8. 0	170
modified		15	60. 0	7. 6	165
		20	59. 0	8. 2	171
MMA/BA	1 : 3	5	66. 0	11. 5	166
powder-		10	64. 0	11. 5	170
modified		15	63. 0	10. 0	172
		20	61. 0	10. 5	173
St/BA	1 : 3	5	68. 0	8. 6	166
powder-		10	67. 0	8. 2	166
modified		15	67. 0	7. 8	168
		20	67. 0	8. 2	170
SBR	1 : 3	5	70. 0	7. 6	166
powder-		10	72. 0	6. 4	170
modified		15	74. 0	5. 4	174
		20	75. 0	5. 2	171

Table 4. Mix proportions of polymer-modified mortars using polymer dispersions

Type of mortar	Cement: sand (by mass)	Polymer-cement ratio (%)	Water-cement ratio (%)	Air content (%)	Flow
Unmodified	1 : 3	0	75. 0	7. 0	170
EVA-modified	1 : 3	5	63. 0	8. 4	170
		10	60. 0	8. 4	174
		15	56. 0	9. 3	168
		20	52. 0	9. 9	169
SBR-modified	1 : 3	5	67. 0	8. 8	168
		10	64. 0	5. 6	170
		15	61. 0	6. 0	170
		20	57. 0	5. 1	170

at 170±5. Beam specimens 40x40x160mm were molded, and then subjected to a 2-day-20℃ -80%(RH)-moist plus 5-day-20℃ -water plus 21-day-20℃ -50%(RH)-dry cure.

3.2 Strength test

The cured beam specimens were heated in a hot-air oven at 150℃ for 28 days. The specimens before and after heating were tested for flexural and compressive strengths at 20℃ according to JIS A 1172 (Method of Test for Strength of Polymer-Modified Mortar). In addition, the retention of strength was calculated by the following formula :

Retention of strength (%)

$$= \frac{\text{Flexural or compressive strength after heating (MPa)}}{\text{Flexural or compressive strength before heating (MPa)}} \times 100$$

4 Test results and discussion

Figs. 1 and 2 represent the polymer-cement ratio vs. flexural and compressive strengths of redispersible polymer powder-modified mortars before and after 28-day heating at 150℃. In general, the flexural strength of the redispersible polymer powder-modified mortars heated at 150℃ increases sharply with increasing polymer-cement ratio, and becomes nearly constant or reaches a maximum at polymer-cement ratios of 10 to 15%. Except for redispersible VA/VeoVa-1 and St/BA powder-modified mortars, the compressive strength of the redispersible polymer powder-modified mortars heated at 150℃ increases with an increase in the polymer-cement ratio, and becomes nearly constant or reaches a maximum at polymer-cement ratios of 10 to 15%. Among eight types of the redispersible polymer powder-modified mortars heated at 150℃, redispersible MMA/BA powder-

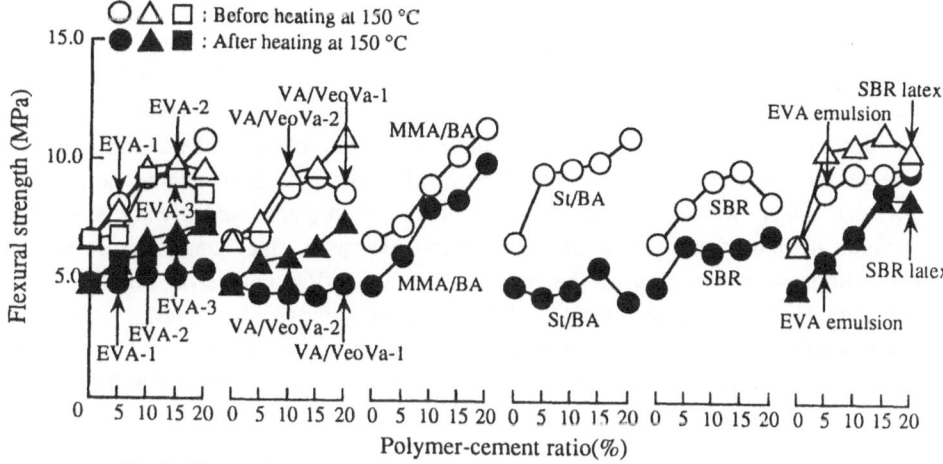

Fig.1. Flexural strength of redispersible polymer powder-modified mortars
before and after heating at 150°C.

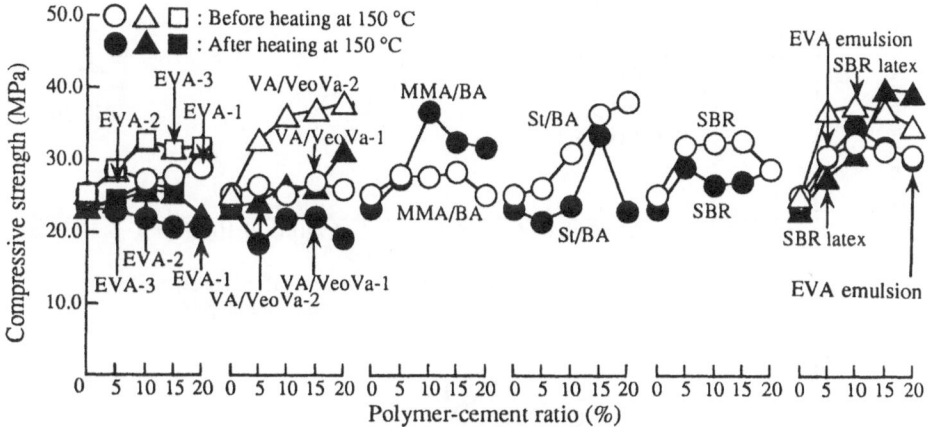

Fig.2. Compressive strength of redispersible polymer powder-modified
mortars before and after heating at 150°C.

modified mortars have the highest flexural and compressive strengths, which are
almost equal to those of the polymer dispersion-modified mortars. In particular, the
compressive strength of the redispersible MMA/BA powder-modified mortars heated
at 150°C is larger than that of redispersible MMA/BA powder-modified mortars
before heating at 150°C. In general, the flexural and compressive strengths of the
redispersible polymer powder-modified mortars heated at 150°C are lower than those
of the polymer dispersion-modified mortars [1].

Figs. 3 and 4 show the polymer-cement ratio vs. retentions of flexural and
compressive strengths of the redispersible polymer powder-modified mortars heated at

150°C for 28 days. Except for redispersible MMA/BA powder-modified mortars, the retentions of flexural and compressive strengths of the redispersible polymer powder-modified mortars heated at 150°C are lower than those of unmodified mortar. Also, the retention of compressive strength of the redispersible polymer powder-modified mortars is about 20% higher than the retention of flexural strength. The reason for this tendency is found to be due to the thermal degradation of the polymer films, which are formed in the modified mortars, and greatly affect the flexural strength development of the mortars. However, the retentions of flexural and compressive strengths of the redispersible MMA/BA powder-modified mortars are somewhat larger than those of the unmodified mortar. It appears that the redispersible MMA/BA powder-modified mortars have an excellent thermal resistance because of the highest glass transition point of 26°C for the redispersible MMA/BA

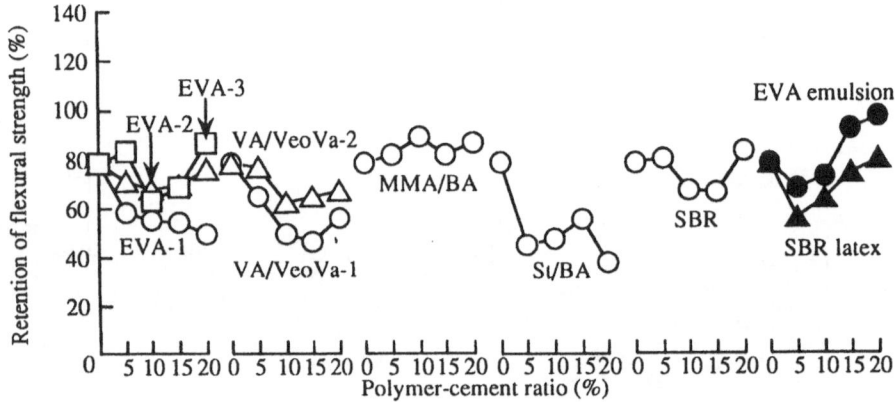

Fig.3. Polymer-cement ratio vs. retention of flexural strength of redispersible polymer powder-modified mortars after 28-day heating at 150°C.

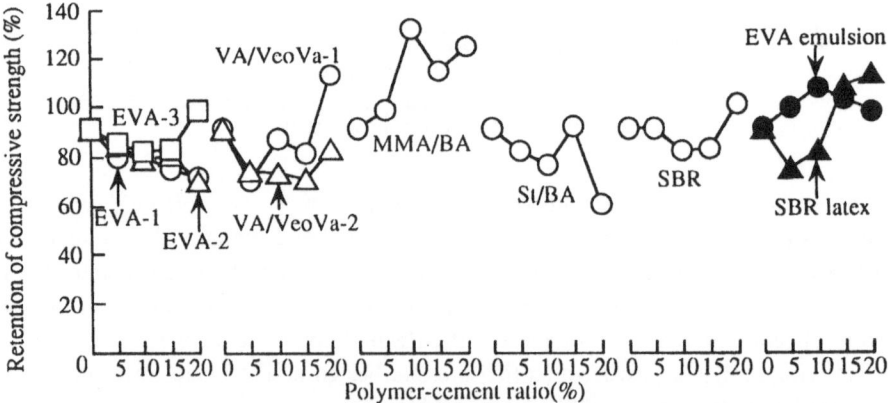

Fig.4. Polymer-cement ratio vs. retention of compressive strength of redispersible polymer powder-modified mortars after 28-day heating at 150°C.

powder. Therefore, it is supposed that the redispersible polymer powder-modified mortars using polymer powders with higher glass transition point have a higher thermal resistance.

Fig. 5 exhibits the relation between the glass transition point of redispersible polymer powders and the loss of flexural strength of redispersible polymer powder-modified mortars after 28-day heating at 150℃. In general, the loss of flexural strength of the redispersible polymer powder-modified mortars after 28-day heating at 150℃ tends to be considerably reduced with a raise in the glass transition point regardless of the polymer-cement ratio [3]. This tendency is remarkable at a polymer-cement ratio of 20%.

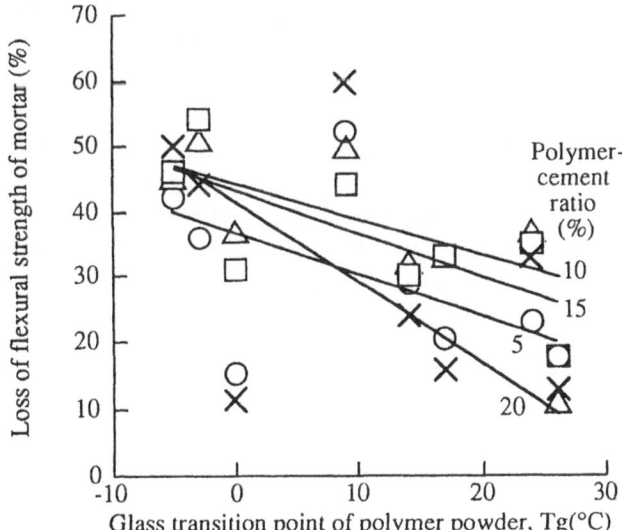

Fig.5. Glass transition point of redispersible polymer powders vs. loss of flexural strength of redispersible polymer powder-modified mortars after 28-day heating at 150°C.

5 Conclusions

The following conclusions can be obtained from the above test results :

(1) In general, the flexural and compressive strengths of the redispersible polymer powder-modified mortars heated at 150℃ tend to be improved with an increase in the polymer-cement ratio, and are somewhat inferior to those of polymer dispersion-modified mortars.

(2) Except for redispersible MMA/BA powder-modified mortars, the retentions of flexural and compressive strengths of the redispersible polymer powder-modified mortars heated at 150℃ are lower than those of unmodified mortar, and the retention of compressive strength of the redispersible polymer powder-modified mortars is about 20% higher than the retention of flexural strength.

(3) From the test results, it is supposed that the redispersible polymer powder-modified mortars using polymer powders with higher glass transition point have a higher thermal resistance.

6 References

1. Ohama, Y., Shiroishida, K. and Miyake, T. (1982) Thermal resistance of polymer-modified mortars, in *Proceedings of the Twenty-fifth Japan Congress on Materials Research*, The Society of Materials Science, Japan, Kyoto, pp. 234-238.
2. Ohama, Y. and Shiroishida, K. (1983) Temperature dependency of strength of polymer-modified mortars, in *Proceedings of the Twenty-sixth Japan Congress on Materials Research*, The Society of Materials Science, Japan, Kyoto, pp. 195-199.
3. Schneider, S. I., DeWacker, D. R. and Palmer, J. G. (1993) Redispersible polymer powders for tough, flexible cement mortars, in *Polymer-Modified Hydraulic-Cement Mixtures*, (eds. L. A. Kuhlmann and D. G. Walters), STP 1176, American Society for Testing and Materials, Philadelphia, pp. 76-89.

PART SEVEN
DURABILITY

EVALUATION OF CHLORIDE PERMEABILITY OF SURFACE COATING MATERIALS BY RAPID CHLORIDE PERMEABILITY TEST

K. Torii and M. Kawamura
Dept. of Civil Engineering, Kanazawa University, Kanazawa, Japan.
S. Tanikawa and M. Achiwa
New Products Research Laboratory, Toagosei Co., Ltd.,
Nagoya, Japan

Abstract
The chloride permeability of ordinary Portland cement mortars treated with and without the surface coating as well as polymer modified mortars was assessed using the rapid chloride permeability test (RCPT, AASHTO T 277). The current vs. time curves and the charge passed in RCPT were compared as a measure of the ability of protective barrier against the chloride ingress into the mortar. After RCPT, the chloride content profiles of the mortars were also determined by the chemical analysis of sliced samples and by spraying 0.1 N $AgNO_3$ solution on splitted surfaces of the mortar, where the chloride diffusivity of the mortar was calculated according to the equation proposed by Tang et al.. From the results, it was found in RCPT that epoxy-type and acrylic rubber-type surface coatings provided a good barrier against the chloride ingress, but silane or silicon penetrant-type treatment was not so effective as expected.
Keywords: AASHTO T 277, diffusion coefficient for chloride ions, electrical resistivity, polymer modified mortar, rapid chloride permeability test, surface treatment

1 Introduction

In the 1980's, a lot of RC and PC bridges on the seashore area facing the Japan Sea were found to be severely damaged by the chloride induced corrosion of steel reinforcement in concrete, where is affected by the strong seasonal wind especially during winter. Also, the use of de-icing salts used on the roads or bridges for traffic safety has rapidly been increasing since the studded tires was prohibited throughout the country in 1991 [1]. Presently, the protecting technique for the steel corrosion has become an issue of concern with respect to the repair and maintenance of damaged concrete structures.

It has well been recognized that the surface treatment with coating materials or polymer modified mortars plays an important role in inhibiting or reducing the chloride ingress into concrete from the surroundings, but there is no standard test to effectively evaluate the chloride permeability of repair materials especially for a short time. The diffusion coefficient of concrete for chloride ions has been estimated on the basis of static diffusion cell experiment or from the chloride ion concentration profiles in concrete, but these methods are time consuming and not applicable for a large variety of concrete

Polymers in Concrete, edited by Y. Ohama, M. Kawakami and K. Fukuzawa. Published in 1997 by E & FN Spon, 2–6 Boundary Row, London SE1 8HN, UK. ISBN: 0 419 22330 4.

mixtures. There is a rapid test method for the chloride permeability of concrete, which was originally developed by Whiting [2] and adopted as a standard test of AASHTO T 277 [3] and ASTM C 1202 [4]. Cabrera et al. [5] has suggested that the RCPT technique is also useful to evaluate the chloride protection resistance for repair materials or surface treatment compounds. However, there are few studies concerning the assessment of effectiveness of surface coating materials against the chloride ingress using RCPT.

The objective of this paper is to evaluate the chloride ingress resistance of the surface treatment with coating materials or polymer modified mortars using the RCPT technique. The applicability of the RCPT technique was also discussed based on the relation between the charge passed in RCPT and the chloride ion penetration in the ponding test into 5 % NaCl or 30 % CaCl$_2$ solution.

2 Experimental Procedure

2.1 Materials and Preparation of Specimens
The ordinary Portland cement with the specific gravity of 3.13 and with the Blaine specific fineness of 3300 cm^2/g was used. The river sand from the Hayatsuki river in Toyama Prefecture with the specific gravity of 2.61 and with the water absorption percentage of 1.3 % was also used to make the mortar. The cement : sand ratio of OPC mortars was 1 : 2, and their water : cement ratios were 0.35, 0.45 and 0.55, where they were labeled OPC35, OPC45 and OPC55. The water : cement ratio of polymer modified mortars was 0.55, the cation-type acrylic polymer dispersion being added with polymer : cement ratios of 5% and 10 % by weight, where they were labeled PC5% and PC10%. The mix proportions of OPC and polymer modified mortars used in this study are presented in Table 1. Five commercially available surface treatment compounds, silane-type and silicon-type penetrant (S1,S2), epoxy-type (E) and acrylic rubber-type coating (A1, A2) were selected. The properties of surface treatment materials are presented in Table 2.

The OPC and polymer modified mortars were cast into the steel mould, which was 500 mm by 500 mm by 50 mm. After demoulding, the slabs were cured in water at 20 °C for 7 days and placed indoors for 21 days, and then surface treatment compounds were provided to the bottom of OPC mortar slabs with the water : cement ratio of 0.55. At the prescribed ages of the test, 100 mm cores were drilled from the slabs.

Table 1. Mix proportions of ordinary Portland cement and polymer modified mortars.

	W/C	Unit content (kg/m^3)			
		Water	Cement	Sand	Polymer*
OPC 35	0.35	244	696	1392	---
OPC 45	0.45	293	651	1302	---
OPC 55	0.55	336	611	1222	---
PC 5%	0.55	265	611	1222	102
PC 10%	0.55	193	611	1222	204

* Solid phase of cation-type acrylic polymer dispersion used is 30 %.

2.2 Procedures of Experiments

2.2.1 Rapid Chloride Permeability Test (AASHTO T 277)
Fig. 1 shows a schematic diagram of electrical circuit of RCPT. As shown in Photo. 1, the 3 % sodium chloride and 0.3 N sodium hydroxide solutions were filled in the cells on the either side of the saturated mortar of 100 mm in diameter and 50 mm thick. In this study, the standard condition of RCPT was modified in order to avoid the increase in

temperature during the test ; the direct voltage supplied reduced from 60 V to 30 V and the time of test prolonged from 6 hours to 24 hours. The chloride permeability of mortars was assessed in terms of total charge passed, coulombs, for 6 hours and 24 hours. After RCPT, distributions of chloride ion concentration from the negative side to the positive one were determined by the chemical analysis of sliced mortar according to JCI SC-5. Similarly, a 0.1 N AgNO₃ solution was sprayed on the splitted surface of mortars, in which the chloride ion penetration depth into the mortar was measured. The diffusion coefficient for chloride ions (D_{cl}) was calculated from the average depth of chloride ion penetration after RCPT according to the following equation proposed by Tang et al. [6].

$$Dcl = \frac{z}{R}\frac{F}{T}E\frac{X_d}{t} \ (cm^2/sec)$$

being: z : electrical charge of chloride, F : Faraday's number, E : voltage applied, R : gas constant, T : absolute temperature, X_d : conversed chloride ion penetration depth, t : time

Table 2 Surface treatment compounds used in this study.

	Description of surface treatment compounds
Silane-type penetrant (S1, T Co. Ltd.)	Water-based silane penetrant AP 40J (0.3 kg/m²)
Silicon-type penetrant (S2, S Co. Ltd.)	Water-based silicon penetrant PC (0.2 kg/m²)
Epoxy-type coating (E, N Co. Ltd.)	Multiple coats, Primer (E), Middle coat (ED 60μm), Top coat (UD 30μm)
Acrylic rubber-type coating (A1, T Co. Ltd.)	Multiple coats, Primer (P200 0.15kg/m²), Middle coat (A100 2kg/m²), Top coat (T300 0.3kg/m²)
Acrylic rubber-type coating (A2, T Co. Ltd.)	Multiple coats, Primer (P200 0.15kg/m²), Middle coat (A300 2kg/m²), Top coat (T300 0.3kg/m²)

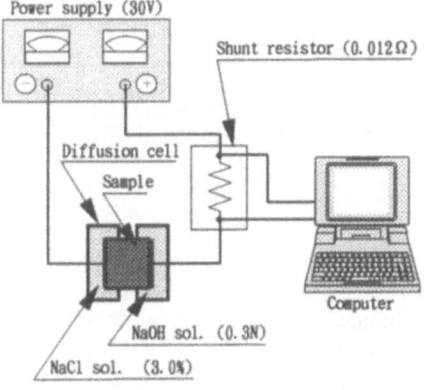

Fig. 1 Schematic diagram of electrical circuit of RCPT.

Photo. 1 External appearance of diffusion cell of RCPT.

2.2.2 Electrical Resistivity Test

The AC impedance, in ohms, is considered to be related to the total charge passed, in coulombs [7].   The electrical resistivity of coated and uncoated mortar specimens was measured by the LCR meter before and after RCPT when 10 mV AC at 1 k Hz was applied with the electrodes of copper plate across the saturated specimens of 100 mm in diameter and 50 mm thick.

2.2.3 Weatherability Test

In order to investigate the influence of ultra-violet ray on the deterioration of surface coating materials, an accelerated weatherability test was carried out using Carbon Arch-type Weather Sun-shine Testing Machine according to JIS B 7753.   After the exposure time of 250 and 1000 hours in the weatherability test, both chloride permeability and electrical resistivity were examined.   Furthermore, the slabs were also exposed to three different environments for 1 year ; outdoors, indoors at 20 °C and 60 % R.H., and in water at 20 °C.

2.2.4 Ponding Test

Cylindrical specimens, 100 mm in diameter and 200 mm high, were prepared for the ponding test.   Coated and uncoated mortar specimens were immersed completely into the 5 % NaCl or 30 % $CaCl_2$ solution for 3 months and 1 year.   After the ponding test, the chloride ion penetration depth into the mortars was determined by the same method mentioned above.

Table 3 Comparison of charge passed, electrical resistivity, chloride ion penetration
depth and diffusion coefficient for chloride ions of coated and uncoated mortars
in RCPT.

	Charge passed (coulombs)	Electrical resistivity (Ω-cm)		Chloride penetration (cm)	Chloride diffusivity* (10^{-8} cm^2/sec)
		before RCPT	after RCPT		
OPC 35	4984	825	871	1.30	0.300
OPC 45	10283	372	404	2.65	3.730
OPC 55	14825	478	315	3.78	7.096
PC 5%	9247	585	427	2.50	3.309
PC 10%	6637	778	562	1.79	1.436
S1	10631	740	362	2.62	3.645
S2	14572	478	298	3.58	6.479
E	46	14402	6234	0	---
A1	8	12062	13232	0	---
A2	0	55640	11948	0	---

* Chloride diffusivity (D_{cl}) was calculated from the average depth of chloride ion
penetration in RCPT according to the equation proposed by Tang et al. [6].

3 Results and Discussion

3.1 Charge Passed through Mortar in RCPT

The comparison of the charge passed, electrical resistivity, chloride ion penetration depth and diffusion coefficient for chloride ions of coated and uncoated mortars in RCPT is summarized in Table 3.   Fig. 2 shows changes in the current passed through coated and uncoated mortars in RCPT.   As shown in Fig. 2, OPC mortars with a very low water : cement ratio of 0.35 show a smaller and more stable current during 24 hours of the test compared with those with higher water : cement ratios of 0.45 and 0.55.   This is due to the increase in the electrical resistivity resulting from the formation of a dense texture

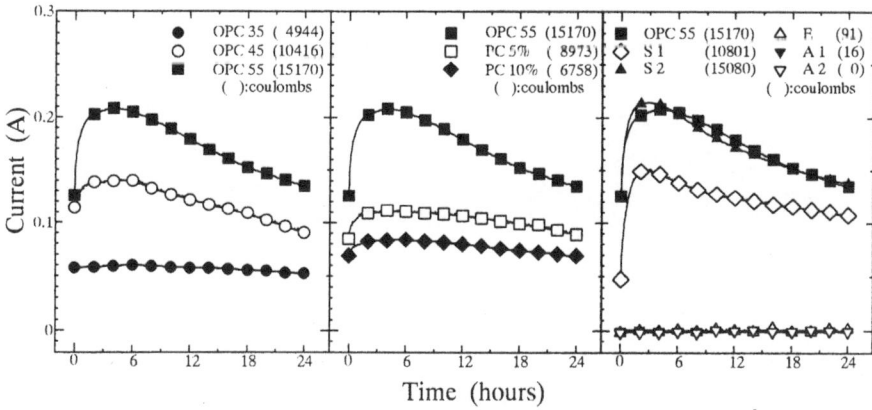

Fig. 2 Changes in current passed through coated and uncoated mortars in RCPT.

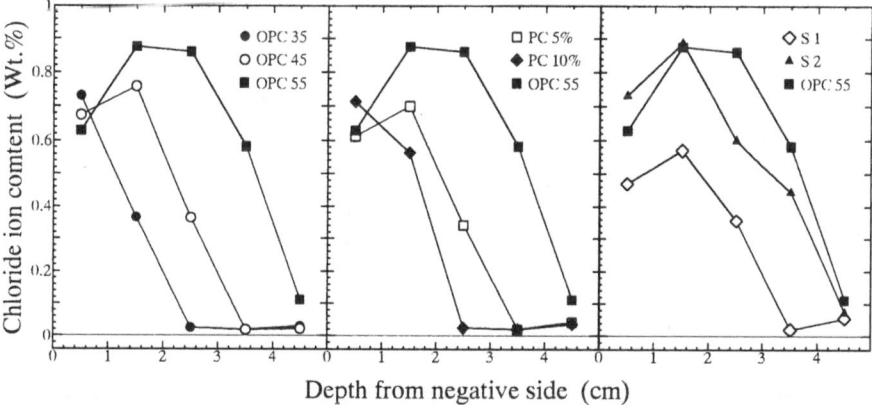

Fig. 3 Changes in chloride ion content with the depth from the negative side in RCPT.

in OPC mortars with a very low water : cement. Again, the current of polymer modified mortars with the polymer : cement ratio of 10 % is almost half that of OPC mortars with the same water : cement ratio, the current passed through polymer modified mortars being reduced in proportion to the amount of the polymer added. It is also observed in the current vs. time curves of some OPC and polymer modified mortars that the current gradually decreases at later stages of the test after it has reached the peak. This phenomenon may be attributed to the difference in the flowing speed of both OH$^-$ and Cl$^-$ ions under the applied electrical field ; OH$^-$ ions in the mortar migrate more rapidly at early stages of the test, followed by a slow migration of Cl$^-$ ions from the negative cell to the positive one. On the other hand, for OPC mortars with surface treatments, only a very small and constant current is passed through the specimens coated with the epoxy-type and acrylic rubber-type resin (E, A1, A2), while a relatively high current is passed through the specimens treated with the silane-type and silicon-type penetrant (S1,S2), which is similar to the current passed of untreated OPC mortars with the same water : cement ratio. Interestingly, a very small charge passed of coated mortars in RCPT results from the high insulation resistance of coating materials. From the results of RCPT, it was confirmed that the surface coating gave a more effective protection against the chloride ion penetration into the mortar under the applied electrical field than the surface treatment with penetrant.

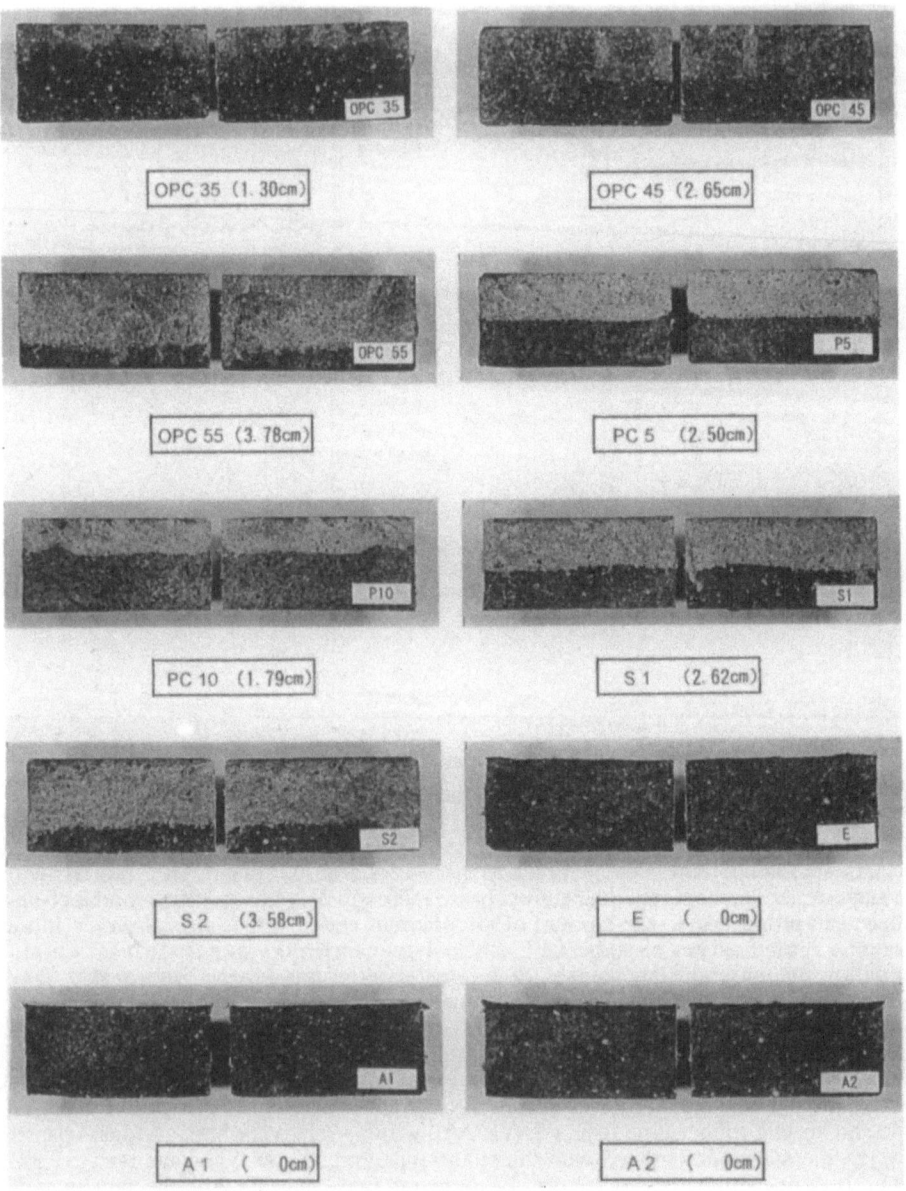

Photo. 2 External appearance of splitted surfaces of coated and uncoated mortars sprayed 0.1 N AgNO3 solution after RCPT.

The ranking order of uncoated and coated mortars in terms of total charge passed of 24 hours can be expressed as follows;
For OPC and polymer modified mortars
OPC35 > PC10% > PC5% > OPC45 > OPC55
For OPC mortars with surface treatments
A2 = A1 > E >> S1 > S2 = OPC55
This ranking order is also in a good agreement with the results of the research by Cabrera et al. [5].

3.2 Chloride Ion Penetration Depth in RCPT

Fig. 3 shows changes in chloride ion content of the mortar with the depth from the negative side in RCPT. As shown in Fig. 3, chloride ions intrude more deeply in OPC mortars with higher water : cement ratio and in polymer modified mortars with low polymer : cement content. Chloride content profiles for polymer modified mortars and OPC mortars treated with the penetrant are similar to those of untreated OPC mortars, while little chloride ion is found in coated mortars with epoxy- and acrylic rubber- type resin.

Photo. 2 shows the external appearance of splitted mortar surfaces in RCPT. It is seen from Photo. 2 that the boundary of the area where chloride ions have penetrated into the mortar in RCPT can be clearly distinguished by spraying a 0.1 N AgNO$_3$ solution on splitted surfaces of the mortar. For OPC and polymer modified mortars, the average depth of chloride ion penetration increases with an increase of the water : cement ratio and with a decrease of the polymer : cement ratio, while the coating with the epoxy-type and acrylic rubber-type resin performs almost as an impermeable barrier.

Figs. 4 and 5 show the relations between the total charge passed and the chloride ion penetration depth in RCPT. It is evident from Figs. 4 and 5 that there is a linear and good correlation between them that higher the charge passed of the mortars is, more deeply chloride ions penetrate into the mortar. This result shows that the measurement of the chloride ion penetration depth as well as the charge passed is very effective to evaluate the chloride permeability of the mortar since it can directly and visually evaluate the effectiveness of protection of surface treatments against the chloride ingress. Diffusion coefficient for chloride ions of OPC and polymer modified mortars estimated from the average depth of chloride ion penetration in RCPT ranges from 10^{-7} to 10^{-8} cm^2/sec, which corresponds in the same magnitude order with the values in the static diffusion cell methods which has already been reported in the literature [8,9].

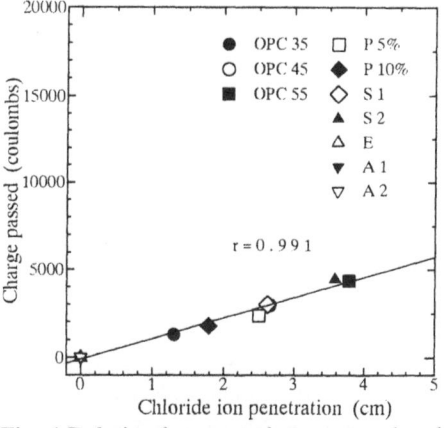 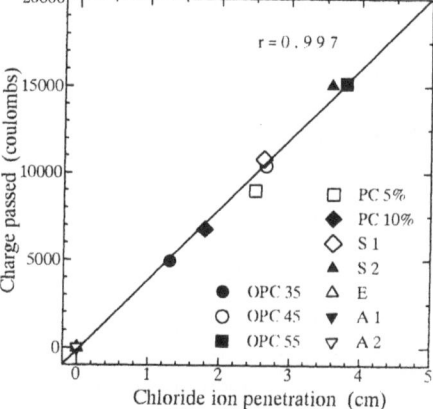

Fig. 4 Relation between charge passed and chloride ion penetration depth in RCPT (6 hours).

Fig. 5 Relation between charge passed and chloride ion penetration depth in RCPT (24 hours).

3.3 Weatherability

Fig. 6 shows changes in the current passed through coated and uncoated mortars after 250 hours of the exposure time in the weatherability test. After the weatherability test, the charge passed of OPC mortars increased by 30 % to 60 %, which corresponded to the decrease in the electrical resistivity of the mortar, although there was not a significant change in the charge passed in polymer modified mortars and OPC mortars treated with the penetrant. It is considered that for the mortars coated with epoxy-type or acrylic rubber-type resin, the current increases and the electrical resistivity decreases if the cracking or peeling off of the coating has occurred. On the other hand, there was no change in total charge passed and electrical resistivity in coated mortars, which maintained their characters of an impermeable barrier. However, a further research is needed to evaluate the weatherability of surface treatment compounds, since it seems that coating materials will normally deteriorate after 1000 hours of the exposure time in the accelerated weatherability test.

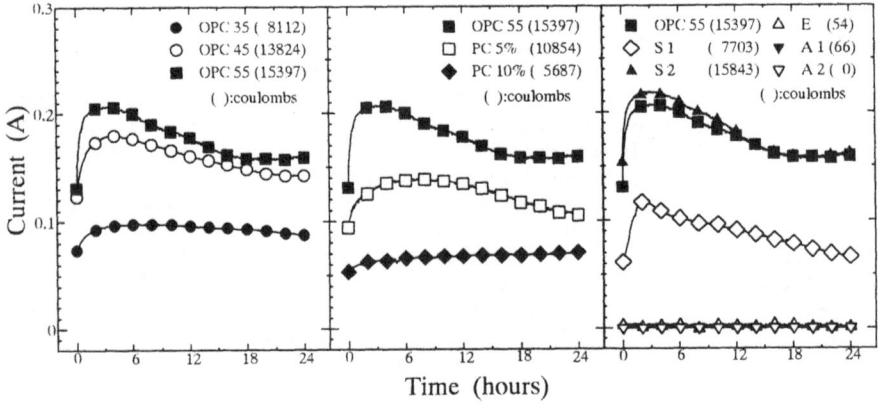

Fig. 6 Changes in current passed through coated and uncoated mortars after 250 hours of exposure time in weatherability test.

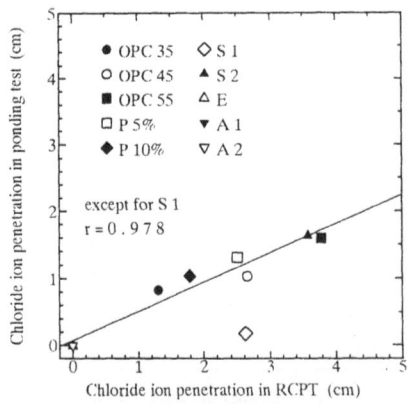

Fig. 7 Relation in chloride ion penetration between RCPT and ponding test in 3 % NaCl sol. for 3 months

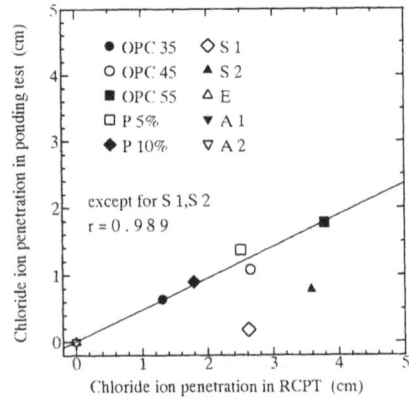

Fig. 8 Relation in chloride ion penetration between RCPT and ponding test in 30 % CaCl2 sol. for 3 months.

3.4 Chloride Ion Penetration in Ponding Test

Figs. 7 and 8 show the relations in chloride ion penetration between RCPT and the ponding test into the 3 % NaCl or 30 % CaCl$_2$ solution. Reports by Whiting [2] and Ozyilidrium [10] have suggested the good correlation between the charge passed in RCPT and the results of the 90 day ponding test using AASHTO T259. The average depth of chloride ion penetration in RCPT is almost doubled when compared with that in the ponding test into the 3 % NaCl or 30 % CaCl$_2$ solution for 3 months, but there is a good correlation between both test methods except for S1 and S2. The mechanism by which silane-type and silicon-type penetrant protects the mortar from the chloride ingress is simply its hydrophobic character. It is considered that under the saturated conditions in RCPT, the hydrophobic character of silane-type penetrant does not fully work to prevent the water from penetrating into the mortar. Also, excepting S1, the results of RCPT gave the almost same ranking order of the chloride permeability as the one found in the ponding test

4 Conclusions

From the experimental results, it was found that the results of RCPT (AASHTO T 277) gave the almost same ranking order of the chloride permeability as the ones found in the ponding test, and that the RCPT technique was very useful to evaluate the chloride protection resistance for repair materials or surface treatment compounds for a short time.

The main results obtained in this study are as follows;

(1) The total charge passed of RCPT correlated with the results of measurement of the electrical resistivity of uncoated and coated OPC mortars and polymer modified mortars.

(2) Both the charge passed and the average depth of chloride ion penetration in RCPT gave a reasonable estimate of the chloride permeability of uncoated and coated OPC mortars.

(3) Epoxy-type and acrylic rubber-type surface coatings provided a good barrier against the chloride ingress in RCPT, but silane or silicon penetrant-type treatment was not so effective as expected.

(4) Within this study, all surface treatments showed a higher weatherability than untreated OPC mortars and polymer modified mortars.

5 Acknowledgments

The authors are grateful to Mr. M. Miyoshi, Civil Engineer of Kobe City, for his contribution to the experimental and computational work .

6 References

1. Torii, K. et al. (1994) Carbonation and steel corrosion in concretes containing mineral admixtures under different environments, Proc. of Inter. Conf. on Corrosion and Corrosion Protection of Steel in Concrete, Vol.2, pp.658-667.

2. Whiting, D. (1981) Rapid determination of the chloride permeability of concrete, FHWA Report No.RD-81/119.

3. AASHTO (1986) Standard method of test for rapid determination of the chloride permeability of concrete (AASHTO T 277-83), Washington DC..

4. ASTM (1992) Standard test method for electrical indication of concrete's ability to resist chloride ion penetration (ASTM C 1202-91), Philadelphia.

5. Cabrera, J.G. et al. (1994) Assessment of the effectiveness of surface treatments against the Ingress of chloride, Proc. of Inter. Conf. on Corrosion and Corrosion Protection of Steel in Concrete, pp.1028-1043.

6. Tang, L. et al. (1992) Rapid determination of the chloride diffusivity in concrete by applying an electrical field, ACI Materials Journal, Vol.89, No.1, pp.49-53.

7. Feldman, R.F. et al. (1994) Investigation of the rapid chloride permeability test, ACI Materials Journal, Vol.91, No.2, pp.246-255.

8. Goto, S. et al. (1981) The effect of w/c ratio and curing temperature on the permeability of hardened cement pastes, Cement and Concrete Research, Vol.11, pp.575-579.

9. Short, N.R. et al. (1982) The diffusion of chloride ions through portland and blended cement pastes, Silicates Indus., Vol.47, No.10, pp.237-240.

10. Ozyildirim, C. et al. (1988) Resistance to chloride ion penetration of concrete containing fly ashes, silica fume or slags, ACI SP 108-3, pp.35-61.

QUANTITATIVE EVALUATION OF SUPPRESSIVE EFFECTS OF POLYMERIC SURFACE COATING MATERIALS ON CARBONATION OF CONCRETE

T. Fukushima
Building Research Institute, Ministry of Construction,
Tsukuba, Japan
I. Fukushi
Housing and Urban Development Corporation, Tokyo, Japan

Abstract
Suppressive effects of polymeric surface coating materials on the progress of carbonation of concrete were successfully affirmed by accelerated carbonation test for concrete specimens treated with many kinds of polymeric coatings. Based upon the experimental results, suppressive effects of polymeric coatings are evaluated by parabolic law ($\sqrt{t}$ law) involving a constant term, derived theoretically by unsteady state dynamic analysis for the diffusion of carbon dioxide in concrete accompanied by carbonation reaction with calcium hydroxide. The influences of the deterioration of polymeric coatings under outdoor exposure on suppressive effects are discussed compared with the case of no deterioration, based upon theoretical consideration. As the result, it was found that the progress of carbonation of concrete treated with a polymeric coating obeys the parabolic law involving a constant term, not depending on whether or not the deterioration of polymeric coating occurs, but that the deterioration of polymeric coating accelerates the progress of carbonation. Concept of equivalent thickness of cover concrete of concrete treated with a polymeric coating has been established, taking account of the suppressive effect, and values of equivalent thickness of cover concrete for many polymeric coatings were calculated from experimental results of accelerated carbonation test.
Keywords : Carbonation, Polymeric surface coating, Suppressive effect, Deterioration, Thickness, Diffusion coefficient, Parabolic law, Equivalent thickness of cover concrete

1 Introduction

The progress of carbonation (neutralization) of concrete, connected with corrosion of reinforcement, affords a basis for the setting of physical service lives of reinforced concrete buildings under the ordinary atmospheric environment, so that the rationalevaluation of the progress of carbonation and pursuit of effective suppressive methods are old, but even now new research subjects[1-5]. If effective methods to stop, retard or suppress the progress of carbonation of concrete are found, we can expect to make best use of them for the establishment of an appropriate durability design for making longer the physical service lives of reinforced concrete buildings.

Polymers in Concrete, edited by Y. Ohama, M. Kawakami and K. Fukuzawa. Published in 1997 by E & FN Spon, 2–6 Boundary Row, London SE1 8HN, UK. ISBN: 0 419 22330 4.

As an effective method to suppress and retard the progress of carbonation of concrete, we can utilize the shielding function against the diffusion of carbon dioxide of various kinds of surface finishing materials, especially polymeric surface coating materials. Polymeric coatings. however. are liable to deteriorate and lose their shielding function during in service under outdoor exposure. In order to make effective use of polymeric coatings for the suppression and retardation of the progress of carbonation of concrete, we should evaluate quantitatively their carbonation preventive functions including the influences of deterioration[6-7].

In this context, this paper describes the prediction of the progress of carbonation of concrete influenced by polymeric coatings, paying special attention to the quantitative evaluation of the mechanisms of carbonation preventive function and the influence of deterioration of polymeric coatings, and further the method of setting the equivalent thickness of cover concrete, taking onto account of the suppressive effect of a polymeric coatings on the progress of carbonation of concrete[8-9].

2 Research Methods and Results

2.1 Evaluation of the suppressive effects of polymeric surface coating materials by accelerated carbonation testing

2.1.1 Experimental methods

Test specimens of concrete ($10 \times 10 \times 40$cm prism) with various kinds of polymeric surface coatings materials(Table 1) treated in advance were acceleratedly carbonated under the condition of $30°C$, 60%R.H., $CO_2 : 5\%$. Accelerated carbonation was carried out only from the two side directions with other four surfaces of concrete specimens shielded against carbonation with epoxy resin(Fig.1). The neutralization depth of acceleratedly carbonated concrete specimens was measured by a 1% phenolphthalein alcohol solution specified by JIS K 8006. In this experiment temperature and humidity were set constant values, and ultraviolet radiation was not done, so that the deterioration of polymeric surface coatings was expected not to occur, and the expectation was affirmed.

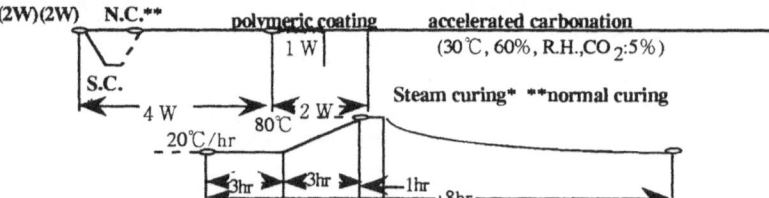

Fig.1 Condition for preparation and curing for concrete specimens

2.1.2 Experimental results

The progress of carbonation was effectively suppressed by polymeric surface coating materials, but the effects depend upon their types. The plotting of neutralization depths versus square root of time shown in Fig. 2 apparently concludes that the progress of carbonation of concrete treated with polymeric coatings can be expressed by the parabolic law involving a constant term. By using the least square method for the experimental data, rates of the progress of carbonation of concrete treated with various types of polymeric coatings were predicted as shown in Table 2. By theoretically calculating the diffusion coefficients of carbon dioxide in various types of polymeric coatings from experimentally observed induction periods(Table 3), the suppressive effects of various polymeric coatings on the progress of carbonation of concrete were successfully evaluated.

Table 1 Type of polymeric surface coating material

Symbols	Type of polymeric coating
a	no finish
b	vinyl wall paper
c	emulsion paint
d	synthetic resin plaster
e	acrylic resin based sprayed
f	epoxy resin based sprayed

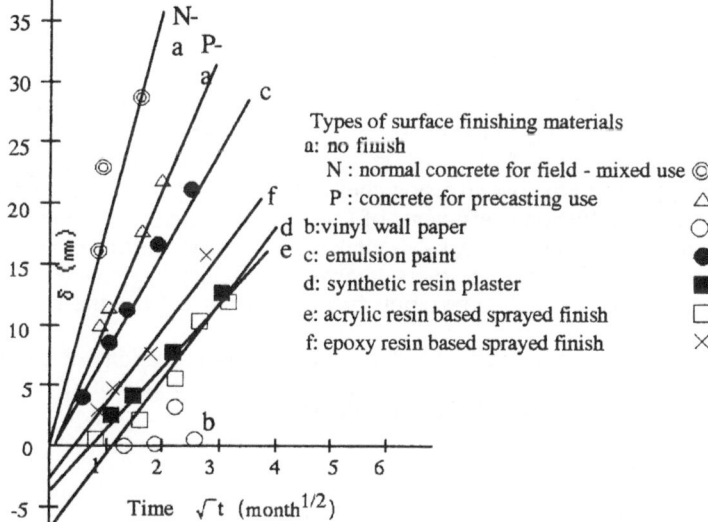

Fig.2 Results of suppressive effects of polymeric coatings materials on the progress of carbonation of concrete by accelerated carbonation test (30 C, 60%R.H., CO_2:5%)

2.2 Theoretical consideration on the progress of carbonation of concrete treated with polymeric surface coatings

2.2.1 Consideration in case that the deterioration of polymeric coatings can be neglected

By assuming the unsteady state diffusion of carbon dioxide in concrete accompanied by the instant irreversible second-order carbonation reaction with calcium hydroxide, taking account of the total mass transfer resistance of polymeric coatings against the diffusion of carbon dioxide from the surface into concrete, the following simultaneous diffusion equations with the initial and boundary conditions are obtained for the dynamic analysis of carbonation process:

[Fundamentals differential equations]

$$\partial C_{a1}/\partial t = D_{a1}\partial^2 C_{a1}/\partial x^2 \quad (-\infty < x \le 0) \tag{1}$$

$$\partial C_{a2}/\partial t = D_{a2}\partial^2 C_{a2}/\partial x^2 \quad (0 < x \le \partial_x N) \tag{2}$$

$$\partial C_{b1}/\partial t = D_{b1}\partial^2 C_{b1}/\partial x^2 \quad (\partial_x N < x \le +\infty) \tag{3}$$

$$\partial C_{p2}/\partial t = D_{a2}\partial C_{a2}/\partial x \quad (x = \partial_x N) \tag{4}$$

[Initial Conditions]

$$t \leq 0, \quad x \to -\infty \quad : C_{a1} = C_{ag} \tag{5}$$
$$0 < x \leq +\infty \quad : C_{b2} = C_{b0} \tag{6}$$
$$0 < x \leq +\infty \quad : C_{a2} = 0 \tag{7}$$
$$0 < x \leq +\infty \quad : C_{p2} = 0 \tag{8}$$

[Boundary conditions]

$$t \geq 0, \quad x \to -\infty \quad : C_{a1} = C_{ag} \tag{9}$$
$$x \to -\infty \quad : C_{b2} = C_{b0} \tag{10}$$
$$x = 0 \quad : N_s = - D_{a1} \partial C_{a1}/ \partial x = - D_{a2} \partial C_{a2}/ \partial x \tag{11}$$
$$x = 0 \quad : - D_{a1} \partial C_{a1}/ \partial x = k_{ag}' (C_{a2} - mC_{a1}) \tag{12}$$
$$x = \partial_x N \quad : C_{a2} = C_{p2} = 0 \tag{13}$$
$$x = \partial_x N \quad : N_f = - D_{a2} \partial C_{a2}/ \partial x = - D_{b2} \partial C_{b2}/ \partial x \tag{14}$$

Where, C_{ag}, C_{b0}: initial molar fractions of CO_2 in gas phase and $Ca(OH)_2$ in concrete; C_{a1}, C_{a2}: molar fractions of CO_2 in gas phase and in concrete as functions of time and depth ; C_{b2}, C_{p2} : molar fractions of $Ca(OH)_2$ and CaO_3 in concrete as functions of time and depth ; D_{a1}, D_{a2}: effecive diffusion coefficients of CO_2 in gas phase and in concrete; D_{b2},: effecive diffusion coefficients of $Ca(OH)_2$ in concrete; m: equilibrium constant concerning CO_2 in the surface; k_{ag}' : total mass transfer coefficient of CO_2 in a polymeric coating including surface mass transfer coefficient k_{ag}; N_s, N_f : molar fluxes of CO_2 at the surface and carbonation front; x: distance from the surface of concrete;t :time.

By using Laplace transformation method and appropriate approximation, the parabolic law involving a constant term expressing the progress of carbonation of concrete treated with polymeric coatings was derived as follows:

$$X = A\sqrt{t} - B \quad : t_0 \equiv (B / A)^2 \leq t \tag{15}$$
$$X = 0 \quad : \quad 0 \leq t \leq t_0 \tag{16}$$
$$A = m\sqrt{\pi} \, D_{a2} \sqrt{D_{a2}/D_{b2}} \{(C_{ag}/2C_{b0})- \sqrt{D_{b0}/D_{a1}}\} \tag{17}$$
$$B = D_{a2}/k_{ag}' \tag{18}$$
$$1/k_{ag}' = 1 /k_{ag} + L_0/D \tag{19}$$

Where, A is the parameter which describes the rate of the progress of carbonation, and it includes the ratio of concentrations of carbon dioxide and calcium hydroxide in the initial stage C_{ag}/C_{b0} , diffusion coefficient of carbon dioxide and hydroxide in concrete D_{a2} , D_{b2} and equilibrium constant of carbon dioxide in the surface of concrete or surface polymeric coatings materials m. The constant term B includes the total mass transfer constant k_{ag}', and describes the retardative effect of polymeric surface coatings materials. The material constants characteristic of finishes are m , D , and together with their thickness L_0, influences the progress of carbonation of concrete treated with polymeric soatings, by determing the two parameters of A and B, though the progress of off-form concrete without polymeric surface coatings can be described by only one parameter A / m as follows:

$$X = (A/m) \sqrt{t} \tag{20}$$

The time t_1 when off-form concrete with thickness Y_0 becomes completely carbonated can be retarded until t_2 by the surface coating, and the retarded physical service life($t_1 - t_2$) can be evaluated as follows:

$$t_1 - t_2 = \{(Y_0 + B)^2 - (Y_{0m})^2 \} / A \tag{21}$$

On the other hand, the progress of carbonation of concrete treated by a polymeric coating shows a certain induction period characteristic of type of coating, and the induction period is connected with the diffusion coefficient of carbon dioxide in the surface coating material by the following equation .

$$D = D_{a2}L_0/\{\sqrt{t_0}m\sqrt{\pi D_{a2}/D_{b2}}[(C_{ag}/2C_{b0})-\sqrt{D_{b2}/D_{a1}}]\} \tag{22}$$

We can analytically evaluate the diffusion coefficients of carbon dioxide in various types of polymeric coatings by inserting the experimentally measured values of the induction periods, which are discussed later in next section. The relationship between the carbonation depth (X_1) of concrete after a time(T_1) and the total thickness of polymeric coatings material (L_0) can be expressed by the following linear equation.

$$X_1 = A\sqrt{T_1} - B = A\sqrt{T_1}-D_{a2}/k_{ag}- (D_{a2}/D)L_0 = X_1'-(D_{a2}/D)L_0 \tag{23}$$

Consequently, the influence of polymeric surface coatings on the progress of concrete can be well described by the parabolic law involving a constant term.

2.2.2 Consideration in case that the deterioration of polymeric surface coatings are taken into account

The deterioration of polymeric surface finishing materials proceeds continuously inwards from the surface with time under outdoor and accelerated aging exposure. The depth of the deteriorated layer increases approximately in proportion to the square root of time[10]. Taking into account of this time dependence of the deteriorated layer(parabolic law), the deterioration of polymers can be expressed by the decrease in their thickness as follows :

$$L(t) = L_0 (1- \beta\sqrt{t}) \tag{24}$$

Consequently, the progress of neutralization of concrete coated with polymeric finishing materials can be expressed as follows :

$$X = \{A + (D_{a2}\beta /D)\} \sqrt{t} - B : t_0 \equiv [B/\{A + (D_{a2} \beta/D)\}]^2 \leq t_0 \tag{25}$$
$$X = 0 \quad : 0 \leq t \leq t_0 \tag{26}$$

This shows that the coefficient expressing the rate of the progress of the neutralization is modified from A to $A + (D_{a2}\beta/D)$, so that the progress of carbonation becomes faster by the deterioration of polymeric coatings, compared with the case of no deterioration. The progress of carbonation, however, obeys also the parabolic law involving a constant term.

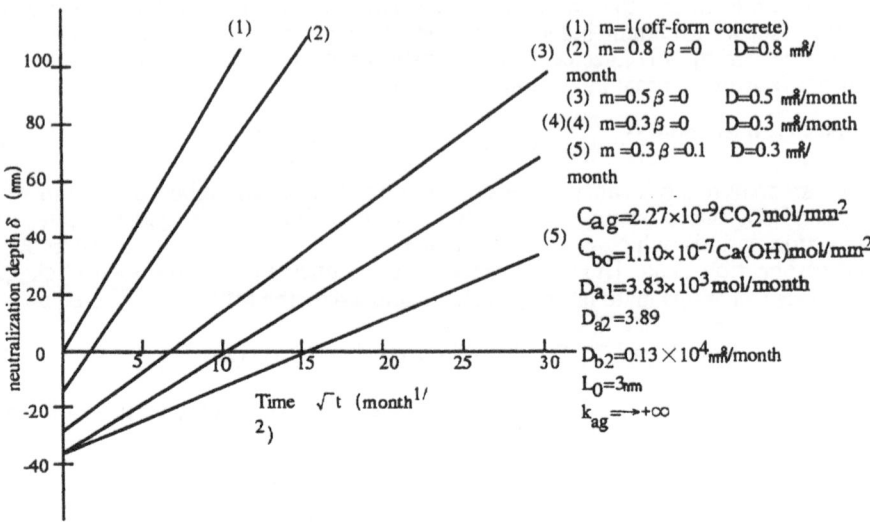

Fig. 3 Theoretical prediction of the progress pf carbonation of concrete with and without polymeric coatings

Table 2 Evaluation of the suppressive effects of polymeric coatings on carbonation of concrete (Part 1 least- square method analysis)

	Type of specimens	Equation obtained by least-square method				Correlation coefficient
	N-R- a	X=	0.65+14.45	$\sqrt{t}$	r=	0.980
	N-R- b	X=	-0.22+0.16	$\sqrt{t}$	r=	0.732
Normal concrete for field-mixed use	N-R- c	X=	-1.27+9.43	$\sqrt{t}$	r=	0.992
	N-R- d	X=	-7.09+6.90	$\sqrt{t}$	r=	0.998
	N-R- e	X=	-3.96+7.06	$\sqrt{t}$	r=	0.959
	N-R- f	X=	0.53+12.36	$\sqrt{t}$	r=	0.994
concrete for precasting use	N-R- a	X=	4.78+8.33	$\sqrt{t}$	r=	0.956
	N-R- a	X=	-0.91+13.03	$\sqrt{t}$	r=	0.996

Table 3 Evaluation of the suppressive effects of polymeric coatings on carbonation of concrete (Part 2 calculation of effective diffusion coefficients of carbon dioxide)

Types of surface finishing materials	parameters		Surface equilibrum m constant	induction Period t (month)	thickness L(mm)	duffusion coefficient D(mm^2/month)
	A (mm/month$^{1/2}$)	B (mm)				
a	14.5	-	1	-	-	3.89
b	-	-	-	-	-	-
c	9.4	1.3	0.65	0.14	1	0.494
d	6.9	7.1	0.48	1.06	3	0.27
e	5.9	4.5	0.41	0.58	5	0.72
f	7.1	4.0	0.49	0.32	5	0.82

3 Equivalent thickness of cover concrete

As described in the previous section, when surface treatment is done on concrete, the time when cover concrete becomes completely neutralized becomes much longer than the time in case no surface treatment is done, so that service life of reinforced concrete becomes lengthened.

The progress of neutralization of off-form concrete is described by the parabolic law not involving a constant term as follows:

$$X = A_0\sqrt{t} \tag{27}$$

The time when cover concrete(cover thickness;X_0) becomes completely neutralized(T_0) is given in the following equation.

$$T_0 = (X_0/A_0) \tag{28}$$

On the other hand, the progress of neutralization of concrete treated with a polymeric surface coatings is described by the following parabolic law involving a constant term, not depending whether or not the deterioration occurs.

$$X = A\sqrt{t} - B \tag{29}$$

In this case, the time when cover concrete becomes completely neutralized(T_1) becomes lengthened obeying the following equation.

$$T_1 = \{(X_0 + B)/A_0\} \tag{30}$$

This indicate that the following equivalent thickness of cover concrete can be set, taking into account the suppressive effects of polymeric surface coatings materials on the progress of neutralization.

$$X_0' = B + (A_0/A)X_0 \tag{31}$$

If we set this equivalent thickness of cover concrete in advance, we can discuss the progress of neutralization of concrete treated with surface finishing materials as if it were that of off-form concrete, obeying the following equation.

$$X_0' = A_0\sqrt{t} \tag{32}$$

The concepts of this equivalent thickness of cover concrete is summarized in Fig.3. On the other hand, the concrete values of equivalent thickness of cover concrete for many polymeric surface coatings materials were calculated from experimental results of accelerated carbonation test, as shown in Table 4.

Table 4 Calculation of the equivalent thickness of cover concrete for various types of polymeric coatings

Types of polymeric coating	parameters		equivalent thickness of cover concrete	
	A (mm/month$^{1/2}$)	B (mm)	X_0 (mm)	X_0' (mm)
no finish	14.5		30	40
vinyl wall paper	-	0	-	-
emulsion paint	9.4	1.3	47	62
synthetic resin plaster	6.9	7.1	70	91
acryl based sprayed finish	5.9	4.5	78	103
epoxy based sprayed finish	7.1	4	65	86

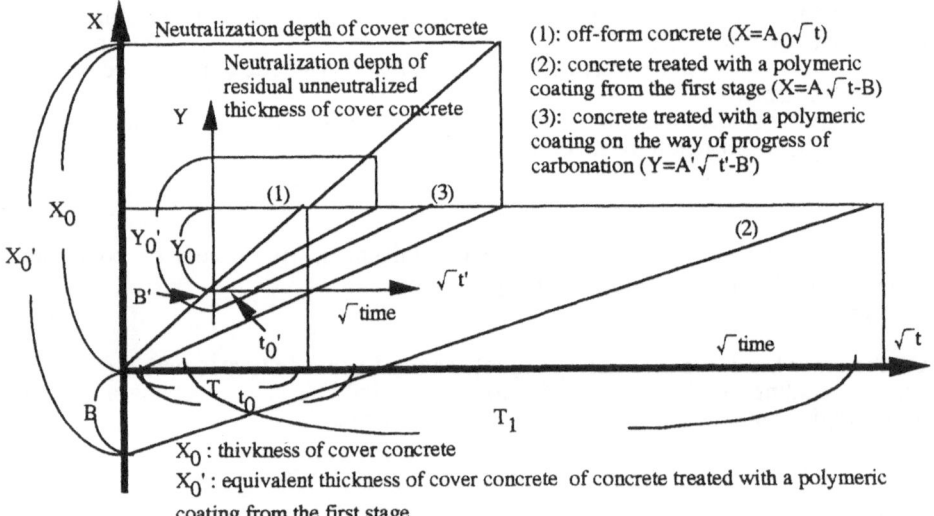

(1): off-form concrete $(X=A_0\sqrt{t})$

(2): concrete treated with a polymeric coating from the first stage $(X=A\sqrt{t}-B)$

(3): concrete treated with a polymeric coating on the way of progress of carbonation $(Y=A'\sqrt{t'}-B')$

X_0 : thivkness of cover concrete

X_0' : equivalent thickness of cover concrete of concrete treated with a polymeric coating from the first stage

Y_0 : unneutralized residual thickness of cover concrete

Y_0' : equivalent unneutralized residual thickness of cover concrete of concrete treated with a polymeric coating on the way of the progress of carbonation

Fig. 4 Schematic diagram of the concept of equivalent thickness of cover concrete taking into account of the suppressive effects of polymeric coatings on the progress of carbonation of concrete

4 Concluding Remarks

This paper deals with the influence of polymeric surface coatings materials on the progress of neutralization of concrete. The main results obtained in the present study are summarized as follows:

(1) The suppressive effects of polymeric surface coatings materials on the progress of carbonation of concrete can be well evaluated by two parameters of the parabolic law involving a constant term.

(2) There is a certain induction period characteristic of a polymeric surface coatings material before the progress of neutralization of concrete starts. The induction period closely correlates with the diffusion coefficient of carbon dioxide in the polymeric surface coatings material.

(3) After a certain period the neutralization depth of concrete decreases linearly with the thickness of polymeric surface coatings material.

(4) Polymeric surface coatings materials such as synthetic resin plaster, acrylic and epoxy based spray finish show the great suppressive effects on the progress of neutralization of concrete.

(5) The influences of the deterioration on the suppressive effects of polymeric surface coatings material are reflected as the increase in the coefficient concerning square root of time, but the progress of concrete treated with polymeric surface coatings materials also obey the parabolic law involving a constant term, not depending whether or not the deterioration occurs.

(6) Equivalent thickness of cover concrete was proposed, and the concrete values of equivalent thickness of cover concrete for many polymeric surface coatings materials were calculated from experimental results of accelerated carbonation test.

5. References

1. Kishitani, K.(1963), *Durability of Reinforced Concrete*. Publication Department of Kashima Corporation, Tokyo, Japan, (in Japanese).
2. Hamada, M.(1968), Neutralization(Carbonation) of concrete and corrosion of reinforcing steel. *Proc. 5th Interntl. Sym. Chemistry of Cement*, 3, Tokyo, Japan, 1968, pp. 343-369.
3. Mori, T. Shirayama, K. Kamimura, K. and Yoda, A. (1972). Carbonation of Portland blast-furnace slag cement concrete, *Cement & Concrete*, **307**, pp. 40-46 (in Japanese)
4. Fukushima, T. and Fukushi, I.(1985). Suppressive on neutralization of concrete and corrosion of reinforcement by surface finishing materials, *ibid.* **463**, pp.74-81 (in Japanese)
5. Fukushima, T.(1990), *Service Lives of Reinforced Concrete Buildings--Focusing on Carbonation of Concrete and Corrosion of Reinforcement--*. Gihodo, Tokyo, Japan, (in Japanese).
6. Fukushima, T. and Motohashi, K.(1985), Deterioration processes of polymeric materials and their influence on durability of reinforced concrete. *Proc. Interntl. Sym. Polymer and Its Control, St. Louis, U.S.A.; American Chemical Society Symposium Series*, **487**,pp.847-861.
7. Fukushima, T., Tomosawa, F., Fukushi, I. and Tanaka, H.(1986), Protective effects of polymeric finishes on the carbonation of concrete and corrosion of reinforcement. *Proc. RILEM Interntl. Sym. Adhesion between Polymers and Concrete--Bonding, Protection, Repair--, Aix en Provance*, France, pp.166-176.
8. Fukushima, T. Fukushi, I. and Tomosawa, F.(1992), Quantitative evaluation of suppressive effective of surface finishing materials on neutralization of concrete considering deterioration and method of establishing equivalent thickness of cover concrete. *JCA Proc. of Cement & Concrete.* **46**, pp.598-603 (in Japanese).
9. Fukushima, T. and Fukushi, I.(1994). Suppressive effects of surface finishing materials on neutralization of concrete and equivalent thickness of cover concrete, *Proc 2nd Joint Finland - Japan Workshop on Service Life and Maintenance of Buildings*, pp.6-1~6-12
10. Fukushima, T(1983). Deterioration processes of polymeric materials and their dependence on depth from surface. *Durability of Building Materials*, **1**, pp. 327-343

EFFECTS OF MOISTURE CONTENT AND PORE STRUCTURE OF SUBSTRATE CONCRETE ON ADHESIVE STRENGTH OF EPOXY COATING

N. Yuasa, Y. Kasai, I. Matsui, Y. Henmi and H. Sato
College of Industrial Technology, Nihon University, Chiba, Japan

Abstract
Effects of the moisture content and pore structure of substrate concrete on the separation of surface finishing, especially floor coatings, have been investigated. The adhesive strength employed to evaluate the separation resistance showed a dependency on the moisture content at a moisture content more than 6 % and on the pore structure at a moisture content less than 6 % . An initiation and development of swelling showed correspondence with the adhesive strength depending on the quality of the substrate concrete.
Keywords: Moisture content, pore structure, floor coatings, adhesive strength, separation, swelling, expoxy coating, substrate concrete

1 Introduction

Damages due to separation or swelling are occasionally observed the finishing on concrete subsequent to its application, which may be attributed to the quality of adhesives and finishing, the thermal environment and the quality of substrate concrete. However, neither the relationship between the quality of concrete and damages nor the concrete itself as a substrate of finishing has been fully investigated whereby the engineering consensus of this problem has not been attained except for a naive recognition that the lower moisture content results less damages.

We have made an extensive research on the variation of strength and durability of structural concrete from the surface to the inside with the help of a newly developed technique capable of measuring local moisture content [1][2][3]. It has been made clear that the moisture content of the skin concrete (the cover part close to the surface) decreases much more rapidly than that of the inside or average, and that the denseness of pore structure of the skin concrete is interfered substantially by drying and coarse

Polymers in Concrete, edited by Y. Ohama, M. Kawakami and K. Fukuzawa. Published in 1997 by E & FN Spon, 2–6 Boundary Row, London SE1 8HN, UK. ISBN: 0 419 22330 4.

pores remain for a long time under an insufficient curing condition.

Based on this knowledge, our previous study dealt with the variations of moisture content and pore structure of concrete as a substrate of finishing and its effect on the adhesive strength of epoxy type floor coatings whereas the influences of moisture content and pore structure cannot be separated [4][5].

This study will deal with the effects of pore structure and moisture content of concrete with different water - cement ratio and curing condition on the adhesive strength of an epoxy type floor coating, and these parameters will be associated with results of the accelerated swelling test.

2 Effects of moisture content and pore structure on adhesive strength

2.1 Effects of water - cement ratio and moisture content

2.1.1 Experiments

Effects of moisture content and pore structure on the adhesive strength of epoxy type floor coatings applied at various intervals on a concrete were reported in the previous paper whereas the effects of moisture content and pore structure cannot be separated [5]. We will reexamine this point in terms of the quality of substrate concrete, which is subjected to drying immediately after casting and has different pore structure associated with the change in water - cement ratio.

(1) Preparation of specimens
• Concrete: Specimen A for the adhesive strength test (15 x 15 x 50 cm) and specimen B for pore structure measurement (Ø10 x 15 cm) were prepared as shown in **Fig.1** on the basis of mix proportions given in **Table - 1**.
• Floor coatings: An epoxy type floor coating without preliminary surface treatment was applied   for 2.0 kg/m²
at the age of 1, 7, 14 and 28 days.

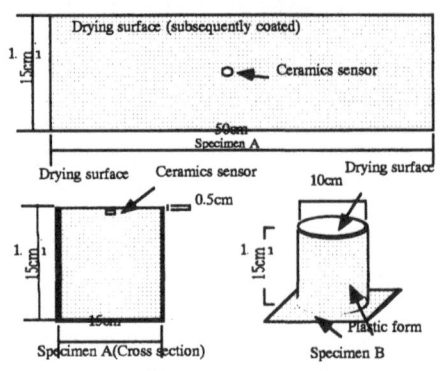

Fig. 1. Details of specimens

Table. 1. Mix proportions and properties of substrate cocretes

W/C (%)	Unit Water (kg/m³)	Unit mass (kg/m³)			Admixture (cm³/m³)			Properties of concrete			
		Cement	Fine aggregate	Coarse aggregate	No.70	SP-8N	No.303A	Mixing Temp. (°C)	Slump (cm)	Air (%)	Comp. strength (MPa)
40	185	463	713	957	-	4630	20	20.0	21.9	5.0	37.1
60	185	308	838	957	770	-	38	20.5	19.8	5.0	26.4
80	185	231	908	957	578	-	40	19.5	21.4	4.7	15.4

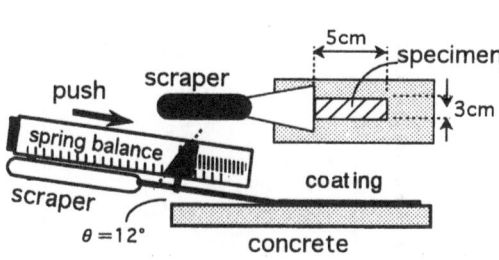

Fig. 2. Testing of adhesive strength

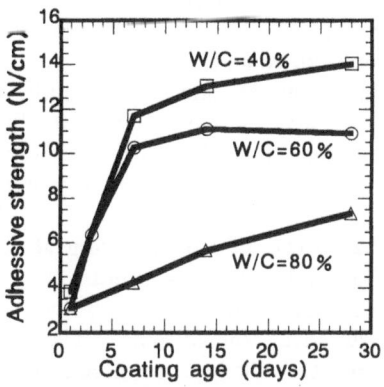

Fig. 3. Adhesive strength and coating age

(2) Method of testing
• Moisture content: A ceramic moisture sensor (Ø10 x 5 mm) embedded at the depth of 0.5 mm from the open surface of the specimen A was used.
• Pore structure: Specimens were taken from a part of 1 cm from the surface of the specimen B, conditioned to a grain size from 2.5 to 5 mm, treated by acetone and D-dried. Mercury injection technique was used for pore structural measurement and the effective pore volume was determined in accordance with the aggregate content measured using muriatic acid.
• Adhesive strength: Maximum load in kgf needed to scrape off the coating as shown in **Fig. 2** was measured for the specimen B at the age of 14 days after

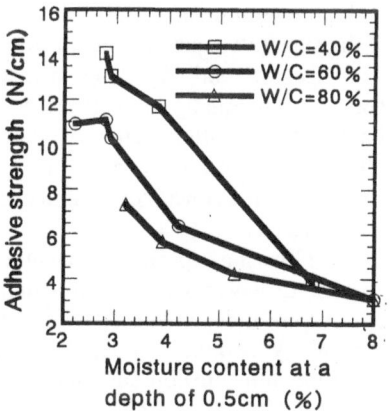

Fig. 4. Adhesive strength and moisture content

coating, and the adhesive strength was calculated by dividing the maximum load by the width of the coating. Recorded adhesive strength is a mean value of 12 measurements.

2.1.2 Results and discussions
Relationship between adhesive strength and ages when coating is applied, is shown in **Fig.3**. The adhesive strength shows greater value as the water - cement ratio becomes smaller, and increases gradually for a specimen with water - cement ratio of 80 % as the coating period after mixing becomes longer while specimens with water - cement ratios of 40 and 60 % show little development.
Relationship between adhesive strength and moisture content of the substrate concrete is shown in **Fig.4**. Moisture content was measured just before coating at a depth of 0.5 cm from the surface. A similar result as reported in the previous paper was obtained, however, it is also shown that the adhesive strength cannot be evaluated solely by moisture content, and water - cement ratio is of substantial importance.
Relationship between adhesive strength and effective pore volume of the substrate concrete is shown in **Fig.5**. The adhesive strength shows quasi linear increase with

decrease of effective pore volume of the substrate concrete even though the moisture content is varied. In order to have a greater adhesive strength, it is necessary to have not only lower moisture content but also well - cured concrete with a low water - cement ratio.

2.2 Effects of sealed curing at early ages and moisture content

2.2.1 Experiments

We have pointed out that adhesive strength is likely to be affected by the pore structure other than moisture content of substrate concrete. However, both pore volume and moisture content of the substrate concrete decreased with their age after mixing as each mix of concrete was equally subjected to drying immediately after setting, whereby effects of moisture content or pore structure on the adhesive strength cannot be separated.

In the following experiments, an attempt is made to have variable pore structure of concrete independent to its moisture content by applying coatings at just specified moisture content which can be monitored by the embedded ceramics sensor. Different pore structure can be obtained by changing the time of drying as reported in the previous study. In this

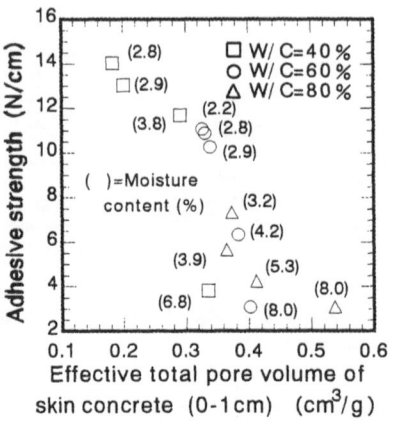

Fig. 5. Adhesive strength and effective total pore volume

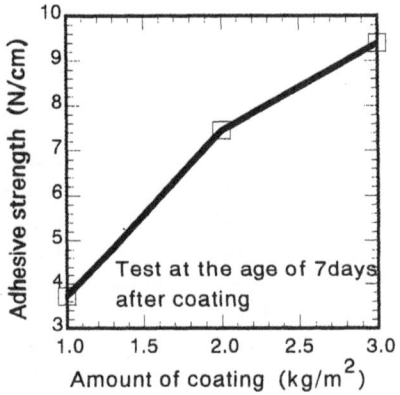

Fig. 6. Adhesive strength and amount of coating

way, the effects of moisture content and pore structure on adhesive strength can be discussed in terms of the significant test of the variance analysis [3].

(1) Preparation of specimens
• Concrete: Specimen A and B were prepared with a water - cement ratio of 60 % on the basis of mix proportion given in Table - 1, and the top surface was subjected to drying from the age of 0, 1, 7, 14 and 28 days under 60 % of relative humidity.
• Floor coatings: An epoxy type floor coating was applied for 1.0 kg/m² without preliminary surface treatment at the moment when moisture content at 0.5 cm below the surface of the substrate concrete becomes 8, 6, 4 and 2 % .

The amount of coating is different from that used in the previous experiment described in section 2.1. Relationship between the adhesive strength and the amount of coating is proportional provided that the quality of the substrate concrete is constant as shown in **Fig. 6**.

(2) Method of testing

All measurement were executed in the same way as the previous experiment described in section 2.1.

2.2.2 Results and discussions

Effects of drying initiation on the relationship between adhesive strength and moisture content are shown in **Fig.7**. Adhesive strength is affected more greatly by the age of drying initiation than moisture content, and increases as the drying initiation becomes later.

Relationship between effective total pore volume and adhesive strength is shown in **Fig. 8**. In spite of the difference in moisture content, the adhesive strength is apparently dependent to the effective total pore volume similar to the results shown in section 2.1 except for specimens with a moisture content of 8 %.

A two - way layout variance analysis of adhesive strength with respect to pore structure parameter and moisture content was executed and the significant difference was analyzed. As shown in the previous study, the pore structure of concrete is affected by the age of initiation of drying whreby the age of initiation of drying was employed as a pore sturcture factor, while the nominal moisture content was employed as a moisture contente factor.[3][5].

The results are shown in **Table. 2**. These two parameters show the significant difference at 1 % of significance level. If the data of 8 % of moisture content are excluded, the moisture content becomes less significant with a significance level of 5 % , while the pore structure is still significant holding 1 % of the significance level. Another analysis executed for moisture range between 8 to 6 % recognized a significance in moisture content at 5 % of the significance level, while pore

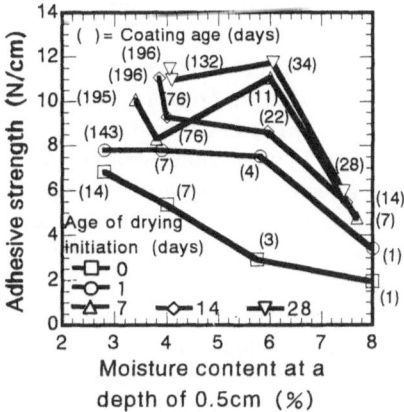

Fig. 7. **Adhesive strength and moisture content**

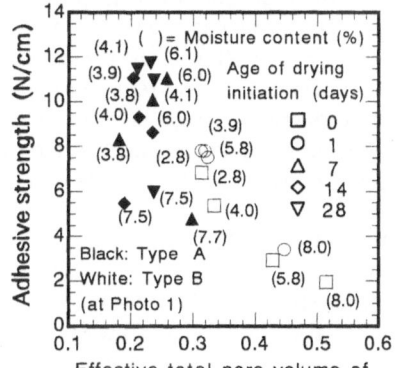

Fig. 8. **Adhesive strength and effective total pore volume**

Table. 2. Results of variance analysis

Case	Cause of variation	Variation	Degree of freedom	Variance	Variance ratio
All level	Moisture	0.800	3	0.267	22.4 **
	Pore	0.825	4	0.206	17.3 **
	Error	0.143	12	0.012	-
	Total	1.769	19	-	-
Moisture content of 8%	Moisture	0.041	2	0.021	1.4
	Pore	0.739	4	0.185	12.5 **
	Error	0.118	8	0.015	-
	Total	0.898	14	-	-
Moisture content of 6 & 8%	Moisture	0.428	1	0.428	18.1 *
	Pore	0.527	4	0.132	5.6
	Error	0.095	4	0.024	-
	Total	1.050	9	-	-

** significant difference at a significant level of 1 %
* significant difference at a significant level of 5 %

structure showed no significant difference.

From the above discussions, it can be made clear that a critical moisture content may exist in a moisture range more than 6 % where adhesive strength can be affected by moisture content whereas pore structure can exhibit a significant influence at moisture content less than 6 %.

Relationships between pore structure and adhesive strength at the moisture range less than 6 % are shown in **Fig.9** taking effective total pore volume and median pore radius as a measure of pore structure. The median pore radius can be defined as a pore radius

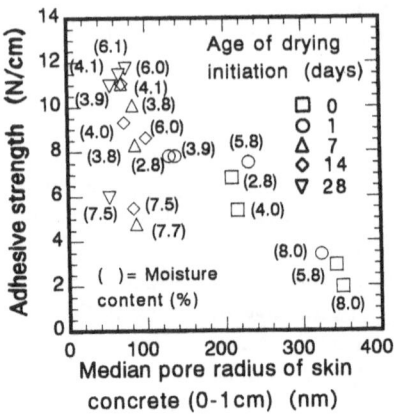

Fig. 9. Adhesive strength and median pore radius

that corresponds to 50 % of the accumulated pore volume.

A set of linear equations representing the relationships between pore structure and adhesive strength is derived in terms of AS: adhesive strength in N/cm, ETP: effective total pore volume in cm³/g and Me: the median diameter in nm. The single correlation coefficient, r, between ETP and Me was 0.94.

$$AS = -35.1 \times ETP + 18.2 \qquad\qquad r = 0.89 \qquad (1)$$
$$AS = -0.0273 \times Me + 12.3 \qquad\qquad r = 0.92 \qquad (2)$$
$$AS = -0.0614 \times (ETP \times Me) + 11.2 \qquad r = 0.91 \qquad (3)$$
$$AS = -10.1 \times ETP - 0.0201 \times Me + 14.1 \qquad r = 0.92 \qquad (4)$$

Interfacial morphology after the adhesive strength test is shown in **Photo. 1**. The fracture type can be grouped in A where separation is made just at interface of a coating and substrate concrete and B where a thin layer of hardened cement paste attached to coatings. These morphology has been observed in practice and is also noted in Fig. 8. This difference may be attributed to the initiation of drying and not to moisture content and the fracture type B can be seen when the sealed curing at early ages is insufficient. Correspondingly all fracture types of experiments presented in section 2.1 were type B as all the specimens were subjected to drying immediately after setting. It may be concluded that the interfacial fracture type is dominated by

Fracture type A   Fracture type B
Photo. 1. Fracture types after adhesive strength test

curing conditions at early ages and independent to water - cement ratio or moisture content of the substrate concrete.

3 Effects of the quality of substrate concrete on the swelling of finishing

3.1 Experiments
As a precursor of separation, swelling will be examined with regards to the quality of the substrate concrete by the accelerated test method. The adhesive strength will also be compared with the accelerated swelling and with the results obtained in section 2 of this study.

(1) Preparation of specimens
• Concrete: Specimen C for the swelling test (Ø20 x 15 cm) was prepared with a water - cement ratio of 60 % on the basis of mix proportions given in Table - 1, and cured in 20 °C under 60 % of relative humidity. The surface was troweled for coating.

• Floor coatings: An elastic epoxy type floor coating, to which a soluble component was admixed to initiate swelling, was applied for 1.0 kg/m^2 at the age of 1, 7, 14 and 28 days without preliminary surface treatment.

(2) Method of testing
• Accelerated swelling test: A method proposed by K. Tanaka et al. [7] was used. Specimens at the age of 14 days after coating were put in water bath of 30 °C with a depth for 14 cm, while all set-up including coated surface was kept in a constant temperature of 20 °C. Swelling was copied, scanned and images were analyzed by a computer.

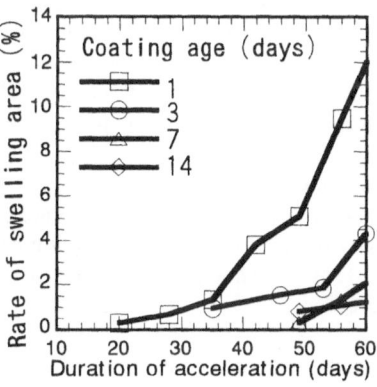

Fig. 10. Swelling area and the duration of acceleration

3.2 Results and discussions
Relationship between the duration of acceleration and swelling area is shown in **Fig. 10**. Swelling could not be observed for specimens with the coating age of 28 days after casting though the acceleration was applied for 60 days. As the coating age becomes earlier, the swelling initiates earlier and its area becomes larger. These results correspond to that of adhesive strength test, which is compared to the swelling area and shown in **Fig. 11**.

It is shown that the smaller the adhesive strength is, the larger the swelling area

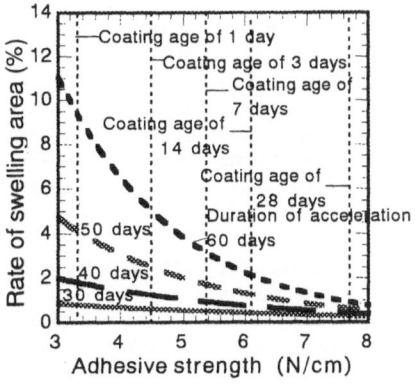

Fig. 11. Swelling area and adhesive strength

is. This tendency is more conspicuous with the longer duration of acceleration. The above findings may lead to a realistic quality control of substrate concrete for floor coatings in terms of prevention of swelling, though more discussion is needed to be an established method.

4 Concluding remarks

Main findings can be summarized as follows:

(1) Adhesive strength becomes greater as the water - cement ratio of the substrate concrete becomes smaller.

(2) Adhesive strength becomes greater as the duration of sealed curing becomes longer.

(3) Adhesive strength decreases at a moisture content of the substrate concrete more than 6 %.

(4) Adhesive strength depends on pore structure of the substrate concrete when coating is applied at a moisture content of the substrate concrete less than 6 %.

(5) The morphology of separation depends on neither water - cement ratio nor moisture content, while it is sensitive to curing condition at early ages.

A specimen subjected to drying at early stage shows a separation of coating material accompanied by thin and weak cement paste layer.

(6) Initiation and development of swelling of coating depends on the quality of the substrate concrete, and corresponds to adhesive strength.

Acknowledgments

The authors would like to acknowledge the advice and support of Prof. K. Tanaka at Tokyo Institute of Technology, members of his laboratory and ABC Shokai Laboratory. This study forms a part of the Grant-in-Aid for Scientific Research (C) in 1994 and 1995 fys.

This paper was also contributed to Proceedings of the Japan Concrete Institute Vol.18.

References

1. Kasai, Y., Matsui, I. and Yuasa, N. (1993) Measurement of moisture content of concrete by embedded ceramics sensors. Proc. 20 the Cement and Concrete Workshop, pp. 7-12.
2. Yuasa, N. Kasai, Y. and Matusi, I. (1994) Study of the quality of the cover concrete (Variation of moisture distribution and pore structure during drying) , Prep. Annual Meeting of AIJ, pp. 449-450.
3. Yuasa, N. Kasai, Y. and Matusi, I. (1994) Study of the quality of the cover concrete (Variation of pore structure of concrete subjected to drying) , Prep. Annual Meeting of AIJ, pp. 199-200.
4. Yuasa, N. Tanaka, K., Asami, T., and Hashida, H. (1994) Moisture distribution and pore structure of the substrate concrete for floor coatings, Prep. Annual Meeting of JCI, Vol. 16, No. 1. pp. 675-680.
5. Yuasa, N. Kasai, Y., Matusi, I. and Henmi, Y. (1994) Moisture content, pore structure and adhesive strength of concrete covered by epoxy coatings, Prep. Annual Meeting of JCI, Vol. 17, No. 1. pp. 695-700.
6. Sato, H., Kasai, Y., Matsui, I., Henmi, Y. and Yuasa, N. (1994) A test method for the adhesive strength of coatings, Proc. 27 th Annual Conference of Nihon Univ. College of Industrial Technology, pp. 13-16.
7. Omori, O., Tanaka, K., and Uchida, M. (1995) Study on the swelling of epoxy floor coatings on the substrate concrete, Proc. 65 th workshop of Kanto branch of AIJ, Structure and Materials division, pp. 169-172.

FROST RESISTANCE OF POLYPROPYLENE FIBER REINFORCED LIGHT-WEIGHT POLYMER MODIFIED MORTAR

Y.S. Soh, S.Y. Soh and H.S. Soh
Dept. of Architectural Engineering, Chonbuk National University, Chonju, Korea

Abstract

The frost resistence of light-weight concrete is lower than that of plain concrete because of the higher water absorption. Therefore, to prevent the frost damage of light-weight concrete, it is necessary to lower the water asorption and to strengthen cement concrete matrixes.

This study is carried out to examine the effects of polymer modification and reinforcing with polypropylene(PP) fiber on the frost resistance of the light-weight concrete. From the test results, the rate of the length change of specimens tends to decrease with an increase in the polymer cement ratio and increase with increasing PP fiber amount. The rate of the weight loss of the polymer modified and fiber mixed specimens increases at a slower rate compare to that of plain specimen, and the number of freezing and thawing cycles of specimens are gradually decreased regardless of polymer modification and reinforced fiber amount.

Therefore, it is found that the frost resistance of mortars improved by polymer modification and reinforcing with PP fiber.

Keywords : Frost resistence, Water absorption, Light-weight concrete, Polymer modification, PP fiber

Polymers in Concrete, edited by Y. Ohama, M. Kawakami and K. Fukuzawa. Published in 1997 by E & FN Spon, 2–6 Boundary Row, London SE1 8HN, UK. ISBN: 0 419 22330 4.

1 Introduction

The frost damage of concrete is one of the general causes of deteriorating the concrete durability[1]. Particulity, the light-weight concrete is weak in freezing and thawing because of the higher absorption.

Recently the light-weight concrete has widely been used for the trend of buildings. The manufacturing processes of light-weight concrete are ; one is made by replacing partial cement with some kinds of admixture(like ash), some others by using of light-weight aggregates(like perlite). In the process, the strength and toughness of light-weight concrete are lower than those of plain concrete. And the resistence of freezing and thawing of light-weight concrete is low compared to that of plain concrete because of the higher absorption due to much air void.

Therefore, to prevent the frost damage of light-weight concrete, it is necessary to lower the water absorption and to strengthen cement concrete matrixes.

In this study, many researchers have attempted to make concrete lighter by using perlite and sludge ash. And Poly-acrylic ester emulsion(PAE) and PP fiber are used to improve the frost resistance of the light-weight concrete. This study is carried out to examine the effects of the polymer modification and reinforcing PP fiber on the frost resistance of the light-weight concrete.

2 Materials

2.1 Cement

Ordinary portland cement as specified in KS(Korean Standard) L 5201 was used for all specimens. The Chemical properties of ordinary portland cement are given in Table 1.

Table 1. Chemical properties of ordinary portland cement

Chemical Component	SiO_2	Al_2O_3	CaO	Fe_2O_3	MgO	SO_3	K_2O	Ig.loss	C_3S	C_2S	C_3A	C_4AF
Component Ratio(%)	21	6	62.1	2.8	3.4	2.0	1.2	1.7	43.1	27.9	11.2	8.5

2.2 Filler

The paper sludge ash used as filler was from the paper industries in Chonbuk, Korea. The chemical properties of the paper sludge ash are shown in Table 2.

Table 2. Chemical properties of ash

Chemical Component	SiO_2	Al_2O_3	Fe_2O_3	MgO	Ig.loss	Total
Component Ratio(%)	26	51.6	6.3	6.9	7.99	99.62

2.3 Fiber

A polypropylene(PP) was used as fiber in this study. The physical properties of fiber are shown in Table 3.

Table 3. Physical properties of polyproplene fiber

Length (mm)	Diameter (μm)	Specific Gravity	Melting point (℃)	Tensile strength (kgf/cm²)
19	0.6–0.8	0.91	160	5.278

2.4 Polymer dispersions

A poly-acrylic ester emulsion(PAE) was as Polymer dispersion. The physical properties of the polymer dispersions are shown in Table 4.

Table 4. Properties of PAE polymer

State	Color Tone	Solid Content (%)	pH	Viscosity (cP. 20℃)	Specific Gravity
Liquid	White	47	8	34.6	1.054

2.5 Aggregate

The perlite was used for light-weight aggregate. The physical and chemical properties of perlite are given in Table 5.

Table 5. Chemical and physical properties of perlite

Chemical Component	SiO$_2$	Al$_2$O$_3$	CaO	Fe$_2$O$_3$	K$_2$O	MgO	Na$_2$O
Component Ratio(%)	71-75	12.5-18	0.5-2	0.5-1.5	4-5	0.1-0.5	2.9-4

State	Softing Point (℃)	Melting Point (℃)	pH	Specific Gravity	Refractive Index
Solid	870-1,090	1,200-1,340	6.6-8.0	0.2	1.5

3. Testing methods

3.1 Preparation specimens

Polymer cement mortar were mixed according to KS F 2476(Making Method of Test Sample of Polymer-Modified Mortar in the Laboratory). Mixing proportions given in Table 6 were a replacement ratio of paper sludge ash to weight of cement and perlite was added to the composites in a ratio of 70%(by volume) of them. Polymer-cement ratios(calculated on the basic of total solids in polymer dispersions) of 0 and 10%, and polypropylene fiber of 0, 1 and 2%(by volume) were added to the composites. Their flow was adjusted to be constant at 180±5. The mixing times were 1 min. for dry blend with water and polymer, 2min. for a secondary blend with PP in the total of 5 min. on the average. Mortar specimens 80×100×400mm were molded and then subjected to a 7-day-20℃-50%R.H-dry cure.

Table 6. Mix proportions

Symbol	Ash (wt.%, by cement)	Perlite (vol.%, by composite)	PAE (wt.%, by cement)	Anti-foaming Agent(%)	PP (vol. %, by composite)	W/C (%)	Flow
UM*-O					0.0	75.4	175
-1			0	0	1.0	79.7	177
-2					2.0	85.8	175
	50	70					
M**-O					0.0	41.9	183
-1			10	2	1.0	44.2	180
-2					2.0	47.9	175

*UM : Un-modified Mortar, **M : Polymer-modified Mortar

3.2 Freezing and Thawing test

Freezing and thawing test of specimens was conducted in a temperature range of -18℃ to 4℃ according to ASTM 666-80(Standard Test Method for Resistance of Concrete to Rapid Freezing and Thawing) and KS F 2456(Standard Test Method for Resistance of Concrete to Rapid Freezing and Thawing). In order to evaluate the freez-thaw durability, the relative dynamic modulus of elasticity, length change and loss were checked at every 30 cycles to 300 cycles, respectivity. Length change, weight loss, relative dynamic modulus of elasticity and durability factor were calculated as following equation.

$L_n = (l_n - l_o)/l_o \times 100$
where,
L_n : Length change of the specimen at N cycles of and thawing, %
l_o : Length comparator reading at 0 cycle, mm
l_n : Length comparator reading at N cycles, mm

$R_n = (W_n - W_o)/ W_o \times 100$
where,
R_n : Weight loss of the specimen at N cycles of freezing and thawing, %
W_o : Weight of the specimen at 0 cycle of freezing and thawing, g
W_n : Weight of the specimen at N cycles of freezing and thawing, g

$P_n = (f_n^2/f_0^2) \times 100$
where,
P_n : Relative dynamic moldulus of elasticity at N cycles of freezing and thawing, %
f_o : Fundamental transverse frequency at 0 cycle of freezing and thawing, Hz
f_n : Fundamental transverse frequency at N cycles of freezing and thawing, Hz

$DF = P \times N/M$
where,
DF : Durabliity fector

P : Relative dynamic moldulus of elasticity at N cycles of freezing and thawing, %

N : Cycles of freezing and thawing below 60% of P., cycle

M : Cycles at end of freezing and thawing.(300cycles in this study), cycle

4 Results and discussion

The length change of specimens through freezing and thawing cycles test are illustrated in Fig. 1. The rate of the length change of specimens tend to decrease with polymer modification and increasing amounts of the PP fiber.

This phenomena may be attributed to effects of polymer and fibers which decrease water-cement ratio, and increase toughness of matrix[2),4)]. In other words, the freezing-thaw resistance of specimens improved by polymer modification and reinforcing with PP fiber.

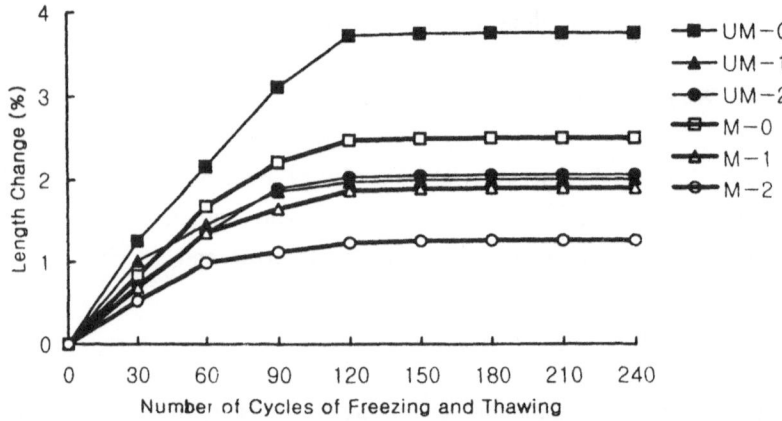

Fig. 1 Relationship between length change and number of freezing and thawing cycles

Fig. 2 Indicates the weight loss of specimens through freezing and thawing cycles test. The weight loss of the polymer-modified and reinforcing with PP fibers specimens incresed at a slower rate compared to that of plain specimens. It is considered that the improvement in the weight loss is due to the effect of co-matrix including the polymer films which formed in the interfaces between the hydrates and aggregates[4].

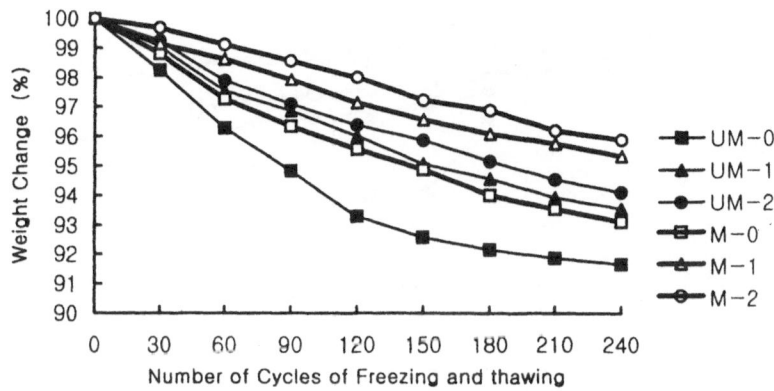

Fig. 2 Relationship between weight change and number of freezing and thawing cycle.

The variations of the relative dynamic moduls of elasticity of specimens are exhibited in Fig. 3. Relative dynamic moduls of elasticity of specimens was gradually decreased regardless of polymer modification and reinforcing with fiber as the increase of the number of freezing and thawing. The relative modulus of dynamic elasticity of PP fiber reinforced light-weight polymer-modified mortar is high range of 85 to 100%, compared to that of plain mortar with 65 or lower at freezing and thawing cycles 240.

It is indicated that the frost resistance of light-weight mortar is improved by polymer modification and reinforcing with PP fiber.

The durability factor of specimens are showed in Fig. 4. The durability factor was increased with increasing the PP fiber content and using of PAE. Generally, It is known that the microstructrures of mortar and concrete are gradually deteriorated by the expansion pressure developed by the frost of capillary pore water. From this point of view, it is concidered that an improvement in the frost resistance of polymer-modified mortar with PP fiber is attributed to excellent watertightness due to the polymer films formed in the matrix, decreased water-cement ratio and increased toughness of matrix[4].

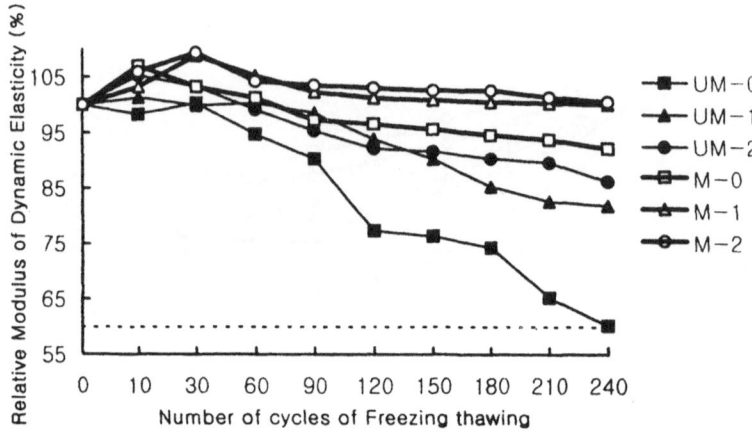

Fig. 3 Relationship between relative modulus of dynamic elasticity
and freezing and thawing cycle

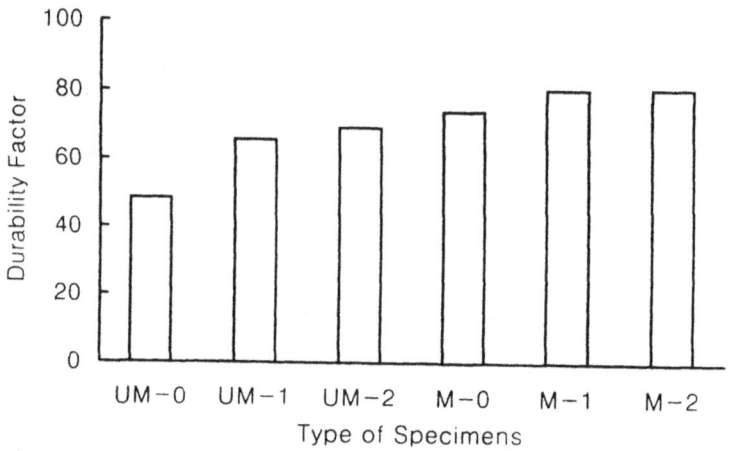

Fig. 4 The durability factor of specimens
(7 day-20℃-50% R.H dry cure)

5 Conclusion

The rate of the length change of specimens by expansion tended to decrease
with polymer modification and increasing of the PP fiber amount. The weight

loss of polymer-modified and fiber mixed specimens increased at a slower rate compared to plain specimens. The relative dynamic moduls of elasticity of polymer modified mortars are decreased regardless of polymer modification and reinforcing with fiber as the increase of the number of freezing and thawing compared to plain specimens. The durability factor was increased with increasing amounts of the PP fiber and polymer modification.   It was found that the frost resistance of mortars are improved by polymer modification and reforcing with PP fiber.

6 References

1.   Okada,K. (1986) Durability of concrete, pp.17-47
2.   Mashima, M.D., Hannant, J. and Keer, J.G. (1990) Tensile properties of polypropylene reinforced cement with different fiber orientations, ACI materials journal, V.87, No.2, pp.172-178.
3.   Kosa, K., Naaman, A.E. and hansen, W. (1991) Durability of fiber reinfoced mortar. ACI materials journal, V.88, No.3, pp.310-319.
4.   Daniel Bordeleau, Michel Pigene, and Nemkumar Banthia, (1992) Comparative study of Latex-modified concretes and normal concretes subjected to freezing and thawing in the presence of a deicer salt solution, ACI materials journal, V.89, No.6, pp.547-553.

PERFORMANCE OF CONCRETE-POLYMER COMPOSITES IN THE HOT CLIMATES OF IRAN

M. Damghani, B. Mashouf and L. Niayesh
Civil Engineering Dept., Tehran-Boston Consulting Engineers,
Tehran, Iran

Abstract
In this paper the climatic conditions in the hot and humid regions of Iran,the state of the ordinary concrete structur -es in these regions, and the diagnostic tests for concrete deteriorations are investigated . As ordinary concrete is undurable in these regions, concrete with polymer compounds has been used in a number of the construction projects.The design and construction of the polymer concrete structures as well as the test results and recommendations made in this connection are discussed in this paper.
Keywords: Chemical attacks, climatic conditions, durability , latex-modified concrete, permeability.

1 Introduction

The environmental conditions of the regions of Iran have caused severe damages in the concrete structures of these regions. From the point of view of the concrete technology, the hot climates of Iran are divided into two categories: 1)hot and dry; 2)hot and humid. Recent investigations have shown that the concrete structures in the hot and humid regions are prone to more severe deteriorations.

Having studied the performance of the concrete structures in the hot regions and carried out a number of tests,the authors of this paper have specified the types of the damages inflicted on these structures . The ingress of the chemical agents such as chlorides and sulphates , which exist in these regions , assisted with the physical conditions and the high permeability of ordinary concrete have caused the deterioration of the structures.

Hence, a study was made of using polymer concrete in order to increase the durability of concrete structures and reduce their permeability.On the completion of the plan for

Polymers in Concrete, edited by Y. Ohama, M. Kawakami and K. Fukuzawa. Published in 1997 by E & FN Spon, 2–6 Boundary Row, London SE1 8HN, UK. ISBN: 0 419 22330 4.

the use of polymer-modified concrete and the related tests , it was applied to a number of concrete structures in the region.The results of the executed works and the tests have demonstrated the high performance of this concrete in the hot and humid regions of Iran [5].

2 The environmental conditions of the regions

The northern coastal region of the Persian Gulf is one of the most severe parts of the world for concrete durability. The climatic condition of the region is characterized by the main environmental factors explained below:

1. The daily and annual temperature fluctuations and the maximum temperature during the summer are rather high; temperature of up to 70 degrees centigrade in the sun has been reported [3].
2. Because of its shallow depth and high rate of transpiration, the salinity of the Persian Gulf water is much higher than that of the free sea water. Affected by the Gulf water, the ground water in the coastal regions of the Gulf contains high amounts of chlorides and sulphates. The top soil is also permanently contaminated with these chemicals due to the capillary rise of the water [1].
3. The humidity of the region is high because of its hot weather and the high rate of transpiration in the Persian Gulf [3].

3 The concrete structures of the region

The concrete structures of the region are exposed to the harsh environmental conditions outlined above. As explained below, these structures can be divided into two groups depending on their locations and the types of deterioration.

1. The structures which are located relatively far from the Gulf shore,but are exposed to chemical attacks because of the permeability of the ordinary concrete they are made of and the ingress of the corrosive agents existing in the soil and the ground water. The corrosion due to chloride attack is the most common type of deterioration in these structures . The diffusion of chlorides has caused the corrosion of the reinforcement in these structures and reduced their stability [1].
2. The structures in the harbour installations,which are in direct contact with the Gulf water.Being continuously in the splash zone , these structures are exposed to the sulphate and chloride ions existing in the Persian Gulf water . The permeability of the concrete in these structures has also been a factor causing the corrosion of the concrete especially in the splash zone.The process of deterioration has ofcourse been intensified by the erosion of concrete due to the impact of waves[1].

4   Preliminary tests, observations and the analyses

The following tests and observations were carried on several damaged structures in the region in order to establish the causes of their deterioration[5]:

1. Classification of the deteriorations in terms of their types and quantities.
2. Measurement of the compressive strength using the core tests ,the schmidt hammer ,and the ultrasonic pulse velocity measurement.
3. Measurement of the depth of the invisible concrete laminations,large cracks and pores using the ultrasonic pulse velocity test.
4. Determination of the density and the porosity of concrete.
5. Determination of the diffusion of sulphates in hardened concrete samples and plotting the results (Fig. 1).
6. Determination of the diffusion of chlorides in hardened concrete samples and plotting the results (Fig. 2).
7. Measurement of the depth of carbonation.
8. Chemical tests on the mix water to determine its PH value and control its sulphate and chloride contents.
9. Chemical tests on aggregates to determine their sulphate and chloride contents.
10. Determination of the degree of the alkali-silica reaction of the aggregates.

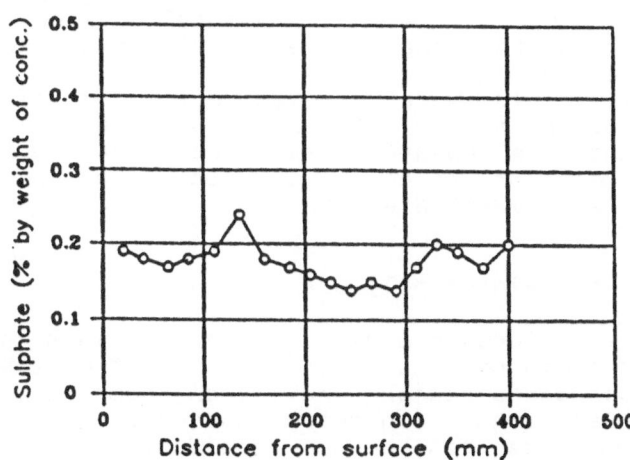

Fig. 1. Penetration of sulphates into the concrete.

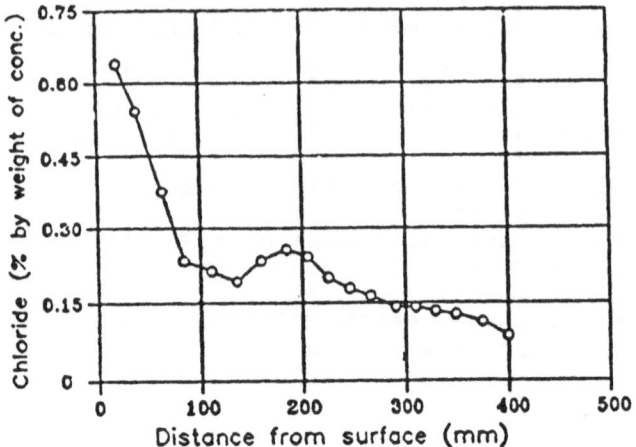

Fig. 2. Penetration of chlorides into the concrete .

It has been concluded from the analyses of the results that: firstly, the compressive strengths of the concretes tested have generally been adequate, and secondly, most deteriorations have been caused by chloride attacks which increase the volume of reinforcement resulting in the spalling of the concrete. The deteriorations have been caused inspite of the sufficient cover to reinforcement, which was up to 10 cm in some cases. Figures (1) and (2) show the diffusion of sulphates and chlorides into the concrete. In fact, as ordinary concrete of high permeability has low resistance to the environmental chemical attacks , it becomes clear that the permeability of concrete should be minimized[5].

Tests 7 to 10 were carried out to investigate and determine other probable factors causing deterioration of concrete, and the results have shown that these factors didn't play a major role in the deterioration process of the structures investigated [5].

5 Concrete containing polymers

As mentioned before, ordinary concrete is not durable in the hot and humid regions of Iran, and most of the concrete structures in these regions have been damaged. Classified as a "High Performance Concrete", the concrete containing polymer compounds has been considered as a suitable building material in the structures of these regions for its interesting properties.

Cement, water, and "Latex" (a polymer) is used in this type of concrete. Polymers used in concrete are made of minute spherical particles suspended in the mix water by the active agents in the mix. They waterproof the hardened concrete by increasing the cohesion in the concrete mix. During the hydration process, the polymer particles are bonded together and form a continuous polymer film which greatly reduces the porosity of the concrete and therefore its permeability. Tests have shown that polymers reduce the voids with a diameter of 0.2 micron and increase those with a diameter of 0.075 micron. The reduction in the porosity of concrete greatly increases its durability and resistance to chloride ions, CO_2 gas, and mineral acids. As the hydration process continues, the polymer particles are binded with one another due to the capillary action and the loss of water and form a polymer film between the aggregates. The elastic properties of the polymer itself increase the cohesion between the concrete constituents and prevent the formation and development of microcracks due to external stresses [2][5].

The electrical conductivity of the polymer concrete is also less than that of ordinary concrete and its electro-chemical reactions in the hardened state are minimized . Table 1 shows the comparison of the permeability and electrical conductivity between various types of concrete investigated [5].

In general, the addition of polymer to concrete reduces its porosity and microcracks, changes its micro-structure, improves the cohesion of the concrete mix ingredients, and decrease the electrical conductivity of the concrete [1].

Table 1. Comparison of permeabilities to chloride ions between different types of concrete in terms of the electric charge flowing through them

Type of concrete	Permeability to chloride ion	Electric charge (coulomb)
Ordinary concrete with water-cement ratio higher than 0.60	High	4000
Ordinary concrete with water-cement ratio of 0.40 to 0.50	Medium	2000-4000
Ordinary concrete with water-cenment ratio less than 0.40	Low	1000-2000
"Latex" polymer-modified concrete	Very Low	100 -1000
Polymer - impregnated concrete and polymer-concrete	Negligible	100

6 Mix design and tests

In order to produce a polymer-modified concrete, preliminary laboratory tests were carried out on samples of different mix proportions resulting in a final mix design, the proportions of which are shown in Table 2 . 4-6 liters of superplasticizer were added to the mix in order to reduce water and achieve adequate workability. Various polymers such as "Poly vynil Astat" (PVA) were used in the test, but the polymer finally chosen was Styrene-Butadiene rubber (SBR) for its availabilty. Before casting the samples, the aggregates were physically and chemically tested to control their hardness, size, and chloride and sulphate contents[5]

The following results were obtained from the tests carried out on the samples of polymer-modified concrete produced in the laboratory[5]:

1. An average compressive strength of 35 MPa on cylindrical samples.
2. An average tensile strength of 1 MPa.
3. A reduction of 10% in the modulus of elasticity compared with ordinary concrete.

Some of the samples were also placed in the sea water in order to test their Long-term resistance to permeability.

The tests were intended for structures such as bridges, reservoirs, harbour walls , etc., which are not required to be constructed of high compressive strength concrete; therefore, the compressive strength obtained in the tests was quite adequate for these structures, and the final concrete mix design was used in several projects[5][1].

Table 2. Proposal concrete mix proportions

Portland cement (type 2 or 5)	400	Kg/m3
Fine aggregates (natural sands)	800	Kg/m3
Coarse aggregates (max. size 12.5 mm)	960	Kg/m3
Water	135	Kg/m3
Polymer (SBR)	80	Kg/m3
Superplasticizer	4-6	Lit/m3

7 Practical specifications

Having taken the climatic conditions and the types of materials available in the region , the following considerations were made in using polymer-concrete composites [4][5]:

1. The chemical properties of the aggregates were tested in order to make sure that they don't chemically react either with one another or with other materials.
2. Considering the high temperature of the region, commonly practised measures were taken to keep the temperatures of water, cement, coarse aggregates, sand, reinforcement,

and formwork low.
3. As concreting had to be carried out at temperatures lower than 30 degrees (C), it was stopped during the summer months.
4. Concrete mixes were made in ordinary mixers which had capacities proportional to the volumes of concrete to be poured and were located at the sites in order to prevent the adverse effects on the quality of concrete caused by carrying it for long distances to bring it to the pouring place.
5. The water-cement ratios were accurately controlled to comply with the values specified, and 4-6 liters of a superplasticizer was added to each one cubic meter of the mix 10 minutes before the start of concreting.
6. The placed concretes were cured for 4 days only as the polymer-modified concrete requires less time for curing than ordinary Portland cement concrete.
7. The "pot life" specified for the additives were controlled, and they were kept at cool and suitable places before being used.
8. The concrete structures in contact with soil or water were coated with one layer of bitumen after the completion of the curing process.
9. Various samples were taken during different stages of concreting for testing the compressive strength, tensile strength, long-term permeability, and durability of the concrete.

8 Conclusions and recommendations

The use of polymer-modified concrete in the structures of the northern coastal regions of the Persian Gulf was started two years ago. The long-term tests on the concrete samples placed in the Gulf water have all yielded satisfactory results , and there have been no reports of deteriorations or low durability of the structures constructed of polymer-concrete composites.

The advantages of the polymer-modified concrete are: easiness of mixing, high workability with low water-cement ratio, little time required for its curing, good adhesion, high impermeability to water and gas, and lower modulus of elasticity (about 10% to 15%) than that of ordinary concrete which makes it more formable and suitable for repair works.

Bearing in mind the high price of polymers, the use of this type of concrete is , at present , limited to the specific orders made by clients and the requirements for repairing deteriorated concrete structures in the region.

9 References

1. Concrete design specifications in hot climates of Iran . (1995) Technical report , Tehran-Boston consulting Engineers, Iran Plan & Budget organization publications.

2. L. A. Kuhlmann (1990) , Styrene-Butadine Latex-modified concrete , ACI concrete international.
3. Publications of Iranian Meteorological organization.
4. Concrete in hot climates , proceedings of the third international Rilem conference ,(1992) ,(ed. Mj. Walker)
5. Tehran- Boston Engineer's documents and reports (1990 – 1995)

WEATHERABILITY OF POLYMER-MODIFIED MORTARS AFTER TEN-YEAR OUTDOOR EXPOSURE IN KORIYAMA AND SAPPORO

Y. Ohama, K. Demura and T. Uchiyama
Dept. of Architecture, Nihon University, Koriyama, Japan

Abstract
This paper deals with the weatherability of polymer-modified mortars after ten-year outdoor exposure in Koriyama and Sapporo, Japan. Polymer-modified mortars using eight types of polymer dispersions and two types of redispersible polymer powders are prepared with various polymer-cement ratios, and tested for appearance change, strength, carbonation and chloride ion penetration after ten-year outdoor exposure. It is concluded from the test results that regardless of the exposure site, the polymer-modified mortars possess an excellent weatherability over unmodified mortars from the view points of surface conditions, strength properties, resistance to carbonation and chloride ion penetration after ten-year outdoor exposure.
Keywords: Appearance change, carbonation depth, chloride ion penetration depth, outdoor exposure, polymer-modified mortars, strengths, weatherability.

1 Introduction

Generally, polymer-modified mortars have excellent flexural strength, adhesion, waterproofness, carbonation resistance, chloride penetration resistance and freeze-thaw durability compared to cement mortars. In recent years, the polymer-modified mortars have widely been used as waterproofing and repair materials in construction works. The history of the polymer-modified mortars is unexpectedly long, and the research and development of the polymer-modified mortars using latexes and emulsions have actively been conducted in many countries for the past 70 years or more. However, there are only a few papers of the weatherability of the polymer-modified mortars through long-term outdoor exposure [1][2].

Polymers in Concrete, edited by Y. Ohama, M. Kawakami and K. Fukuzawa. Published in 1997 by E & FN Spon, 2–6 Boundary Row, London SE1 8HN, UK. ISBN: 0 419 22330 4.

In this paper, polymer-modified mortars using ten commercial cement modifiers are prepared with polymer-cement ratios of 0, 10 and 20%, and tested for appearance change, strength, carbonation and chloride ion penetration after ten-year outdoor exposure in Koriyama and Sapporo.

2 Materials

2.1 Cement and fine aggregate
Ordinary portland cement and Toyoura standard sand specified in JIS (Japanese Industrial Standard) were used for all the mortar mixes.

2.2 Cement modifiers
Commercial cement modifiers used were three styrene-butadiene rubber (SBR ; SBA, SBB and SBC) latexes, two ethylene-vinyl acetate (EVA ; EVA and EVB) emulsions, three polyacrylic ester (PAE ; AEA, AEB and AEC) emulsions, one ethylene-vinyl acetate (EVA ; PEV) and vinyl acetate-vinyl carboxylate (VA/VeoVa; PVV) redispersible polymer powders. Their basic properties are listed in Table 1. Before mixing, a silicone emulsion-type antifoamer was added to the polymer dispersions in a ratio of 0.7% of the silicone solids of the antifoamer to the total solids of the polymer dispersions. The same amount of the antifoamer was added to the redispersible polymer-modified mortars during mixing.

Table 1.   Properties of polymer dispersions and redispersible polymer powders

Type of polymer	Specific gravity (20 ℃)	pH (20 ℃)	Viscosity (20 ℃, mPa·s)	Total solids (%)
SBA	1.016	8.0	165	45.3
SBB	1.012	8.4	1100	45.2
SBC	1.006	10.6	21	48.6
EVA	1.078	5.7	2205	53.7
EVB	1.076	5.2	1200	54.3
AEA	1.074	9.4	36	45.6
AEB	1.028	8.9	240	49.9
AEC	1.016	9.0	100	44.8
PEV	0.450*	6-7**	30000**	-
PVV	0.400*	5-6**	10000**	-

Notes, * : Apparent specific gravity. , **: pH and viscosity of 50% water dispersion.

3 Testing procedures

3.1 Preparation of specimens
According to JIS A 1171 (Method of Making Test Sample of Polymer-Modified Mortar in the Laboratory), polymer-modified mortars were mixed with the mix proportions given in Table 2, and their flow was adjusted to be constant at 170±5. Beam specimens 40×40×160mm were molded, and then subjected to a 2-day-20 ℃ - 80%(RH)-moist plus 5-day-20 ℃ -water   plus   84-day-20 ℃ - 50%(RH)-dry cure.

Table 2. Mix proportions of polymer-modified mortars

Type of polymer	Cement: sand (by mass)	Polymer-cement ratio (%)	Water-cement ratio (%)	Flow	Air content (%)
Plain	1:3	0	77.2	172	7.0
SBA		10	65.6	170	7.0
		20	61.4	170	6.6
SBB	1:3	10	63.6	168	8.3
		20	65.0	168	6.7
SBC		10	58.8	170	7.8
		20	50.6	170	7.3
EVA		10	62.1	172	8.7
	1:3	20	57.8	170	8.8
EVB		10	63.2	172	8.3
		20	57.3	170	6.9
AEA		10	55.2	170	15.3
		20	51.4	170	10.0
AEB	1:3	10	54.8	170	12.7
		20	45.8	168	13.5
AEC		10	46.1	170	26.4
		20	37.4	170	28.8
PEV	1:3	10	70.0	170	8.9
		20	71.2	169	10.6
PVV	1:3	10	69.2	167	8.4
		20	68.5	170	8.5

3.2 Outdoor exposure test

After curing, beam specimens were exposed to the outdoors or indoors for ten years at the following exposure sites:

(1) Koriyama exposure site : roof floor of the main building of College of Engineering, Nihon University in Koriyama, Japan.
(Long. 140°23′E, Lat. 37°21′N)

(2) Sapporo exposure site : roof floor of the main building of the Hokkaido Prefectural Cold Region Building Research Institute in Sapporo, Japan.
(Long. 141°20′E, Lat. 43°03′N)

(3) Indoor exposure site : room controlled at 20 °C and 50%(RH).

Annual weather observation data at two outdoor exposure sites are shown in Table 3.

3.2.1 Appearance observation test

The appearance of beam specimens after ten-year outdoor exposure was visually observed.

3.2.2 Strength test

Beam specimens after ten-year outdoor exposure were stored at 20 °C and 50%(RH) for 7 days for conditioning. After conditioning, the beam specimens were tested for flexural and compressive strengths in accordance with JIS A 1172 (Method of Test for Strength of Polymer-Modified Mortar), and then the relative flexural and compressive strengths are calculated by the following equations:

Table 3. Annual weather observation data at outdoor exposure sites

Observation item	Exposure site	Jan.	Feb.	Mar.	Apr.	May	June	July
Average temperature (°C)	Koriyama	0.9	1.3	4.1	10.5	15.4	19.3	22.5
	Sapporo	-4.6	-4.0	-0.1	6.4	12.0	16.1	20.2
Average relative humidity [%(RH)]	Koriyama	75	73	69	66	70	77	80
	Sapporo	72	71	69	64	67	75	78
Rainfall (mm/month)	Koriyama	29.9	54.9	83.9	81.4	94.6	132.5	137.4
	Sapporo	107.6	94.1	81.8	62.3	54.8	66.4	68.7
Sunshine duration (h/month)	Koriyama	159	164	194	219	217	165	154
	Sapporo	99	112	159	183	202	192	179

Observation item	Exposure site	Aug.	Sept.	Oct.	Nov.	Dec.	Annual average or sum total
Average temperature (°C)	Koriyama	24.6	20.3	14.1	8.4	3.9	12.1*
	Sapporo	21.7	17.2	10.8	4.3	-1.4	8.2*
Average relative humidity [%(RH)]	Koriyama	79	80	80	78	77	75*
	Sapporo	78	74	69	68	71	71*
Rainfall (mm/month)	Koriyama	145.2	170.2	115.9	53.9	26.6	1126.4**
	Sapporo	142.0	137.7	115.6	98.5	100.1	1129.6**
Sunshine duration (h/month)	Koriyama	191	140	157	146	144	2050**
	Sapporo	169	167	156	100	86	1805**

Notes, * : Annual average. , **: Annual sum total.

$$\text{Relative flexural strength (\%)} = \frac{\text{Flexural strength after exposure (MPa)}}{\text{Flexural strength before exposure (MPa)}} \times 100$$

$$\text{Relative compressive strength (\%)} = \frac{\text{Compressive strength after exposure (MPa)}}{\text{Compressive strength before exposure (MPa)}} \times 100$$

3.2.3 Carbonation test

After strength test, the crosssections 40×40mm of beam specimens were sprayed with a 1.0% phenolphthalein alcoholic solution. The depth of the uncolored rim of each crosssection after phenolphthalein solution spraying was measured as a carbonation depth with slide calipers. In this measurement, the maximum carbonation depth was 20mm.

3.2.4 Chloride ion penetration test

After strength test, the crosssections 40×40mm of beam specimens were sprayed with a 0.1% sodium fluorescein solution and a 0.1N silver nitrate solution. The depth of the white-colored rim of each crosssection after sodium fluorescein solution and silver nitrate solution sprayings was measured as a chloride ion penetration depth with slide calipers. In this measurement, the maximum chloride ion penetration depth was 20mm.

4 Test results and discussion

Fig.1 shows the visual observation results of the appearance of the polymer-modified mortars exposed to the outdoors in Koriyama and Sapporo for ten years. The surface conditions of the polymer-modified mortars are classified into four types : (A) having rough surfaces, (B) having a few cracks, (C) having many cracks, and (D) having molds or mosses. Some of the mortars have complex surface conditions. In general, most polymer-modified mortars have the rough surfaces. However, the surface conditions of the polymer-modified mortars are sounder than those of unmodified mortars with many cracks and molds or mosses. The surface conditions of the polymer-modified mortars are improved with an increase in the polymer-cement ratio. PAE-modified mortars have good surface conditions in comparison with other mortars. This may be due to the excellent weatherability of

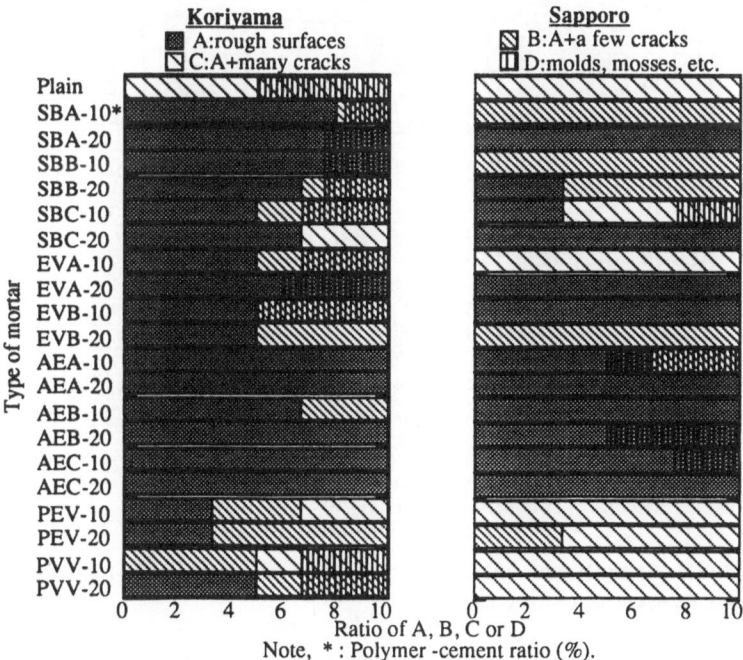

Fig.1. Appearance change distribution of polymer-modified mortars after ten-year outdoor exposure.

PAE films formed on the surfaces of the mortars. The surface conditions of the redispersible polymer powder-modified mortars exposed in Sapporo give almost the same as those of the unmodified mortar.

Fig.2 represents the flexural strength and relative flexural strength of the polymer-modified mortars exposed to the outdoors in Koriyama and Sapporo, and the indoors for ten years. Most polymer-modified mortars exposed to the indoors provide relative flexural strengths of 100% or more. On the other hand, the relative flexural strength of most polymer-modified mortars exposed to the outdoors in Koriyama and Sapporo is less than 100%, and is smaller than that of outdoor-exposed unmodified mortars. This may be due to the degradation of the polymers used as cement modifiers. However, the flexural strength of the polymer-modified mortars is still higher than that of unmodified mortars after ten-year outdoor exposure because the degradation of the polymer phases may occur on the surface layers of the mortars. The flexural strength and relative flexural strength of the polymer-modified mortars with a polymer-cement ratio of 20% are larger than those of the mortars with a polymer-cement ratio of 10% regardless of polymer type. PAE-modified mortars provide a higher relative flexural strength than other mortars because of the superior weatherability of PAE polymers.

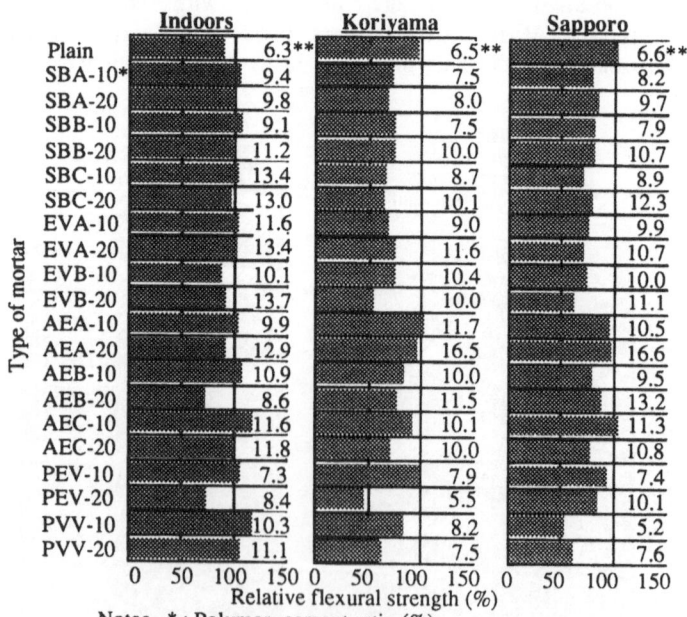

Fig.2. Flexural strength and relative flexural strength of polymer-modified mortars after ten-year outdoor exposure.

Fig.3 illustrates the compressive strength and relative compressive strength of the polymer-modified mortars exposed to the outdoors in Koriyama and Sapporo, and the indoors for ten years. Almost the same tendency as the relative flexural strength is recognized for the relative compressive strength of the polymer-modified mortars exposed to the outdoors and indoors. Generally, the polymer modification for cement mortar does not improve the compressive strength of the cement mortar[3]. However, the compressive strength of the polymer-modified mortars is higher than that of unmodified mortars regardless of exposure site. It is considered that the hydration of cement in the polymer-modified mortars progresses during outdoor exposure because the water retention of the polymer-modified mortars is improved by the polymer films formed in the mortars[3].

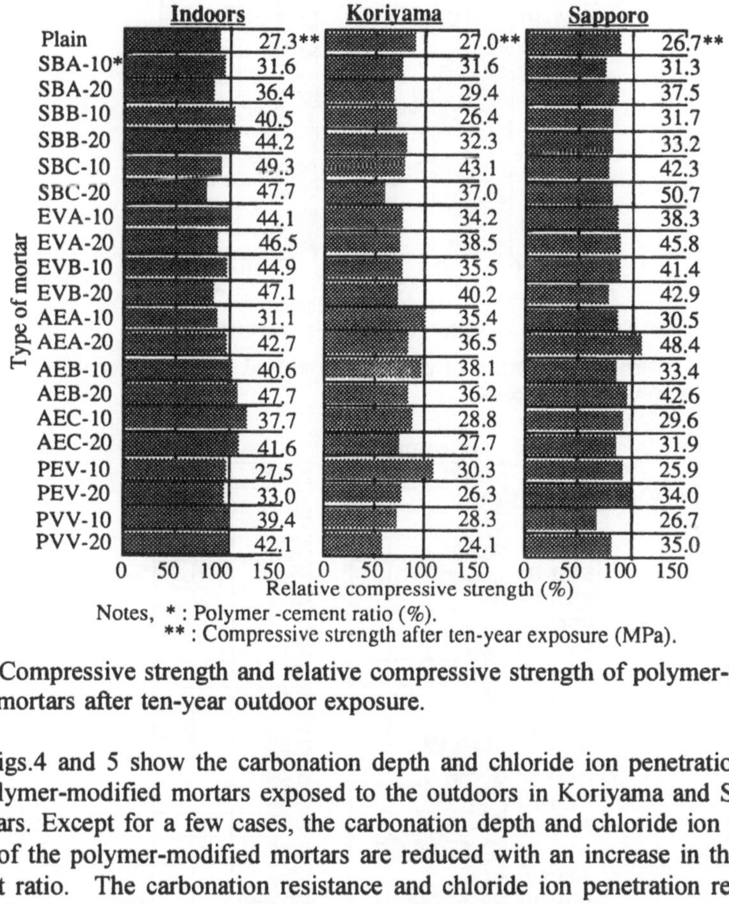

Notes, * : Polymer -cement ratio (%).
** : Compressive strength after ten-year exposure (MPa).

Fig.3. Compressive strength and relative compressive strength of polymer-modified mortars after ten-year outdoor exposure.

Figs.4 and 5 show the carbonation depth and chloride ion penetration depth of the polymer-modified mortars exposed to the outdoors in Koriyama and Sapporo for ten years. Except for a few cases, the carbonation depth and chloride ion penetration depth of the polymer-modified mortars are reduced with an increase in the polymer-cement ratio. The carbonation resistance and chloride ion penetration resistance of the polymer-modified mortars are affected by the type of polymer used. EVA-modified mortars show superior carbonation resistance and chloride ion penetration resistance. The carbonation and chloride ion penetration depths of PAE-modified mortars having high air content are almost the same as those of unmodified mortars.

From the annual weather observation data, the number of freeze-thaw cycles in Sapporo may be lager than that in Koriyama with higher average temperature and longer sunshine duration. The freeze-thaw cycles are generally considered to be the

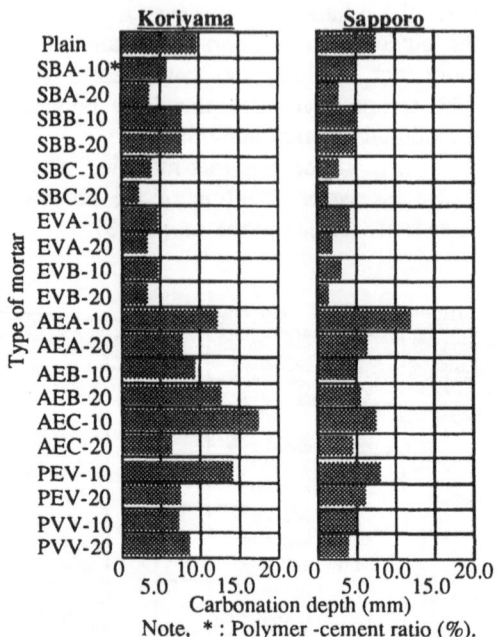

Fig.4. Carbonation depth of polymer-modified mortars after ten-year outdoor exposure.

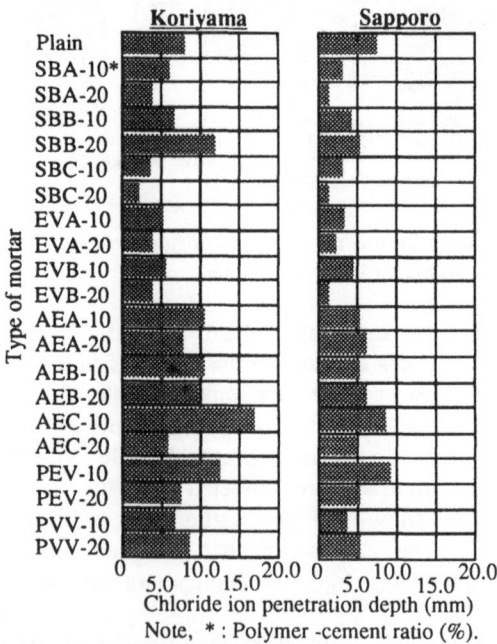

Fig.5. Chloride ion penetration depth of polymer-modified mortars after ten-year outdoor exposure.

most important deterioration factor for cement mortar and concrete. Nevertheless, the flexural and compressive strengths, resistance to carbonation and chloride ion penetration of Sapporo-exposed polymer-modified mortars are somewhat superior to those of Koriyama-exposed ones except for some cases. The polymer-modified mortars have basically a good freeze-thaw resistance. It is considered that factors such as the temperature and sunshine duration are much important to discuss the weatherability of the polymer-modified mortars because the temperature and ultraviolet rays from the sun attack the polymer phases in the polymer-modified mortars.

5 Conclusions

The conclusions obtained from the above-mentioned test results are summarized as follows:
(1) After ten-year outdoor exposure, the relative flexural and compressive strengths of polymer-modified mortars are smaller than those of unmodified mortars. However, the outdoor-exposed polymer-modified mortars provide a higher flexural strength and compressive strength than the outdoor-exposed unmodified mortars.
(2) In comparison with unmodified mortars, polymer-modified mortars have a superior weatherability from the view points of surface conditions, strength properties, resistance to carbonation and chloride ion penetration after ten-year outdoor exposure. The weatherability of the polymer-modified mortars is affected by the type of polymer used.
(3) The flexural and compressive strengths, resistance to carbonation and chloride ion penetration of Sapporo-exposed polymer-modified mortars are somewhat superior to those of Koriyama-exposed ones except for some cases in this study.

6 References

1. Ohama, Y. (1982) Adhesion durability of polymer-modified mortars through ten-year outdoor exposure, in *Proceedings of the Third International Congress on Polymers in Concrete Vol. 1*, College of Engineering, Nihon University, Koriyam, Japan, pp.209-221.
2. Ohama, Y., Moriwaki, T. and Shiroishida, K. (1984) Weatherability of polymer-modified mortars through ten-year outdoor exposure, in *Proceedings of the Fourth International Congress on Polymers in Concrete*, (ed. H. Schulz), Institut für Spanende Technologie und Werkzeumaschinen, Technische Hochschule Darmstadt, Darmstadt, West Germany, pp.67-71.
3. Ohama, Y. (1995) *Handbook of Polymer-Modified Concrete and Mortars*, Noyes Publications, Park Ridge, New Jersey, USA, pp.45-85.

DURABILITY OF HIGH FLEXURAL STRENGTH ALUMINA CEMENT–PHENOL RESIN COMPOSITE

G.K.D. Pushpalal, T. Kawano, T. Kobayashi and S. Miura
Maeta Techno-Research, Inc., Sakata, Japan
N. Maeda
Maeta Concrete Industry Ltd., Sakata, Japan
M. Hasegawa and T. Takata
Dept. of Materials Science and Technology, Toin University of Yokohama, Yokohama, Japan

Abstract

Polymer cement composite with very high flexural strength was investigated here in view of its durability. This newly developed composite is made by combining alumina cement, phenol resin precursor, a small amount of high viscous modifier and plasticizer under high shear mixing to produce viscoelastic cement paste through a twin roll mill. This material is known as a water disuse cement paste in which hardening takes place together with the solidification of phenol resin precursor. The flexural strength of the composite is between 120-220 MPa while those with the usual cement pastes are lower than 10 MPa.

The strength loss after immersion in water at 20°C and after outdoor exposure for one year are only 9% and 6%, respectively, and maintains the elastic modulus almost unchanged. Linear expansion and weight increase of alumina cement - phenol resin composite are 0.12% and 0.82%, respectively after immersion in water at 20°C for one year. High durability is specific property of alumina cement - phenol resin composite but not to other cement/phenol resin composites or other fillers/phenol resin combinations. However, this composite contains a large amount of unhydrated cement because basic hardening takes place with a very small quantity of water which is released from the phenol resin during the polycondensation reaction to form crosslinking structure. Present paper discusses the influence of unhydrated cement particles on durability, based on the results of several durability tests and some other experiments.

Keywords: Alumina cement, calcium aluminate hydrates, durability, high flexural strength, immunization, linear expansion, phenol resin, weight change.

Polymers in Concrete, edited by Y. Ohama, M. Kawakami and K. Fukuzawa. Published in 1997 by E & FN Spon, 2–6 Boundary Row, London SE1 8HN, UK. ISBN: 0 419 22330 4.

1   Introduction

We introduced a system of very high strength cement paste with water insoluble phenol resin precursor in 1992 [1][2]. This innovation is directed to the cement composition that contains a hydraulic cement and a water insoluble polymer precursor that is substantially anhydrous but generates water during the heating process. However, the amount of generated water is limited to 6-7% of the nonvolatile component of the precursor and no additional free water is added to the composition in order to hydrate the cement. The nonvolatile content is rather depend on commercially available phenol resin precursors and it was found to be 58 to 62% by weight of the precursor. Accordingly, the water utilized in hydration is nearly 1% by weight of the cement in standard mix proportion of present study. Such a small amount of water might be enough only hydrating the surface of the cement particles, but not the inert body of the particles. Hence, this manner of hydration should leave large part of the cement body unreacted.

Phenol resin - alumina cement composite demonstrated favorable stability in water, but post hydration of unreacted high alumina cement particles in the composite cast doubt on long term stability of the material. This paper aims to provide an extensive study of long term stability properties in water immersion and outdoor exposure and to suggest the methods to improve their stability.

2   Materials

Mainly two types of mix proportions were used and these are given in Table 1. One is 'standard,' which consists of alumina cement(AC) and the other one is 'comparison,' in which cementitious components are not contained but consist of equal volume fractions of Al_2O_3 and $Al(OH)_3$. Alumina cement (r.d. 3.01) used is composed of 54.3wt% Al_2O_3, 37.0wt% CaO, 4.5wt% SiO_2 and 1.5wt% Fe_2O_3. The main mineral constituent of alumina cement is monocalcium aluminate (CA), but CA_2, $C_{12}A_7$, C_2AS and αAl_2O_3 are also identified by X-ray diffraction. Alumina powder (r.d. 3.95) contains 99.6wt% Al_2O_3. Reagent grade pure $Al(OH)_3$, of which relative density is 2.42 was used. Commercially available resole type phenol resin precursor was used as the main binder. The precursor is essentially anhydrous and soluble in methanol, and contains 58 to 62wt% of nonvolatile matter. Specific gravity is 1.06 and viscosity is 250cps. N-methoxymethyl 6-nylon was incorporated to modify the phenol resin and to develop the plasticity of the paste. Glycerol was used as a plasticizer.

Table 1. Mix proportions.

Mix type	Resin+ solvent+ modifier/ Solid (%)	Mix proportions (by weight)						
		Solid substances			Resin	Solvent	Modifier	Plasticizer
		AC	Al_2O_3	$Al(OH)_3$				
Standard	23	100			13.06	8.24	1.70	2.30
Comparison	23		66	40	13.06	8.24	1.70	2.30

3 General methods

3.1 Preparation of specimens

The materials shown in Table 1 were processed by the method described by Hasegawa et al. [3]. Heat pressing after calendering was excluded for present study. To make test specimens, processed sheets were cut into strips, 25x2.5x180 mm before heat curing. Mainly two types of specimens were prepared. One was prepared according to 'standard' mix proportion and the other was according to 'comparison' (preparation of wholly Al_2O_3 based specimens were unable due to reason has been given elsewhere[4]). This 'comparison' specimens (based on non cementitious compounds) were prepared to examine the effects of alumina cement on durability related performances. Some of the 'standard' specimens were heat treated at 300°C for 24 hours and were used for the purpose described in paragraph 4.3.

3.2 Measurements

The heat cured specimens were tested for flexural strength. Half length of each specimen was subjected to a three point bend test in which the original strength and the modulus of elasticity of the specimen was measured. Deflection of the specimens was measured by a displacement transducer and the span/depth ratio of the three point bend test was 30 to 40 for strength and elastic modulus test. The remaining part of specimens, in sets of more than six, were tested in the following conditions:

- water at 20°C
- water at 40°C
- Outdoor exposure conditions

The residual strength and modulus of the specimens in each of the above storage conditions were again measured in the same manner at prescribed time intervals.

The change of length and weight were also measured. The change of length was measured in accordance with the comparator method of JIS A 1129 (Methods of Test for Length Change of Mortar and Concrete). 15 mm diameter milky glasses with fine straight lines perpendicular to each other were stuck to surface of the specimens using epoxy resin before placed in each storage conditions. Distance between the marks before placing was about 100 mm. Linear expansion and change of weight were expressed as the ratio of the change of length and weight to the value before the placing.

3.3 Outdoor exposure test

The specimens were installed on a wooden stand at an outdoor exposure site located at Sakata, Yamagata, Japan. Fig. 1 shows a view of the specimen bearing stand. The stand and its positioning were made in accordance with the recommended values given in JIS A 1410 (Recommended Practice for Outdoor Exposure of Plastics Building Materials). The specimens were removed from the stand at prescribed time intervals and performances were evaluated.

Fig. 1. Disposition of the specimens at the outdoor exposure site.

4 Results

4.1 Original strength and modulus

The original flexural strength for the 'standard' specimens was 123±7 MPa (i.e. 2.5 mm thick specimens), except for the 3 mm thick specimens for which the flexural strength is 113±7 MPa. The flexural strength highly depended on the specimen thickness probably due to the differences of stress distribution in the cross section, and also due to the lamination defects which are bound to occur in fabrication. The average flexural strength of the 'comparison' mixture was 88±5 MPa. One of our previous studies [4] showed that porosity and pore size of the specimens made by the same mix proportion was lower than that of 'standard,' still, strength is lower than 'standard.'

4.2 General durability performances

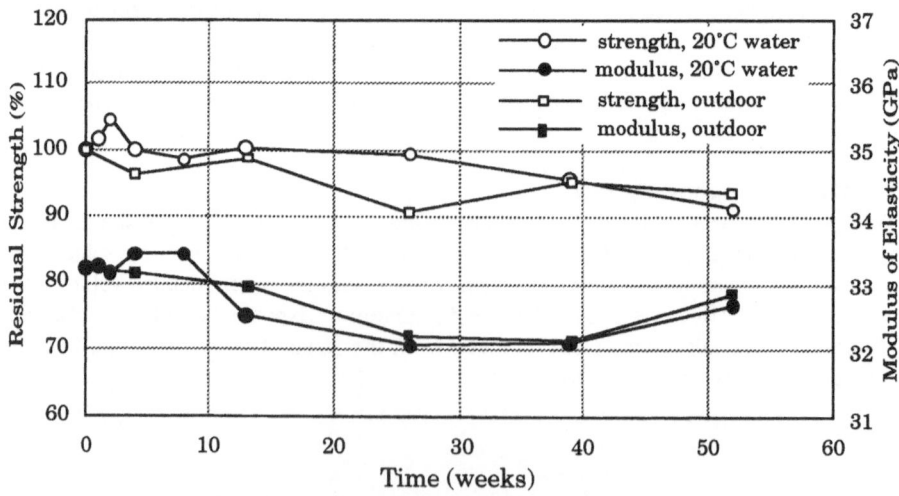

Fig. 2. Durability characteristics of standard specimens for one year at 20°C water immersion and outdoor exposure.

Durability data of standard specimens for one year at 20°C water immersion and outdoor exposure are presented in Fig. 2. The strength was decreased by both placing conditions. Strength loss for one year at 20°C water immersion and outdoor exposure were still only 9% and 6%, respectively, and maintains the elastic modulus relatively unchanged. These strength and modulus data ensured considerable stability of phenol resin cement system for structural uses.

4.3 Linear expansion and change of weight

The 'standard' specimens and heat treated specimens at 300°C were immersed in water at 20°C. Heating at 300°C significantly reduced the strength to 43 MPa, since the phenol resin was partially burnt off by heating. This was done to evaluate the responsibility of polymer matrix on expansion and weight gain. Linear expansion and weight change of standard and treated specimens over one year in 20°C water are shown in Fig. 3. Both expansion and weight gain of standard specimens are lower than that of the heat treated specimen as can be seen in the Fig. 3. The standard specimens have reached 0.12% expansion level after one year in water. It has been shown experimentally[5] that sintered high alumina cement samples expand 0.03%, after 28 days at 20°C water. None of the specimens made from standard mix proportion reached this expansion level during the first 56 days.

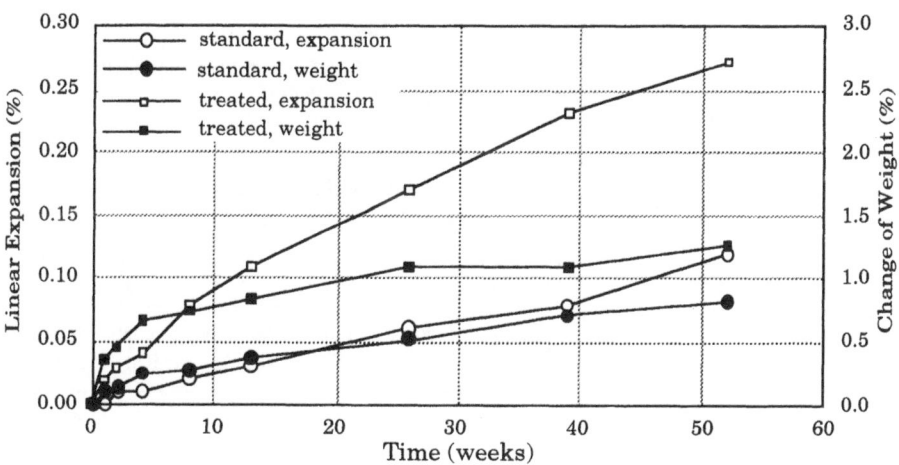

Fig. 3. Linear expansion and weight change of standard and
treated specimens over one year in 20°C water.

4.4 The effect of water at 40°C

The 'standard' and 'comparison' specimens were immersed in water at 40°C. Two sets of each type were taken out at prescribed time intervals; one set was tested for strength and modulus; other set was dried at 80°C in an oven until constant weight had been achieved. Then the specimens were tested for strength and modulus. Fig. 4 shows that the strength of standard specimens tends to increase or remain unchanged within first 14 days of immersion and further immersion tends to decrease gradually. An interesting aspect of these results is that the both wet and dry strength of standard specimens marked highest residual strength at 14th day over this time scale. The strength of 'comparison' specimens dropped immediately within first 21 days and maintained its flexural strength until test termination. The results (Table 2 and Fig. 4) show that, unlike standard, comparison specimens when dried regain its flexural strength up to more than 80% of initial strength.

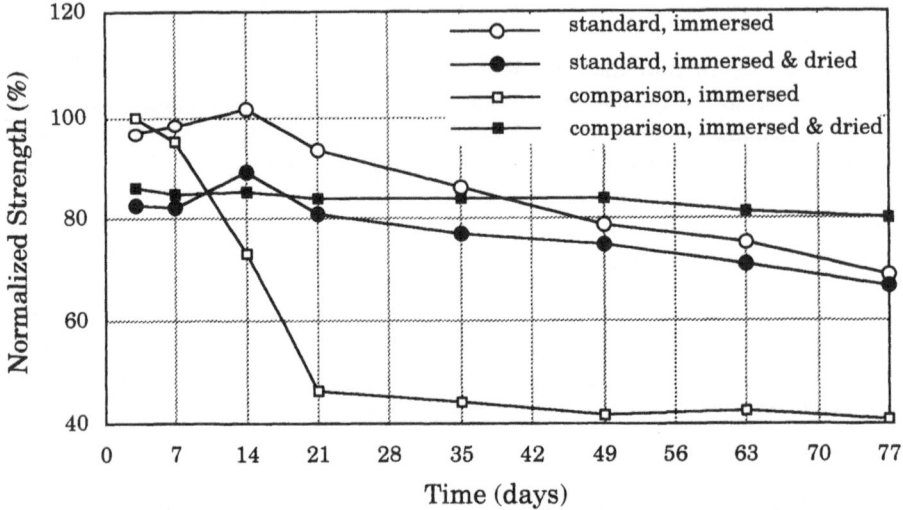

Fig. 4. The residual strengths of the standard and comparison specimens after storage at 40°C water and dried at 80°C.

Table 2. The bending modulus of elasticity of the standard and comparison specimens after storage in water at 40°C and dried at 80°C.

Type of Specimens	State	Original		14 days		77 days	
		Strength (MPa)	Modulus (GPa)	Strength (MPa)	Modulus (GPa)	Strength (MPa)	Modulus (GPa)
Standard	immersed	123	33.2	125	32.4	85	31.9
	immersed & dried			109	32.0	82	32.1
Comparison	immersed	88	28.1	64	23.5	36	21.7
	immersed & dried			75	26.4	71	25.6

4.5 The effect of 'immunization' and specimen size

Experiments were also carried out to study the effects of prolonged hydration and accompanied mineralogical changes on flexural strength. Some of the standard specimens were immunized before storage at 20°C water. Immunization was done in following manner. The standard specimens were cured in water at 20°C for 14 days and then heat cured in an oven at 80°C for 24 hours. The purpose of immunization was to restrain the disadvantage of making unstable alumina cement hydrates when immersed in water at ambient temperature. Curing specimens in water at 20°C, produces the metastable phases CAH_{10} and alumina gel; following heat curing, causes the stable phases C_3AH_6 and AH_3 to be produced [6], but with 15% loss in strength. As reflect in Fig. 5, immunized specimens show higher resistivity to water at 20°C than that of standard specimens over the measured time scale.

Fig. 5 also shows the effect of specimen size on resistivity in water at 40°C. 2.5 mm thick (the normal specimen size used through out this study) and 3 mm thick specimens were compared; the effect of hot water on the thicker specimens were less harmful. Initial strength was lower in 3 mm thick specimens, but final residual strength after 35 days was higher.

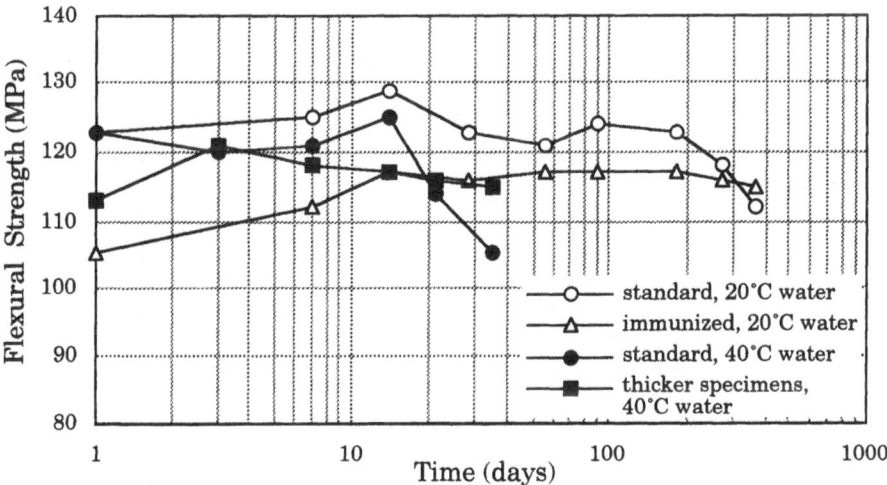

Fig. 5. The effect of immunization and specimen thickness on durability.

5 Discussion

From the results in present study it is obvious that the high strength of phenol resin - alumina cement composite is due to a strong organic/inorganic interphase and physical adhesion between particles and polymer. The lower strength of non cement based material conforms this idea. The knowledge gained from an earlier study suggests that the resole type phenol and calcium aluminate interact to make a strong composite with high water resistivity [7]. This organic/inorganic interphase considerably obstructs the water ingress into the structure; therefore, restrains the negative influences of water on mechanical properties. In absence of the interphase, water ingress into the microstructure through the interfacial capillary paths, becomes more rigorous (Fig. 6); thus, the mechanical breaking of weak van der Waals bonds occurs progressively.

As we have seen, the bulk polymer region rarely transport water into the structure and also does not alter with moisture, since the bending modulus remains unchanged after immersion in water at both 20 and 40°C. As Fig. 2 demonstrates, a longer time (one year) is required to showing a small decrement of strength; thus degradation is a much slower process. In addition, when the standard specimens soaked in water at 40°C, were followed by drying strength was not restored; meanwhile the comparison specimens when dried regained their flexural strength up to more than 80% of their initial strength [Fig. 4]. This result informs that certain irreversible changes have been induced on the cement during the water immersion. It appears that over longer periods of immersion, the strength of immersed and immersed/dried specimens approach the same final value, this substantiates standard specimens have reached equilibrium within the 77 days by complete hydration and conversion of hydrates products to stable state.

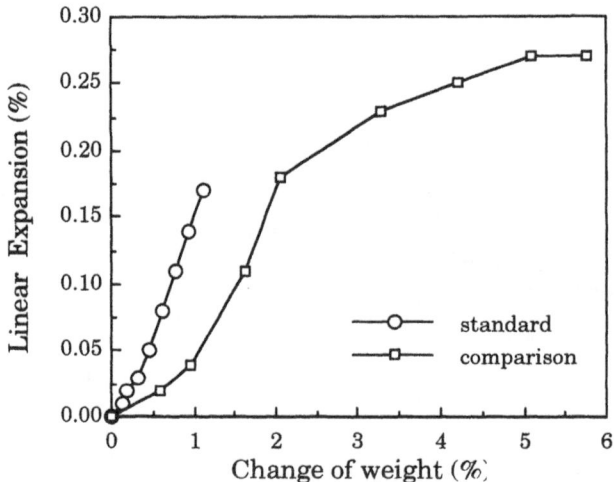

Fig. 6. Relation between change of weight and linear expansion in water at 40°C for prescribed time intervals up to 77 days.

Fig. 7 shows the hydration products detected on the surface of a standard specimen which was immersed in water at 40°C. However, from X-ray diffraction any peaks of calcium aluminate hydrates were not detected in fine powder made by pulverizing these specimens. This is probably due to the hydration products is formed only on the surface of the specimen. The hydration in the material might have occurred on progressing from the surface to the bulk since lower porosity restrains water penetration into its interior. This assumption is in good agreement with higher resistivity demonstrated by thicker specimens. The effect of the surface deterioration on strength of the material decreases with longer the distance between the neutral axis and the surface of specimens.

An interesting observation is that the linear expansion of the standard specimens is higher than that of the comparison for each water content penetrate into the material, as seen in Fig. 6. This reveals that alumina cement is more susceptible to water in expansion point of view, in contrast with the non cementitious compounds Al_2O_3 and

Al(OH)$_3$. Thus, it is easily inferred that the linear expansion of the alumina cement based specimens is a function of the degree of hydration.

Further important achievement of present study is that, the flexural strength of 'immunized' samples is apparently unaffected by immersion in water at 20°C water. It must therefore be deduced that there is a small but significant enhancement of long term durability due to stabilization the surface of specimens, which we have already concluded as most susceptible to hydration. In addition, we have recently found more effective immunization system, which is based on hydrating for 14 days in water at 40°C, followed by drying in an oven at 80°C; this effectively utilized knowledge gained by results demonstrated in Fig. 4.

Fig. 7. Standard specimen after 77 days, 40°C water immersion (SEM)

6 Conclusions

Phenol resin - alumina cement composite showed considerable long term durability. However, a small deterioration in water at 20°C and a large deterioration in hot water was observed. This weakening of the composite was attributed to further hydration of alumina cement through absorption of water, allowing split growth in the polymer cement interface, leading to a lower fracture surface energy of the material [8].

Stabilization of alumina cement in the surface layer by hydrating at high temperatures for a specified time interval (the immunization) improves long term durability in ambient temperature.

Acknowledgment

The authors wish to thank Mr. Seiichi Hashimoto for assistance in SEM analysis.

7   References

1.Kobayashi, T., Pushpalal, G.K.D., and Hasegawa, M., (1992) Japanese Patent Application No. JP 301514 / 92   to Maeta Concrete Industry Ltd., Japan. (in Japanese)
2.Kobayashi, T., Pushpalal, G.K.D., and Hasegawa, M., (1993) European Patent Application to Maeta Concrete Industry Ltd., Japan, Publication No. 0590948 A1.
3.Hasegawa, M., Kobayashi, T., and Pushpalal, G.K.D., (1995) A new class of high strength, water and heat resistant polymer - cement composite solidified by an essentially anhydrous phenol resin precursor, Cement and Concrete Research, Vol. 25, No. 6, pp. 1191-1198.
4.Pushpalal, G.K.D., Maeda, N., Kawano, T., Kobayashi, T., Hasegawa, M., Takata, T., (1995) Properties and Flexural Failure Mechanism of a High Strength Phenol Resin-Cement Composite, Proceedings of the VIIIth ICPIC Congress, Belgium, pp. 631-636.
5.Poon, C. S., Wassell, L. E., and Groves, G. W., (1987) "Stability of macrodefect-free cement", Mater. Sci. and Technol., Vol. 3, pp. 993-996
6.Kosmac, T., Lahajnar, G., and Sepe, A., (1993) Proton NMR relaxation study of calcium aluminate hydration reactions, Cement and Concrete Research, Vol. 23, pp.1-6.
7.Hasegawa, M., Pushpalal, G.K.D., Takata, T., Maeda, N.,Kobayashi, T., (1995) Mat. Res. Soc. Symp. Proc., Vol. 385, Materials Research Society: Pittsburgh, pp. 167-172.
8.Pushpalal, G.K.D., Doctoral Dissertation, Toin University of Yokohama, in preparation.

HYDROPHOBIC IMPREGNANT FOR DAMAGED CONCRETE STRUCTURE

T. Miyagawa, Y. Kubo, A. Hattori and M. Fujii
Dept. of Civil Engineering, Kyoto University, Kyoto, Japan
K. Hori and S. Kurihara
Sho-Bond Construction Co. Ltd., Osaka, Japan

Abstract

This paper deals with various kinds of silanes, which are typical hydrophobic impregnants, as repair materials.   In the deterioration mechanism of concrete structures, water control is one of the most important factors.   Thus, many kinds of surface treatments are applied to control the water content in concrete.   Surface treatments can be classified into two types from the viewpoint of how to control the water in concrete. One type is a water proof treatment.   The other type is a hydrophobic treatment which restricts liquid water penetration into concrete, but allows water vapour to move out. The former system may promote concrete deterioration.   In this study, the effect of the molecular size (molecular weight: 120 - 416), the type (alkyl, alkoxyl) and the number of alkoxyl groups of silane on the hydrophobicity of concrete were investigated.
Keywords: hydrophobic impregnant, molecular structure, repair, water control, weight change, alkali-silica expansion, chloride induced corrosion

1   Introduction

Many reports of premature deterioration of concrete structures caused by alkali-silica reaction and/or chloride induced corrosion of reinforcing steel have been published. Water plays one of the most important roles in these deterioration mechanisms. Therefore, in order to prevent the deterioration, many kinds of surface treatments which can control water content in concrete are applied to concrete.

These surface treatments can be classified into two categories from the viewpoint of the way how they control the water in concrete.   One type, water proof treatment, permits no water ingress into concrete, and no water to get out.   Another type, the hydrophobic treatments, restricts liquid phase water to penetrate into concrete, but allows vapour phase water to get out.   Since the former may cause deterioration by the water left in concrete, the latter must be a superior methods [1].

Polymers in Concrete, edited by Y. Ohama, M. Kawakami and K. Fukuzawa. Published in 1997 by E & FN Spon, 2–6 Boundary Row, London SE1 8HN, UK. ISBN: 0 419 22330 4.

In the latter system, silanes are commonly used as typical hydrophobic impregnants. Silanes are silicone-based products with low molecular weight and the alkyl alkoxysilanes [2] are commonly used.   When a silane is applied to concrete surface, a series of chemical reactions between the silane and the silicate structure of concrete occurs in two steps, which are hydrolysis and condensation.   During the hydrolysis, the moisture provided from concrete produces unstable silanol molecules.   During the condensation, the unstable silanol molecules shake hands with available hydroxyl groups in the silicate structure and some crosslinkings occur.   In this way, the silane treated concrete becomes water repellent [3].

This paper deals with the hydrophobic surface treatment of concrete using some types of silanes.

2   Outline of experiment

2.1   Molecular structure of silanes

By changing the kind and number of the alkyl and alkoxyl groups of silanes, nine kinds of silanes shown in Table 1 were prepared as 1 mol solutions in isopropyl alcohol.

Table 1. Silanes

Name	Molecular formula (alkyl) (alkoxyl)	Molecular weight	Note
dimethyldimethoxy silane	$(CH_3)_2Si(OCH_3)_2$	120	two alkoxyls
methyltrimethoxy silane	$CH_3Si(OCH_3)_3$	136	
ethyltrimethoxy silane	$C_2H_5Si(OCH_3)_2$	150	alkoxyl is
iso-buthyltrimethoxy silane	$C_4H_9Si(OCH_3)_3$	178	methoxy.
n-octhyltrimethoxy silane	$C_8H_{17}Si(OCH_3)_3$	234	
n-deciletrimethoxy silane	$C_{10}H_{21}Si(OCH_3)_3$	262	
n-octadeciletrimethoxy silane	$C_{18}H_{37}Si(OCH_3)_3$	374	
methyltriethoxy silane	$CH_3Si(OC_2H_5)_3$	178	alkoxyl is
n-octadeciletriethoxy silane	$C_{18}H_{37}Si(OC_2H_5)_3$	416	ethoxy.

Table 2. Factors for Series 2

concrete	non-reactive aggregate without chloride, non-reactive aggregate with chloride reactive aggregate without chloride, reactive aggregate with chloride
silane	no treatment, 234, 262, 374, 416 (molecular weight)
condition	outdoors, dry and wet chamber, partially immersing in chloride solution (NaCl:3.13wt.%)

2.1.1 Series 1: Preliminary test

Preliminary test were conducted on the nine kinds of silanes to select silanes used in Series 2. The specimens were small concrete prisms (W/C = 0.60, 40 x 40 x 160 mm) with non-reactive aggregate. After being cured in water for three months, they were dried in air for a week and then impregnated with the silanes. The amount of impregnating silane solution was 400 cm^3/m^2 in all test series. Two days after impregnation, the specimens were exposed to four different conditions - indoors, underwater, in the dry and wet chamber (20 °C, 60 %RH, 12 h - 40 °C, 100 %RH, 12 h), or outdoors. The weight change of specimens impregnated with various silanes were measured to evaluate the hydrophobic performance and to select silanes for Series 2.

2.1.2 Series 2: Alkali-silica expansion and corrosion test

The specimens were concrete prisms (W/C = 0.70, 100 x 100 x 400 mm). Table 2 shows the types of concrete, silanes and environmental conditions. For corrosion series, three deformed bars (D = 10 x 300 mm) were embedded in the concrete with the cover thickness of 20 mm. The Sc/Rc of reactive aggregate was 4.63. The total amount of equivalent alkali of the concrete was 8 kg/m^3. After being cured at 20 °C, 80 %RH for two weeks, each specimen was impregnated with the silane selected from the preliminary test and exposed to the condition shown in Table 2. The weight change, strain and half cell potential were measured. The long term performance of prisms impregnated with typical silanes were examined.

2.1.3 Series 3: Quasi-actual scale test

Finally, by using the quasi-actual size concrete specimens (non-reactive, W/C = 0.51, 1000 x 1000 x 150 mm, Fig. 1) placed outdoors, the effects of two types of hydrophobic treatment system were evaluated in the state of quasi-actual condition. The applied hydrophobic systems were the silane which was selected based on the results of Series 2, and the same silane with water vapour permeable, flexible mortar top lining with the thickness of 1.2 mm. In Japan, a mortar lining is usually adopted as top coat from an aesthetic viewpoint and to control of carbonation. The water vapour permeability of the mortar lining was 12 $g/m^2 \cdot$ day and the elongation was 75 %. PVC tubes of various lengths were embedded in the concrete to measure the relative humidity in the concrete.

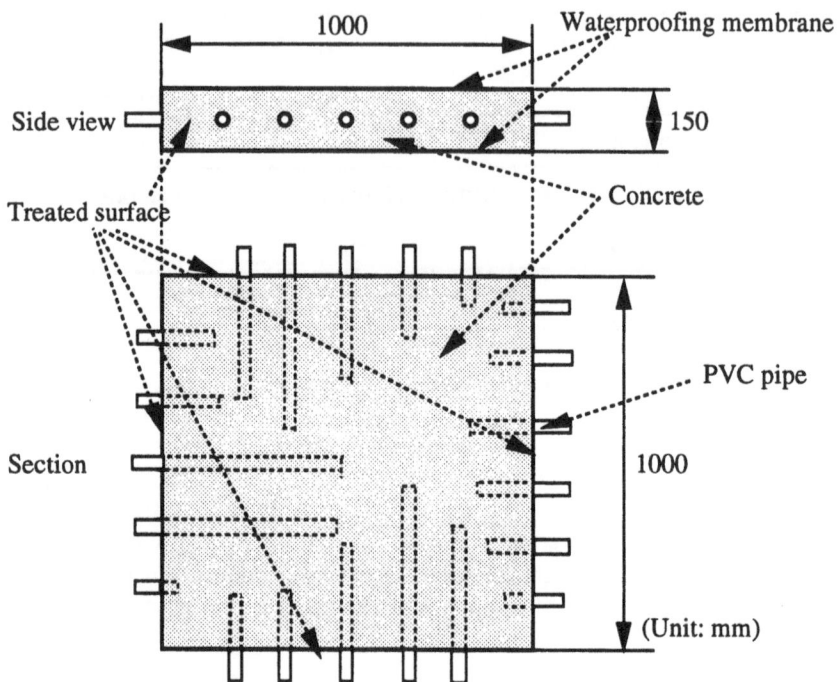

Fig. 1. Quasi-actual specimen.

3 Results and discussions

3.1 Series 1: Preliminary test

Surface treatments are expected to have the ability not only to permit little liquid phase water to penetrate into concrete but also to permit a lot of water vapor in concrete to get out. As the weight changes of specimens exposed to indoor and underwater can be regarded as indexes of "water vapour permeability" and "water liquid permeability" respectively, a larger ratio of "water vapour permeability/water liquid permeability" corresponds to better hydrophobic performance.

3.1.1 Water vapour permeability

Fig. 2 shows the relationship between molecular weight and water vapour permeability. After 9 days of exposure, the silanes with smaller molecular weight in the methoxy series had the larger water vapour permeability. The same tendency is observed in the ethoxy series. The tendency is still observed after 30 days of exposure, although the influence of the molecular weight on the water vapour permeability is reduced. These results indicate that the water vapour permeability of the concrete treated by silanes with smaller molecular weight is generally larger than that with larger ones during the above period.

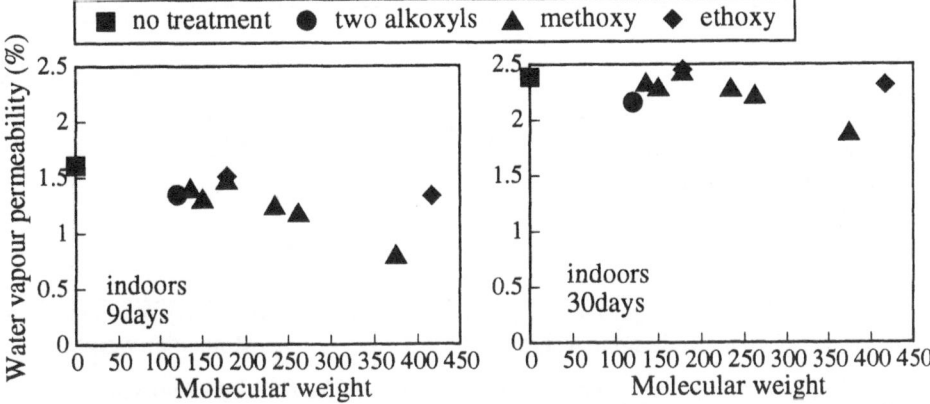

Fig. 2. *The relationship between water vapour permeability of concrete treated with silanes and molecular weight.*

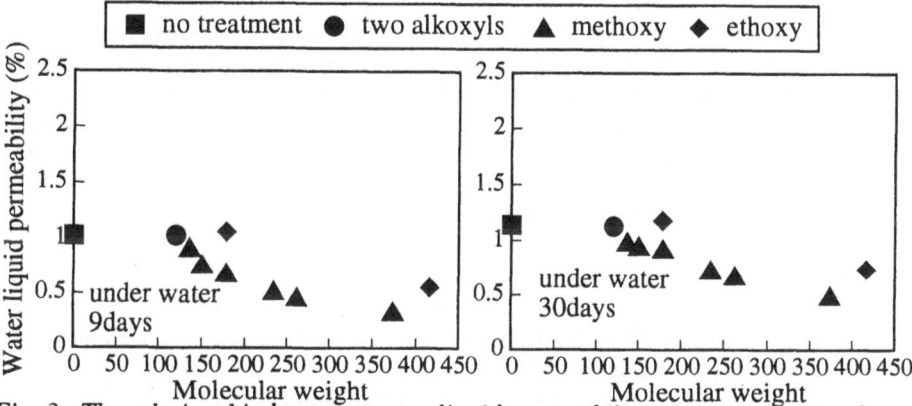

Fig. 3. *The relationship between water liquid permeability of concrete treated with silanes and molecular weight.*

3.1.2 Water liquid permeability

Fig. 3 shows the relationship between molecular weight and water liquid permeability. In the methoxy series, the silanes of larger molecular weight showed the lower water liquid permeability. Since the size of the alkoxyls are the same, this might be due to longer hydrophobic alkyls corresponding to larger molecular weight. The same tendency is observed in ethoxy series as well. The silanes of larger molecular weight had better resistance against penetration of water.

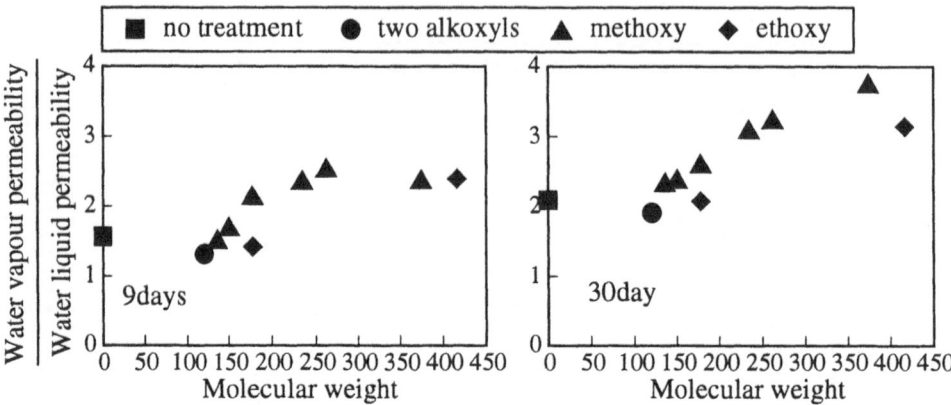

Fig. 4. The effect of molecular weight of silane on the ratio of water vapour permeability/ water liquid permeability.

3.1.3 "Water vapour permeability /water liquid permeability"

Fig. 4 shows the effect of molecular weight on the ratio of "water vapour permeability / water liquid permeability". After 9 days of exposure, the silanes with molecular weight of 120, 136 and 178 show smaller ratio as compared with those of the non-treated specimen, 262, 234, 374 and 416 of larger molecular weight. Almost similar results are obtained after 30 days of exposure, that is, the larger molecular weight resulted in the larger ratio. The silanes which showed a large ratio also indicated good hydrophobicity under dry and wet chamber and outdoor conditions. From these results, the four types of silanes with molecular weight of 234, 262, 374 and 416 are selected for Series 2.

3.2   Series 2: Alkali-silica expansion and corrosion

3.2.1   Weight changes

In all conditions, the weight changes of exposed specimens impregnated with the silanes were smaller than that of non-treated specimens. From the viewpoint of weight change, this indicates that the silanes could also control the water content of the specimens used in Series 2, even under the condition of partial immersion in chloride solution.

3.2.2   Effect on expansion caused by alkali-silica reaction

Figs. 5 and 6 show the expansion of concrete. Under outdoor conditions, the expansion was very small. In the dry and wet condition, non-treated specimens expand significantly due to alkali-silica reaction. On the other hand, the specimens impregnated with the silanes expand much less than non-treated specimens. Among them, the silane with molecular weight of 262 is regarded to have the best effect. The result indicate that silanes control the water content in the specimens and silane treatment is effective against alkali-silica expansion. However, the strains of treated specimens with reactive aggregate are much larger than those of the specimens without reactive aggregates. It should be noted that unless the alkali content of concrete is low, silanes would fail to restrain reactive concrete from excessive expansion in the long term.

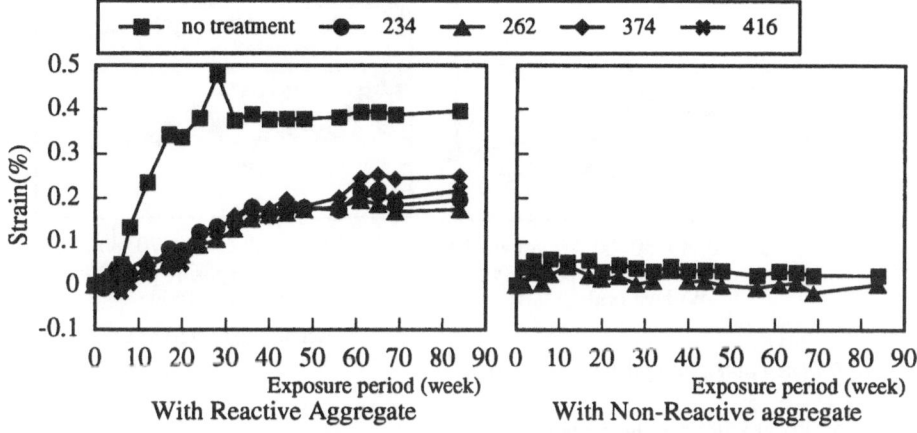

Fig. 5. Expansion of concrete exposed to dry and wet conditions.

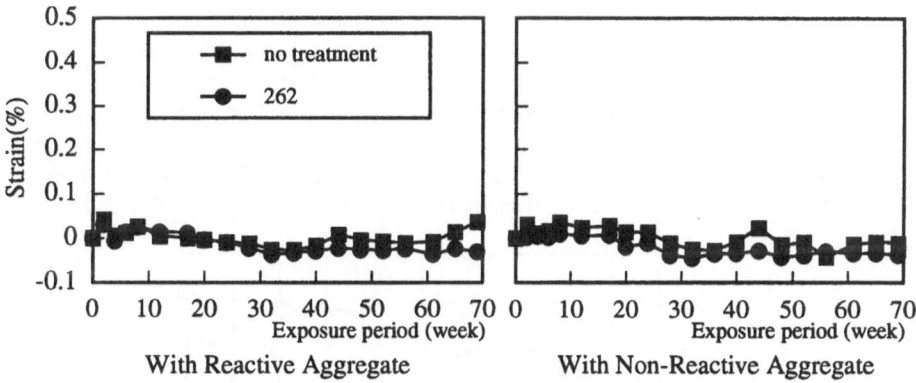

Fig. 6. Expansion of concrete exposed to outdoor.

3.2.3 Effect on reinforcement corrosion

Fig. 7 shows some results of the half cell potential obtained from the concrete specimens without reactive aggregates in the condition of partially immersing in sodium chloride solution (3.13 %). While the half cell potential of the non-treated specimen without chloride is in the corrosion area, that of the specimen impregnated with silane 262 is still in the uncertain zone. This indicates that the silane work effectively also against corrosion of reinforcement caused by the external chloride solution. However, when concrete contaminated with much chloride, although the half cell potentials of the impregnated specimens are less negative than that of non-treated specimens in a short term, they gradually became negative with the time until some of them are more negative than those of non-treated specimens.

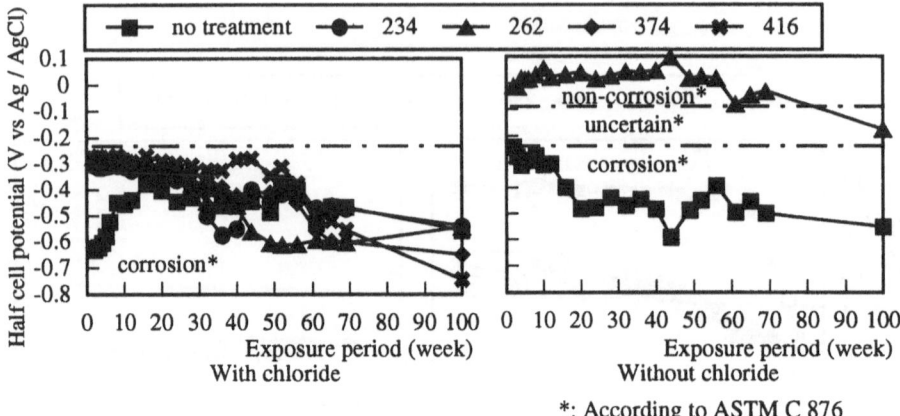

Fig. 7. Half cell potential - partially immersion.

3.2.4   Series 3: Quasi-actual scale test

Silane with molecular weight of 262, which showed the best effect in the alkali-silica expansion series, was used in the quasi-actual scale test.   Fig. 8 shows the typical profiles of relative humidity in concrete of quasi-actual scale specimens after about 6 months of exposure.   The clear effects of silane systems are recognized in Fig. 8.   In the surface 10 cm layer, the relative humidity of the treated specimens is lower than that of the non-treated one.   Particularly, the silane without the mortar system gave a significantly good effect.   However, in the zone deeper than 20 cm of the specimen, the effect of the silane treatment could not be recognized clearly.   Therefore, in this series, it is thought that the effective depth of silane treatment ranged between 10 and 20 cm.

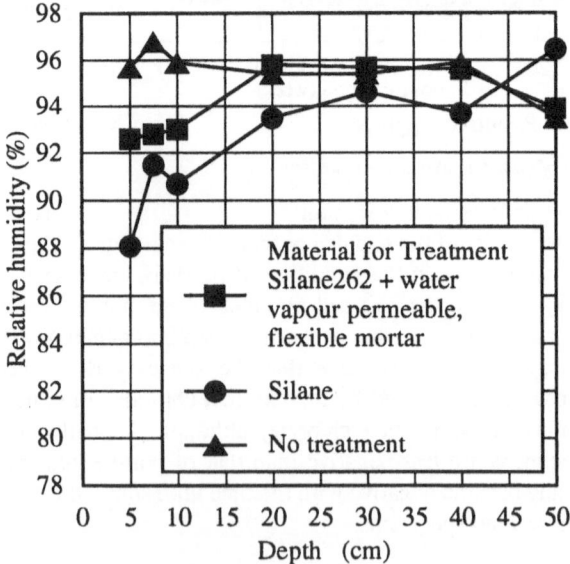

Fig. 8. Profile of relative humidity.

4   Conclusion

The main results obtained in this study are summarized as follows.

1. The silanes of small molecular weight allow the large amount of water vapour to get out.   On the other hand, the silanes of large molecular weight which have long alkyl groups have better resistance against ingress of water.
2. Silane treatment work effectively against the expansion of concrete caused by alkali-silica reaction and the reinforcement corrosion, the expansion or half cell potential is reduced or even restrained.   However, if concrete has excessively reactive potential and/or chloride, it is difficult to restrain deterioration in the long term.
3. Under outdoor conditions in Japan, the effective depth to which silane lowers the relative humidity of concrete, is between 10 - 20 cm.

5   Acknowledgment

The authors wish to express their sincere gratitude to Prof. C.L. Page for his helpful advice.

6   References

1. Miyagawa, T., Hisada, M., Inoue, S. & Fujii, M. (1991) Effect of concrete surface treatment on expansion due to alkali-silica reaction, *Concrete Library International*, No.18, 237-261
2. Edwards, S.C. (1987) Surface coatings, in *THE REPAIR OF CONCRETE STRUCTURES*, BLACKIE ACADEMIC & PROFESSIONAL, Glasgow, 122-148
3. Mailvaganam, N.P., Deans, J.J. & Cleary, K. (1992) Sealing and water proofing materials, in *Repair and protection of concrete structures*, CRC Press, Boca Raton, 87-116
4. Miyagawa, T., Tanaka, H., Hattori, A., Fujii, M. & Hori, K. (1995) Effect of hydrophobic agent as repair material for concrete structures, in *Proc. of IABSE Symposium San Francisco 1995*, 281-286

INFLUENCE OF PHYSICAL PROPERTIES OF WATER BARRIER PENETRANTS ON PERMEABILITY INTO MORTAR

T. Numao, K. Fukuzawa and T. Chen
Dept. of Urban and Civil Engineering, Ibaraki University,
Hitachi, Japan

Abstract
Twelve kinds of commercially available water barrier penetrants, including nine silane systems and three acrylic systems, were subjected to testing. Multiple regression analysis was performed to obtain the relationships between permeability and physical properties of the penetrants, namely surface tension, specific gravity, viscosity, rate of evaporation, and content of non-volatile matter. Consequently, the influence of the properties of the penetrants on penetration depth was evaluated quantitatively.
Keywords: Mortar, multiple regression analysis, permeation coefficient, specific gravity, surface tension, viscosity, water barrier penetrants,

1 Introduction

In recent years, premature deterioration of concrete structures has become an object of public concern. As one countermeasure, water barrier penetrants have been applied to prevent or retard deterioration. After being painted on the surface of concrete, the penetrants permeate into concrete, fill the pores and form a protective layer near the surface. These penetrants are considered to prevent deterioration of the concrete, because the protective layer prevents water from permeating into the concrete. Therefore, during evaluation of the effectiveness of water barrier penetrants, the penetration depth must be taken into account. Although effectiveness of the penetrants has been studied[1][2][3], very little attention has been paid to the relationship between penetration depth and effectiveness. Furthermore, the influence of physical properties of penetrants on penetration depth has not been reported.

The objective of this study was to examine the relationship between physical

Polymers in Concrete, edited by Y. Ohama, M. Kawakami and K. Fukuzawa. Published in 1997 by E & FN Spon, 2–6 Boundary Row, London SE1 8HN, UK. ISBN: 0 419 22330 4.

properties and permeability of twelve commercially available water barrier penetrants.

2   Experimental methods

2.1   Outline of experiment

Nine kinds of silane (S1-S9) and three kinds of acrylic water barrier penetrants (A1-A3) were selected from commercially available penetrants. Multiple regression analysis was performed in order to determine the relationship between permeability and physical properties of the penetrants, including surface tension, specific gravity, viscosity, evaporation rate and content of non-volatile matter. The relationships are compared with a theoretical equation developed by Washburn.

2.2   Specimens and mix proportion

Mortar specimens measuring 40x40x160mm and having a sand-cement ratio (S/C) of 1.0 and a water-cement ratio of 0.5 were used in the experiments. Ordinary Portland cement that has a specific gravity of 3.15 was used as binder, and Toyoura standard sand was used as fine aggregate. In order to eliminate the influence of form-oil on permeability, the oil was not applied to the mould's surface so that the penetration surfaces (40x40mm) of specimens are oil free. A forced mixer was used to produce the mortar matrix. Cement and fine aggregate were mixed for approximately one minute, water was added, and the resultant mixture was mixed for one additional minute and subsequently placed in molds.

The mortar specimens were removed from the molds two days later, and were then cured in water for four weeks. Subsequently, the specimens were dried in a oven at 110℃ for two days, and then cooled in air (20℃, 60%RH) for one day.

On each specimen, penetration surface was polished with a wire brush in order to remove a film of laitance. All other surfaces, except the surface opposite the penetration side, were sealed with silicone resin to restrict the direction of penetration to the axial direction. The specimen was then fixed to the permeation test apparatus by the same silicone resin, and the permeation test was performed the following day.

2.3   Permeation test method

Permeation tests were carried out as shown in Fig.1. In this test, pressure head on the penetrants was controlled to as low a level as possible so as to avoid the effect of hydraulic pressure on the penetration of penetrants.

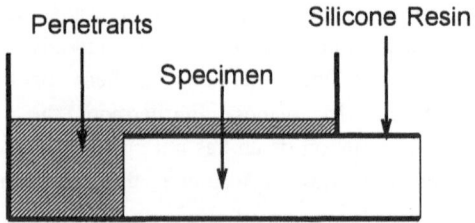

Fig. 1   Method of permeation test

2.4 Method of measuring permeation depth

Each specimen was taken out of the testing apparatus after a certain time and was bisected in the longitudinal direction. Permeation depth was measured by a caliper at the center of the area where the water barrier penetrant permeated. Measurements were carried out on three specimens to obtain one datum of permeation depth.

2.5 Methods of measuring physical properties of water barrier penetrants

Surface tension of water barrier penetrants was measured by du Noulli's method, specific gravity by a method prescribed in JIS B 7525, and viscosity by a method employing the B type viscometer. The evaporation rate of volatile ingredients was obtained by measuring the change in weight of the penetrants upon exposure to air at 20°C and 60%RH. The non-volatile content of the penetrants was obtained by measurement of the weight of residue remaining after the penetrants were heated to 110°C and maintained at this temperature for 3 hours. Measurement of evaporation rate and non-volatile content was carried out two times for each type of penetrant, and the mean values of measurements were used for analysis.

3 Results and discussion

3.1 Results of permeability test

Fig. 2 shows the relationship between time and permeation depth for penetrant S1 and water. This figure shows that permeation velocity is relatively high at the time of application and slows down with the elapse of time.

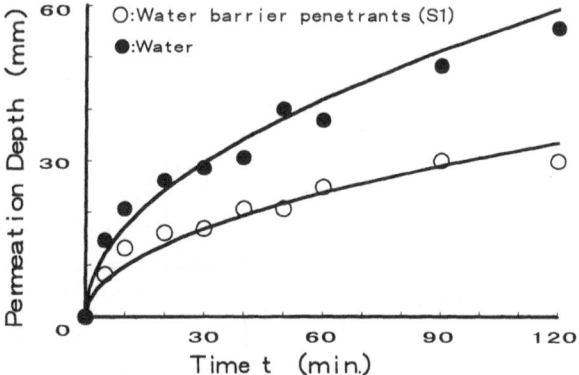

Fig. 2 Relationship between permeation depth and elapsed time

The relation between permeation depth and time can be derived from equation (1), which was developed by Washburn [4] on the basis of capillary permeation theory.

$$\ell = \frac{r}{2}\sqrt{\left(2\frac{\gamma_L \cos\theta}{r} + \Delta p\right)\frac{t}{\eta}} \tag{1}$$

where ℓ = permeation depth, r = capillary radius, γ_L = surface tension, θ = angle of contact, Δp = difference in pressure between both ends of the capillary tube, t = time, and η = viscosity of liquid.

In equation (1), the permeation depth is proportional to the square root of time t. Therefore, equation (1) could be rewritten as equation (2).

$$\ell = a\sqrt{t} \qquad\qquad (2)$$

where, ℓ = permeation depth (mm), a = coefficient of permeation (mm·min$^{-1/2}$), and t = time (min).

The permeation coefficient (a) was determined empirically by plugging the test results into equation (2) and applying the least squares method. In Fig. 2, permeation depth, as estimated by equation (2) with the empirically determined coefficient (a), is represented by a solid line. In the present paper, permeability of water barrier penetrants is evaluated in reference to permeation coefficient (a).

3.2 Relationship between physical properties of water barrier penetrants and coefficient of permeability.

The physical properties and permeation coefficients of water barrier penetrants are shown in Fig. 3. Multiple regression analysis was performed in order to clearly express the relationships between properties of the penetrants and their permeation coefficients. In this analysis, the values of the properties were treated as predictive variables (Xi), and the permeation coefficient (a) was treated as criterion variable (Y).

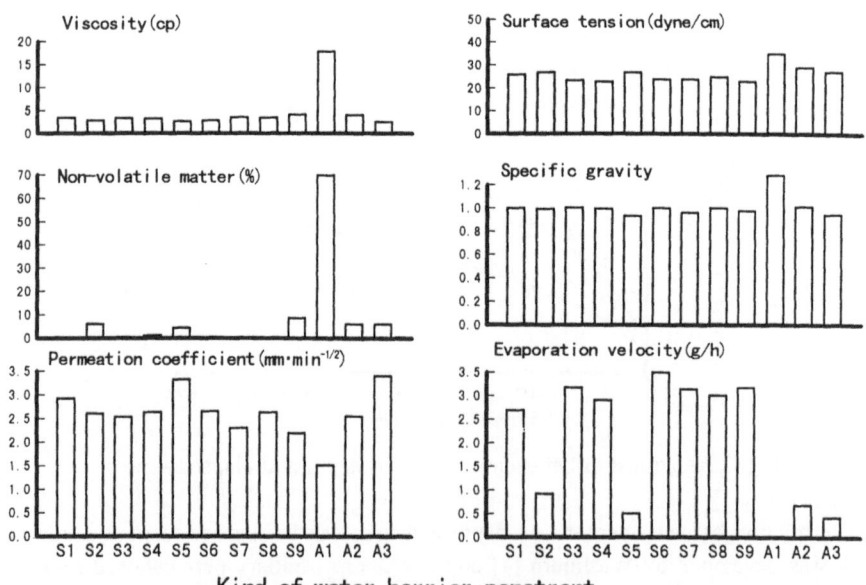

Kind of water barrier penetrant

Fig. 3 Properties of water barrier penetrants and permeation coefficients

A multiple regression model of equation (3) was assumed.

$$Y = 1 \, 0^{K_0} X_1^{K_1} \sim X_i^{K_i} \tag{3}$$

where, K_i = coefficient which shows the effect of each predictive variable (X_i) on criterion variable (Y).

Table 1 shows each of the variables. Table 2 shows the correlation coefficients between each of the predictive variables (X_i) and the criterion variable (Y). This table also shows the mutuality of each X_i. These results show that the permeation coefficient increases with increasing surface tension and specific gravity, whereas the coefficient decreases with increasing viscosity and content of non-volatile matter.

On the basis of the mutuality of each coefficient X_i as shown in Table 2, the predictive variables X_3 and X_4 were struck off the predictive variable in order to control the influence of the mutual relation between each predictive variable X_i and to obtain a more reliable regression equation. The multiple regression analysis was performed on the basis of variance selection method in which a variable was selected

Table 1. Variables

	X_1 Surface Tension (dyne/cm)
Pre-dictive Variable	X_2 Specific Gravity
	X_3 Viscosity (cp)
	X_4 Evaporation Rate (g/h)
	X_5 Non-Volatile Matter (%)
Criterion Variable	Y Permeation Coefficient ($mm \cdot min^{-1/2}$)

Table 2. Correlation Coefficient Matrix

	X_1	X_2	X_3	X_4	X_5
X_1	1.00				
X_2	0.73*	1.00			
X_3	0.70*	0.96**	1.00		
X_4	-0.94**	-0.73*	-0.73*	1.00	
X_5	0.61*	0.41	0.47	-0.74*	1.00
a	-0.41	-0.84**	-0.90**	0.45	-0.33

Notes   *:Significance level=5%
         **:Significance level=1%

Table 3. Analysis of Variance

Source of Variation	Sum of Square	Degrees of Freedom	Variance	Variance Ratio	F 0.01
Regression	0.08	2	0.0400	36.36**	8.02
Residual	0.01	9	0.0011	-------	-----
Total	0.09	11	-------	------	------

Notes **:Significance level = 1%

if its variance ratio F was greater than 2.0. The regression equation obtained from this analysis is shown in equation (4). The coefficient of determination R^2 was 0.898, and this equation was judged as statistically significant at the significance level of 1% on the basis of the analysis of variance shown in Table 3.

$$Y = 1 0^{-0.375} \cdot X_1^{0.757} \cdot X_3^{-0.484} \qquad (4)$$

where, Y = permeation coefficient (mm·min$^{-1/5}$), X_1 = surface tension (dyne/cm), and X_3 = viscosity (cp).

On the assumption that the pressure (Δp) and angle of contact (θ) are equal to zero, equation (1) could be rewritten as equation (5).

$$a = K \cdot \gamma_L^{1/2} \cdot \eta^{-1/2} \qquad (5)$$

where, K = constant.

The variables of equations (4) and (5) coincide with each other and when we compare the empirically derived equation (4) with the theoretical equation (5) obtained from Washburn's equation (1), we find that the exponents of the two equations are almost identical. Consequently, the influence of the physical properties of water barrier penetrants on permeation depth can be fundamentally presumed on the basis of capillary permeation theory under the conditions that no pressure is applied and that the penetrants are supplied in a quantity sufficient to permeate the specimens.

4 Conclusions

The following conclusions were obtained in the present study.
(1) Permeation depth is proportional to the square root of time. The permeation coefficient increases with increasing surface tension and specific gravity and decreases with increasing viscosity and content of non-volatile matter.
(2) An equation for estimating permeation depth from surface tension and viscosity of the penetrants was derived from multiple regression analysis.
(3) The influence of the physical properties of water barrier penetrants on permeation depth can be fundamentally presumed on the basis of capillary permeation theory.

5 References

1.   Numao, T., Fukuzawa, K., Iwamatsu, S. and Okuzawa, T. (1987) Permeability of Mortar Painted with Impregnating Paints, Trans. of JCI, Vol. 9, pp. 63 - 70.
2.   Ono, H., Ohgishi, S. and Itoh, S. (1990) Improvement Effect of Strength and Durability of Cement Mortar by Inorganic Polymer Impregnation, Journal of Structural and Construction Engineering, Trans. of AIJ, No. 407, pp. 13 - 22.
3.   Wada, T., Yamashita, T. and Shindo, T.(1987) Fundamental study on Silane water barrier penetrants, Proc. of the 42nd annual conference of JSCE, pp. 422 - 423.
4.   Washburn, E. W. (1921) The Dynamics of Capillary Flow, Physical Review, Vol. 17, No. 3, pp. 273 - 283.

THE FATIGUE BEHAVIOR OF R/C BEAM REPAIRED WITH CONCRETE-POLYMER COMPOSITES

J.S. Sim and I.H. Bae
Dept. of Civil & Environmental Engineering, Hanyang University, Ansan, Korea
Y.K. Hong
Dept. of Architectural Engineering, Hongik University, Seoul, Korea
E.S. Hwang
Dept. of Civil Engineering, Kyunghee University, Yongin, Korea

Abstract
The purpose of the study is to investigate the fatigue behavior of the reinforced concrete beams repaired with concrete-polymer composites. The beams, which were produced with the tension part defected, were repaired with polymer composites such as epoxy and polyester, polymer made cement such as latex and premix, and cement grout. Then static and dynamic tests were performed.

Bonding of repair materials, relationship between the load and midspan deflection, crack growth, failure mode, and S-N curves to present the fatigue strength of the repaired beams were investigated throughout the tests.

From the results of tests, the polymer material was the most excellent on the ability of bonding to concrete beams. In case of the polymer-cement materials and cement grout, cracks occurred on the layer between the concrete and repair material. It was shown that the midspan deflection of the repaired beams were similar to those of the unrepaired beam in case of the stress level of 60%. However, at higher stress levels, only beams repaired with polymer-cement materials showed similar trend to the unrepaired beam. According to the S-N curves, the fatigue characteristics of beams repaired with polymer-cement were the most similar to those of the unrepaired.

Keywords: composite material, concrete, fatigue, repair

1 Introduction

For reinforced concrete structures, cracks develop due to several causes including impact, repeated loads, shrinkage, temperature change, and creep under permanent loads. They cause the corrosion of the reinforcement by chemical action and then spalling occurs since the corrosion increase the volume of material by 8 times. Therefore, structures deteriorated by cracks need repair to prevent functional deficiencies. In addition, especially for bridges subjected to repeated fatigue loading,

Polymers in Concrete, edited by Y. Ohama, M. Kawakami and K. Fukuzawa. Published in 1997 by E & FN Spon, 2–6 Boundary Row, London SE1 8HN, UK. ISBN: 0 419 22330 4.

durability of the repair material is needed.

The purpose of this study is to establish a fundamental basis of the optimum repair method for damaged reinforced concrete members subjected to dynamic loading. Dynamic tests were performed on reinforced concrete beams repaired with concrete-polymer composites to investigate the fatigue characteristics of the various repair materials.

2 Test Plan

2.1 General
The fatigue tests were performed on simply-supported reinforced concrete beams. Damage (spalling) on beams was artificially made at the tension part and was repaired with three types of material: polymer, polymer-cement and cement based materials. Repair work was done by injecting or patching.

2.2 Test parameter
Each specimen has constant parameters such as the ratio of repair thickness to beam depth (0.24), the ratio of repair length to span (0.75) and the ratio of repair volume (0.15). 18 specimens were tested with different repair materials (two polymer based types, two polymer-cement based types and one cement based type) and various stress levels (60%, 70% and 80% of static strength) and they are summarized in Table 1.

Table 1. Test specimen and parameters

Class	Repair material	Specimen	
		Name	Stress level (%)
		CON-1	80
Unrepaired		CON-2	70
		CON-3	60
		DT5E1	80
	Epoxy	DT5E2	70
Polymer		DT5E3	60
based		DT5P1	80
	Polyester	DT5P2	70
		DT5P3	60
		DT5L1	80
	Latex	DT5L2	70
Polymer-cement		DT5L3	60
based		DT5R1	80
	Premix	DT5R2	70
		DT5R3	60
		DT5G1	80
Cement based	Grout	DT5G2	70
		DT5G3	60

2.3 Materials
Characteristics of the materials used in the test including the reinforcements, concrete, and repair materials as follows.

2.3.1 Concrete and reinforcement
Ordinary portland cement, natural sand, and 25mm maximum size crushed granite were used and their mix proportion were 1:2.88:3, with a water-cement ratio of 0.55. Design strength of the concrete is 210 kgf/cm^2 (20.6 MPa) and compressive strength at 28days is 220 kgf/cm^2 (21.6 MPa) and slump is 22 cm. The reinforcing bar was D10 and D13 with yielding stress of 4,000 kgf/cm^2 (372 MPa).

2.3.2 Physical properties of the repair materials
Physical properties of the repair materials from the standard cylinder test are shown in Table 2.

Table 2. Physical properties of the repair materials

Physical properties (kgf/cm^2) (MPa)	Polymer based		Polymer-cement based		Cement based
	Epoxy	Polyester	Latex	Premix	Grout
Compressive strength	797 (78.1)	704 (69.0)	286 (28.0)	383 (37.5)	597 (58.5)
Elastic modulus (*10^5)	2.17 (0.213)	1.19 (0.117)	1.52 (0.149)	2.01 (0.198)	4.67 (0.458)
Flexural strength	238 (23.3)	222 (21.8)	62 (6.08)	42 (4.11)	109 (10.7)
Tensile strength	73 (7.15)	71 (6.96)	29 (2.84)	33 (3.23)	47 (4.61)
Adhesive strength	98 (9.60)	63 (6.17)	24 (2.35)	17 (1.67)	21 (2.06)

2.4 Fabrication of specimen
Dimensions of the specimen are shown in Fig.1. Total length of the beam is 240cm and distance between the supports are 200cm. Reinforcing bars are consisted of 2-D10 in compression side, 2-D13 in tension side, and D10 stirrups in 10cm spacing. 33% of the balanced reinforcement ratio is used. To simulate spalling of concrete, styrofoam

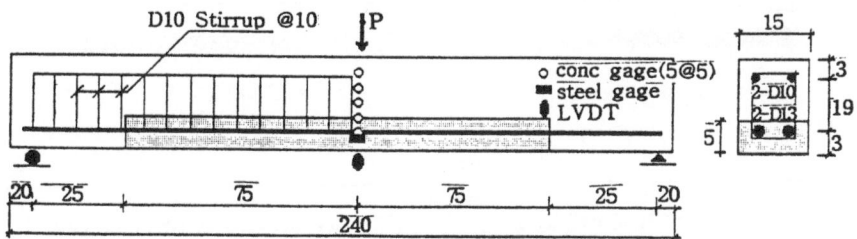

Fig. 1. Dimensions of the specimen

was used. Its size is 5cm deep and 150cm long. After 28 days curing, the specimens were repaired with 5 different materials by patching(for premix) or injecting.

2.5 Loadings and measurement
Fatigue test machine, made by Shimadzu Co. with a capacity of 50tonf and 50mm stroke, was used in the tests. The beams were simply supported and loaded at the midspan. Up to 10 cycles, the specimens are loaded statically then repeated loads were applied with 2Hz speed. Crack development and midspan deflection were checked and measured with the data acquisition system.

3 Test results

3.1 Static tests
Ultimate strength of the specimen varied from 4.37ton for premix specimen to 5.92ton for polyester specimen and there is not much difference between repaired and unrepaired specimen. Debonding between the concrete and repair material occurs slightly after yielding of tension rebars except for the epoxy specimen. However, for premix specimen repaired with the patching method, large cracks occurred between the concrete and repair material due to the low bonding strength of the material.

3.2 Fatigue tests
3.2.1 Crack and failure mode by number of loading
Typical flexural crack growth tendency was found in unrepaired, epoxy, polyester, and grout specimens regardless of the stress level. However, for latex and premix, crack grew at the interface between the concrete and repair material. Figure 2 shows the crack growth for the stress level of 70%. N in the figures shows the number of loading until failure. Specimen with 80% and 90% of stress level showed similar results.

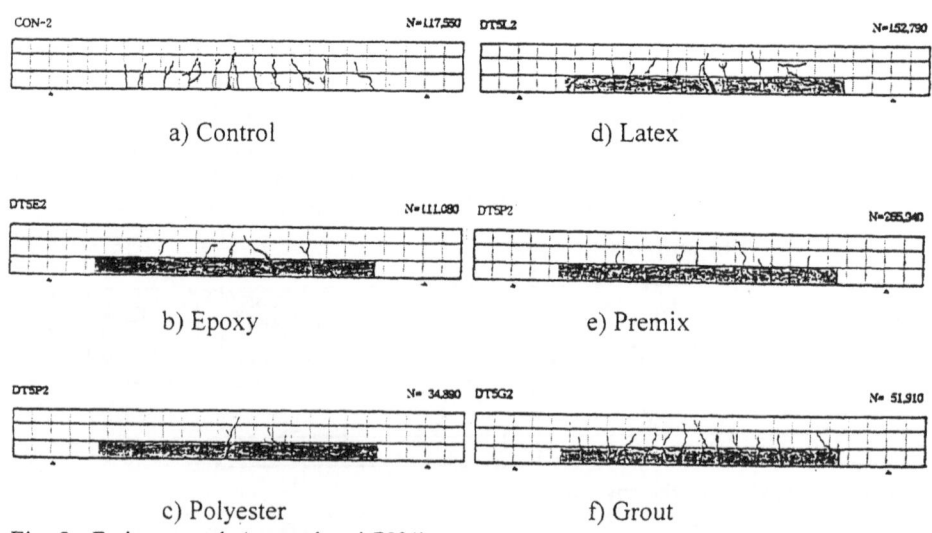

a) Control   d) Latex

b) Epoxy   e) Premix

c) Polyester   f) Grout
Fig. 2. Fatigue crack (stress level 70%)

3.2.2 Maximum midspan deflection by number of loading

Maximum midspan deflection occurred in the polyester specimen regardless of stress level and minimum in epoxy specimen. Fig. 3 shows the midspan load-deflection diagram at the stress level of 60% and 70% until 40,000 loadings. As the number of loading increases, epoxy, latex, premix and grout specimens show deformation similar to the unrepaired specimen at the stress level of 60% (service load state). At the stress level of 70%, where crack has developed, only latex and premix specimens show deformation similar to the unrepaired specimen. Grout specimen showed larger deflection than the unrepaired specimen at the stress level beyond the service load and latex specimen showed the most similar deflection to unrepaired specimen at the stress level of 80%.

Largest permanent deformation occurred at the first cycle of loading regardless of repair material and stress level, especially for grout specimen. The rate of deflection increase dimished rapidly as the number of loading increased.

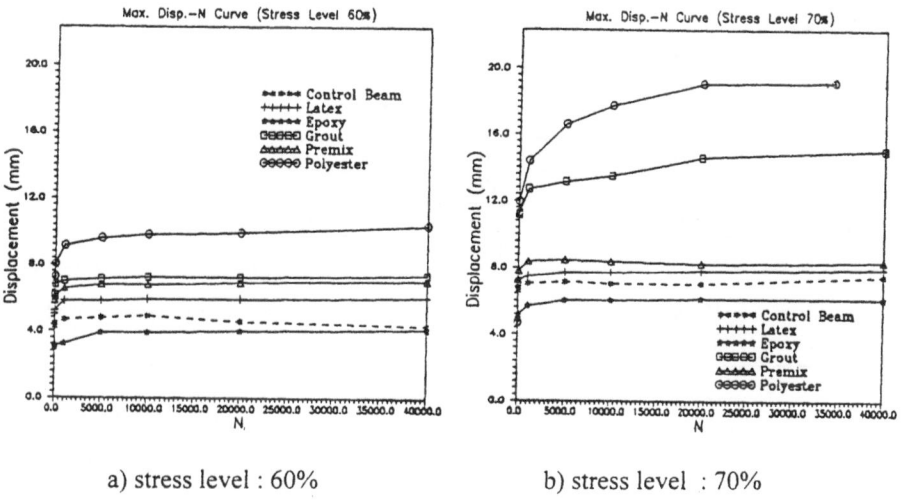

a) stress level : 60%   b) stress level : 70%

Fig. 3. Load-deflection curve

3.2.3 Fatigue strength by repair material

S-N curves, which show the relationship between the applied loads and number of loading, were used to forecast the fatigue strength, i.e. stress level at the certain number of loading. Though most metals have their own fatigue strength, concrete use the fatigue strength at the certain number of loadings. Many researchers show different fatigue strengths at one million cycles of loading but, in general, fatigue strength is known to 50% to 70% of the maximum static strength[1].

Table 3 shows the fatigue strength and fatigue strength ratio of the repaired specimen to unrepaired specimen. The fatigue strength at one million cycles of loading varies from 50% to 60% and the ratio varies 0.6 to 1.2. Latex and premix specimens show the most similar fatigue strength to unrepaired specimen and highest strength for grout specimen. Table 4 shows the regression analysis results for various repair materials.

Table 3. Fatigue strength at one million cycles of loading

	Unrepaired	Epoxy	Polyester	Latex	Premix	Grout
Fatigue Strength (%)	50.4	31.1	54.9	47.3	50.9	57.6
Repair/Unrepaired	1.0	0.62	1.09	0.94	1.01	1.14

Table 4. Regression analysis results

	Stress level (%)	Number of cycle at failure (N)	Regression equations
Unrepaired	80	69,840	
	70	117,550	$S=-10.4347*Ln(N)+194.511$
	60	425,880	
Epoxy	80	51,120	
	70	111,080	$S=-16.702*Ln(N)+261.837$
	60	162,040	
Polyester	80	340	
	70	34,890	$S=-3.200*Ln(N)+99.106$
	60	60,100	
Latex	80	57,390	
	70	152,790	$S=-11.592*Ln(N)+207.423$
	60	318,490	
Premix	80	107,140	
	70	265,940	$S=-13.270*Ln(N)+234.272$
	60	471,420	
Grout	80	460	
	70	51,910	$S=-3.014*Ln(N)+99.200$
	60	176,390	

4 Conclusions and remarks

In this study, the following conclusions were reached from the fatigue tests of the specimens repaired with polymer and polymer-cement type repair materials.

- Under fatigue loading, specimens repaired with polymer and cement type materials showed typical bending failure regardless of the stress level. On the other hand, for specimens repaired with polymer-cement type material, crack started at the layer between the concrete and repair material and bending failure occurred.
- For midspan deflection under fatigue loading, the polyester specimen showed the largest deflection and epoxy showed smallest. Under service load state(stress level of 60%), all specimen showed the similar tendency to the unrepaired specimen but as the stress level increased, only specimen repaired with polymer-cement type material showed similar results.

- From the regression analysis on the fatigue strength at one million cycled, specimen repaired with polymer-cement type material showed similar results to the unrepaired specimen. However, due to less bonding strength, crack occurred at the layer between the concrete and repair material. Further research is needed to solve this problem.

5 References

1. Hsu, T.C. (1981) Fatigue of Plain Concrete, *ACI Journal*, Vol.78, pp. 292-305.
2. Shi, X.P., Fwa, T.F. and Tan, S.A. (1993) Flexural Fatigue Strength of Plain Concrete, *ACI Materials Journal*, Vol. 90, No. 5, pp. 435-440.
3. Basunbul, I.A., Gubati, A.A., Al-Sulaimani, G.J. and Baluch, M.H. (1990) Repaired Reinforced Concrete Beams, *ACI Materials Journal*, Vol. 87, No. 4, pp. 348-354.
4. Vipulanandan, C. and Paul, E. (1990) Performance of epoxy and polyester Polymer Concrete, *ACI Materials Journal*, Vol. 87, No. 3, pp. 241-251.
5. ACI Committee 215 (1986) *Considerations for Design of Concrete Structures Subject to Fatigue Loading*, pp. 1-25.
6. Yeon, K.S., Jung, Y.S., Han, M.Y., Lee, J.Y., Jang, T.Y. and Jung K.H. (1995) Repair of Reinforced Concrete Structures - Repair Materials and Methods, *Proceeding of Korea Concrete Institute Spring Conference*, Vol. 7, No. 1, pp. 212-218.
7. Sim, J.S., Hong, Y.K., Hwang, E.S., Bae, I.H. and Lee, E.H. (1995) Repair of Reinforced Concrete Structures - Static and Dynamic Flexural Characteristics, *Proceeding of Korea Concrete Institute Spring Conference*, Vol. 7, No. 1, pp. 225-230.

STRENGTHENING OF TIMBER STRUCTURES USING POLYMER CONCRETE

D. Van Gemert and A. Beeldens
Civil Engineering Dept., Katholieke Universiteit Leuven,
Heverlee, Belgium

Abstract
The structural restoration of deteriorated beam ends by means of anchored epoxy mortar prostheses is presented. The design method is based on the results of an experimental program concerning the bond strength between wood and epoxy mortar as a function of the degree of humidity of the wood, the anchoring strength of glass-fibre reinforced polyester rods and steel reinforcement bars glued in wood or in epoxy mortar, and the structural and rupture behaviour of restored real-size beams. A case-study presents the application of the technique for the repair of a broken beam in an historic castle in Horst, Belgium.
Keywords: epoxy, mortar, strengthening, timber.

1 Introduction

Structural restoration of wooden beams can be necessary due to deterioration of the beam ends, at which they are supported in the masonry walls. Due to entering humidity the wood material becomes a preferred food for fungi and beetles. In most cases the middle part of the beam is still in excellent condition, and can carry the loads as before. However, the end parts are loaded by shearing forces, and a real danger of severe damage to the structure exists. Depending on the degree of deterioration of the beam end, there can arise a serious sagging of the beam and the part of the structure resting upon it, or the beam can collapse and cause partial collapse of the building. In these cases the beam ends have to be restored as soon as possible. Until now this has been done by replacing the

Polymers in Concrete, edited by Y. Ohama, M. Kawakami and K. Fukuzawa. Published in 1997 by E & FN Spon, 2–6 Boundary Row, London SE1 8HN, UK. ISBN: 0 419 22330 4.

decayed portion with a new piece of wood. The new wooden beam end was connected to the sound part of the original beam using scarfed and glued lap joints or glued joints with special geometries. To obtain a solid connection these joint constructions extend over an important length, and they generally include the removal of a considerable part of still sound wood. As a consequence this classical repair method involves a great amount of skilled hand work. In many cases it would be easier to replace the entire beam, but this drastic intervention can be prevented by artistic, emotional or even economic reasons. In such cases the deteriorated parts of wood can be removed and epoxy mortars cast in situ to reintegrate them.

In special cases the technique can be extended to other damage or strengthening cases. The epoxy mortar has a good bond to the wood, its modulus of elasticity is nearly equal to the one of wood, and plastic behaviour after first cracking can be assured by glued-in anchoring rods. With these characteristics the epoxy mortar can also be used in the zones, loaded in bending. Such an application will be discussed in the case-study.

2 Principles of repair method

The affected part of the wooden beam is thoroughly inspected in order to determine the extension of the wood attack into the beam. The beam end will now be denuded completely by removing the surrounding brickwork, or by making a hole through the wall. From now on the beam has to be supported at the decayed end by a scaffolding. The decayed part is sawn off as far as it extends. No affected wood will be left in place, because it is unable to transfer forces, and because it is a possible source of future deterioration.

A casing is now installed which reconstituates the original shape of the beam end. If the reconstruction has to meet aesthetic requirements, the casing must be made of the same wood as the original beam, and it will be kept in place to cover the epoxy mortar. Holes are now bored into the beam and reinforcement bars are inserted into these holes. Normally the holes are bored from the top surface of the beam at a distance of some 500-600 mm from the sawcut, slightly sloping downwards towards the wall. Depending on local circumstances the holes are also bored horizontally, or inclined in the other direction. The reinforcement bars are ribbed steel bars or pultruded glassfibre reinforced polyester bars. The space between the reinforcement bar and the wall of the hole is filled with an epoxy resin grout. This grout can be poured or injected under pressure. It cures after a short time and so it anchors the reinforcement rod in the sound wood. The casing is now filled with an epoxy mortar. The mechanical characteristics of this epoxy mortar are excellent, it has a strong bond to wood and to the reinforcing bars, and it is not affected by moisture.

The required mechanical properties for the epoxy mortar are: bending strength 20 N/mm² and compressive strength 40 N/mm² (NBN B12-208), modulus of elasticity 8.000 N/mm² (NBN B15-230). The epoxy resin grout should minimally have a tensile strength of 20 N/mm² and a modulus of elasticity of 2500 N/mm² (ASTM D638). All the repair materials are subjected to a continuous quality control system, according to UEAtc, the European Union for Technical Agrement in construction.

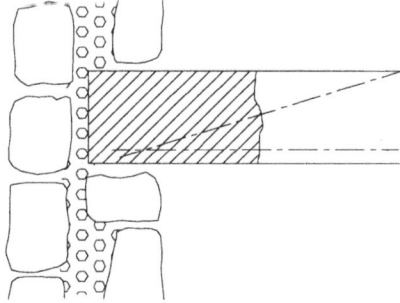

Fig. 1.   Principle of repair method

After some 24 hours of cure, the epoxy mortar reaches its final strength, and the supports can be removed. The restoration can be completed by matching the original surface colour and relief. An important element is the relatively low cost of this chemical restoration procedure, as compared to other more classical methods. An experimental program was executed in the Reyntjens Laboratory to obtain realistic experimental data about the structural behaviour of chemically restored beams, and to set up a reliable design and calculation method [1].

3 Experimental program

The experimental program contained three chapters. Firstly the bond between wood and epoxy mortar was studied. As this bond revealed to be lower than the cohesion strength, it becomes necessary to take into account the strengthening effect of the reinforcing bars. Therefore we need data about the anchoring strength of the bars in the wood and in the mortar. The last part of the experiments comprised loading tests on real scale oak beams, with one beam end built up of epoxy mortar, which was attached to the wooden beam by means of reinforcing bars. The epoxy mortar was DELTAPOX RM, kindly provided by De Neef Engineering, Belgium.

3.1 Bond strength between wood and epoxy mortar
The bonding strength between timber and epoxy mortar was determined using a four-point bending test on a prismatic test sample, with dimensions 70 x 70 x 280 mm, composed of a wooden part and an epoxy-mortar part (Fig. 2). From the rupture load P_R, the ultimate tensile strength at rupture between wood and epoxy-mortar was calculated. The wooden parts of the test samples were sawed from a larger oak beam, so that the fiber direction of the sample coincides with its longitudinal direction. The wooden samples have been kept at different environments so as to vary their relative humidity, which is determined as the weight percentage of moist content to dry weight. Some of the wooden blocs were treated with the pure resin of the epoxy-mortar, which was applied on the contact surface as a primer, before the casting of the epoxy-mortar itself.

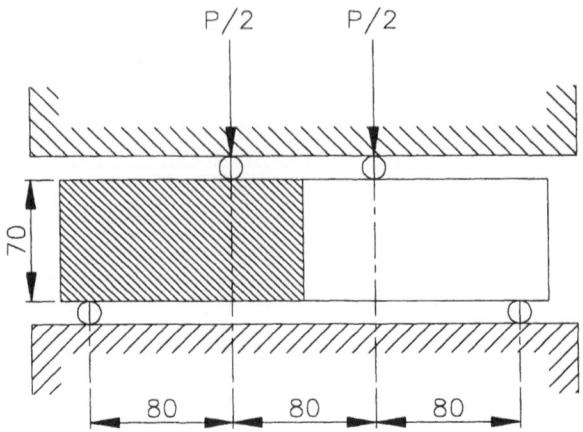

Fig. 2. Bending test set-up

The bond strength clearly decreases as the relative humidity increases. It is possible to find an explicit formula for this relationship, using linear regression. Linear regression also shows that the bond strength is not depending on the use of the pure resin as a primer (Fig. 3). The relation between the relative humidity and the bond strength is characterised by the equation: bond strength (N/mm^2) = -0,14 x R.H. (%) + 10,24.

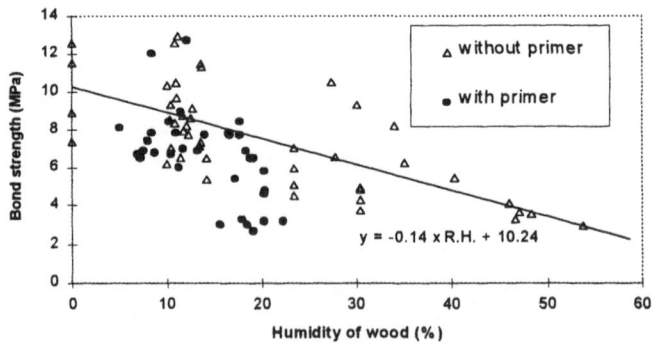

Fig. 3. Relation bond strength - humidity of wood

3.2 Anchoring strength of reinforcement bars

The test program started with a comparison of three types of reinforcement bars : smooth glass-fibre reinforced polyester, notched GRP-bars and ribbed steel bars [1]. The test results showed the poor performances of smooth glass fibre polyester bars, and to a lesser extent of notched glass fibre polyester bars. Therefore, the tests were continued only with ribbed steel bars of 20 mm diameter, quality BE 400.

The anchoring strength of the steel reinforcement bar to the wood determines the length, needed for gluing the steel bar to obtain rupture of the bar before slipping of the connection under a tensile load on the bar.

Two different types of tests were executed to determine this anchoring length:

- pull-out tests, in which a steel bar is glued centrally in a wooden block, and then submitted to a tensile test.
- beam-tests, in which a tensile test is executed on a steel bar, wich is glued excentrically in a wooden block, which involves a bending moment. In this test the constraining action of the jack in the pull-out test is avoided.

Figure 4 shows schematically the two test set-ups.

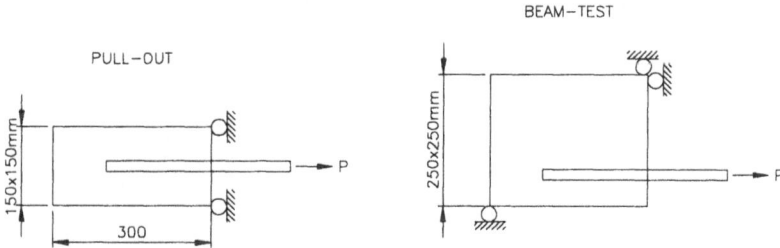

Fig. 4. Schematical representation of the pull-out-test and the beam test

By gluing bars over different lengths in the wood, the anchoring length which satisfies the above criterion can be determined. However, the anchoring length is a function of the tensile strength of the steel bar. The mean shear stress at rupture is a strong simplification of the stress distribution in the anchoring zone, but is used as a design tool.

During the first loading step, shear stresses will vary exponentially alongside the steel bar. During the second loading step, after fissuration in the first step, the curve of the shear stresses will have a triangular shape. This mean shear stress of rupture also tends to be depending on geometrical factors as the anchoring length itself, diameter of the bore hole and probably diameter of the steel bar.

The following tests were executed :

10 pull-out-tests. Steel bars of quality BE400, diameter 20 mm, were glued in wooden blocks of 150 x 150 x 300 mm, with bore holes of diameter 32 mm and different lengths.

7 beam-tests. 3 steel bars of quality BE400, and diameter 20 mm, were glued in wooden blocks of 600 x 250 x 250 mm, with bore holes of diameter 32 mm (3 samples) and of diameter 25 mm (4 samples), and gluing over different lengths.

The results are represented in table 1. From table 1 it can be seen that:

- the mean shear stress at rupture increases slightly with decreasing gluing length. The characteristic value (P > 0,95) is more then 8,0 MPa for the three tested lengths;
- the diameter of the bore hole and the gluing length have an influence on the mean shear strength at rupture in the beam tests;
- influence of diameter of steel bar was not studied;
- the constraining action of the jack in the pull-out test seems to be very limited, probably due to the limited stiffness of the wood in the transverse direction.

Table 1. Results of pull-out and beam tests

Sample number	Type	Length	Force of rupture	Mean shear stress of rupture	Type of rupture
		mm	kN	MPa	
P0	pull-out φ32	300	161	8,5	slipping of bar
P1	pull-out φ32	300	190	10,1	slipping of bar
P2	pull-out φ32	300	183	9,7	slipping of bar
P3	pull-out φ32	300	197	10,5	slipping of bar
P4	pull-out φ32	250	147	9,4	fracture of wooden block
P5	pull-out φ32	250	159	10,1	fracture of wooden block
P6	pull-out φ32	250	170	10,8	slipping of bar
P7	pull-out φ32	200	166	13,2	fracture of wooden block
P8	pull-out φ32	200	158	12,6	fracture of wooden block
P9	pull-out φ32	200	135	10,7	slipping of bar
B1	beam φ32	600	188	>5,0	rupture of steel
B2	beam φ32	500	173	>5,5	rupture of steel
B3	beam φ32	400	171	6,8	slipping of steel
B4	beam φ25	400	192	>7,6	rupture of steel
B5	beam φ25	350	176	>8,0	rupture of steel
B6	beam φ25	300	161	8,5	slipping of steel
B7	beam φ25	200	139	11,1	slipping of steel

Full scale tests have been executed on oak wood beams, to which the epoxy mortar prostheses were anchored using different configurations of the steel anchoring rods [2]. If only inclined bars can be placed, they should cut the bonding face as low as possible, because the plastic reserve after cracking of the bond face is limited in this case. Such layouts of anchoring bars should be avoided. All the beams were tested in an asymmetric three point bending test. Based on the real scale experiments the adopted design philosophy is as follows. First a calculation in bending is made, at which the bond stress is limited to the appropriate bond strength in Fig. 3. The number of reinforcing bars is calculated, taking into account the geometrical disposition and the feasible gluing length, by comparing the necessary anchoring force with the experimental values in table 1. Here a suitable safety factor is applied, because in the real scale tests the force in the bars at rupture of the beams was very variable, and always smaller than in the pull-out tests. This is probably due to bending of the steel bars, by which the wood fibres are splitted longitudinally.

4 Case-study

The oldest parts of the Horst Castle in St. Pieters Rode at 20 km distance from Leuven date back to the 13th century. The castle was drastically restored in the 16th century. Important decorations were executed in the 17th century. In 1655 stucco cealings with scenes from the Roman writer Ovidius were installed. They are attributed to the artist J. Christian Hansche, the most famous stucco artist of the 17th century.

Due to imprudent actions of the manager, one of the floors nearly collapsed, as shown in fig. 5. This was partly due to overloading, but the wooden beam was also thoroughly penetrated by the death-watch beetle. The only possibility to save the worthful stucco consisted in removing the old timber as much as possible, and to replace it with epoxy mortar [3]. At first the loads on top of the wooden floor were carefully removed. These loads consisted of the concrete and the reinforcements, that had caused the collapse. Of course this concrete had hardened during the period of negotiations with the authorities about the necessary actions that had to be taken to save the floor and the stucco. After removal of the concrete the broken floor was lifted with hydraulic jacks to its original position. During the lifting operation the stucco was supported by strongly deflected thin glass fibre reinforced polyester rods. They could straigthen during the lifting operation and thus provide a continuous support for all the loose stucco parts, preventing them from falling off, and keeping them in place for later fixing with epoxy adhesive.

During the lifting operation the rupture zone was carefully observed and if necessary, some wooden splinters were removed to allow the two parts of the beam to fit perfectly together again. After reaching the horizontal position the beam was blocked, and the floor was made accessible for the strengthening works. A large groove was than sawed in the top side of the beam, as shown in fig. 6. In this groove a strong high grade steel reinforcement has been placed, as shown in fig. 7. The bending reinforcement has been given a parabolic shape, to limit the amount of original wood that had to be removed. The hollow beam served as a lost casing at refilling of the groove with epoxy mortar, fig. 8. At the top side some dowels are placed already now, to allow for a later strengthening of the beam by applying an additional layer of mortar on top of the actual one. The dowels will than act as shear connectors. The dowels are shown in fig. 8. After hardening of the epoxy mortar, the supports under the beam were removed, and the stucco was repaired and fixed again to the wooden supports. Fig. 9 shows the repaired beam and the saved stucco.

Fig. 5. Broken and collapsed beam

Fig. 6. Groove at top side of beam

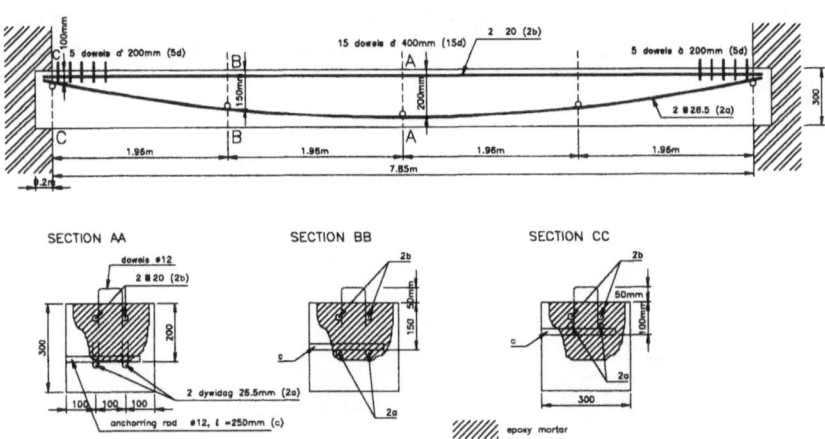

Fig. 7. Reinforcement of beam

Fig. 8. Epoxy mortar placed in groove

Fig. 9. Repaired beam and saved stucco

5 Conclusions

Structural restoration of wooden beams by means of epoxy mortar is a strengthening technique which can offer an elegant solution for the repair of deteriorated beam ends and other damages to timber beams. The tests showed that the bond between wood and epoxy mortar is limited and generally does not reach the cohesion strength of the wood. The placement of anchoring bars, especially steel bars, provides an improvement of the plastic reserve of the beam. The layout of the anchoring bars should always include horizontally placed anchors. The anchoring forces can be taken from the experiments. Innovative solutions are possible, as shown in the case-study, if the composite action of the wood and the polymer based repair mortar is taken into account.

6 References

1.   Van Gemert, D., Vanden Bosch, M. (1986) Structural restoration of wooden beams by means of epoxy resin, RILEM Materials and structures, pp. 165-170.
2.   Van Gemert, D., Horckmans, J. (1990) Concrete-Polymer composites in restoration of ancient monuments, Proceedings of the International Seminar on Concrete-Polymer Composites, Polymers-in-Concrete Committee, Japan Technology Transfer Association, Sept. 19.
3.   Van Gemert, D. (1995) Horst Castle - Repair of Cealing, Reyntjens Laboratory, Report 28263.

PART EIGHT
HIGH PERFORMANCE

MECHANICAL PROPERTIES OF POLYMER CONCRETE AND FIBER-REINFORCED POLYMER CONCRETE

B.H. Oh, S.H. Han, Y.S. Kim, B.C. Lee and H.S. Shin
Dept. of Civil Engineering, Seoul National University, Seoul, Korea

Abstract
Concrete material has some weak points in the sense of tensile strength as well as durability. Great effort has been made to enhance the basic performance of concrete. The polymer concrete may be one of the solution to improve this tensile and durability problems. Many studies have been conducted so far to explore the properties of plain polymer concrete. However there exist only a very few studies on fiber-reinforced polymer concrete.

The purpose of the present paper is therefore to explore mechanical properties of plain polymer concrete as well as fiber reinforced polymer concrete. The major test variables are the amount of steel fibers. The effects of curing temperature have been also studied. The stress-strain properties of polymer concrete and fiber-reinforced polymer concrete have been extracted. The present study indicates that a certain optimum value for the polymer-filler ratio exists for each mix proportion. The effect of curing temperature have been quantatively measured. The increase of durability and tensile resistance has been also obtained for fiber reinforced polymer concrete. The effects of fiber volume on the mechanical behavior of fiber reinforced polymer concrete have been presented.

The present study allows more realistic design and application of fiber reinforced polymer concrete to actual structures.
Keywords : polymer concrete, fiber reinforced polymer concrete, steel fibers, synthetic polypropylene fibers, unsaturated polyester resin.

Polymers in Concrete, edited by Y. Ohama, M. Kawakami and K. Fukuzawa. Published in 1997 by E & FN Spon, 2–6 Boundary Row, London SE1 8HN, UK. ISBN: 0 419 22330 4.

1. Introduction

As concrete is considered to be a two-phase composite material whose constituents are binder(or cement) and aggregate, the use of polymers in concrete for these two constituents dramatically increases its strength, stiffness and durability. But the inherent brittleness of this material poses serious limitations on its ready acceptance for many applications.

The purpose of the present paper is therefore to explore mechanical properties of polymer concrete as well as fiber reinforced concrete.

2. Materials

2.1 Polymer resin

Unsaturated polyester resins are used in this study. The characteristics of unsaturated polyester resins(ortho type) which used in this study are listed in Table 1. And we used MEXPO(Methyl-Ethyl Kethone Peroxide) as an intaitor by 1% addition of resin volume.

Table 1. The characteristics of unsaturated polyester

vicosity(25℃) (PS)	acidity (mg KOH/g)	hardening condition (%)	gell time(25℃) (minute)	SM content (%)	resin type
2~4	20	M1.0	8~15	38	low viscosity medium activity

unit of vicosity(PS) : Poise
SM content : Stylene Monomer content(weight) ratio in resin

2.2 Filler

The silica rock powder is used as a filler. Specific gravity of this powder is 2.75. By sieve analysis, it passes 100% through No.30 sieve(size:0.6mm) and 85% through No.100 sieve(size:0.15mm).

2.3 Aggregates

F.A. has a specific gravity of 2.64 and a fineness modulus of 2.18. And C.A. has specific gravity of 2.66, finess modulus of 7.12, and maximum size of 9.52mm. F.A. is river sand and C.A. is granite crushed gravel, respectively. The river gravel C.A. is used in some specimens for comparison with crushed gravel C.A.

2.4 Fibers

Steel fibers are blended in polymer concrete mix by 0%, 1.5%, 3.0% of total volume, and fiber lengths are varied 20mm to 40mm. Synthetic polypropylene fiber with the length of 19mm was also used by 0%, 1.5%, 3.0% volume.

3. Mechanical characteristics of polymer concrete

3.1 Variables and specimens

In order to determine optimum contents of polymer and filler, the filler-to-polymer ratio was varied from 0.0 to 1.7 as shown in Table 2. The polymer contents varied from 9.7% to 18.4% by weight for the total weight.

Table 2. Summary of test series for polymer concrete

		Test series No.					
		ZR1	ZR2	ZR3	ZR4	ZR5	ZR6
	resin	232	279	329	387	442	477
mix proportion	C.A.	712	715	724	744	756	771
(kg/m^3)	F.A.	1064	1068	1081	1112	1129	1152
	filler	391	338	266	156	73	0
resin content (by weight, %)		9.7	11.6	13.7	16.1	18.4	19.9
filler content (by weight, %)		16.3	14.1	11.1	6.5	3.0	0
filler/resin ratio		1.7	1.2	0.8	0.4	0.2	0

3.2 Experimental procedure

F.A., C.A. and filler are oven dried for 24hrs at 105℃ and those are slowly cooled down to room temperature 24℃. The dry-mixed aggregates and filler are blended with polymer resin in which 0.5% accelerator and 0.5% initiator of resin volume are added respectively.

The cylinder specimens are demoulded at 1 day after casting and cured in the oven at 45℃ for 7 days, and at the room temperature for 14 days.

3.3 Experimental results

3.3.1 Workability and setting time

Consistence test is used to estimate the workability of polymer concrete. The initial setting time to variable initiator(MEKPO) contents is explored. The workability of polymer concrete decreased with increasing polymer content as in following Fig 1. Therefore initail setting time and the content of initiator are properly selected by considering the batch time as shown in Fig. 2.

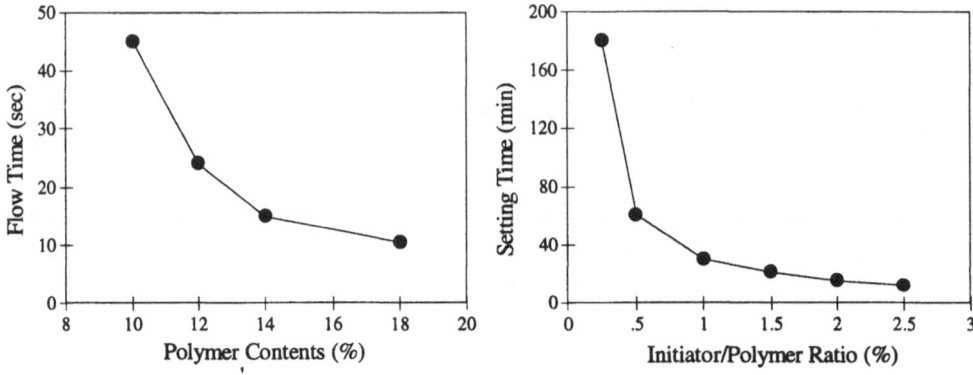

Fig 1. Flow Time(V-B test) vs
polymer contents for polymer concrete

Fig 2. Setting time vs MEKPO
contents for polymer concrete

3.3.2 Static compressive strength

The static compressive strength and modulus of elasticity of polymer concrete were examined by compressive test.

As shown in Fig. 3, the compressive strength increased as the filler/polymer ratio increased up to 0.8. However it started to decrease above the ratio 1.2. Therefore, it may be adequate to maintain the filler/polymer ratio 0.8~1.2.

The relations between compressive strength and modulus of elasticity were also examined. This results are skipped here due to the length limitation of the paper.

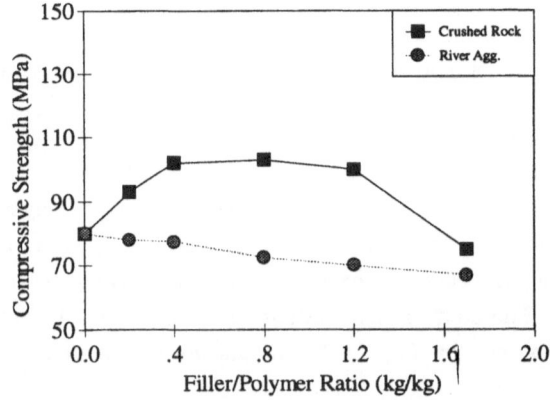

Fig. 3 Compressive strength for polymer concrete

4. Mechanical characteristics of fiber reinforced polymer concrete

4.1 Variables

The test variables of fiber reinforced polymer concrete are listed in Table 3. Among the variables are fiber type, fiber length and fiber contents. The curing temperature and filler/resin ratios are also varied to evaluate their effect on fiber reinforced polymer concrete.

Table 3. Mix proportions and test variables of fiber reinforced polymer concrete

specimen No.	fiber type	fiber length	fiber content(%)	resin (kg/m³)	filler (kg/m³)	F.A. (kg/m³)	C.A. (kg/m³)	resin/filler	temperature (℃)
AV0			0	398.3	398.3	747.4	1115.9		
AV1			1.0	405.6	405.6	722.1	1081.7		20±2
AV2			2.0	413.4	413.4	699.0	1047.2	1:1	
AHV0			0	398.3	398.3	747.4	1115.9		
AHV1			1.0	405.6	405.6	722.1	1081.7		60
AHV2		30mm	2.0	413.4	413.4	699.0	1047.2		
BV0	steel		0	399.2	599.0	665.7	998.5		
BV1	fiber		1.0	406.6	609.9	642.3	962.1		20±2
BV2			2.0	414.4	621.6	617.8	925.4	1:1.5	
BHV0			0	399.2	599.0	665.7	998.5		
BHV1			1.0	406.6	609.9	642.3	962.1		60
BHV2			2.0	414.4	621.6	617.8	925.4		
C1V		20mm	1.5	336	216	1056	672		
C2V		30mm	1.5	336	216	1056	672	1:0.6	
C3V		40mm	1.5	336	216	1056	672		20±2
DV0			0	336	216	1128	720		
DV1	synthetic fiber	19mm	1.5	336	216	1128	720	1:0.6	
DV2			3.0	336	216	1128	720		

4.2 Experimental procedure

The specimen manufacturing procedure is same as that in article 3.2. In order to see the effects of loading rate, the rate of loading varied from 1.67×10^{-5} *mm*/ sec to 6.67×10^{-3} *mm*/ sec. The splitting tensile tests, flexural strength tests(3rd point method) as well as compressive strength tests were performed under these loading rate.

4.3 Test results

4.3.1 Compressive strength
(1) Fiber content

Static and dynamic compressive strengths increased corresponding to the fiber content increase. The specimens which have 2% steel fiber volume, 60℃ curing temperature and 1.5 resin/filler ratio, showed 17% larger compressive strength than the plain polymer concrete specimens as shown in Fig. 4(a). And the specimens which have same mix propotios but 20℃ curing temperature showed 13% larger compressive strength in Fig. 4(b).

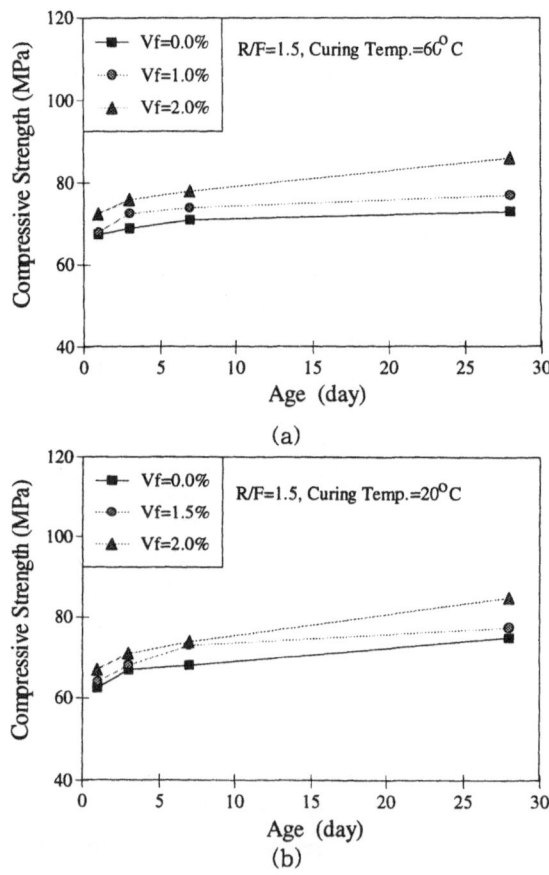

(a)

(b)

Fig. 4   Fiber content vs compressive strength of fiber reinforced polymer concrete

(2) Curing temperature

The effect of suring temperature on the compressive strength is shown in Fig. 5. High curing temperature increases the strength. This is because the high temperature accelerate the polymerization reaction.

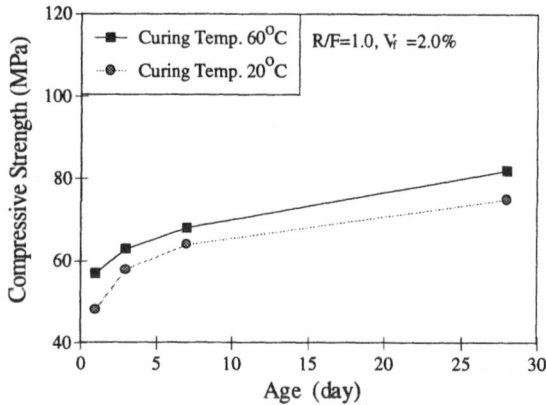

Fig. 5 Effect of curing temperature on the compressive strength of fiber reinforced polymer concrete

(3) Filler content

The compressive strengths according to filler content amount are shown in Fig. 6. It is seen from this figure that the specimens with ratio(Filler:F.A.:C.A.) of 1:1.7:2.5 showed smaller compressive strength than those with ratio of 1:1.0:1.5.

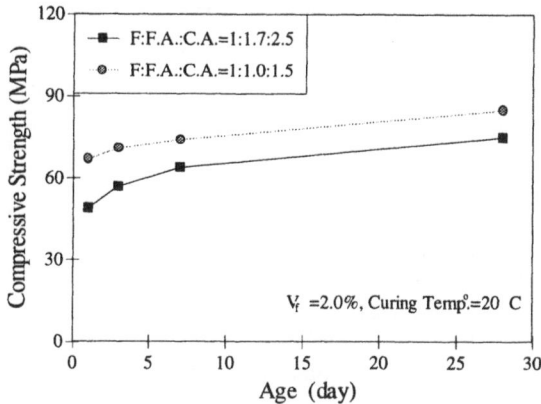

Fig. 6 Compressive strength of fiber reinforced polymer concrete for different filler amounts.

(4) Strain rate

Dynamic compressive strength tests are performed by different stroke rates, i.e. 0.1, 1.0, 10 and 40 mm/min, respectively. These correspond to the strain rates of 1.67×10^{-5} *mm*/sec, 1.67×10^{-4} *mm*/sec, 1.67×10^{-3} *mm*/sec, and 6.67×10^{-3} *mm*/sec, respectively.

As the strain rate increases, the compressive strength of fiber reinforced polymer concrete increases as shown in Fig. 7. In these Figures, the relative strength is defined that the ratio of compressive strength to that of plain polymer concrete with 1.67×10^{-5} *mm*/sec strain rate. This relation shall be represented as following Eq. (1).

$$\sigma_{dc} / \sigma_{oc} = a + b \ln(\dot{\varepsilon}) \tag{1}$$

where, σ_{dc} : dynamic compressive strength
σ_{oc} : static compressive sterngth
a, b : empirical constant (a=82~104MPa, b=1.515~2.790)

Irrespective of the fiber length, as the fiber contents increase, the compressive strength and the modulus of elasticity increase. But the longer fiber length is, the less the workability of fiber reinforced polymer concrerte is.

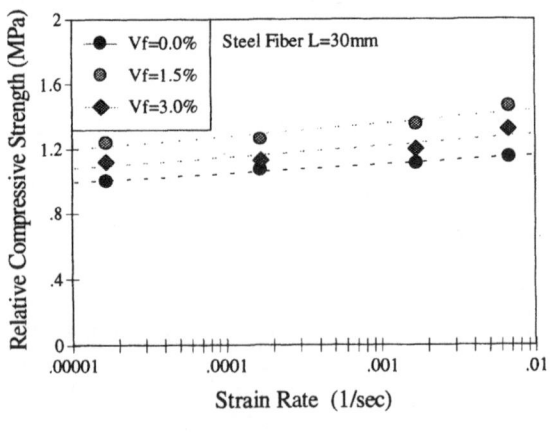

(a)

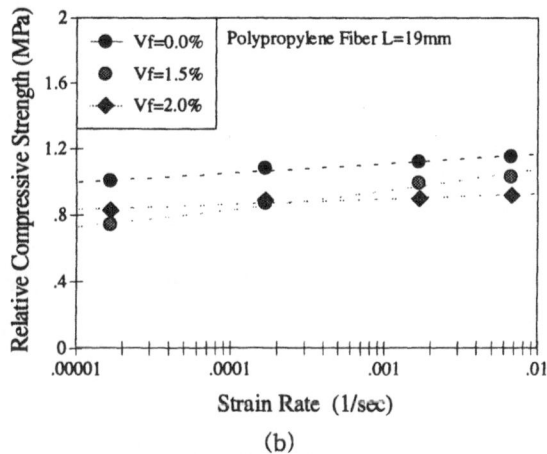

Fig. 7 Compressive strength as a function of strain rate for fiber reinforced polymer concrete

4.3.2 Flexural strength

Flexural strength characteristics of fiber reinforced polymer concrete are very similar to the compression characteristics. The flexural strength of polymer concrete with 2.0% fiber content is 50% larger than control specimen strength as shown in Fig. 8.

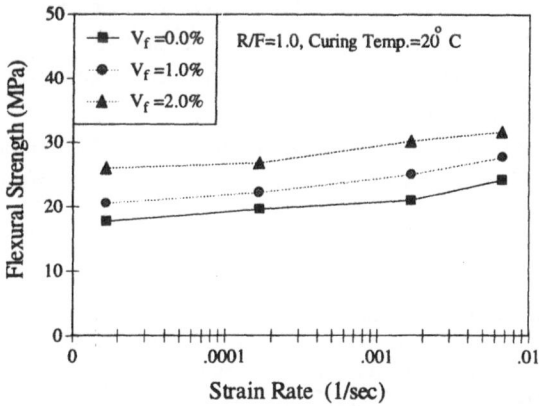

Fig. 8 Flexural strength as a function of strain rate for fiber reinforced polymer concrete

4.3.3 Splitting tensile strength

The splitting strength of polymer concrete with 2.0% fiber content increase up to about 220% as shown in Fig. 9. This results indicate that the addition of fiber contributes to the tensile resistance of polymer concrete and curbs the crack occurrence.

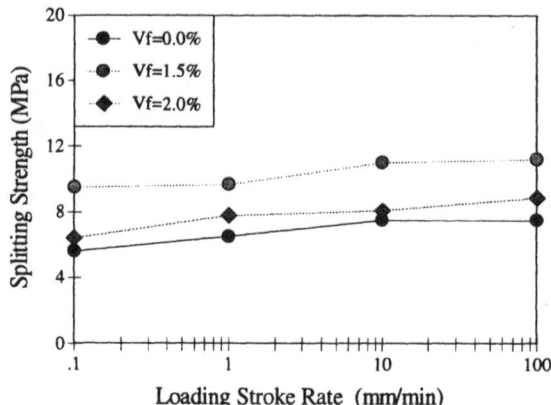

Fig. 9 stroke rate vs splitting strength of fiber reinforced polymer concrete

5. Conclusions

Polymer concrete is one of the relatively new construction materials. The present study is conducted to explore the mechanical characteristics of plain polymer concrete and fiber reinforced polymer concrete. It is found from this study that the addition of fibers in polymer concrete enhances greatly the tensile resistance and thus curbs the occurrence of cracks. The effects of loading rate are also studied in the present study and the test results indicate that the strengths increase with loading rate. This behavior is very important, for example, for foundations of vibrating machines under high loading rates.

The present study provides useful information to perform more realistic design and application of fiber reinforced polymer concrete to actual structures.

References

1. Koyanagi, W., Rokugo, K., and Hayashi, F., "Evaluation of Toughness of Resin Concrete and Its Improvement by Steel Fibers," Polymers in Concrete, ICPIC '84, Darmstadt, pp. 93-98.
2. Huges, B. P. and Guest, J. E., "Polymer Modified Fibre-Reinforced Cements Composites," Proceedings of the First International Congress on Polymer Concretes, May5-7, 1975, pp.85-92.
3. Aguado, A., Martinez, A., and Salla, J. M., "Effects of Different Factors in Mixing and Placing of Polymer Concrete," Polymers in Concrete, ICPIC '84, Darmstadt, pp.299-303.
4. "Polymers in Concrete", Reported by ACI Committee 548, 1977.

BEHAVIOUR OF CONCRETE BEAMS WITH EXTERNALLY BONDED POLYMER-IMPREGNATED HIGHLY REINFORCED FERROCEMENT PLATES

M. Neelamegam and J.K. Dattatreya
Structural Engineering Research Centre, Madras, India

Abstract
The technique of bonding plates/laminates to the tension faces of the reinforced concrete flexural members is now used worldwide to both strengthen and stiffen existing RC members. Until now, steel plates have been the most popular external reinforcing material for strengthening of the reinforced concrete flexural elements. However, the difficulties in handling, cutting and placement, besides the problems of corrosion, have made researchers and engineers to look for alternative reinforcing materials, such as, glass fibre reinforced plastic (GRP), carbon fibre reinforced plastic (CFRP) and ferrocement (FC) plates for strengthening of RC flexural elements. An attempt has been made by the authors to use polymer impregnated highly reinforced ferrocement(PIHRFC) laminates, which possess high strength and excellent durability properties, for strengthening RC beams. Flexural behaviour of RC beams strengthened with ferrocement laminates, with and without polymer impregnation, were investigated at the Structural Engineering Research Centre(SERC), Madras, India. The performance of plated beams with regard to cracking, service load, ultimate load, deflection, ductility, and failure mode have been discussed in this paper.
Keywords: Crack width, bond, deflection, ductility ratio, external reinforcement, ferrocement, laminate, polymer impregnation, strengthening, , serviceability, strain

1 Introduction

It often becomes necessary to strengthen existing concrete flexural members that might have developed some form of distress due to reinforcement corrosion, unanticipated heavy overloading, poor quality of construction, or due to a combination

Polymers in Concrete, edited by Y. Ohama, M. Kawakami and K. Fukuzawa. Published in 1997 by E & FN Spon, 2–6 Boundary Row, London SE1 8HN, UK. ISBN: 0 419 22330 4.

of one or more of these factors. Several techniques are used for strengthening purposes viz., bonding of thin plates to concrete surfaces, providing additional rebars with projected mortar/concrete, external prestressing etc. The behaviour of strengthened concrete elements is a world wide subject of research.

Until now, steel plates have been the most popular external reinforcing material for strengthening of RC flexural elements[1][2][3]. However, difficulties in handling, cutting, splicing, and placement, besides the problem of corrosion, especially in aggressive environs have necessitated the development of alternative reinforcing materials of which FRP plates have been more widely investigated[4][5]. At the Structural Engineering Research Centre, Madras, India, extensive studies have been carried out on alternative reinforcing materials for strengthening of RC beams[6][7][8]. This paper discusses relative performance of RC beams strengthened with highly reinforced ferrocement(HRFC) and polymer impregnated ferrocement (PIHRFC) laminates in respect of cracking, service load, ultimate load, deflection, ductility, and failure mode.

2 Highly reinforced ferrocement(HRFC)

Highly reinforced ferrocement plates, consisting of different number of layers of chicken wire mesh were prepared from a rich cement mortar of proportion 1:1.5 (cement : sand) by weight using a specially fabricated wooden mould. A water cement ratio of 0.34 was adopted and a superplasticizer (1% by weight of cement) was used as a water reducing admixture. Mould-Shutter vibration was resorted to for achieving proper compaction. The HRFC laminates were provided with 2.5%, 3.75%, and 5%(by volume) of chicken- wire-mesh reinforcement. The laminates were 100mm wide x 2200mm long and the thickness was 16mm for lower reinforcement percentages and 27mm for higher percentages of reinforcement. The laminates were moist cured for 28 days before use.

3 Polymer impregnated highly reinforced ferrocement(PIHRFC)

The cured HRFC specimens were thoroughly dried at 110°C to eliminate moisture present in the plates. After cooling to room temperature, the plates were placed in an impregnation chamber of size 250 X 200 X 2500mm and a suction pressure of 3MPa was applied for period of 4 hours. Then methyl methacrylate monomer solution mixed with 3% by weight of benzoyl peroxide was pumped into the impregnation chamber. The plates were soaked in the solution for eight hours under a pressure of 0.2 MPa. After soaking, all the plates were immersed in hot water at a temperature of 80° for about 8 hours to facilitate complete polymerization.

4 Reinforced concrete beams

Ordinary Portland cement concrete of proportion 1:2.5:3.5 (cement : fine aggregate : coarse aggregate) by weight with a water-cement ratio of 0.46 was used for casting the RC beams. Natural river sand passing 2.36mm IS sieve and crushed blue granite stone chips (maximum size 12.5mm) were used as fine and coarse aggregate respectively. A naphthalene-based superplasticizer was added to the concrete mix to achieve the desired workability.. The average compressive strength of the mix after 28 days of *moist curing varied from 40 to 48 MPa. High-yield strength deformed bars with an average yield strength of 475 MPa and an ultimate strength of 550 MPa(2 nos. of 8mm diameter at top and bottom) were used as the main tensile reinforcement in the test beams. The shear reinforcement consisted of mild steel links with a yield stress of 275 MPa and an ultimate strength of 360 MPa provided at 100mm c/c over the shear span. After curing the test beams for a minimum period of 28 days, the laminates were affixed to the tension faces of the beams using an epoxy resin based adhesive formulation. The details of adhesive mix and the procedure followed for plate gluing are discussed elsewhere[8].

5 Testing of beams

The beams were tested under two- point static loading over a span of 2300mm. The load was applied in increments of 2kN. The deflections at midspan and under the load points were measured using dial gauges, while the longitudinal strains in the beams and in the plate/laminates were measured using a Pfender type of mechanical strain gauge. A hand held microscope was used for measuring the width of the cracks. All the measurements were taken at regularly intervals of load until the beams failed. Fig. 1 shows the location of strain gauges and the test set up used while Table 1 gives the strength and stiffness characteristics of control and plated beams.

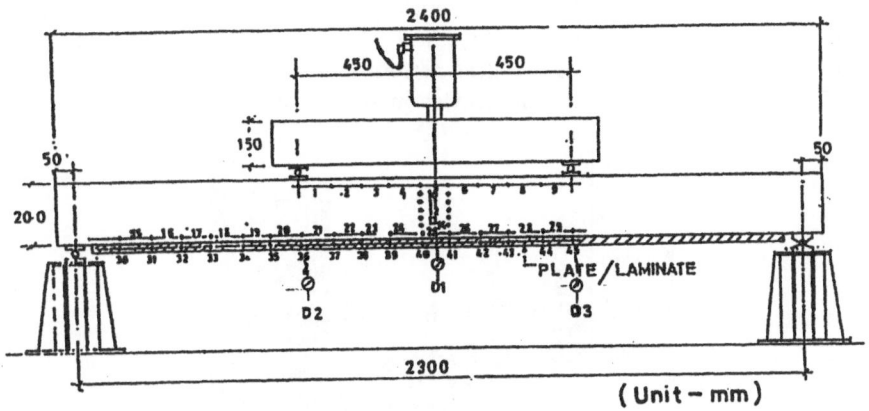

Fig. 1 Location of Pfender Points, Dial Gauges, and Load Points for the Test Beam

Table 1 Strength and Stiffness Characteristics of Plated beams

Laminate type	V_f	t_p	P_{cr}	P_{SL}	P_U	d_{cr}	Properties d_{20}	d_U	EI_{cr}	EI_{20}	EI_U	R
CONTROL	0.00	---	8	20	30	1.2	7.2	42.0	1.04	0.46	0.15	35.0
HRFC1	2.50	16	20	26	34	3.5	3.5	70.0	1.18	1.12	0.12	20.0
HRFC2	3.75	27	20	32	36	3.0	3.0	26.0	1.43	1.43	0.34	8.6
HRFC3	5.0	27	16	44	46	1.8	2.9	17.0	2.00	1.53	0.41	9.4
PIHRFC1	2.5	16	20	32	34	2.6	2.6	14.0	2.00	1.54	0.69	5.3
PIHRFC2	3.75	27	30	36	42	3.0	3.9	30.0	2.07	2.07	0.34	7.7
PIHRFC3	5.0	27	32	41	50	3.7	3.7	24.0	1.52	1.94	0.51	6.4

V_f = Volume fraction(%) of reinforcement, t_p =Plate thickness(mm), P_{cr} = First crack load (kN), P_{SL} = Service load(kN) , P_U = Ultimate load (kN), d_{cr} = Deflection(mm) at first crack, d_{20} = Deflection(mm) at a load of 20kN, d_U= Deflection (mm) at failure, EI_{cr} , EI_{20} , EI_U =Flexural rigidity(MJm) at first crack, 20kN, and failure, R= Ductility ratio

6 Test results and discussions

6.1 First crack load
As seen in Table 1, all the plated beams exhibited delayed appearance of first visible crack when compared with the control beam. HRFC and PIHRFC plated beams exhibited good crack control mechanism with first cracking load registering increase of 100 to 150% in the case of HRFC plates and 150 to 300% in the case of PIHRFC plates, owing to their improved stiffness and superior tensile strength. Fig. 2 shows the percentage improvement in first crack load of plated beams compared to that of control RC beam. The slight reduction in the first crack load of HRFC2 plated beams could be attributed to the lower matrix strength of the laminates.

6.2 Service load
In the present study, service load was reckoned as the load corresponding to a deflection of span/350 or max. crack width of 0.2mm, whichever is less. HRFC and PIHRFC plated beams showed excellent performance under service load conditions. Fig. 3 shows the percentage improvement in the service load of plated beams in

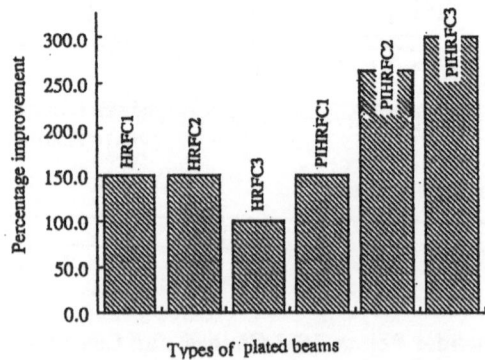

Fig. 2 First Crack Load Improvement for Plated Beams

comparison with that of control RC beam. The service load improvements in HRFC and PIHRFC plated beams are 30 to 120% and 60 to 105% respectively. The slightly lower service load of PIHRFC3 plated beam could be attributed to the local variations in thickness of the RC beam and the laminate used.

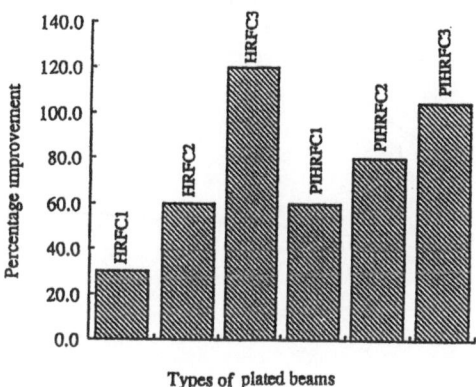

Fig. 3 Service Load Improvement for Plated Beams

6.3 Ultimate load

The HRFC plated beams registered only 13 to 53 % increase in ultimate load capacity over control RC beam while the PIHRFC plated beams showed 13 to 67% increase. Fig. 4 shows the percentage increase in ultimate strength of plated beams.

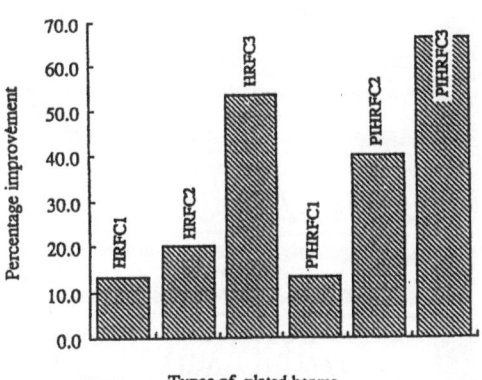

Fig. 4 Ultimate Strength Improvement for Plated Beams

6.4 Crack width and Failure pattern

Fig. 5 shows the variation in average crack width with load for all the test beams. PIHRFC plated beams developed smaller crack widths compared to HRFC plated beams and control beam. Both HRFC and PIHRFC plated beams showed reduction in crack width with increase in the reinforcement content of the plates.

The failure pattern of all HRFC and PIHRFC plated beams and control beams was almost identical. First, a few cracks appeared in the flexural zones of concrete followed by cracking of the laminate. Subsequently, there was profuse cracking of both the laminate and the concrete beam leading to failure of the beam by crushing of concrete, after the yielding of main tension reinforcement. The "post-plate-cracking" behaviour was not very much different from that of the control RC beam. Fig. 6 shows the failure pattern of tested beams.

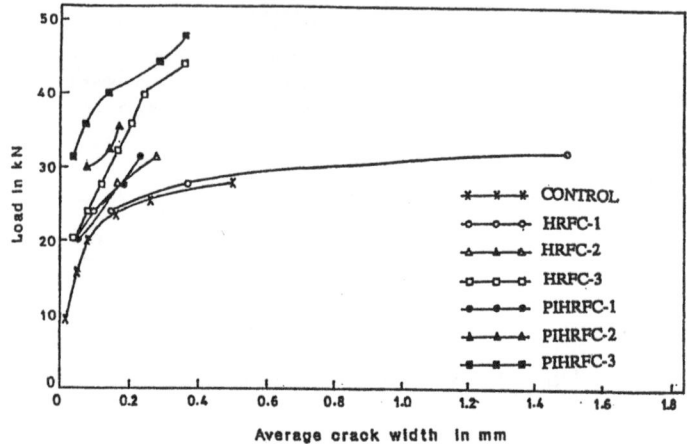

Fig. 5 Variation of Average Crack Width with Load

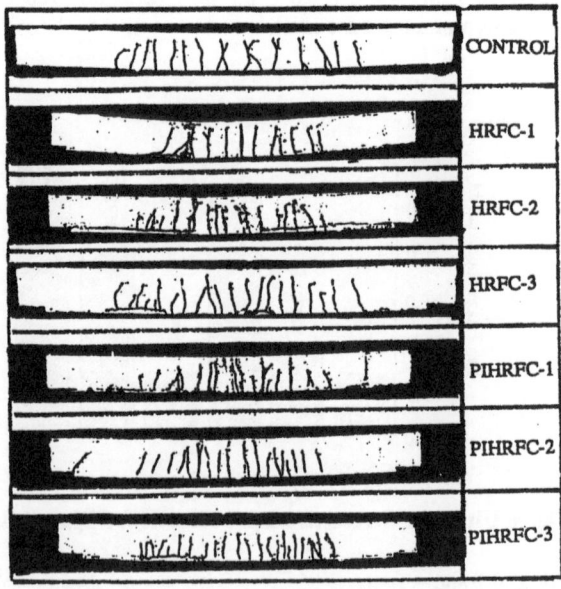

Fig.6 Failure Pattern of Control, HRFC, and PIHRFC Plated Beams

6.5 Strain variation

Fig. 7 shows the variation of plate tensile strain with load for all the plated beams. PIHRFC plated beams registered lesser strain than HRFC-plated beams with the same reinforcement content. Also, increasing the reinforcement content in the plates results in reduction in tensile strain. Fig. 8 shows the typical strain profile across the cross the section at midspan under 20 kN load. It clearly shows the reduction in compressive and tensile strains in all the plated beams compared with the control beam.

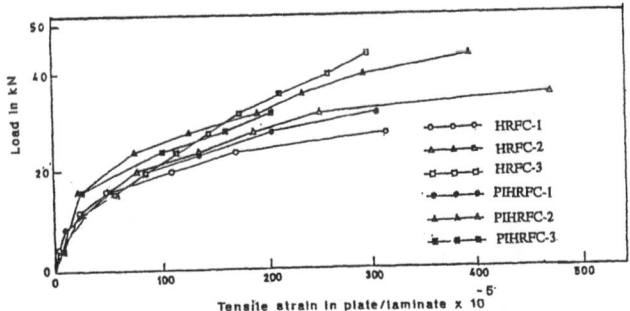

Fig. 7 Variation of Bonded Plate Tensile Strain with Load

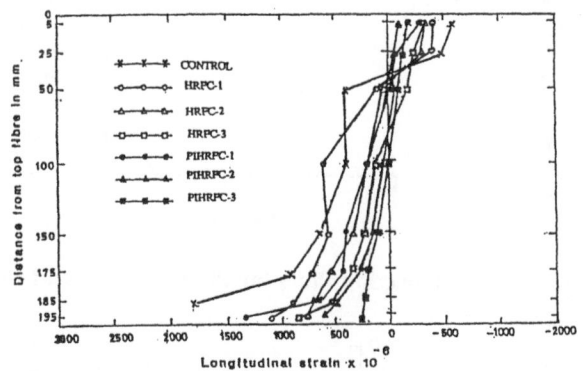

Fig. 8 typical Strain Profile at Midspan at 20kN Load

6.6 Deflection

Fig. 9 shows the load deflection characteristics of the HRFC and PIHRFC plates. The load deflection characteristics of the control and plated beams are given in Fig. 10. Plated beams registered substantial reduction in deflection during the service load stages. Fig. 11 shows the percentage reduction in deflection at a load of 20 kN(which corresponds to the service load of companion beam) for all the plated beams compared to the control RC beam.

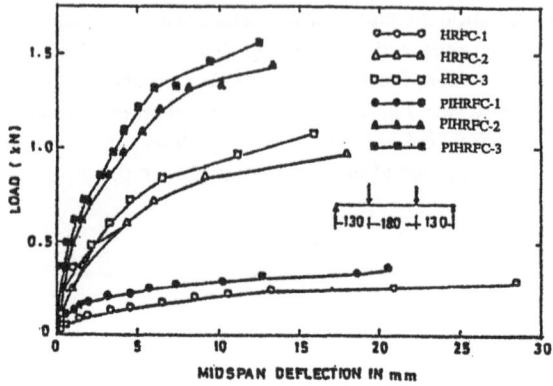

Fig. 9 Load-deflection Plots for HRFC and PIHRFC Plates Under Flexure

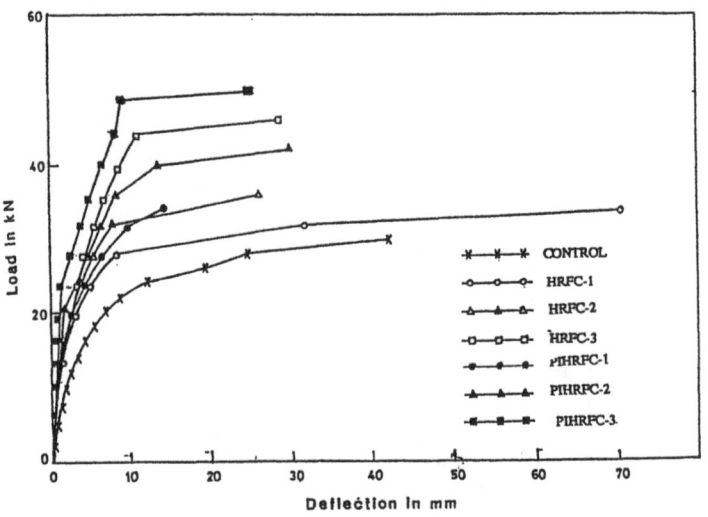

Fig. 10 Load-deflection Plots for Plate Bonded and Control Beams

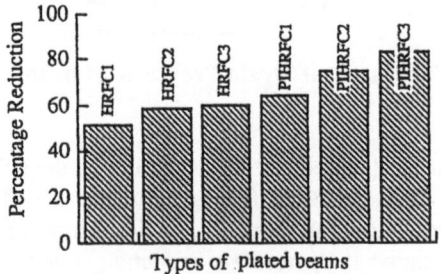

Fig. 11 Reduction in Deflection of Plated Beams at 20kN Load

6.7 Flexural rigidity

The flexural rigidity was found to be increase for all the strengthened beams (Fig. 12). Table 2 gives the comparison of experimental and theoretically predicted strength and stiffness characteristics of plated beams. The theoretical cracking load land flexural rigidity were computed based on simple beam theory assuming no bond slip.

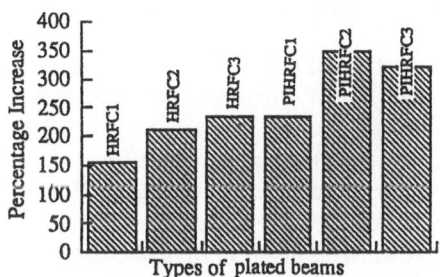

Fig. 12 Flexural Rigidity Improvement for Plated Beams

Table 2 Comparison of Experimental and Theoretical Strength and Stiffness Characteristics of Plated Beams

Laminate type	P_{cre} (kN)	P_{crt} (kN)	P_{Ue} (kN)	P_{Ut} (kN)	EI_e (MJm)	EI_t (MJm	Failure Mode
CONTROL	8	9.3	30	32.0	1.04	0.82	Primary tension
HRFC1	20	13.7	34	32.0	1.18	1.62	Primary tension
HRFC2	20	15.6	36	36.0	1.43	2.31	Primary tension
HRFC3	16	17.2	46	44.0	2.00	2.63	Primary tension
PIHRFC1	20	13.6	34	36.0	2.00	1.94	Primary tension
PIHRFC2	30	15.1	42	40.0	2.07	2.74	Primary tension
PIHRFC3	32	16.5	50	48.0	1.52	3.12	Primary tension

P_{cre}, P_{crt} = Experimental and theoretical first crack load (kN), P_{Ue}, P_{Ut} = Experimental and theoretical ultimate load (kN), EI_e, EI_t = Experimental and theoretical flexural rigidity(MJm)

7.0 Conclusions

1. Polymer impregnated highly reinforced ferrocement possesses good mechanical and durability properties.
2. All the plated beams exhibited improvement in first crack load, service load, ultimate load and flexural rigidity indicating that PIHRFC and HRFC laminates can be satisfactorily used for strengthening of RC flexural elements.
3. PIHRFC and HRFC plated beams did not show any peeling or debonding of plates.
4. In view of their simple process technology and low cost of production, HRFC plates can be a suitable choice for most common situations, while PIHFRC plates may be preferred in extremely aggressive environs.

8 Acknowledgements

This paper has been published with the kind   permission of Director, Structural Engineering Research Centre, Madras

9 References

1. Irwin, C A K(1975), "Strengthening of Concrete Beams by Bonded Steel Plates", TRRL Supp. Rep. 160 UC , TRRL, Dept. of Environment, Crowthorne, UK, 18pp

2. MacDonal & Calder, A D(1982),"Bonded Steel Plating for Strengthening of Concrete Structures", Int. Jnl. of Adheseon and Adhesives, Vol.2, No.2, April, pp.119- 127

3. Swamy R N, Jones, R, & Bloxham, J W(1987)," Structural Behaviour of Reinforced Concrete Beams Strengthened by Epoxy Bonded Steel Plates", ibid, Vol.68a, No.2, Feb., pp.59-68

4. Ritchie et al(1991),"External Reinforcement of Concrte Beams Using FRPS ", ACI Structural Jnl, Vol. , No. , July-Aug., pp.490-500

5. Saadatmanesh, H & Ehsani, M R(1990), "Fiber Composite Plates can Strengthen Beams", Concrete International, Mar., pp 65-71

6. Neelamegam,   M., Dattatreya, J K., and   Parameswaran,   V S.(1993),   " Strengthening of RC Beams with Externally Bonded Plates/Laminates", SERC Technical Report, CCL/94/1, Structural Engineering Research centre,Madras

7. Parameswaran V S, Neelamegam, M , and Dattatreya J K(1993), "Use of Non-Ferrous Externally Bonded Reinforcement for Strengthening of Concrete", Proc. Concrete 2000, Int. Conf. , Univ. of Dundee, Vol. 1, pp 239-253

8. Neelamegam M., Dattatreya J.K. and Parameswaran V.S.(1992) "Studies on Bonding Materials and Laminates for Strengthening of Concrete   Flexural Members", Civil Engineering & Construction Review, Feb., pp.29-37

INFLUENCE OF POLYMER SPECIES AND ADDITIVES ON HIGH-STRENGTH POLYMER-CEMENT COMPOSITE

T. Takata, M. Hasegawa and G.K.D. Pushpalal
Dept. of Materials Science and Technology, Toin University of
Yokohama, Yokohama, Japan
N. Maeda
Maeta Concrete Industry Ltd., Sakata, Japan
Q. Huang, T. Kawano and T. Kobayashi
Maeta Techno-Research, Inc., Sakata, Japan

Abstract
The influence of polymer species, such as phenol resin, melamine resin, and additives, such as diphenolic compounds and some other hardeners, to flexural strength of polymer-cement composite, has been investigated. Analysis by Fourier Transform Infrared Spectrophotometer(FT-IR), Differential Scanning Calorimeter and Thermogravimetric Analyzer(DSC/TG) indicated that interactions or chemical bonding exist between hydroxyl groups of diphenolic compound moieties and cement components in the interface area.

In present study, composite with flexural strength higher than 100MPa was obtained by means of aqueous melamine resin and alumina cement(AC). By comparing the Scanning Electron Microscope(SEM) images of fracture surfaces between melamine-ordinary portland cement(OPC) and resol resin-OPC composites, it is suggested that the flexural strength is influenced by pore features of the composite.

Addition of a slight amount of diphenolic compounds such as resorcinol raised the flexural strength of the resol resin-AC or -OPC composite by approximately 45% in comparison with that without addition of resorcinol. Addition of maleic anhydride into alcohol soluble melamine resin-OPC composite has raised the flexural strength by 55%.
Keywords: Heat and water resistance, high flexural strength composite material, polymer-cement composite, resol resin, resorcinol.

Polymers in Concrete, edited by Y. Ohama, M. Kawakami and K. Fukuzawa. Published in 1997 by E & FN Spon, 2–6 Boundary Row, London SE1 8HN, UK. ISBN: 0 419 22330 4.

1 Introduction

Macro-defect-free(MDF) cement has been studied in several research laboratories[1] since first report by J.D.Birchall et al. in 1981[2]. The MDF cement consists of, for example, a mixture of cement and aqueous polyvinyl alcohol with a few additives, and shows extremely high flexural strength. It has been reported that in the MDF cement an unusual micro structure was created by chemical interactions between the organic-inorganic components and greatly contributed to the flexural strength. However, one general drawback of this material is its instability in water with swelling of polymer matrix and serious decrease in strength after immersion in water because of hydrophilic nature of polymeric binders used.

On the other hand, a new class of polymer cement composite with high flexural strength(>200MPa) and with good water- and heat-resistant properties has been developed in our group[3]. The new polymer cement composite consists of cement, thermo-setting resin precursor, such as formaldehyde resin, and small amounts of a few additives.

In previous papers we have reported that mechanism study indicated the existence of interaction in the interface area between phenolic resin and AC by means of EPMA, SEM, TEM, X-ray diffraction, FT-IR and DSC/TG. From the model reaction between phenol moiety and N-methoxymethyl 6-nylon, it was presumed that not only in the phenol resin precursor itself, cross linking also occurred between phenol resin and the polyamide[3-6]. The flexural strength higher than 100MPa was readily obtained from the composite consisting of alumina cement and the polymer which contained phenol and amide linkage in repeating unit[7].

In this paper, the influence of polymer species (such as resol resin, melamine resin) and additives (diphenolic compounds, such as resorcinol, catechol and some other hardeners) on properties of these polymer-cement composites is studied.

2 Experimental

2.1 Materials

Alumina cement is the product of Denki Kagaku Kogyo Corporation(Denka Alumina No.1). Ordinary portland cement is the product of Titibu Onoda Cement Corporation. Phenol resin is the product of Showa Highpolymer Co. Ltd.(resol type). The precursor, which contains less than 2.0% of water, is soluble in methanol and contains approximately 60wt% nonvolatile component.

Three types of melamine resins (products of Mitsui Toatsu Chemicals Inc.) were used; Type I is 88.0% solution in iPrOH/iBuOH, methoxymethyl type, Type II is 82.6% solution in water, methylol type, and Type III is 82.0% solution in iBuOH,methoxymethyl and methylol/imino type. Maleic anhydride is the product of Wako Pure Chemical Industry Ltd.

2. 2   Preparation of the resin and cement composites

Polymer-cement composite was prepared by mixing alumina cement, resin precursor (resole type or melamine)-methanol solution, alcohol soluble polyamide and glycerol in a twin roll mill. Alcohol soluble polyamide was used as the modifier to improve the viscoelastic property of the mixture. The sheet thus obtained was re-rolled several times. The samples were cut into 100.0mm long and 25.0mm wide, before heat curing. The proportions and curing conditions are showed in Table 1 and Table 2.

2.3   Test and Analysis

The three point flexural strength test has been carried out on the samples. Thermal behavior of the material was measured by DSC/TG analysis by Shimadzu TA-50 thermal analyzer (Shimadzu TGA-50 and DSC-50). SEM pictures were taken by JEOL JSM-5300 scanning electron microscope. The crystalline structure was determined by X-ray diffraction analysis by Shimadzu XD 610 X-ray diffractometer.

3   Results and Discussion

3.1 The relation between resin   species and flexural strength

Starting compositions, curing conditions and flexural strength of resulting composites are shown in Table1. The highest flexural strength(>200MPa) was attained from the heat pressed composite which consists of AC, phenol resin, N-methoxymethyl 6-nylon and glycerol. Composite with flexural strength higher than 100MPa (109.4MPa) was obtained from the mixture of melamine resin(type II) and alumina cement(Table1).

In contrast with the relative flexural strength(2.5) of resol resin-AC to resol resin-OPC composites[7], the composite from the melamine resin(type II) with either AC or OPC have showed almost the same flexural strength (76.7MPa and 74.3MPa). Nearly the same strength of these two composites, independent of the type of cement in present composites, indicates that the interaction between melamine resin and cements is, if any, different from that between phenolic resin and cements. SEM analysis indicated that the pores in melamine resin (type II)-OPC composite are fewer and smaller than those in resol resin-OPC composite(Fig.1). The result of pore features may explain the higher strength of melamine resin(typeII)-OPC composite than resol resin-OPC composite. .

Flexural strength of the composite from methylol type melamine resin(type II) was the highest among the three types of melamine-cement composites. One of the possible reasons is that the higher methylol group percentage makes the more smooth rolling process, to result in higher strength.

Table 1   Starting compositions, curing conditions and flexural strength of resulting composites.

| Cement | Starting composition | | | | Curing condition | Flexural strength |
| | Phenol [a] | Melamine | Polyamide [b] | Glycerol | | |
g	g	g	g	g	°C/h	MPa
100[c]	21.2		1.8	2.3	200/18[h]	220.0
100[c]	21.2		1.8	2.3	180/21	117.6
100[d]	27.9		2.1	3.0	180/21	46.4
100[c]		16.0[e]	2.0		180/3	56.6
100[d]		16.0[e]	2.0		180/3	54.9
100[c]		17.0[f]	1.8	2.4	180/3[i]	109.4
100[d]		15.8[f]	1.8	2.4	180/3[i]	96.0
100[c]		17.0[f]	1.8	2.4	180/3	76.7
100[d]		17.0[f]	1.8	2.4	180/3	74.3
100[c]		17.0[g]	2.4		180/3	60.6
100[d]		17.0[g]	2.4		180/3	61.5

a)   Resol type(approximately 60% methanol solution)
b)   N-methoxymethyl 6-nylon
c)   Alumina cement
d)   Ordinary portland cement
e)   Melamine resin precursor(Type I)
f)   Melamine resin precursor(Type II)
g)   Melamine resin precursor(Type III)
h)   Heat pressed under 6MPa at 80°C for 30min., followed by heat cured at 200°C for 18 hours.
i)   Heat pressed under 2MPa at 154°C for 10min., then under 12MPa at 154°C for 10min. followed by heat curing at 180°C for 3 hours.

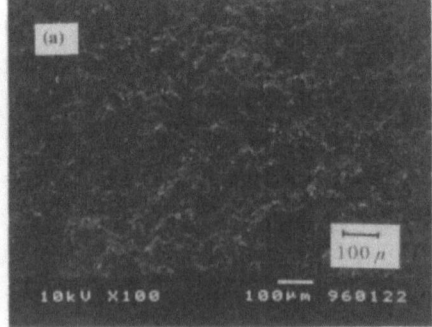 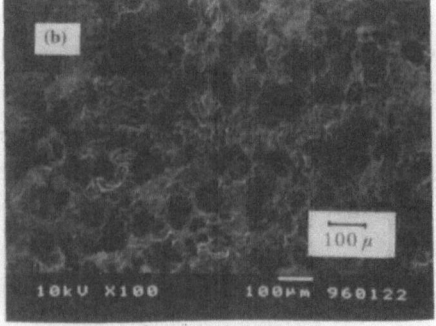

Fig.1   SEM images of fracture surface of (a) melamine resin(type II)-OPC composite, cured at 180°C for 3 hours, (b) resol resin -OPC composite, cured at 180°C for 21 hours.

3.2 X-ray diffraction analysis

The X-ray diffraction patterns of AC, OPC, and melamine resin (type II)-cements (AC and OPC) composite are shown in Fig.2. Any patterns assigned to hydration products of AC, such as C_3AH_6, AH_3 or OPC, such as $Ca(OH)_2$, are not seen in Fig.2(b) and (d), which are nearly identical with the patterns in Fig.2(a) and (c), respectively. Neither X-ray diffraction of the composite is different from that of the respective cement means that no hydration reaction of AC or OPC ever happened. The X-ray analysis indicates that no hydration reaction of the cements has occurred, and that the composites have been hardened by certain other mechanisms, as was the case in the hardening of phenol resin-cement (AC or OPC) composite.

3.3 The influence of hardener

Maleic anhydride, which is known to be effective as a cross linking reagent of melamine resin, was applied to the composite as hardener in melamine resin (type I)-OPC composite. By the addition of 2.5wt% maleic anhydride (vs. melamine resin), the flexural strength of the composite increased by approximately 55%(Table 2).

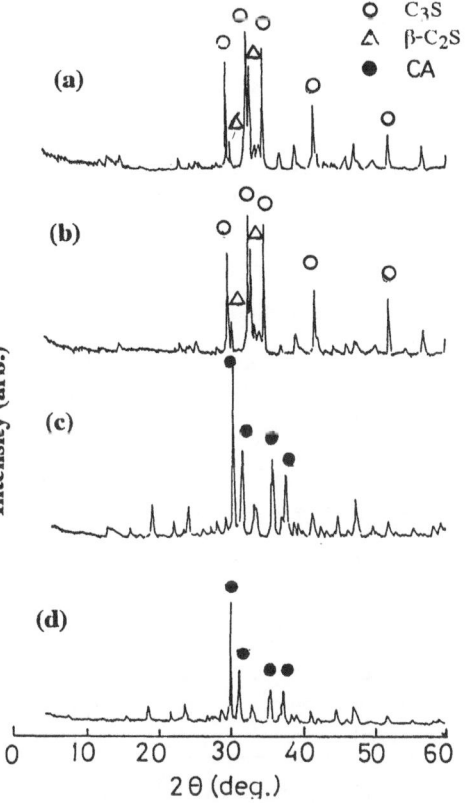

Fig.2 X-ray diffraction patterns of (a) OPC powder, (b) melamine resin (type II)-OPC composite, cured at 180°C for 3 hours, (c) AC powder, (d) melamine resin (type II)-AC composite, cured at 180°C for 3 hours.

C : CaO
S : SiO$_2$
A : Al$_2$O$_3$

Table 2. Influence of hardener to flexural strength of melamine-cement composite

Cement g	Melamine a) g	Polyamide b) g	Maleic anhydride g (wt%) c)	Curing condition °C/h	Flexural strength MPa
100d)	16.0	2.4	0.0(0.0)	180/3	550
100d)	16.0	2.4	0.352(2.5)	180/3	855
100d)	16.0	2.4	0.704(5.0)	180/3	784

a) Melamine resin precursor(Type I)   b) N-methoxymethyl 6-nylon
c) vs.melamine resin precursor   d) Ordinary portland cement

From the result, it is assumed that appropriate amount of maleic anhydride made the polymer matrix stronger by accelerating the cross linking reaction in melamine resin(type I) and, as the result, enhanced the flexural strength of the composite.

3.4   Influence of resorcinol or catechol on flexural strength of polymer-cement composite

Table 3 shows starting composition of the composite(AC or OPC, alcohol soluble polyamide, glycerol, resorcinol or catechol), curing conditions of the composites and flexural strength of resulting composites.

Table 3   Influence of resorcinol and catechol on flexural strength of polymer-cement composite.

Cement g	Phenol [a)] g	Polyamide [b)] g	Resorcinol g	Catechol g	Paraform g	Glycerol g	Curing condition °C/h	Flexural strength MPa
100[c)]	22.5	1.5				2.4	200/12	98.9
100[c)]	22.5	1.5		2.26	0.66	2.4	200/12	113.9
100[c)]	22.5	1.5	2.26		0.50	2.4	200/12	107.5
100c)	22.5	1.5	2.26		1.32	2.4	200/12	109.7
100[c)]	22.5	1.5	2.26		0.66	2.4	200/12	143.2
100[d)]	28.2	1.8				3.0	200/12	46.4
100[d)]	28.2	1.8	0.84			3.0	200/12	54.6
100[d)]	28.2	1.8	1.68			3.0	200/12	67.1

a) 60% resole/MeOH solution.   b) N-methoxymethyl 6-nylon.
c) Alumina cement.   d) Ordinary portland cement.

The addition of resorcinol results in prominent increment of the flexural strength. In comparison with the samples without addition of resorcinol, the addition of 2.26 wt% resorcinol(by comparing with AC)to the resol resin-AC composite has raised the flexural strength from 98.9MPa to 143.2MPa, increment of 44.8%; the addition of 1.68wt% resorcinol(by comparing with OPC)to the resol resin-OPC composite has raised the flexural strength from 46.4MPa to 67.1MPa, increment of 44.6%.

However, the addition of catechol into the composites has not increased the flexural strength very much(from 98.9MPa to 113.3MPa). The reason may be that due to its low polymerization reactivity of catechol.

For mechanism study, FT-IR and DSC /TG have been applied to the mixture of esorcinol and AC.

By comparing the FT-IR of the AC and resorcinol mixture heated at   180 °C for 48 hours with the IR of pure resorcinol, it was confirmed that the -OH stretching vibration absorption at 3,300 cm^{-1} of resorcinol disappeared in the IR spectrum of the heat-treated product of the mixture of AC and resorcinol (Fig. 3).

Also, in DSC/TG curve of the heat-treat product of the mixture of AC and resorcinol, the endothermic peak at melting point of pure resorcinol(110.9 °C) disappeared, whereas the exothermic peak at 379°C due to the thermal degradation of the heat-treated product of AC and resorcinol mixture newly appeared(Fig. 4).

The results of IR and DSC/TG indicates that in the interface area of the composite of phenolic resin-cement with the addition of resorcinol, besides the interaction between phenol resin and cements, certain interaction have existed between resorcinol and AC or OPC, and contributed to the enhancement of flexural strength of the composites.

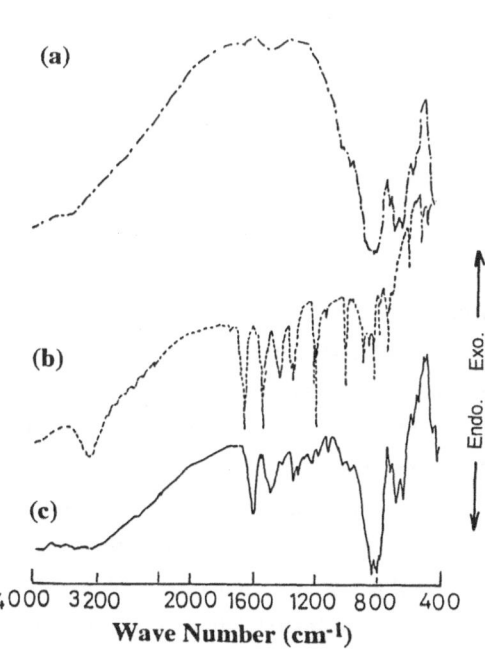

Fig.3   FT-IR spectra of (a)AC powder,
(b) mixture of AC and resorcinol,
(c) heat treatment product of
mixture of AC and resorcinol, at
180°C for 24 hours.

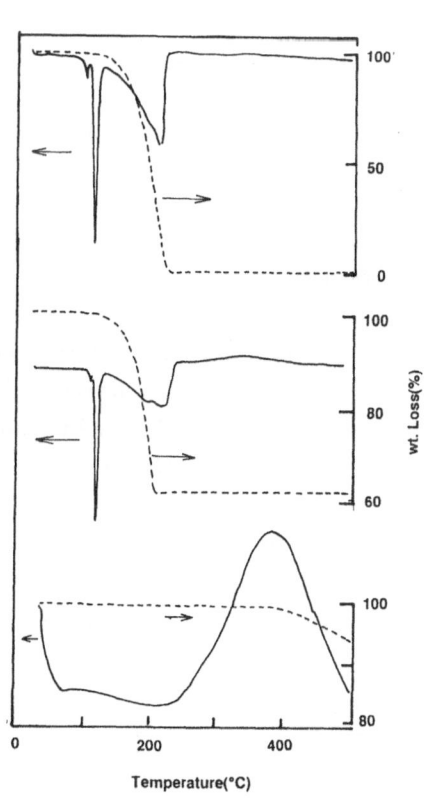

Fig.4   DSC/TG curves of (a) resorcinol,
(b) mixture of AC and resorcinol, (c)
heat treatment product of AC and
resorcinol mixture, at 180°C for 24
hours.

4 Conclusion

1) Addition of resorcinol to the resol resin-cement composite resulted in increment of flexural strengths by approximately 45% for both cases of AC and OPC.

2) The IR and DSC/TG of the mixture of AC and resorcinol, indicated the existence of interaction or chemical bonding formation between hydroxyl group and cement composite in the interface area.

3) Composites with flexural strength higher than 100 MPa have been obtained from the mixture of melamine resin and alumina cement. Furthermore, that the higher flexural strength of melamine-OPC composite than that of resol resin and OPC was explained by fewer and smaller pores of the former composite.

4) The composite from melamine resin containing more methylol group shows higher flexural strength.

References

1. J.F.Young, et al., (March 1993) MRS Bull. Research on cement-based materials. Vol.33, pp.33-4, J.A.Lewis et al., (March 1993) MRS Bull. Micro structure property relationships in Macro-Defect-Free cement, Vol.33, pp.35-77.
2. J.D.Birchall, A.J.Howard, K.Kendall, (1981) Flexural strength and porosity of cements, Nature, Vol. 289, No.29, pp.388 -9.
3. M.Hasegawa, et al., (1993) Development of a new class of high strength polymer cement composite, Abstract of MRS-JAPAN Annual Symposium, 1-P5
4. M.Hasegawa, et al., (1994) Development of a new conceptual polymer cement composite with very high flexural strength, Cement Concrete proceedings, Vol. 48, pp.820 -5.
5. M.Hasegawa, et al.(1995) A new class of high strength water and heat resistant polymer-cement composite solidified by an essentially anhydrous phenol resin precursor, Cement and Concrete research Vol.25, No.6, pp.1191-8.
6. M.Hasegawa, et al., (1995) Interraction of phenol resin precursor and calcium aluminates, MRS Symposium Proceedings Vol.385, pp.167-71.
7 T. Takata, et al., (1995) Mechanisms of enhanced strength of high strength polymer-cement composites, Cement Concrete proceedings, Vol. 49, pp.892 -7.

WATER RESISTANCE OF POLYMETHYL METHACRYLATE MORTARS

Y. Ohama and K. Demura
Dept. of Architecture, Nihon University, Koriyama, Japa 1
M.A.R. Bhutta
Maeta Techno-Research, Inc., Sakata, Japan

Abstract

Polymethyl methacrylate (PMMA) mortar specimens are prepared using two types (MMA-1 and MMA-2) of methyl methacrylate monomers with various silane contents in the air, and immersed in tap water at 20°C for water resistance. Furthermore, PMMA-1 mortar specimens are separately molded underwater, and immersed in tap water at 20°C for one year in view of underwater construction. The water absorption, flexural strength and compressive strength of PMMA mortar specimens are determined after 1, 3, 7, 28, 90, 180 and 360 days. The effects of the silane coupling agent content on the water resistance and strength properties of PMMA mortars are discussed. As a result, a slight strength reduction of PMMA mortars is found by water absorption, but will cause no problem in their practical applications. The addition of the silane coupling agent is considered to be most effective for improving the water resistance of PMMA mortars and for reducing their water absorption during 20°C water immersion for one year.

Keywords: Polymethyl methacrylate mortars, water resistance, water absorption, silane coupling agent, underwater placement, strength properties.

1 Introduction

Polymethyl methacrylate (PMMA) mortars and concretes have actively been developed, and widely used in the construction industry in Japan. Recently, the polymethyl methacrylate mortars have been developed for underwater placement in Japan [1]. Because of their wide applications in the underwater construction work, their water resistance has become most important property. In this paper, two types of PMMA (PMMA-1 and PMMA-2) mortars

Polymers in Concrete, edited by Y. Ohama, M. Kawakami and K. Fukuzawa. Published in 1997 by E & FN Spon, 2–6 Boundary Row, London SE1 8HN, UK. ISBN: 0 419 22330 4.

are prepared with proper binder formulations and mix proportions, and a silane coupling agent is used for improving their durability. PMMA mortars are placed, dry- or heat -cured in air, and immersed in water at 20°C for 1 to 360 days (1 year). PMMA-1 mortars are separately molded underwater, and immersed in water at 20°C for 1 to 360 days (1 year) in view of underwater construction. The water absorption, flexural strength and compressive strength of PMMA mortars are determined after 1, 3, 7, 28, 90, 180 and 360 days. The effects of the silane coupling agent on the water resistance and strength properties of PMMA mortars are discussed.

In general, polymer concretes or mortars have high water resistance compared to conventional cement concretes or mortars. In particular, the data on the water resistance of PMMA mortars for a long time have not been reported till now [2,3]. The objective of this study is to obtain such data which will be effective and useful to clarify the water resistance of PMMA mortars in various long-term underwater applications in construction work.

2 Materials

2.1 Materials for binder systems
Binder systems for polymethyl methacrylate (PMMA-1) mortar were based on methyl methacrylate (MMA) monomer, together with trimethylolpropane trimethacrylate (TMPTMA) as a crosslinking agent, unsaturated polyester resin (UP) and polyisobutyl methacrylate (PIBMA) as shrinkage-reducing agents, benzoyl peroxide (BPO) as an initiator, and N, N-dimethyl-p-toluidine (DMT) as a promoter. Commercially available prepackaged materials which were composed of binder systems, premixed fillers and fine aggregates were used for the preparation of PMMA-2 mortars. The binder systems were prepared by mixing the above materials with γ– methacryloxypropyltrimethoxy silane (Silane) as a coupling agent.

2.2 Filler and fine aggregates
In the preparation of PMMA-1 mortars, commercially available ground calcium carbonate (size; 2.5μm or finer) was used as a filler, and silica sands (sizes; 0.04-0.30mm and 0.21-1.19mm) were done as fine aggregates. As stated in 2.1, the premixed fillers and fine aggregates were used for the preparation of PMMA-2 mortars.

3 Testing procedures

3.1 Preparation of PMMA mortars
According to JIS A 1181 (Method of Making Polyester Resin Concrete Specimens), polymer mortars were mixed with the binder system formulations and mix proportions as shown in Tables 1 to 3.

Table 1 Formulations of MMA-1 binder system for PMMA-1 mortars

Formulations by mass						
(%)				(phr*)		
MMA	TMPTMA	UP	PIBMA	DMT	BPO	Silane
67.40	1.80	23.10	7.70	0.50	2.00	0
						0.50
						1.00
						2.00

Note, *: Parts per hundred parts of resin.

Table 2 Mix proportions of PMMA-1 mortars

Mix proportions by mass				
Binder	Filler	Silica sand		Binder-filler ratio, B/F
		No.4	No. 7	
15.00	15.00	35.00	35.00	1.00

Table 3 Formulations of MMA-2 binder system for PMMA-2 mortars

Formulations by mass		
(%)	(phr*)	
Resin	BPO	Silane
11.00	2.00	0
		0.50
		1.00
		2.00

Note, *: Parts per hundred parts of resin.

Table 4 Mix proportions of PMMA-2 mortars

Mix proportions by mass		
Binder	Fine aggregate	Binder-filler ratio, B/F
11.00	89.00	0.12

3.2 Preparation of specimens

PMMA mortars using MMA-1 and MMA-2 binder systems with Silane contents of 0, 0.5, 1.0 and 2.0 phr were mixed as shown in Tables 1 to 3. In the preparation of specimens for flexural and compressive strength tests, PMMA mortars were placed into molds 40x40x160 mm at 20°C and 50% R.H., and then cured as follows: (1) 1-day dry cure at 20°C and 50% R.H., (2) 7-day dry cure at 20°C and 50% R.H., and (3) 15-hour heat cure at 70°C. PMMA mortars with MMA-1 binder systems were also placed into molds 40x40x160 mm underwater, and kept in water at 20°C.

3.3 Water resistance test

The specimens placed underwater were demolded, their mass was measured, and again immersed in water at 20°C. The heat- and dry-cured specimens were also immersed in water at 20°C. After water immersion for 1, 3, 7, 28, 90, 180 and 360 days, the appearance of the specimens was visually checked, and then their mass was measured. The water absorption of the specimens was calculated by the following equation:

$$\text{Water absorption } (\%) = [(W_n - W_o) / W_o] \times 100$$

where W_o is the mass (g) of the specimens before water immersion or the mass (g) of the underwater-placed specimens after demolding, and W_n is the mass (g) of the specimens after water immersion for the respective periods.

3.4 Strength tests

According to JIS A 1172 (Method of Test for Strength of Polymer-Modified Mortar), the flexural strength test of cured specimens was conducted by use of the Amsler-type universal testing machine. After the flexural strength test, the broken portions were tested for compressive strnegth by using the same testing machine according to JIS A 1172. The relative flexural and compressive strengths were calculated by the following equation:

$$\text{Relative flexural strength or compressive strength } (\%) = (\sigma_n / \sigma_o) \times 100$$

where σ_o is the flexural strength (MPa) or compressive strength (MPa) of the specimens before water immesion, and σ_n is the flexural strength (MPa) or compressive strength (MPa) after water immersion for the respective periods.

4 Test results and discussions

Fig. 1 shows the relation between the 20°C water immersion period and water absorption of 7-day dry-cured PMMA mortars with Silane contents of 0, 0.5, 1.0 and 2.0 phr. Generally, the water absorption of PMMA mortars with Silane is smaller than that of PMMA mortars without Silane. The water absorption of PMMA-1 mortars increases with additional water immersion period regardless of Silane contents. However, the water absorption of PMMA-2 mortars increases till a water immersion period of 28 days, and becomes nearly constant at a water immersion period of 90 days. The water absorption of PMMA-1 mortars placed underwater at 20°C is almost half of that of PMMA-1 mortars placed in air regardless of Silane contents. The reason for this may be explained to be due to the insolubility of MMA monomer in the binder systems in water at 20°C. The binder systems provide a protective covering for the aggregates and filler, and help to prevent water absorption by them. A noticeable difference in the water absorption between the Silane contents is recognized.

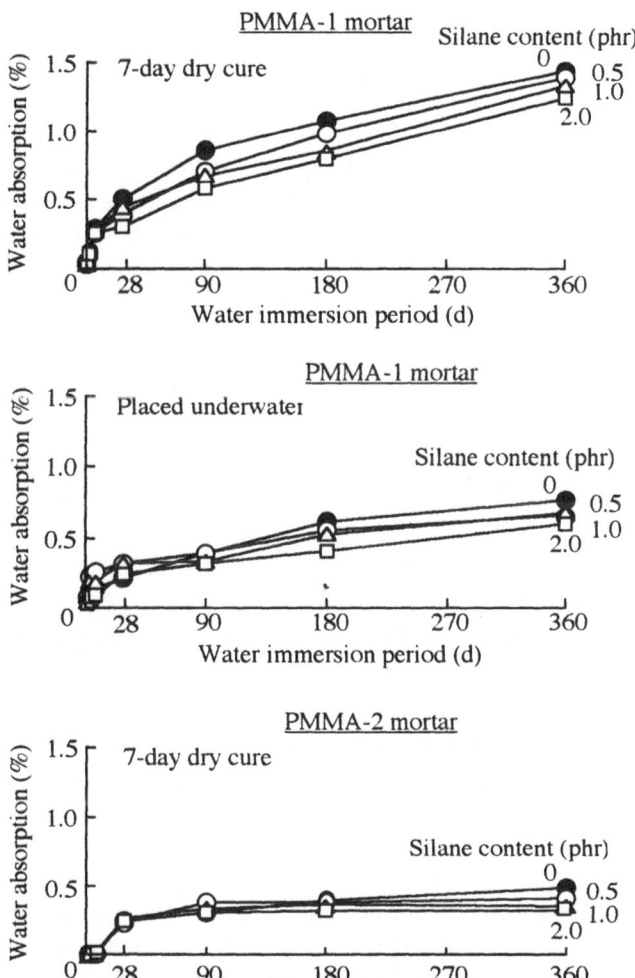

Fig. 1 Water immersion period vs. water absorption of PMMA
mortars with various Silane contents.

Fig. 2   represents the relation between the 20°C water immersion period and water
absorption of 7-day dry-, 1-day dry- and heat-cured or underwater-placed PMMA mortars
with a Silane content of 1.0 phr. As mentioned in Fig.1, the water absorption of PMMA
mortars increases with additional water immersion period regardless of the curing conditions
and the type of the binder systems. The water absorption of PMMA-1 mortars is larger than
that of PMMA-2 mortars. This is considered to be due to the type of the binder systems and

aggregates. A noticeable differnce in the water absorption due to the curing conditions of PMMA-1 mortars is seen at a water immersion period of 360 days. The water absorption of the heat-cured PMMA-1 mortars is smaller than the dry-cured PMMA-1 mortars at a water immersion period of 360 days because of the almost complete polymerization of the MMA-based binder systems. On the other hand, the effects of the curing conditions on the water absorption of PMMA-2 mortars are hardly recognized. At a water immersion period of 360 days, the water absorption of the 7-day dry-, 1-day dry- and heat-cured PMMA mortars is less than 1.5%, and the water absorption of PMMA-1 mortars placed underwater is 0.67% at water immersion period of 360 days which is half of those PMMA-1 mortars placed in air. No appearance changes such as cracks, swellings and color changes of the water-immersed PMMA mortars were observed.

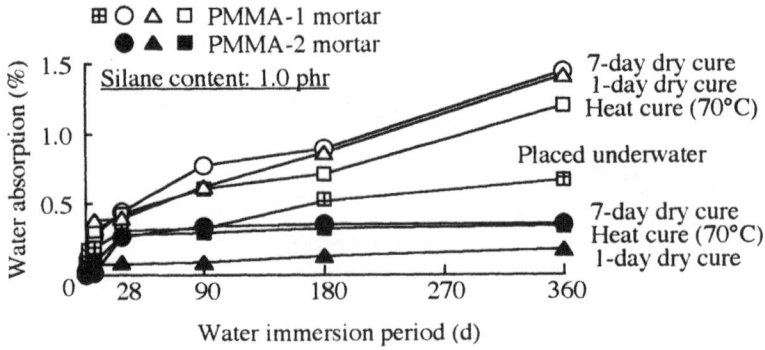

Fig. 2 Water immersion period vs. water absorption of PMMA
mortars with a Silane content of 1.0 phr.

Figs. 3 and 4 exhibit the effects of different curing conditions and water immersion periods on the flexural and compressive strengths of 7-day dry-, 1-day dry- and heat-cured or underwater-placed PMMA mortars with a Silane content of 1.0 phr . Generally, the flexural and compressive strengths of PMMA mortars after 360-day-20°C water immersion are slightly reduced by water absorption irrespective of the curing conditions. The flexural and compressive strengths of the heat-cured PMMA mortars is slightly higher than those of 7-day dry- and 1-day dry cured PMMA mortars. The flexural and compressive strengths of PMMA-1 mortars are somewhat higher than those of PMMA-2 mortars. The reasons for this were mentioned above. The flexural and compressive strengths of PMMA-1 mortars placed underwater are 60% of those of PMMA-1 mortars placed in air, and show also a slight reduction in the strengths due to the water absorption regardless of the water immersion period and Silane content. The addition of Silane and the heat cure casue increases in the flexural and compressive strengths by reducing the water absorption.

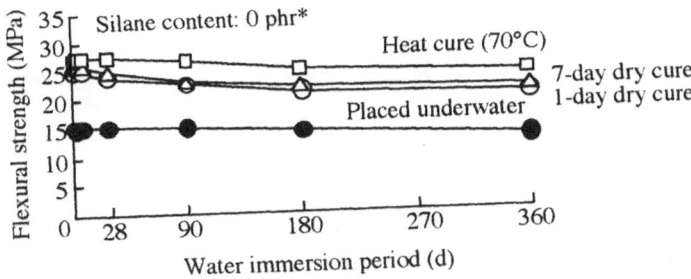

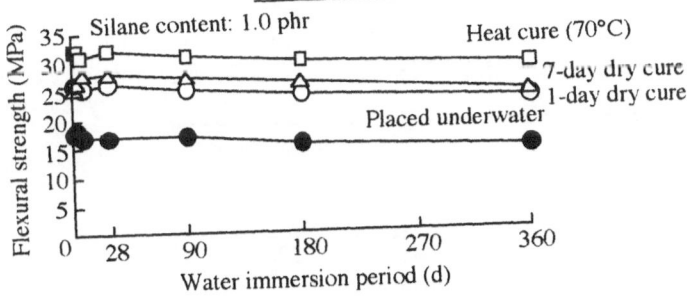

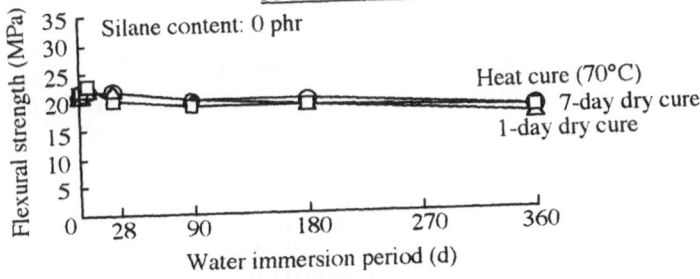

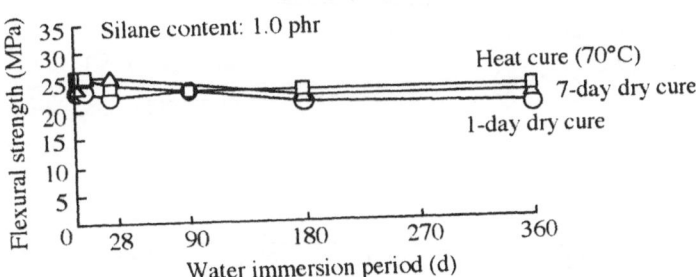

Fig. 3 Flexural strength vs. water immersion period of PMMA mortars placed with a Silane content of 1.0 phr, cured under different conditions.

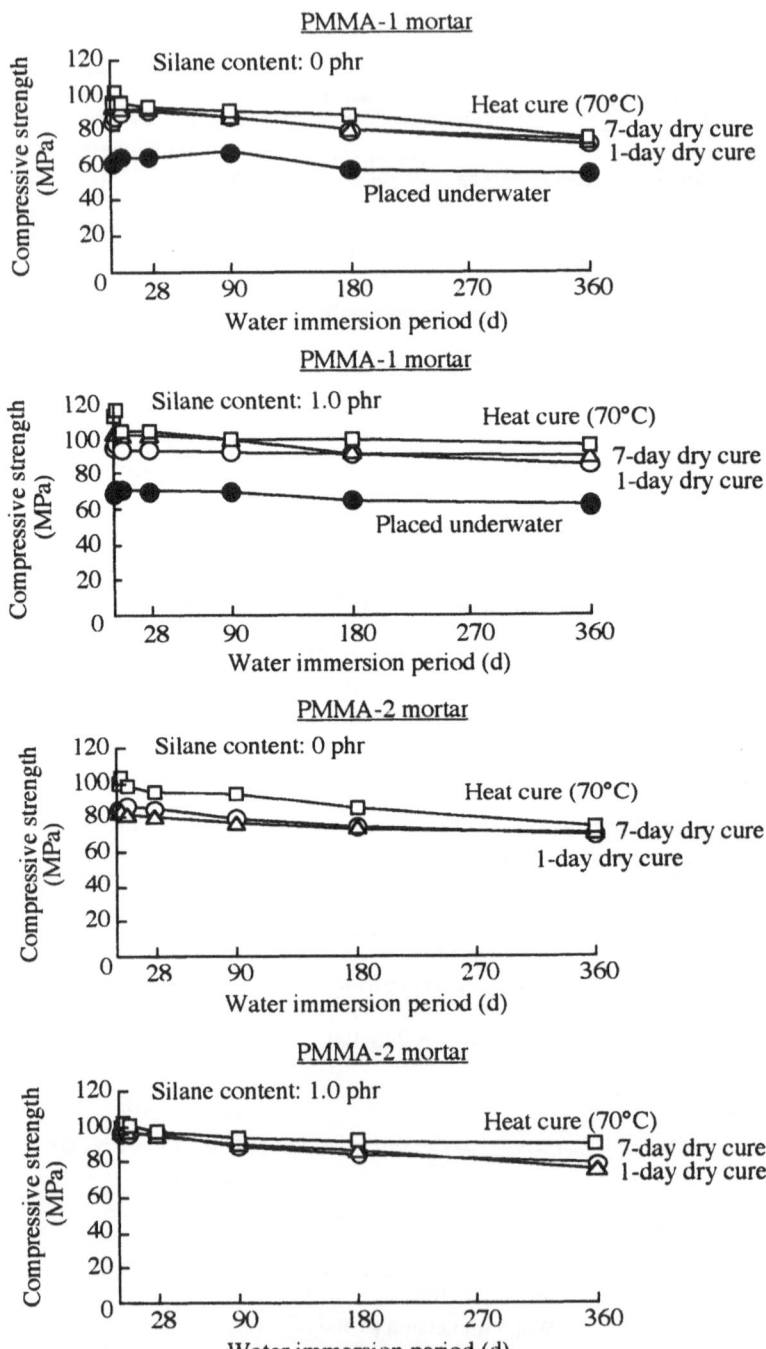

Fig. 4 Compressive strength vs. water immersion period of PMMA mortars with a Silane content of 1.0 phr, cured under different conditions.

Fig. 5 exhibits the relation between the Silane content and flexural and compressive strengths of dry-, heat-cured and underwater-placed PMMA mortars before and after a water immersion period of 360 days at 20°C. The flexural and compressive strengths of PMMA mortars with a Silane content of 1.0 phr tend to be higher than those of PMMA mortars with a Silane content of 0 phr. In general, the addition of Silane to PMMA mortars provides a good water resistance, developing relative strengths of 85 and 82%

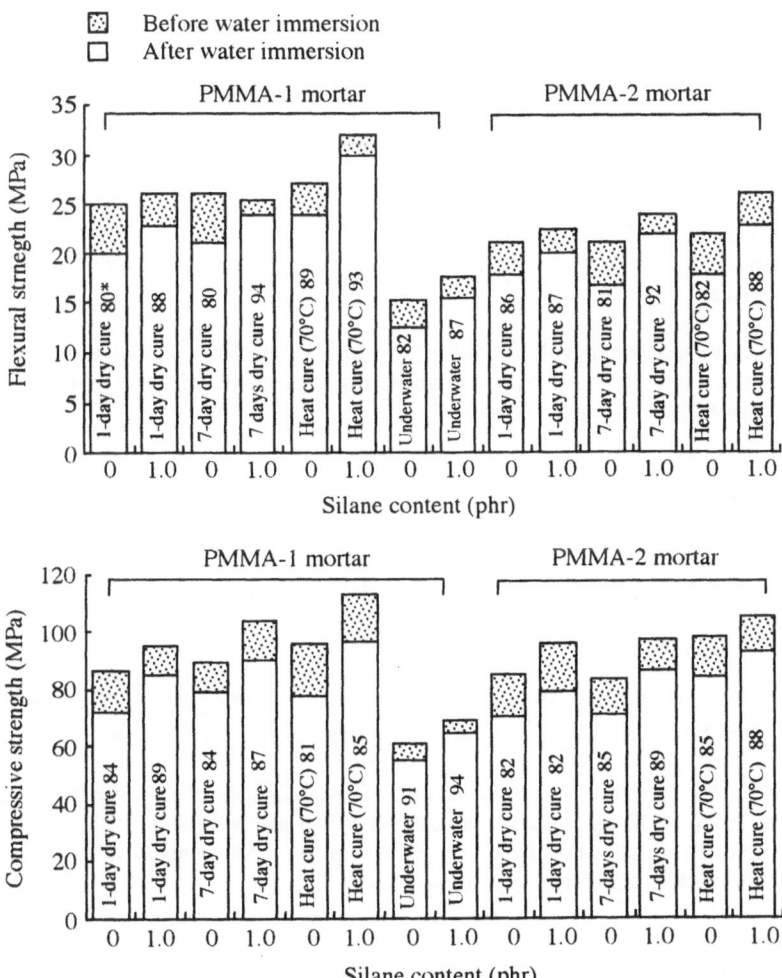

Fig. 5 Flexural and compressive strengths of PMMA mortars with Silane contents of 0 and 1.0 phr, immersed in water at 20°C for 360 days. Note, *: Relative strength

or higher for PMMA-1 and PMMA-2 mortars respectively. As a result, the addition of Silane to the binder systems and the heat cure are recommended for the long-term water resistance of PMMA mortars.

5   Conclusions

The conclusions obtained from the above test results are summarized as follows:

(1) At a water immersion period of 360 days, the water absorption of 7-day dry-, 1-day dry- and heat-cured PMMA mortars is less than 1.5% regardless of the type of binder systems, and their flexural and compressive strengths tend to reduce slightly. It is found that the water absorption of PMMA-1 mortars placed underwater is 0.67% at a water immersion period of 360 days which is half of those of PMMA-1 mortars placed in air. However, such water absorption and slight strength reduction are found to cause no problem in the practical applications of PMMA mortars.

(2) The addition of Silane is considered to be most effective for improving the water resistance of PMMA mortars and for reducing their water absorption during 20°C water immersion.

(3) The above test results will be useful and effective to clarify the water resistance and long-term durability of PMMA mortars in the practical applications of polymethyl methacrylate mortars in construction work.

6   References

1. Bhutta, M.A.R., Ohama, Y., and Demura, K. (1993) Polymethyl methacrylate concrete for underwater construction, Proceedings of the International Congress on "Concrete 2000", Economic and Durable Construction through Excellence, E & FN Spon, London, pp.1061-1070.

2. Ohama, Y., (1977) Hot Water Resistance of Polyester Resin Concrete, Journal of the College of Engineering of Nihon University, Series A-18, pp.33-37.

3. Ohama, Y., Demura, K., Kobayashi T, and Nawata, K. (1987) Temperature Dependency of Flexural Behavior and Water Resistance of Polymethyl Methacrylate Concretes, Proceedings of the 13th Japan Congress on Materials Research, The Society of Materials Science, Japan, Kyoto, pp.157-161.

Acknowledgement

The authors wish to thank the late undergraduate student, Mr. T. Hoshi for his useful help and valuable assistance in carrying out this experimental work.

DEVELOPMENT OF MDF CEMENTITIOUS COMPOSITES BASED ON PACKING THEORY

H. Nakamura and H. Mihashi
Dept. of Architecture, Tohoku University, Sendai, Japan
T. Kobayashi, Y. Ohama and K. Demura
Dept. of Architecture, Nihon University, Koriyama, Japan

Abstract
In order to develop cementitious composites with lightweight, high-strength and ductility, high-strength and fine lightweight aggregates and short fibers were used. Furthermore reduction of micro- voids by packing theory and removal of the macro-voids by applying manufacturing techniques of Macro-Defect-Free(MDF) cement were tried. Thus effectiveness of these methods was experimentally investigated for the target composites.

As a result of the experiments, flexural strength of 15~30 MPa was obtained with the specific density of 1.5~1.8 t/m³. It was also shown that steel fibers and polypropylene fibers are effective to increase the ductility after water immersion.
Keywords:lightweight, high-strength, ductility, packing theory, MDF cement, flexural strength.

1. Introduction

Cementitious composites with both lightweight and high-strength are required earnestly. Materials with not only high-strength but also high ductility are required, too. However it is a very difficult subject to make such a high performance material because the lightweight and high-strength are usually the opposite features. In particular, the cementitious composites have usually high density and the fracture behavior is brittle. Therefore it is difficult to be used in various purposes.

To solve these problems, it is important to consider the selection of materials and the

Polymers in Concrete, edited by Y. Ohama, M. Kawakami and K. Fukuzawa. Published in 1997 by E & FN Spon, 2–6 Boundary Row, London SE1 8HN, UK. ISBN: 0 419 22330 4.

manufacturing process. In short, usage of high-strength lightweight aggregate and reinforcing fiber in conjunction with the reduction of macro- and micro- voids are needed. For this purpose, it was tried to develop the cementitious composites by combining the packing theory with MDF cement.

For the purpose of developing lightweight, high-strength and high-ductility cementitious composites, combination of following items may be effective:
1. to use lightweight and high-strength hollow micro-ceramics as aggregates;
2. to use micro-fiber with high elastic modulus and strength;
3. to make dense microstructure based on packing theory;
4. to make dense microstructure by using water-soluble polymers and by applying manufacturing techniques with kneader and twin-roll mill.

While MDF cement has been studied since around 1980[2], study of MDF cement with lightweight aggregate has not been tried. In this paper, the material development technique combined with these ideas is proposed, and the effectiveness is experimentally investigated.

2.   Experimental procedures

2.1 Materials
Properties of materials used in this experimental study are shown in Table 1.

Table 1   Properties of materials used in this experiment

Kinds of used Materials	Specific gravity	Size	Mechanical properties
high-early-strength Portland cement	3.14	13.68μm	fc:39.7 MPa,fb:6.5 MPa(seven days)
Silica fume	2.20	0.15μm	
Ceramics aggregate	0.70	115μm	
Steel fiber	0.91	φ40μm×6mm	ft:0.98GPa, E:210GPa
Polypropylene fiber	7.85	0.10×0.10×6mm	ft:482MPa, E:2.61GPa
Polyacrylamide	1.30	125μm	
Super plasticizer	1.1	Main component:aromatic aminosulfonic acid macromolecular compound	

where values of size of cement, aggregate,silica fume and polyacrylamide in the Table mean their average diameters; fc is compressive strength; fb is flexural strength; ft is tensile strength; and E is Young's modulus of elasticity

2.2 Mixtures and specimens
By using the Lee's packing model[1],packing density(bulk density/solids density) of three components which are cement,silica fume and aggregate was investigated.The result is shown in Fig.1. This figure shows packing ternary diagram of all possible combinations of the three components, that is cement,aggregate and silica fume. In Fig.1, solid lines show the packing density and dashed lines show the weight per unit volume which is calculated on the assumption that voids between particles are filled

with water.

As a result of trials, it was found that mixing proportions described with points *b* and *c* shown in Fig.1 were impossible to form a sheet. Therefore, mix proportion described with the point *a* shown in Fig.1 was selected as a base of mortar matrix.

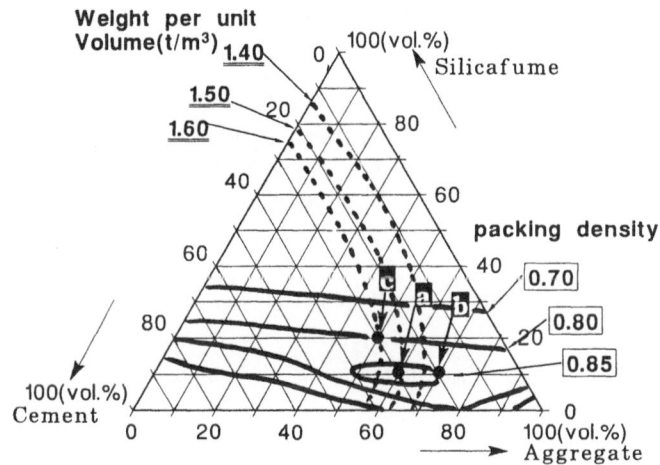

Fig.1 Estimated Packing Densities of Ternary Mixtures.

2.3 Preparation of specimens

Cement, polymer, silica fume, water and superplasticizer were premixed with the mix proportions indicated in Table 2 by use of a kneader for about 10 minutes. The premixed cement paste was fed onto a twin-roll mill 9-12 times to form a sheet. The sheet was cut

Table 2　Mix proportions of base mortal matrix(unit:kg/m³)

Water	Cement	Aggregate	Silica fume	Polyacryl-amide	Super-plasticizer
104	816	342	190	40	27

where Cement:Aggregate:Silica fume=30:60:10vol.% from Fig.1,
Polyacrylamide/(Cement+Silica fume) is 4wt.%; Superplasticizer/(Cement+Silica fume) is 3wt.%.

Table 3　Series of specimens

Symbol of specimen	Type of fiber	Volume fraction of fiber(%)
Plain	-----	0
ST-2	steel	2
ST-4		4
PP-2	polypropylene	2
PP-4		4
PP-6		6

into strip-like specimens(2 x 25 x 100mm) and then subjected to a moist cure atmosphere at 20℃ and 80%R.H. for 7 days. Then they were cured in the following conditions, (1)1day-45℃ heat cure, or (2)1day-45℃ heat cure plus autoclave cure at a temperature of 180℃ under a pressure of 0.98MPa for 7 hours. During the moist cure, specimens were pressed between two glass plates.

2.4 Bend test

Cured specimens were immersed in water at 20℃ for 48 hours. The specimens before and after 48-hour water immersion were tested in three-point bend load at a crosshead speed of 2 mm/min and a span of 80mm by use of Instron universal testing machine (load cell capacity :500kgf). The flexural strength was calculated from the following equation:

$$\sigma_b = 1.5 \times P_{max} \times l / bd^2 \tag{1}$$

where σ_b:flexural strength, P_{max}:maximum load, l:span, b,d:width and depth of the specimen, respectively.

3 Experimental results and discussion

Results of the density and the flexural strength are shown in Fig.2 and Fig.3. Relations between load and displacement of the crosshead are shown in Fig.4 and Fig.5.

3.1 Density and flexural strength

In general , cementitious composites have a tendency that higher density achieves higher strength. However mixing high volume percentage of high-strength and high-modulus fiber prevents the extension of crack in matrix and it may be capable to increase the strength of composites[3]. Consequently the mix of fiber has a possibility to not only increase the ductility but also strengthen without increasing the density. In this experimental study, two dimensional random orientation of short fibers because of the shape of sheet, has the high resistance to flexural and tensile deformations. In addition improvement of bond between high-strength hollow ceramics aggregates and polymeric co-matrix phase in conjunction with removal of voids by means of kneader and twin-roll mill was effective to increase high flexural strength with a low density.

In the series mixed with steel fibers, the flexural strength was decreased than that of plain mortar series regardless of volume fraction of fiber. In case of the volume fraction of polypropylene fiber 2% and 4% , the flexural strength was increased to 8% and 24% higher than that of plain mortar series, respectively. In case of polypropylene fiber 6%, however , the flexural strength was decreased. Consequently, in the range of specific density 1.5~1.8 t/m³, the flexural strength showed a large variation of 15~30MPa.

Although the steel fibers used in this experimental study are very fine in comparison with those usually used in civil engineering and architectural fields, these steel fibers are large in comparison with the size of specimens. Therefore flaws after peeling off the surface might cause the strength reduction.

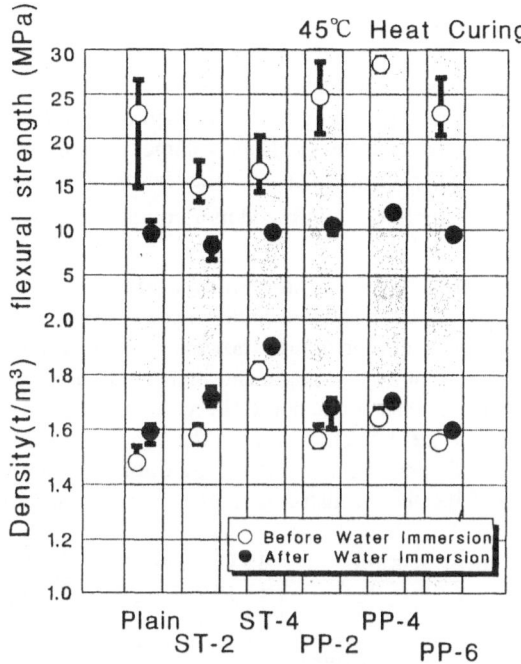

Fig.2 Density and flexural strength after 45 ℃ heat curing.

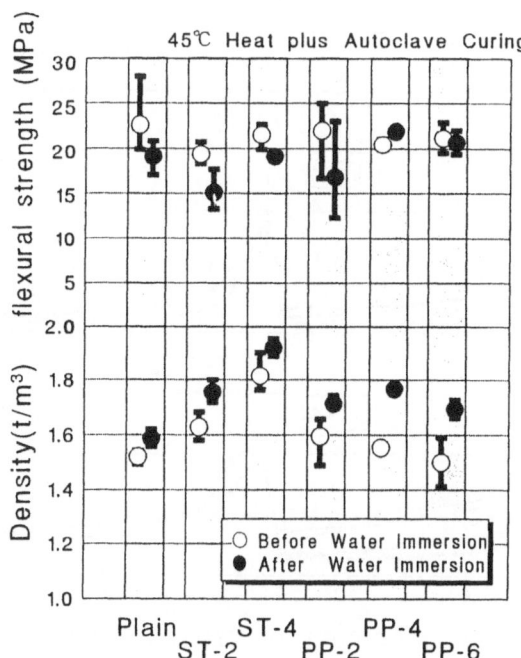

Fig.3 Density and flexural strength after 45 ℃ heat curing plus autoclave curing.

In case of the polypropylene fiber, favorable specimens without such flaws were made because of the small diameter and the low modulus of the fiber up to the volume fraction of 4%. In case of 6%, however, creases and flaws which existed near the surface of specimens might cause the reduction of strength.

After water immersion, the flexural strength was declined over 50% and shown about 10MPa without autoclave curing regardless of fiber volume fraction and fiber types. On the other hand, autoclave cured specimens showed smaller strength reduction of 15%,11 ~22% in plain materials and steel fiber reinforced mortar, respectively.

3.2 Deformation properties before water immersion

The specimens mixed with steel fiber showed lower flexural strength than that of plain mortar series. After cracking in mortar matrix, however, the load-displacement curve drew the upper convex because of high-strength and high-modulus of steel fiber. As the volume fraction of steel fiber increased, the curve shifted upward. In the large deflection range, crack resistance due to steel fibers increased the ductility. In case of the mortar mixed with polypropylene fibers, the low elastic modulus of polypropylene fibers caused the brittle fracture of the mortar matrix just after the peak load. Nevertheless the

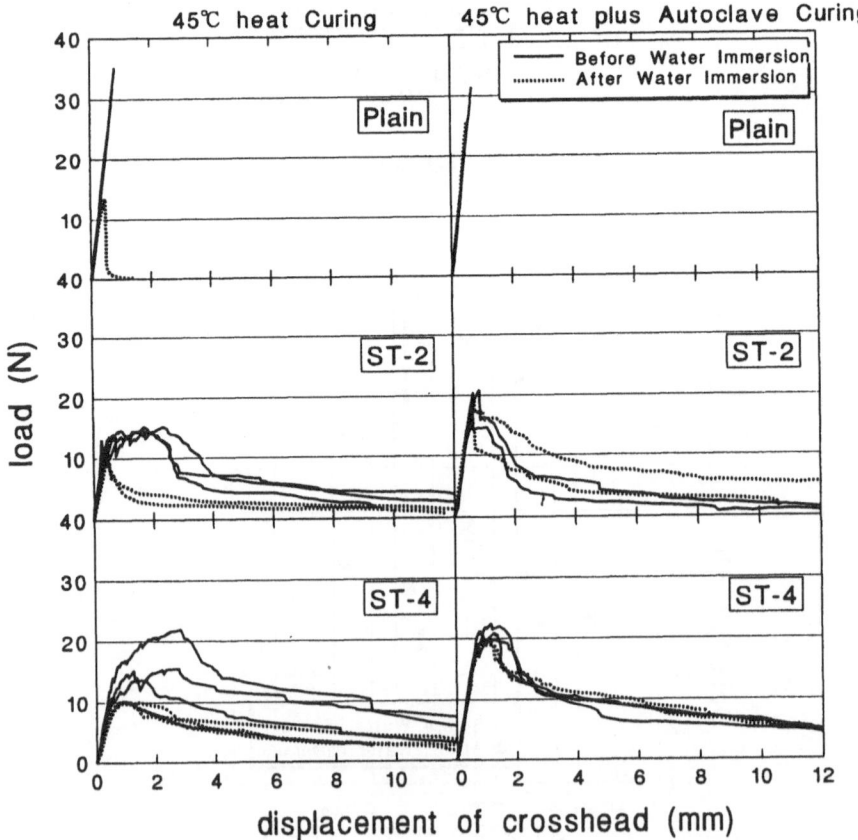

Fig.4 Load-displacement of crosshead in case of the steel fiber reinforced mortar.

high-strength and high-elongation of fibers give the high ductility in the range of large deformation. As the volume fraction of fibers increased , the ductility increased.

In the mortar mixed with the steel fiber, autoclave curing has caused poor ductility after the maximum load. In general, autoclave curing develops hydration and makes the microstructure denser. Therefore the bond property between cement matrix and inclusion is expected to be improved. In this experiment, however, the chemical properties of polymer might influence the strength of interface between the steel fiber and the mortar matrix.

3.3 Deformation properties after water immersion

Influence of different type of fiber was investigated about the series subjected to 45℃ heat curing. After water immersion, the deformation property after the maximum load showed high-ductility in case of over 4% volume fraction of polypropylene fiber. In that case, the multiple cracking were observed on the surface of specimens subjected to tensile stress because of the decrease of matrix strength and high interface strength

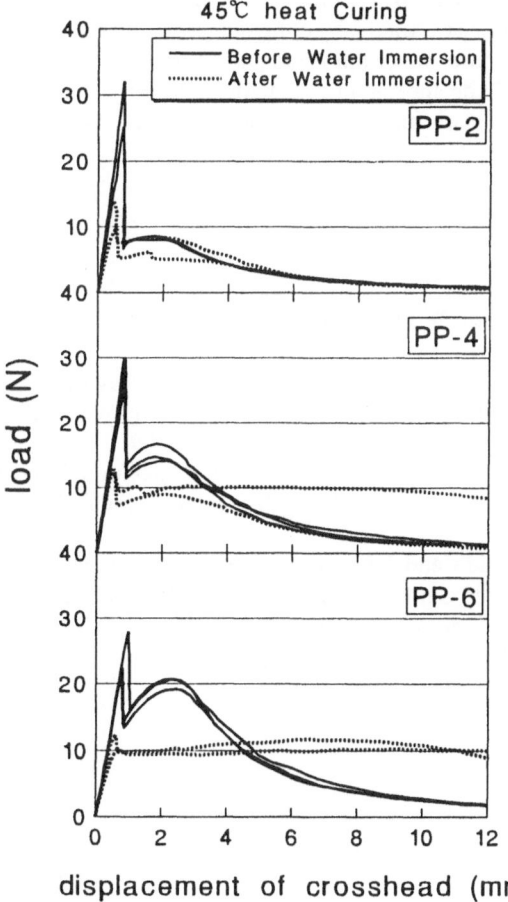

Fig.5 Load-displacement of crosshead in case of polypropylene fiber reinforced mortar.

between polypropylene fiber and mortar matrix.

Influence of different curing methods was investigated about the steel fiber reinforced mortar. It is said that the autoclave curing improves the water-resisting due to the decomposition of a water-soluble polymer and the accelerated hydration of cement [4]. Deformation properties of the mortar mixed with steel fibers are very similar before and after water immersion if the autoclave curing was subjected. Therefore,the autoclave curing is effective to increase not only flexural strength but also the deformation property after water immersion.

4 Conclusions

The following conclusions were obtained.

1. Flexural strength of 15~30 MPa was achieved in the range of specific density 1.5~1.8 t/m³.

2. It is difficult to increase the flexural strength by mixing steel fiber since making flaws on the surface of specimens. After subjecting autoclave curing, however, the flexural strength and the ductility after water immersion are improved.

3. Flexural strength of the mortar mixed with polypropylene fiber is increased in comparison with that of plain mortar. Polypropylene fiber also improves the ductility after water immersion.

5   References

1.   D.J.Cumberland and R.J.Crawford(1987) *The Packing of Particles*, Elsevier, pp.41-62.

2.   J.F.Young(1991) Macro-Defect-Free Cement :A Review,*Materials Research Society Symposium*, Volume 179, Specialty Cements with Advanced Properties, Materials Research Society, Pittsburgh,pp.101-121.

3.   V.C.Li and C.K.Y.Leung(1992)Steady-State Cracking of Short Random Fiber Composites, *Journal of Engineering Mechanics*, ASCE, Volume 118, No.11, pp.2246-2264.

4.   Y.Ohama, K.Demura and T.Kobayashi(1993)Improvement in Water Resistance of Macrodefect-Free Cements Using Ordinary Portland Cement, *Proceedings of the Thirty-Sixth Japan Congress on Materials Research*, The Society of Materials Science, Japan, Kyoto, pp192-195.

FACTORS AFFECTING FLEXURAL STRENGTH OF MDF CEMENTS

T. Kobayashi, Y. Ohama, K. Demura and K. Ochiai
Dept. of Architecture, Nihon University, Koriyama, Japan

Abstract
Macrodefect-free (MDF) cements using an ordinary portland cement, a polyacrylamide and a high-range water-reducing agent are prepared with various mix proportions, and tested for flexural strength. The effects of the mix proportions on the flexural strength of MDF cements are examined. The microstructures of MDF cements are also observed by use of scanning electron microscope. As a result, the flexural strength of MDF cements tends to increase with a decrease in the water-cement ratio and an increase in the polymer-cement ratio. The flexural strength development of MDF cements is found to depend on the polymer cohesion governed by the water-polymer ratio. An empirical equation obtained from the flexural strength-water-polymer ratio relationships is proposed for the flexural strength prediction.
Keywords: Flexural strength, MDF cements, ordinary portland cement, polyacrylamide, polymer-cement ratio, water-cement ratio, water-polymer ratio.

1 Introduction

In general, macrodefect-free (MDF) cements are prepared by using ordinary portland cement or alumina cement and water-soluble polymers, and by applying mechanochemical processing techniques at very low water-cement ratios (8.0 to 20.0%). The flexural strengths of MDF cements attain 70 to 150MPa under dry conditions[1], [2]. Such high flexural strengths are achieved by the removal of relatively large pores from their microstructures, and by an adhesive effect of the polymers at the interfaces between interacting calcium silicate hydrate gels[3], [4]. Many papers on MDF cements have been published since Birchall's work in 1981. However, the detailed effects of process conditions on the flexural strength of MDF

Polymers in Concrete, edited by Y. Ohama, M. Kawakami and K. Fukuzawa. Published in 1997 by E & FN Spon, 2–6 Boundary Row, London SE1 8HN, UK. ISBN: 0 419 22330 4.

cements have scarcely been discussed.

In this paper, MDF cements using an ordinary portland cement, a water-soluble polyacrylamide and a high-range water-reducing agent are prepared with various mix proportions, and tested for flexural strength. The microstructures of MDF cements are also observed by scanning electron microscopy. The effects of the mix proportions on the flexural strength of MDF cements are examined to find out the factors affecting flexural strength of MDF cements.

2 Materials

2.1 Cement

An ordinary portland cement specified in JIS (Japanese Industrial Standard) R 5210 (Portland Cement) was used in all the mixes. The chemical compositions and physical properties of the cement are listed in Table 1.

Table 1. Chemical compositions and physical properties of ordinary portland cement

Chemical Compositions (%)							
CaO	SiO_2	Al_2O_3	Fe_2O_3	SO_3	MgO	ig.loss	Total
64.0	22.0	5.2	2.6	2.2	1.7	1.8	99.50

Specific gravity	Blaine's specific surface (cm^2/g)	Setting time (h-min)		Compressive strength of mortar (MPa)		
		Initial set	Final set	3d	7d	28d
3.16	3250	2-13	3-06	16.1	26.0	43.1

2.2 Admixtures

A water-soluble polyacrylamide and a commercial polyalkyl aryl sulfonate-type high-range water-reducing agent were employed. The properties of the polyacrylamide are given in Table 2.

Table 2.   Properties of polyacrylamide

Chemical formula	Molecular weight	Particle size (μm)	Specific gravity (20°C)
$\begin{array}{c} -\!\!\left[\!\!\begin{array}{c} CH_2-CH- \\ \mid \\ C=O \\ \mid \\ NH_2 \end{array}\!\!\right]_n \end{array}$	1300×10^4	125	1.30

3 Testing procedures

3.1 Preparation of specimens
MDF cements were mixed with water-cement ratios (W/C) of 10.0, 11.0, 12.0, 13.0, 14.0, 15.0, 16.0, 17.0 and 20.0%, polymer-cement ratios (P/C) of 4.0, 6.0 and 8.0% and a water-reducing agent content of 1.0% by use of a kneader for 10 min. The mixed MDF cements were fed onto a twin-roll mill 10 times to form sheets. The sheets were cut into strip-like MDF cement specimens 25x2x100 mm, and then subjected to a 3-day-20°C-80%(RH)-moist plus 1-day-45°C-heat cure. During the moist cure, MDF cement specimens were pressed between two glass plates.

3.2 Flexural strength test
The cured MDF cement specimens were tested in three-point flexure at a crosshead speed of 2mm/min and a span of 80mm by use of the Instron universal testing machine.

3.3 Observation of microstructures of MDF cements
The microstructures and polymer films of the MDF samples taken from the specimens after flexural strength test were observed by using a scanning electron microscope.

4 Test results and discussion

Fig.1 shows the relation between the water-cement ratio and flexural strength of MDF cements. Fig. 2 represents the relation between the polymer-cement ratio and flexural strength of MDF cements. The flexural strength of MDF cements with a polymer-cement ratio of 8.0% becomes nearly constant at water-cement ratios of 11.0 to 15.0%, and decreases at water-cement ratios of 16.0% or more. The flexural strength of MDF cements with a polymer-cement ratio of 6.0% becomes nearly constant at water-cement ratios of 10.0 to 14.0%, and decreases at water-cement ratios of 15.0% or more. Such

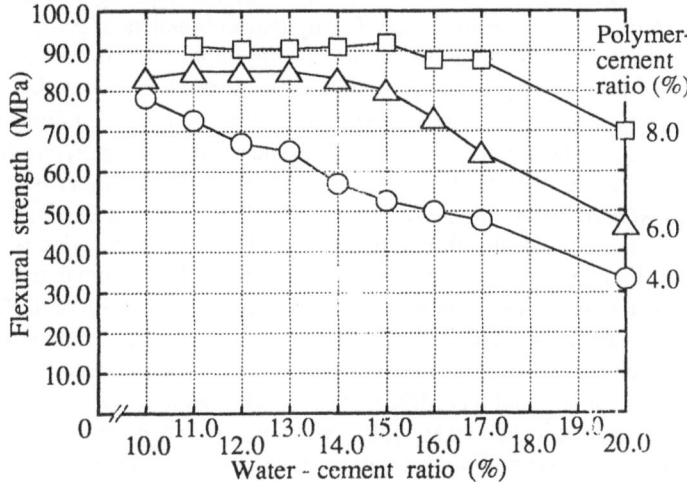

Fig.1.   Water-cement ratio vs. flexural strength of MDF cements.

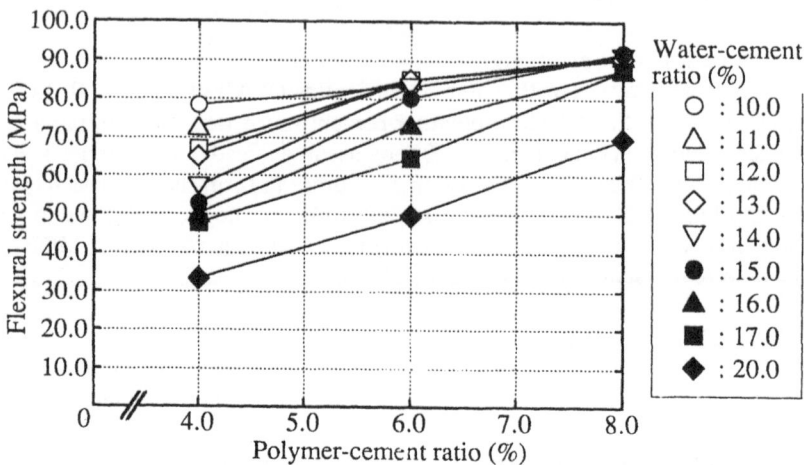

Fig.2. Polymer-cement ratio vs. flexural strength of MDF cements.

nearly constant flexural strengths at water-cement ratios of 11.0 to 15.0% and 10.0 to 14.0% of MDF cements with polymer-cement ratios 8.0 and 6.0% are attributed to the high cohesion of polyacrylamide as an admixture[5]. The flexural strength of MDF cements with a polymer-cement ratio of 4.0% decreases with an increase in the water-cement ratio. Such a decrease in the flexural strength of MDF cements with a polymer-cement ratio of 4.0% is due to an excess in the hydrolysis of the polyacrylamide [5]. The flexural strength of MDF cements increases sharply with an increase in the polymer-cement ratio irrespective of the water-cement ratio.

Photo 1 represents the microstructures of MDF cements with water-cement ratios of 11.0 and 15.0%, and polymer-cement ratios of 4.0 and 8.0%. Photo 2 exhibits the polymer film networks formed in MDF cements. The macrodefects, namely large pores, are not observed in the microstructures of MDF cements regardless of the water-cement ratio and polymer-cement ratio. The hydration products of cement hardly exist in the microstructures of MDF cements as shown in Photo 1 because of the low water-cement ratio and the short period of moist cure. In general, it is considered that MDF cement is a two-phase composite material consisting of the discontinuous phase of cement particles and the continuous phase of polymer films. The polymer film network may act as a binder for the cement particles. Therefore, mixing water is consumed for dissolving powdered water-soluble polymers in MDF cement. In Photo 2, the pores of 20 to 30 µm in diameter are observed, and are almost the same size of the cement particles. MDF cements with a polymer-cement ratio of 8.0% give densified polymer film networks in comparison with ones with a polymer-cement ratio of 4.0% regardless of the water-cement ratio. The polymer films in MDF cements become massive with decreasing water-cement ratio and increasing polymer-cement ratio. From the above-mentioned things, it is found that MDF cements with high polymer-cement ratio have a higher flexural strength than ones with low polymer-cement ratio, and the flexural strength of MDF cements with low polymer-cement ratio is reduced with an increase in the water-cement ratio. It is considered from the above test results that the flexural

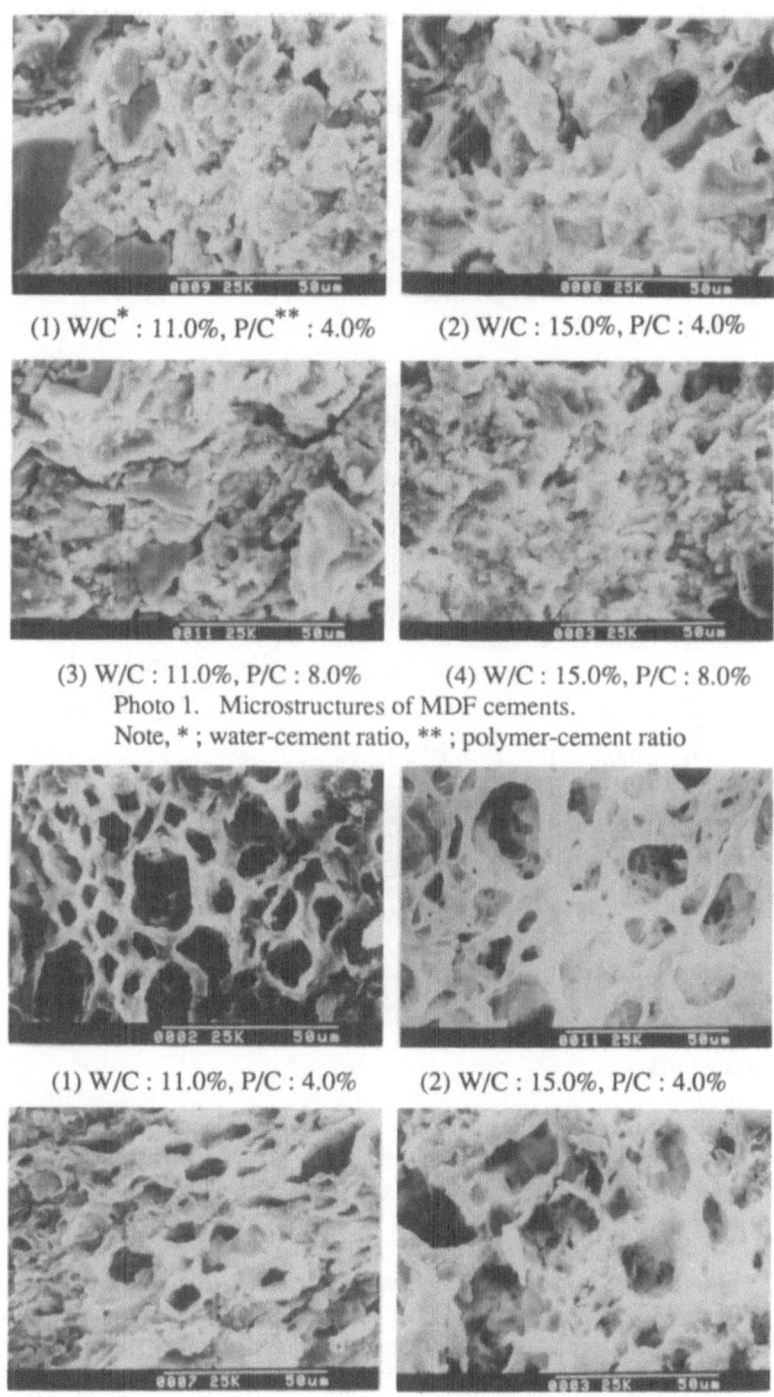

(1) W/C* : 11.0%, P/C** : 4.0%   (2) W/C : 15.0%, P/C : 4.0%

(3) W/C : 11.0%, P/C : 8.0%   (4) W/C : 15.0%, P/C : 8.0%

Photo 1. Microstructures of MDF cements.

Note, * ; water-cement ratio, ** ; polymer-cement ratio

(1) W/C : 11.0%, P/C : 4.0%   (2) W/C : 15.0%, P/C : 4.0%

(3) W/C : 11.0%, P/C : 8.0%   (4) W/C : 15.0%, P/C : 8.0%

Photo 2. Microstructures of polymer films in MDF cements.

strength of MDF cements is greatly related to the strength of the polymer films formed in the MDF cements, and the mixing water is an important factor affecting the polymer film formation and the strength of the polymer films. As a result, the water-polymer ratio is proposed as a factor influencing the flexural strength of MDF cements.

Fig.3 shows the relationship between the water-polymer ratio and flexural strength of MDF cements. The flexural strength of MDF cements is decreased with an increase in the water-polymer ratio. Close correlation between the water-polymer ratio and flexural strength of MDF cements is observed, and the flexural strength can be predicted by using the following equation:

$$\sigma_f = -0.18(W/P) + 120 \quad (\gamma = 0.98)$$

where σ_f and W/P are the flexural strength(MPa) and water-polymer ratio(%) of MDF cements, respectively, and γ is coefficient of correlation.

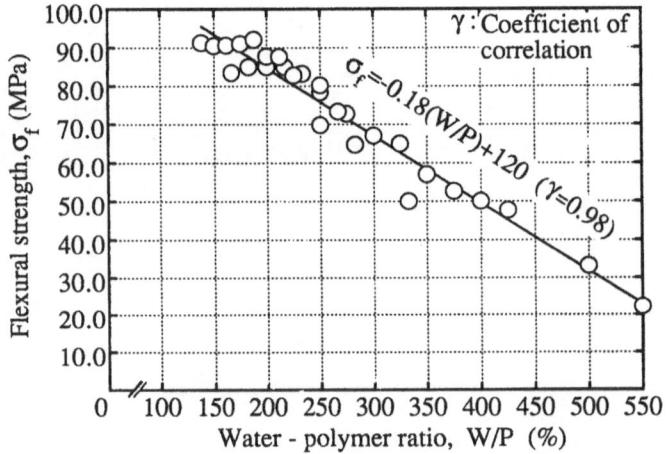

Fig.3.   Water-polymer ratio vs. flexural strength of MDF cements.

5   Conclusions

The conclusions obtained from the above test results are summarized as follows:

(1) The flexural strength of MDF cements decreases with an increase in the water-cement ratio, or becomes nearly constant, and then decreases with an increase in the water-cement ratio. In addition, the flexural strength of MDF cements increases with an increase in the polymer-cement ratio.

(2) The flexural strength of MDF cements is greatly related to the strength of the polymer films formed in MDF cements, and the mixing water is an important factor affecting the polymer film formation and the strength of the polymer films. Therefore, the water-polymer ratio is proposed as a factor influencing to the flexural strength of MDF cements.

(3) The flexural strength of MDF cements can be predicted by an empirical equation obtained from flexural strength-water-polymer ratio relationships.

6 References

1. Birchall, J.D., Howard, A. J. and Kendall, K. (1981) Flexural strength and porosity of cements. *Nature*, Vol.289, No.5796. pp.388-389.
2. Goto, S., Koyata, E. and Kimura, S. (1983) Hardened cement paste with high bending strength (in Japanese), in *Sement Gijutsu Nempo* 37, The Cement Association of Japan, Tokyo, pp.109-111.
3. Birchall, J.D. (1983) Cement in the context of new materials for an energy-expensive future. *Philosophical Transactions of the Royal Society of London*, Series A, Vol.310, No.1511, pp.31-42.
4. Beaudoin, J.J. and Feldman, R.F. (1985) High-strength cement pastes-A critical appraisal. *Cement and Concrete Research*, Vol.15, No.1, pp.105-116.
5. Michaels, S.S. (1954) Aggregation of suspensions by polyehectrolytes. *Industrial and Engineering Chemistry*, Vol.46, No.7, pp.1485-1490.

KEYWORD INDEX